Carbon and Nitrogen Cycling in Soil

Rahul Datta • Ram Swaroop Meena
Shamina Imran Pathan • Maria Teresa Ceccherini
Editors

Carbon and Nitrogen Cycling in Soil

Springer

Editors
Rahul Datta
Department of Geology and Pedology
Mendel University in Brno
Brno, Czech Republic

Ram Swaroop Meena
Department of Agronomy
Institute of Agricultural Sciences, BHU
Varanasi, Uttar Pradesh, India

Shamina Imran Pathan
Department of Agrifood and
Environmental Science (DISPAA)
University of Florence
Florence, Italy

Maria Teresa Ceccherini
DAGRI – Department of Agriculture,
Food, Environment and Forestry
University of Florence
Florence, Italy

ISBN 978-981-13-7263-6 ISBN 978-981-13-7264-3 (eBook)
https://doi.org/10.1007/978-981-13-7264-3

This Springer imprint is published by the registered company Springer Nature Singapore Pte Ltd.
The registered company address is: 152 Beach Road, #21-01/04 Gateway East, Singapore 189721, Singapore

Foreword

This book, *Carbon and Nitrogen Cycling in Soil*, deals with the hot topic of carbon and nitrogen interactions in soil, that is, it deals with two elements that have a crucial role in soil functioning and fertility. Soil is a living system and the decrease in soil functionality can generate negative effects on functionality and thus production of agricultural and forest systems and in general in the functionality of all terrestrial ecosystems. This book has 13 chapters contributing to increase our knowledge on processes and mechanisms of carbon and nitrogen in soil. Particular attention is given to microbial processes of the carbon and nitrogen dynamics in soil, also considering the impact of climate changes. One chapter deals with the role and dynamics of the two elements in the rhizosphere soil. Other chapters deal with the impacts of agronomical management, including the use of biochar, on both carbon and nitrogen transformations. The sustainability of agricultural management is also discussed. This book represents an essential resource for soil scientists, biologists, and biogeochemists, and for all individuals in the fields of agronomy, ecology and environmental sciences.

University of Firenze
Livorno, Italy

Paolo Nannipieri
Professor Emeritus

Preface

The problem of carbon (C) and nitrogen (N) cycling and its impact on climate has been attracting attention for many decades. The last few decades have seen tremendous changes in agriculture and the food chain of the world. New and modern agriculture techniques result in more depletion of C and N from the soil and cause a remarkable increase in C and N concentration in the atmosphere. Increased demand for food and energy are the two main anthropogenic factors affecting C and N cycles in a climate change era. While global trade in agricultural commodities has increased inter-connectivity among food resources in developed and developing countries, it has also contributed and exacerbated the challenges related to malnutrition, food security, environmental degradation and large-scale soil sustainability, making it harder to achieve the targets of Sustainable Development Goals (SDGs) of eliminating poverty and hunger. The C and N cycles are playing a crucial role in the earth's biochemistry at the global level under the changing climate. Essential plant nutrients like C and N have a vital role in biomass production and ultimately in crop yield in agriculture production systems. The N content in soil enhances the metabolic activities in the plant, whereas the C element in the plant acts as a basic nutrient and the primary platform for microbial reaction in the soil. The rapidly growing human population is placing an increasing demand for food and nutritional security. Intensive cultivation of soils is increasing soil sickness and decreasing soil functionality due to the reduction in organic C and N content in the soils. It is therefore imperative to understand sustainable cycling for primarily preventing the losses of C and N content in the soils. This is critical to soil biodiversity. Therefore, there is a need to adopt sustainable management practices to restore soil fertility and sustainable crop production. Terrestrial C sequencing through biotic processes is a possible option to reduce the rates of CO_2 emissions. Transfer of C and N from the soils to the atmosphere not only affects the atmosphere but also the soil processes.

Soil organic matter (SOM) is the primary pool of C, either sink, in case of sequestration, or source, when mineralization is favoured. Keeping this in mind, the objectives of this book are: (1) to discuss the role of C and N cycling in sustaining agricultural productivity, and its importance to sustainable soil management; (2) to explain organic and inorganic remediation technologies to treat heavy metal-contaminated soils; (3) to elucidate the main agronomical benefit of inoculating AMF in cropped soils; (4) to throw light on the abiotic and biotic factors that manage plant–microbe interactions in the rhizosphere; and (5) to discuss the effects of excess metals on N-fixation in plants and provide guidelines for further studies. In this book, the editors/authors contributed a broad range of information on C and N cycling with a global perspective. Highly qualified researchers from different countries such as Australia, India, Japan, Mexico China, Nigeria, Italy, Pakistan and Bangladesh have contributed to the cutting-edge scientific information on topics covering management of C and N cycling in soil. The editors have provided a roadmap for sustainable approaches for C and N cycling in soil systems for food and nutritional security, and soil sustainability in agricultural operations. In general, the book is suitable for teachers, researchers, and policymakers, undergraduate and graduate students of soil science, soil microbiology, agronomy, ecology and environmental sciences.

Brno, Czech Republic — Rahul Datta
Varanasi, India — Ram Swaroop Meena
Florence, Italy — Shamina Imran Pathan
Florence, Italy — Maria Teresa Ceccherini

Contents

About the Editors

Rahul Datta is a soil microbiologist and enzymologist. Dr. Datta holds an MSc and PhD from the Department of Geology and Pedology at Mendel University in Brno, Czech Republic, where he is currently a researcher. He has acquired global experience in soil science research by working with a number of prominent experts, including as a visiting scientist with both Dr. Richard Dick (President of the SSSA) at OSU, Ohio, USA, and Prof. Paolo Nannipieri at the University of Florence, Italy. Dr. Datta has published numerous research articles and review papers in peer-reviewed journals and is currently a reviewer for thirty major scientific journals. Dr. Datta is at present a member of Soil Science Society of America.

Ram Swaroop Meena PhD, is an agronomist and has completed his postdoctoral research on soil carbon sequestration under Prof. Rattan Lal, Distinguished Scientist and Director, CMASC, OSU, USA. With 9 years of research and teaching experience, he is currently working on three ongoing, externally funded projects, including one for the DST (GOI). His current research interests are in soil sustainability, crop productivity and resource-use efficiency.

Dr. Meena has published more than 100 research and review papers in peer-reviewed journals, as well as 25 book chapters. He has worked as an expert at the NCERT, MHRD, GOI, to develop two books for school education for standards XI and XII. A member of various national and international societies, he has also been honoured with several awards, including an RGNF award from the UGC, GOI, and a Raman Fellowship from the MHRD, GOI.

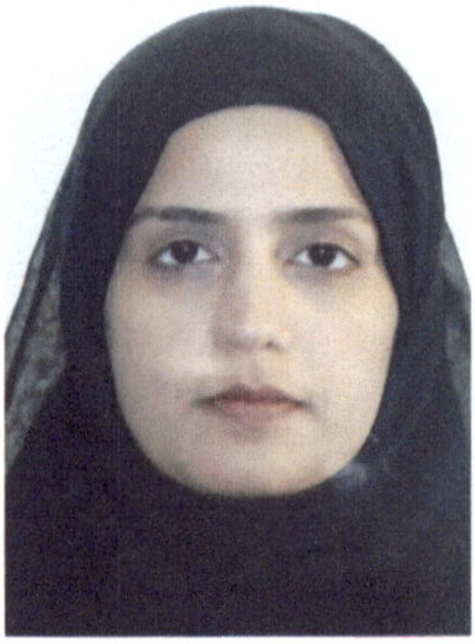

Shamina Imran Pathan is a soil microbial ecologist and is currently working as a postdoctoral researcher at the University of Florence, Italy. Previously, she worked at the Mediterranean University of Raggio Calabria, Italy, where she studied how different management practices such as tillage systems and digestate amendment affect soil microbial communities. She holds a PhD from the University of Florence, Italy. Her doctorate dissertation was conducted in consultation with Prof. Paolo Nannipieri. She has published seven peer-reviewed papers in high-impact journals (e.g. *Soil Biology and Biochemistry*, *Plant and Soil*), including four as lead author. She has been a reviewer for the journals *BFOS*, *Plant* and *Soil*, *Scientific Reports*, etc.

Maria Teresa Ceccherini PhD., is Assistant Professor at the DISPAA – Department of Agrifood and Environmental Science, University of Florence, Italy. She holds a PhD in Soil Science and Climatology and completed her postdoctoral research on environmental metagenomics under Prof. Paolo Nannipieri, Distinguished Scientist and Director of the DISPAA, University of Florence. She is currently researching soil sustainability, soil plant pathogens, use of industrial biomass as fertilizers, and metagenomic approaches. She has been the project manager and coordinator of several research projects and has published more than 60 papers and book chapters. She is member of the editorial advisory board of the journal *Applied Soil Ecology* and serves as a reviewer for various other international journals.

Carbon and Nitrogen Cycling in Agroecosystems: An Overview

Tariq Shah, Sadia Lateef, and Mehmood Ali Noor

Abstract Soil organic carbon (SOC) is the key constituent of soil organic matter (SOM). As a sign for soil health, SOC plays significant contributions to food production, extenuation, and variation to climate change. High SOM content delivers nutrients to plants and enhances water availability, both of which improve soil fertility and eventually enhance food production. Additionally, SOC enhances soil structural firmness by encouraging aggregate development which, together with porosity, warrants adequate aeration and water infiltration to support plant growth. Soils are also a source of greenhouse gas (GHG) emissions into the atmosphere. While the general effect of climate change on SOC stocks is very inconsistent according to the region and soil type, increasing temperatures and enhanced frequency of extreme events are likely to lead to enhance SOC losses. Global nitrogen fixation is the main contributor of the reactive nitrogen (Nr) to terrestrial and marine ecosystems, for which the anthropogenic activities are responsible. Majority of the conversions of anthropogenic Nr are on land within soils and vegetation. Emissions of ammonia (NH_3) and nitrogen oxides (NOx) in the atmosphere lead to the productions of secondary pollutants such as ozone and other photochemical oxidants and aerosols (specially ammonium nitrate and ammonium sulfate). This chapter aims to deliver an outline to decision-makers and specialists of the key scientific details and information concerning the existing knowledge and knowledge gaps on carbon and nitrogen cycling in the agroecosystem.

Keywords Agroecosystem · Climate change · Greenhouse gas emissions · Nitrogen cycling · Soil organic carbon

---

T. Shah (✉)
Oil Crops Research Institute, Chinese Academy of Agricultural Sciences, Wuhan, China
e-mail: tariqshah@aup.edu.pk

S. Lateef
Faculty of Biological Sciences, Quaid-e-Azam University, Islamabad, Pakistan

M. A. Noor
Institute of Crop Sciences, Chinese Academy of Agricultural Sciences, Beijing, China

R. Datta et al. (eds.), *Carbon and Nitrogen Cycling in Soil*,
https://doi.org/10.1007/978-981-13-7264-3_1

Abbreviations

BNF	Biological nitrogen fixation
C	Carbon
DOC	Dissolved organic carbon
FAO	Food and Agriculture Organization
GHG(s)	Greenhouse gas (es)
GWP	Global warming potential
HONO	Nitrous acid
N	Nitrogen
N_2O	Nitrous oxide
NH_3	Ammonia
NH_4	Ammonium
NO	Nitric oxide
NO_2	Nitrogen dioxide
NO_x	Oxidized forms of nitrogen
NPP	Net primary productivity
N_r	Reactive nitrogen
PAN	Peroxyacetyl nitrate
SOC	Soil organic carbon
SOM	Soil organic matter
UNCBD	Convention of biological diversity at the United Nations

1 Introduction

The importance of carbon (C) and nitrogen (N) in agriculture and its wide-ranging implications to environmental aspects bring the C and N cycles to the centerpiece of agroecosystems, especially against the backdrop of climate change and its impacts on life on earth. While the C and N cycles are central to the earth's biogeochemistry in all kinds of ecosystems, the cycling of C and N in agroecosystems is of enormous importance due to the two elements' key status as major nutrients. Apart from being two key essential elements for plants, C and N are also key elements of greenhouse gases causing global warming. Agroecosystems are not only challenged by sustainable resource use but also by climate change, to which they themselves are contributors as well. Therefore, better understanding of C and N cycling in agroecosystems is essential to any possible improvements in functioning of agroecosystems.

Global N cycle is largely affected by human activities, as most of the fixed N comes from the anthropogenic activities, thus having vast environmental, agricultural, and health implications. For example, the huge production and consumption of synthetic fertilizers is causing groundwater pollution by the accumulation of nitrate and nitrite through leaching process. The excessive deposition of N forms in soils also causes environmental pollution through N_2O emissions, and the groundwater contamination poses serious threats to human health causing gastric cancer by drinking contaminated water. A better understanding of key processes involved in N cycle and the contribution of each process as source or sink of N_2O can enable to improve agroecosystem management, and various options can be explored to ameliorate the negative effects of relevant anthropogenic activities.

Cycling of elements in the ecosystem is important for sustainability, in which the nutrient cycling is essential for living organisms on globe as the cycling prevents accumulation of elements, thus avoiding toxicity. Along with N cycling, the cycling of C is also of great importance as in recent decades both the C and N cycling have gained attention, because they are capable to produce greenhouse gases (GHGs). In C cycling, CO_2 and CH_4 gases have significantly contributed to increased atmospheric CO_2 and the global warming due to GHGs emission effects. Various ecosystem processes are responsible for GHG emissions, in which the release of CO_2 and CH_4 gases by plants and soil respiration is prominent. Therefore, recent research efforts are focusing on increasing the carbon sequestration rates as a mitigation strategy to global warming and most importantly to improve the soil health to sustain the soil microbiota and for increasing crop yields.

This chapter briefly outlines the key components and processes of C and N cycling in agroecosystems and presents the understanding of vital processes and their interrelationship involved in C and N cycling. Using this knowledge, management options and strategies can be employed to improve the anthropogenic activities responsible for negative effects due to the excessive N-inputs on the agroecosystem.

2 Soil Organic Carbon (SOC): A Crucial Component of Carbon Cycle

Soil organic carbon (SOC) is one of the major parts of global carbon cycle which basically involves in the cycling of C (carbon) from and between the soil, ocean, vegetation, and atmosphere. The SOC pool is estimated to store 1500 PgC in the upper layer of soil (FAO and ITPS 2015). The phonemical SOC reservoir is continually cycling with different molecular forms of C between various global carbon pools (Kane 2015). As CO_2 and CH_4 are vital C-based gases in the atmosphere and also crucial component of autotrophic organisms as well as chemo- and photoautotrophic microbes which are involved in the conversion of CO_2 into organic forms of C. Plant exudates and residues (dead organic matter) are incorporated into the soil from plants by heterotrophic microorganisms through transformations of organic materials. The transformation process of organic matter results in the formation of complex biochemical mixtures and microbial decomposition products from plant litter (Paul 2014). These products form complexes with soil minerals in the form of aggregates, thus ensuring the SOC persistence in the soil for decades and in some cases for centuries or millennia (Schmidt et al. 2011). A small amount of C is also transported to rivers and the ocean in the form of dissolved organic carbon (DOC) or along with erosion material from soil.

At different stages of decomposition, soil organic matter includes dead tissues of plants and animals which are less than 2 mm in size. Transformation of SOM plays a crucial role in global warming and soil ecosystem. SOM is vital not only for the stabilizing soil structure but also for the release and retention of plant nutrients and the maintenance of soil water holding capacity. Thus, it is a key factor in environmental resilience and agricultural productivity. SOM decomposition releases mineral nutrients in the soil, thus ensuring plant growth by increasing their availability (van der Wal and de Boer 2017). While improved plant vigor and higher yield ultimately ensures food security.

3 Carbon-Based GHGs

Soil serves as a double-edged sword for carbon fluxes. Soil can turn into either a net source or a net sink of GHGs due to anthropogenic activities. As a source, it involves in the emission of GHGs into the atmosphere which causes greenhouse effect as these GHGs trap thermal radiations and ultimately contribute to global warming. CO_2 and CH_4 are two leading carbon-based GHGs which are emitted from the soil due to human activities (Pachauri et al. 2014).

Nitrous oxide (N_2O) is another important form of GHGs emitted from agricultural soils and livestock, and its emission has also increased due to human activities. The inclusion of all these gases in soil carbon dioxide budgets is crucial due to the interconnection of all processes in ecosystem cycling such as C-N cycle. Global warming potential (GWP) of all these gases is different as it depends on their relative greenhouse efficiency.

CO_2 is a carbon containing gas that is normally present in the atmosphere. Before the industrial time period, the concentrations of CO_2 in the atmosphere ranged between 180 and 290 ppm by 2.1 million years (Honisch et al. 2009). Collectively considered, the atmospheric CO_2 was increased to 240 Pgc between 1750 and 2011 and was abundant at 397 ppm in 2014 (Le Quéré et al. 2016), and this was 40 times more than that before the industrial era.

Carbon dioxide discharges into the environment occur due to the oxidization of SOM or organic remnants. Soil respiration is the discharge of respired CO_2 by soil animals and also due to roots being the second largest discharge on earth (Raich and Potter 1995).

According to its GWP, CH_4 as a greenhouse gas is 28 times more vigorous (Solomon et al. 2007). It is released from the soils in a process called methanogenesis, occurring under anaerobic conditions due to the decay of organic matter. Methanogens in addition to fermenting acetate utilize carbon dioxide as an electron accepting agent instead of oxygen for their metabolism, and CH_4 is released as a by-product. So, the main origins of CH_4 discharges are waterlogged soils and most importantly peatlands, wetlands, and paddy fields (FAO and ITPS 2015). Distinctively, these greenhouse gases are substantially stored in soils (importantly carbon, a process called C sequestration). In the presence of oxygen, methanotrophic bacteria in soil flourish and use CH_4 as a carbon source – a process named methanotrophy in which CH_4 is oxidized. So, soils in forests may be a good CH_4 sinks because their water tables are low and bacteria can grow in these (Serrano-Silva et al. 2014).

4 SOC Sequestration

The process in which carbon fixation takes place through organic matter and plant material and then stored in soil is known as SOC sequestration. It involves three stages, when we deal with CO_2: (1) through the process of photosynthesis, CO_2 removal from the environment, (2) carbon transfer from CO_2 into biomass of plants, and (3) carbon transfer from biomass of plants into the soil as SOC stock. More rate of turnover is characteristic of this pool, enclosing freshly taken residues of plants and is easily decomposed by soil animals, normally CO_2 discharges back into the environment are caused. So, this SOC program needs much more besides accumulating atmospheric CO_2, and it is important to look into the methods of retaining carbon in a pool of SOC. In another way, it is demonstrated by research that there is a minor power for sequestration of carbon because of its resisting property to change, and so in management, it is not responsive (Kane 2015).

A number of mechanisms are involved in the stabilization of newly incorporated carbon in the soil (Six et al. 2006; Kane 2015). In a chemical process, carbon is strongly adsorbed by clay through chemical bonds preventing carbon consumption by other organisms. In a biochemical way, resynthesis of carbon is possible into more complex structural molecules, and decomposition may be prevented. Carbon stabilization efficiency is shaped by the three main procedures that rely on a multitude of biotic and abiotic factors (Six et al. 2006; Kane 2015).

4.1 SOC and Biodiversity

The various living organisms found collectively in soil is termed as soil biodiversity. The organisms that make up soil diversity interact with other plants and animals as well as one another and ultimately form a web of intense biological activity that can be detected as well as studied (Orgiazzi et al. 2016). Soil biodiversity not only greatly contributes to the creation of SOM using the organic litter of the soil, but it also contributes to the augmentation of SOC content in the soil to some extent. Additionally, the quality as well as the amount of SOM and SOC indirectly determines the density along with the activity of plant roots and their interaction with soil biota. In this manner, the structure of the microbial community in the soil is greatly influenced both by the quantity and the quality of SOC as well as the diversity of plant present in the soil (Thiele-Bruhn et al. 2012).

4.2 Importance of Soil Biodiversity

In 1992, during the Convention of Biological Diversity at the United Nations (UNCBD), the vitality of biodiversity was made official. It stated that biodiversity has a key role to play by ensuing that the ecosystem continues to function at an optimal level and every single organism found in the ecosystem, has its place in the ecosystem regardless of its mass and density. The World Soil Charter, released in 2015, referred to soil as the key to global diversity, which includes both flora and fauna present in the ecosystem. This biodiversity plays an essential part in sustaining the various functions of the soil and indirectly supports the properties and the services of the soil. Thus, the maintenance of soil biodiversity is critical to its various purposes (FAO 2015).

The soil biodiversity, i.e., the organisms found in the soil like the protozoa, fungi, bacteria, worms, and insects along with the various invertebrates and mammals, merge with the SOC and form the metabolic capacity of the soil. This capacity is believed to play an important role in the production of food and increasing the resilience of soil in response to climate change. The complex communities of soil organism are responsible for (i) the determination of the size and the course of C fluxes occurring between the soil and the atmosphere, (ii) cycle the SOC that deeply impact the nutrient availability, i.e., the acquisition of nutrients in plants is helpful especially when it works in tandem with associations formed by soil microorganisms, (iii) with the help of aggregation, the physical structure of the soil is somewhat improved, and (iv) boost crop pollination along with biological pest control (FAO and ITPS 2015).

4.3 Soil Biodiversity Losses

Various functions of the ecosystem are seen to have an impact on the density of losses faced by the biodiversity of soil. This mainly includes nutrient retention in soil, SOC decomposition, and cycling of nutrients (FAO and ITPS 2015). Inadequate land management methods and environmental modifications affect various soil communities and rapid decline in the biodiversity of soil (Fig. 1) (Wall et al. 2012). The fragile web of various interactions existing between the communities is damaged by poor agricultural practices that disrupt the agronomic ecosystems. The interactions existing between pests and their natural enemies ultimately have a negative impact on SOC stocks within the soil. Since physical properties of the soil are unable to explain the ratio of loss faced by SOC, it can be assumed that the SOC stability relies on the diversity as well as the activity of the soil organisms (Gardi and Jeffery 2009).

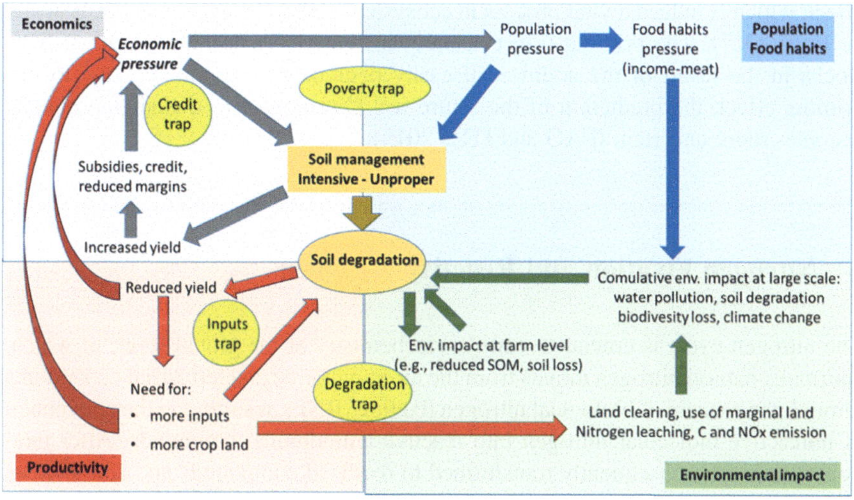

Fig. 1 Impact of land use decisions on soil biodiversity (Cited from Gomiero 2016)

5 SOC Status Under Changing Climate

According to current studies, the range of soil carbon responses varies from small losses to modest gains. On the basis of soil types and region, it is difficult to predict the multiple effects of climatic change on the soil as it gives the complex interaction between increased decomposition and productivity besides moisture and temperature (FAO and ITPS 2015; Keesstra et al. 2016).

The major factors that control the SOC dynamics include precipitation and temperature (Deb et al. 2015). The microbial decomposition of SOC increases with the increase in the temperature (Keesstra et al. 2016). Indeed, there is a strong support for the idea that increasing temperature accelerates the climatic change (Crowther et al. 2016). The ecosystem dynamic is greatly affected by the climatic change than the combined effect of temperature and carbon dioxide (Pachauri et al. 2014).

The increase in net primary productivity is due to increased anthropogenic activity in atmospheric carbon dioxide. This increased NPP (net primary productivity) has a negative feedback on atmospheric CO_2 due to increased SOC input (van Groenigen et al. 2014; Amundson et al. 2015). According to progressive nutrient limitation theory, the response of NPP is limited by the soil nutrient supply specifically of nitrogen. However, it is not clear that either increased NPP will translate into increased SOC storage.

The consequences of the human actions on the global climate are still not obvious, which partly owes to the limited knowledge and understanding about soil respiration and its depiction in various Earth systems (Gougoulias et al. 2014). For instance, on SOM decomposition, a high uncertainty relates to so-called priming effect, which is indeed a vital process in ecosystem carbon balance (van der Wal and de Boer 2017). The priming effect actually increased the decomposition of SOC stocks in the result of the addition of easily degradable compounds. Due to this priming effect, the prediction of the future soil C responses to a changing climate becomes more uncertain (FAO and ITPS 2015).

6 Nitrogen Fixation and Reactive Nitrogen

The nitrogen cycle is crucial to earth's biochemistry at the global level, in which, enormous natural nitrogen moves from the air into marine and terrestrial ecosystems through the process of biological nitrogen fixation (BNF), resulting in the conversion of unreactive molecular nitrogen into reactive ammonium compounds. After fixation, nitrogen is subsequently transformed to oxidized compounds and amino acids by microorganisms and then finally returned as molecular nitrogen to atmosphere from marine and fresh waters, soils, and sediments through the process of microbial denitrification (Galloway et al. 2004). Initial steps of BNF result in compounds that contain N_r (reactive nitrogen) including PAN, NH_4, NO_2, NH_3, N_2O, NO, and HONO as well as other organic nitrogen compounds which are widely distributed in the atmosphere, specifically in cryosphere and play a vital role in marine and terrestrial ecosystems. Galloway et al. (2003) discussed that N_r being component of this earth element basically serves the role of tracer for the biogeochemical cascade. Oxidized form of nitrogen (NO_x) performs a primary role in the production of radical species photochemical ozone and other oxidants (Wayne 1991), and it is also vital for increasing oxidizing capacity of the atmosphere (Isaksen et al. 2009). Emission of N_2O, as a result of denitrification process, also plays an important role in radiation balance on earth and ozone layer chemistry specifically of stratosphere by lysis of N_2O in the presence of light (Wayne 1991). Mere sources of N_r production are BNF and NOx photolysis in the absence of anthropogenic activities.

6.1 Natural Sources of Fixed Nitrogen

The main natural sources of new N_r in the environment are biological nitrogen fixation and lighting. Fowler et al. (2013) estimated that among global natural sources of N_r (203 Tg N $year^{-1}$), lightning just comprised approximately 2.4%. Actually, biological nitrogen fixation in terrestrial ecosystems has been largely influenced by human activities after industrialization. Looking at the pre-industrial BNF estimates, terrestrial ecosystems constituted approximated 1/3 of total BNF as compared to 2/3 of marine ecosystems (Fowler et al. 2013).

6.1.1 Biological Nitrogen Fixation

In purview of global nitrogen cycle, biological nitrogen fixation is the most important source of N_r whether it is in terrestrial or marine ecosystem. Biological nitrogen fixation offers imperative reference while enumerating the value of human inputs in conformity with the nitrogen cycle at the global level because this is the important natural input regarding reactive nitrogen (Vitousek et al. 2002). The magnitude from lightning, discussed later, as this magnitude is smaller than other estimates of BNF, although significant for ozone formation and maintenance regarding the oxidation potential of the atmosphere at the global level.

The procedure of biological nitrogen fixation was first recognized in last decades of the nineteenth century and since then considered as a focal point of ecological interest. There are many limits in understanding why microorganisms are not able to fix all available N as it can benefit them compared to competitor non-nitrogen-fixing organisms, although there is vast capacity in ecosystems to fix available atmospheric N_2. In various ecosystems, the frequently available Nr in the soil generally downregulates BNF, so perhaps N_r application on cultivated soil and deposition into the seminatural region has lowered nonagricultural BNF (Galloway et al. 2004). The present knowledge about strategies has not yet delivered clear solutions according to these queries. The analysis on this subject by Vitousek et al. (2013) presents an estimation on annual pre-industrial BNF into terrestrial ecosystems of 58 Tg N, with an extent from 40 to 100 Tg, with a discussion on modern perception and restrictions. The doubtful range is wide as it represents the difficulty in estimation of the constituent terms. The range inferred by Vitousek et al. (2013) is lesser than mostly available approximations, mostly previous values proposing prior to industrial BNF in range of 100–290 Tg N $year^{-1}$. Nevertheless, this recent estimate is based on calculations of hydrological nitrogen losses in terrestrial ecosystems, and nitrogen portion that is denitrified in rivers or oceans might also be overemphasized.

N cascade in the waters along with BNF and denitrification process is reviewed by Voss et al. (2013). Estimates regarding each term propose also surplus of denitrification process than BNF or equilibrium of both processes (Gruber and Galloway 2008). Nevertheless, doubts among the specific terms prevent a vibrant consent. The sustaining mechanism that maintains the balance between BNF and denitrification at global scale has yet not been verified. However, iron and phosphorus (P) availability may be one of the contributors. By covering approximately three quarters on the Earth's surface and oceans simply take over surface while relatively small fluxes per unit area hold the potential of substantial contribution to overall N fluxes of the atmospheric molecular nitrogen reservoir and marine reactive nitrogen forms. One of major uncertainties in the rates regarding marine BNF is spatial variability, as these are associated with the supply of other essential nutrients too like Fe and P are essential for many biological processes because their supply is also spatially variable (Berman-Frank et al. 2007). Large BNF rates have basically recommended for the Atlantic oceans than the Pacific because of more nutrient availability (Voss et al. 2013).

6.1.2 Lightening

In addition to BNF, nitrogen is also fixed as NO_x by the natural process of lightning, which presents N_r to rather far off areas of the troposphere. The process has been scrutinized via direct measurements and also spacecraft remote sensing of lightning activity has supported this process. Accessible data and models are used in the estimation of global production, but with considerable unreliability to some extent due to strain in up-scaling, by Brasseur et al. (2006), who also examined viable effects of changing climate on NO_x production rate via lightning. These authors enumerate increasing NO_x production rate with an increase of global temperature ranging 3–12% per 8 K. Overall, worldwide informant strength estimates vary from 2 to 10 Tg N year^{-1} (Tie et al. 2002; Cape et al. 2004), with contemporary values or at 5 Tg nitrogen annually. For this evaluation 5 Tg nitrogen per year, value is assumed.

6.2 Impact of Anthropogenic Activities on N Fixation

Nitrogen fixation of compounds based on anthropogenic activities is dubious but well known than natural process of fixation. Because, more substantial measurements have been made to original sections and these are also subjected to considerable scientific study, with continuous supervision of some considerable industrial origins. The gases produced from the transport and industry are oxidized compounds of N, e.g., nitrogen oxide (NO) and nitrogen dioxide (NO_2), burning of biomass, and reduction of nitrogen (N) as ammonia (NH_3) via Haber-Bosch process.

Biological N exists widely in the world as amines in considerable amount (Cape et al. 2004; Jickells et al. 2013). Nevertheless, there is no evidence that these are reactive nitrogen representing compounds, which is acquired from the BNF or commercial sources of ammonia or nitrogen oxides. The involved compounds and mechanisms in the discharge of biological nitrogen and their streaming into the environment are not well known to allow the upscaling for territorial or worldwide estimation of their original strength. Consequently, a substantial involvement to the manmade discharge of Nr may be lacking from the worldwide reactive N budgets raised to date, comprising the one depicted here, as these compounds are dubious to depict more key origins of Nr.

Anthropogenic fixing of reduced nitrogen, i.e., ammonia, is via nitrogen fixing plants and mainly through the process of Haber-Bosch where hydrogen and nitrogen react together in the presence of catalyst and extreme temperature and pressure (Erisman et al. 2008). During the early years of the twentieth century, the process was evolved and supplying 120 Tg nitrogen as ammonia per year, of which 80% is used as fertilizer in agriculture and 20% as raw material for commercial purposes (Galloway et al. 2008). The outcome of N usage in the production of crops differs, with only 17% utilized in crops, milk, and meat productions by humans, the rest of this is lost into the soils, freshwater bodies, and the atmosphere (Leach et al. 2012; Noor 2017). In decades to centuries, the process of denitrification returns most of the Nr to atmosphere as dinitrogen, but the lifetime in dissimilar repositories come back to the atmosphere allows slots for movement in reactive form in freshwater bodies or the atmosphere. Some of the Nr applied in the form of fertilizer is discharged to the atmosphere as ammonia relying on the comparative balance between ambient volumes of ammonia and the equilibrium volumes with the concentration of ammonium within intercellular fluids (Sutton et al. 1995). The total per year production of 120 Tg nitrogen as ammonia via the process of Haber-Bosch depicts the major single involvement to Nr formation via human activities. The total contribution by nitrogen fixing crops to agricultural crops is approximately 60 Tg per annum (Herridge et al. 2008), while the total production of Nr due to anthropogenic activities is 180 (+20) Tg N per annum.

6.3 Nitrogen Fixation in Cropland

N-fixing crops present considerable volumes of Nr to soils. In the evaluation of crop BNF, Herridge et al. (2008) assembled data from the uninterrupted computation of BNF from an array of agroecosystems worldwide and upscaled N fixation rates per year via land use and clipped data to compute a worldwide total. The present worldwide BNF from crops and ruminating savannahs computed by Herridge et al. (2008) is 50–70 Tg N per year. For the sake of condensing data, a principal value of 60 Tg nitrogen per year as the worldwide per year Nr streaming for BNF in cropland is included in this evaluation. The BNF value for cropland is near to the pre-commercial BNF and is identical within the recent array of unreliability.

7 Overview of Nitrogen Cycle

The chemical process of naturally fixed reactive nitrogen takes place in the atmosphere, earth, and oceanic environments. As a result of chemical processing, a large variety of inorganic and organic products are formed. The quantities of Nr cycling in each of these environments and their magnitudes show that worldwide component processes 240 Tg nitrogen in different forms, while 230 Tg nitrogen per year processing takes place in marine, and probably 100 Tg nitrogen per year processing takes place in the atmosphere, neglecting the discharges into and out of the huge environmental nitrogen source. Studies showed that the environmental occupancy time is small as compared to oceanic and earth environments, and with the exception of earth environments, the normal lifespan of Nr is a few decades (Galloway et al. 2013). Such an average over worldwide biosphere is confusing as long-lived sources are there such as peat lands contain organic matter and ice sheets, deserts, and aquifers contain reactive nitrogen. In oceanic and earth environments, the processing of the Nr generation takes place by microbial activity and metabolic activities of plants and their conversion into plant and proteins of microbes. The Nr in agriculture can have direct use as food for human or for feeding purpose for livestock and its conversion, fairly in an ineffectual way to protein and used by humans as food. The conversions resulting from initially fixed nitrogen are important for discharges into groundwater or the environment.

The transfer of reactive form of nitrogen (Nr) from soils to fresh water and atmosphere has a broader application in understanding the leakages of a different form of Nr in soils, water, and vegetation. Galloway et al. (2003) defined the concept of the N cascade to demonstrate many Nr interconversions as well as leakages of Nr during processes of original fixation and denitrification ultimately return back to the atmosphere as N_2. It is helpful to consider an example which describes the fate of nascent NH_3 molecule production to explain the point: consider nitrogen (N) atom converted into NH_3 molecule during the Haber-Bosch process. As it is applied as N fertilizer to the soil, while transformed in many forms before returning as N_2 into the atmosphere.

8 Conclusion

The global N cycle has been greatly modified by human activity, and it is the most perturbed biogeochemical cycle on the planet on which ecosystems depend. Many components of the global budget have been quantified over the last 20 years. However, many fluxes are subjected to large uncertainties and require extensive measurements to constrain the current range of values. Approximately half of the global human population is dependent on the increased yields of agricultural crops owing to fertilizer N usage. These interventions have substantially enhanced the carbon sequestration resulting from N_r deposition to forests and other seminatural terrestrial ecosystems. The full scope of the global carbon cycle and its interdependency on SOC dynamics and relation to other biogeochemical cycles is yet to be fully understood. More accurate SOC measurement, mapping, monitoring, and reporting can contribute to achieving progress in this regard. Land management practices and systems that foster SOC sequestration should aim to retain carbon in the soil over the long term.

References

Amundson R, Berhe AA, Hopmans JW, Olson C, Sztein AE, Sparks DL (2015) Soil and human security in the 21st century. Science 348(6235):1261071–1261071

Berman-Frank I, Quigg A, Finkel ZV, Irwin AJ, Haramaty L (2007) Nitrogen-fixation strategies and Fe requirements in cyanobacteria. Limnol Oceanogr 52(5):2260–2269

Brasseur GP, Schultz M, Granier C, Saunois M, Diehl T, Botzet M, Roeckner E, Walters S (2006) Impact of climate change on the future chemical composition of the global troposphere. J Clim 19(16):3932–3951

Cape JN, Anderson M, Rowland AP, Wilson D (2004) Organic nitrogen in precipitation across the United Kingdom. Water Air Soil Pollut Focus 4(6):25–35

Crowther TW, Todd-Brown KEO, Rowe CW, Wieder WR, Carey JC, Machmuller MB, Snoek BL, Fang S, Zhou G, Allison SD, Blair JM, Bridgham SD, Burton AJ, Carrillo Y, Reich PB, Clark JS, Classen AT, Dijkstra FA, Elberling B, Emmett BA, Estiarte M, Frey SD, Guo J, Harte J, Jiang L, Johnson BR, Kröel-Dulay G, Larsen KS, Laudon H, Lavallee JM, Luo Y, Lupascu M, Ma LN, Marhan S, Michelsen A, Mohan J, Niu S, Pendall E, Peñuelas J, Pfeifer-Meister L, Poll C, Reinsch S, Reynolds LL, Schmidt IK, Sistla S, Sokol NW, Templer PH, Treseder KK, Welker JM, Bradford MA (2016) Quantifying global soil carbon losses in response to warming. Nature 540(7631):104–108

Deb S, Bhadoria PBS, Mandal B, Rakshit A, Singh HB (2015) Soil organic carbon: towards better soil health, productivity and climate change mitigation. Clim Chang Environ Sustain 3(1):26

Erisman JW, Sutton MA, Galloway J, Klimont Z, Winiwarter W (2008) How a century of ammonia synthesis changed the world. Nat Geosci 1(10):636–639

FAO (2015) Revised World Soil Charter Rome. Food and Agriculture Organization of the United Nations.

FAO, ITPS (2015) Status of the World's Soil Resources (SWSR)–technical summary. Food and Agriculture Organization of the United Nations and Intergovernmental Technical Panel on Soils, Rome, Italy

Fowler D, Coyle M, Skiba U, Sutton MA, Cape JN, Reis S, Sheppard LJ, Jenkins A, Grizzetti B, Galloway JN, Vitousek P, Leach A, Bouwman AF, Butterbach-Bahl K, Dentener F, Stevenson D, Amann M, Voss M (2013) The global nitrogen cycle in the twenty-first century. Philos Trans R Soc B Biol Sci 368(1621):20130164–20130164

Galloway JN, Aber JD, Erisman JW, Seitzinger SP, Howarth RW, Cowling EB, Cosby BJ (2003) The nitrogen cascade. Bioscience 53(4):341

Galloway JN, Dentener FJ, Capone DG, Boyer EW, Howarth RW, Seitzinger SP, Asner GP, Cleveland CC, Green PA, Holland EA, Karl DM, Michaels AF, Porter JH, Townsend AR, Vosmarty CJ (2004) Nitrogen cycles: past, present, and future. Biogeochemistry 70(2):153–226

Galloway JN, Townsend AR, Erisman JW, Bekunda M, Cai Z, Freney JR, Martinelli LA, Seitzinger SP, Sutton MA (2008) Transformation of the nitrogen cycle: recent trends, questions, and potential solutions. Science 320(5878):889–892

Galloway JN, Leach AM, Bleeker A, Erisman JW (2013) A chronology of human understanding of the nitrogen cycle. Philos Trans R Soc B Biol Sci 368(1621):20130120–20130120

Gardi C, Jeffery S (2009) Soil biodiversity, vol 27. EUR-OP, Brussels

Gomiero T (2016) Soil degradation, land scarcity and food security: reviewing a complex challenge. Sustainability 8(3):281

Gougoulias C, Clark JM, Shaw LJ (2014) The role of soil microbes in the global carbon cycle: tracking the below-ground microbial processing of plant-derived carbon for manipulating carbon dynamics in agricultural systems. J Sci Food Agric 94(12):2362–2371

Gruber N, Galloway JN (2008) An Earth-system perspective of the global nitrogen cycle. Nature 451(7176):293–296

Herridge DF, Peoples MB, Boddey RM (2008) Global inputs of biological nitrogen fixation in agricultural systems. Plant Soil 311(1–2):1–18

Honisch B, Hemming NG, Archer D, Siddall M, McManus JF (2009) Atmospheric carbon dioxide concentration across the mid-pleistocene transition. Science 324(5934):1551–1554

Isaksen ISA, Granier C, Myhre G, Berntsen TK, Dalsøren SB, Gauss M, Klimont Z, Benestad R, Bousquet P, Collins W, Cox T, Eyring V, Fowler D, Fuzzi S, Jöckel P, Laj P, Lohmann U, Maione M, Monks P, Prevot ASH, Raes F, Richter A, Rognerud B, Schulz M, Shindell D, Stevenson DS, Storelvmo T, Wang WC, van Weele M, Wild M, Wuebbles D (2009) Atmospheric composition change: climate–chemistry interactions. Atmos Environ 43(33):5138–5192

Jickells T, Baker AR, Cape JN, Cornell SE, Nemitz E (2013) The cycling of organic nitrogen through the atmosphere. Philos Trans R Soc B Biol Sci 368(1621):20130115–20130115
Kane D (2015) Carbon sequestration potential on agricultural lands: a review of current science and available practices. Association with: National Sustainable Agriculture Coalition Breakthrough Strategies and Solutions, LLC
Keesstra SD, Bouma J, Wallinga J, Tittonell P, Smith P, Cerdà A, Montanarella L, Quinton JN, Pachepsky Y, van der Putten WH, Bardgett RD, Moolenaar S, Mol G, Jansen B, Fresco LO (2016) The significance of soils and soil science towards realization of the United Nations sustainable development goals. Soil 2(2):111–128
Le Quéré C, Andrew RM, Canadell JG, Sitch S, Korsbakken JI, Peters GP, Manning AC, Boden TA, Tans PP, Houghton RA, Keeling RF, Alin S, Andrews OD, Anthoni P, Barbero L, Bopp L, Chevallier F, Chini LP, Ciais P, Currie K, Delire C, Doney SC, Friedlingstein P, Gkritzalis T, Harris I, Hauck J, Haverd V, Hoppema M, Klein Goldewijk K, Jain AK, Kato E, Körtzinger A, Landschützer P, Lefèvre N, Lenton A, Lienert S, Lombardozzi D, Melton JR, Metzl N, Millero F, Monteiro PMS, Munro DR, Nabel JEMS, Nakaoka S-i, Brien K, Olsen A, Omar AM, Ono T, Pierrot D, Poulter B, Rödenbeck C, Salisbury J, Schuster U, Schwinger J, Séférian R, Skjelvan I, Stocker BD, Sutton AJ, Takahashi T, Tian H, Tilbrook B, van der Laan-Luijkx IT, van der Werf GR, Viovy N, Walker AP, Wiltshire AJ, Zaehle S (2016) Global carbon budget 2016. Earth Syst Sci Data 8(2):605–649
Leach AM, Galloway JN, Bleeker A, Erisman JW, Kohn R, Kitzes J (2012) A nitrogen footprint model to help consumers understand their role in nitrogen losses to the environment. Environ Develop 1(1):40–66
Noor MA (2017) Nitrogen management and regulation for optimum NUE in maize – a mini review. Cogent Food Agric 3(1)
Orgiazzi A, Bardgett RD, Barrios E, Behan-Pelletier V, Briones MJI, Chotte JL, De Deyn GB, Eggleton P, Fierer N, Fraser TJ (2016) Global soil biodiversity atlas. European Commission, Publications Office of the European Union, Luxembourg
Pachauri RK, Allen MR, Barros VR, Broome J, Cramer W, Christ R, Church JA, Clarke L, Dahe Q, Dasgupta P (2014) Climate change 2014: synthesis report. Contribution of Working Groups I, II and III to the fifth assessment report of the Intergovernmental Panel on Climate Change. IPCC
Paul EA (2014) Soil microbiology, ecology, and biochemistry in perspective. In: Paul E (ed) Soil microbiology, ecology and biochemistry. Academic Press, New York, pp 3–24
Raich JW, Potter CS (1995) Global patterns of carbon dioxide emissions from soils. Glob Biogeochem Cycles 9(1):23–36
Schmidt MWI, Torn MS, Abiven S, Dittmar T, Guggenberger G, Janssens IA, Kleber M, Kögel-Knabner I, Lehmann J, Manning DAC, Nannipieri P, Rasse DP, Weiner S, Trumbore SE (2011) Persistence of soil organic matter as an ecosystem property. Nature 478(7367):49–56
Serrano-Silva N, Sarria-Guzm ÁNY, Dendooven L, Luna-Guido M (2014) Methanogenesis and methanotrophy in soil: a review. Pedosphere 24(3):291–307
Six J, Frey SD, Thiet RK, Batten KM (2006) Bacterial and fungal contributions to carbon sequestration in agroecosystems. Soil Sci Soc Am J 70(2):555
Solomon S, Qin D, Manning M, Averyt K, Marquis M (2007) Climate change 2007-the physical science basis: working group I contribution to the fourth assessment report of the IPCC, vol 4. Cambridge University Press, Cambridge
Sutton MA, Schjørring JK, Wyers GP (1995) Plant—atmosphere exchange of ammonia. Philos Trans R Soc Lond A Phys Eng Sci 351(1696):261–278
Thiele-Bruhn S, Bloem J, de Vries FT, Kalbitz K, Wagg C (2012) Linking soil biodiversity and agricultural soil management. Curr Opin Environ Sustain 4(5):523–528
Tie X, Zhang R, Brasseur G, Lei W (2002) Global NO x production by lightning. J Atmos Chem 43 (1):61–74
van der Wal A, de Boer W (2017) Dinner in the dark: illuminating drivers of soil organic matter decomposition. Soil Biol Biochem 105:45–48

van Groenigen KJ, Qi X, Osenberg CW, Luo Y, Hungate BA (2014) Faster decomposition under increased atmospheric CO_2 limits soil carbon storage. Science 344(6183):508–509

Vitousek PM, Cassman K, Cleveland C, Crews T, Field CB, Grimm NB, Howarth RW, Marino R, Martinelli L, Rastetter EB, Sprent JI (2002) Towards an ecological understanding of biological nitrogen fixation. Nitro Cycle Reg Glob Scal:1–45

Vitousek PM, Menge DNL, Reed SC, Cleveland CC (2013) Biological nitrogen fixation: rates, patterns and ecological controls in terrestrial ecosystems. Philos Trans R Soc B Biol Sci 368 (1621):20130119–20130119

Voss M, Bange HW, Dippner JW, Middelburg JJ, Montoya JP, Ward B (2013) The marine nitrogen cycle: recent discoveries, uncertainties and the potential relevance of climate change. Philos Trans R Soc B Biol Sci 368(1621):20130121–20130121

Wall DH, Bardgett RD, Behan-Pelletier V, Jones TH, Herrick JE (2012) Soil ecology and ecosystem services. Oxford University Press, Oxford

Wayne RP (1991) Chemistry of atmospheres, by Wayne RP. Clarendon Press, Oxford, 460 p, ISBN 0-19-855571-7

Rhizosphere as Hotspot for Plant-Soil-Microbe Interaction

Shamina Imran Pathan, Maria Teresa Ceccherini, Francesco Sunseri, and Antonio Lupini

Abstract In the last decades, many studies were addressed to focus the interplay between plant and microbial community into the soil and especially in the small soil zone in contact to plant root, called rhizosphere, which can be considered as a hotspot for interactions and therefore is a major target for improving nutrient use efficiency in crops. In this regard, unraveling the microbial activities that can be used to improve nutrient use efficiency may be the major challenge considering a sustainable agricultural contest. However, although using different approaches (metabolomics and transcriptomic) it has made it possible to characterize many interaction mechanisms, more remains largely unknown. Here, we summarize and discuss the abiotic and biotic factors that may manage plant-microbe interactions in the rhizosphere as well as in those parts of the soil furthest from the root, focusing on root architecture and nitrate as well.

Keywords Metabolomics · Transcriptomic · Rhizosphere · Plant-microbe interaction

Abbreviations

AM	Arbuscular mycorrhizal
ECM	Ectomycorrhizae

S. I. Pathan
Department of Agrifood Production and Environmental Sciences, University of Florence, Florence, Italy
e-mail: shamina.pathan@unifi.it

M. T. Ceccherini
DAGRI – Department of Agriculture, Food, Environment and Forestry, University of Florence, Florence, Italy
e-mail: mariateresa.ceccherini@unifi.it

F. Sunseri · A. Lupini (✉)
Department of AGRARIA, Mediterranean University of Reggio Calabria, Reggio Calabria, Italy
e-mail: francesco.sunseri@unifi.it; antonio.lupini@unirc.it

R. Datta et al. (eds.), *Carbon and Nitrogen Cycling in Soil*,
https://doi.org/10.1007/978-981-13-7264-3_2

GS	Glutamine synthetase
HMW	High molecular weight
JA	Jasmonic acid
LMW	Low molecular weight
N	Nitrogen
NiR	Nitrite reductase
NR	Nitrate reductase
NUE	Nutrient use efficiency
P	Phosphorus
PGPR	Plant growth-promoting rhizobacteria
PLFA	Phospholipid fatty acid analysis
SOM	Soil organic matter
T-DNA	Transmission of DNA
UpE	Uptake efficiency
UtE	Utilization efficiency

1 Introduction

To meet the food needs of the population, an agricultural model based on the overuse of fertilizers has been adopted in recent decades. This model has created severe problems for both environmental quality and human health (Good et al. 2004). In addition, this approach has involuntarily led to the selection of genotypes with a low nutrient use efficiency. Given forecasts that the world population is increasing, around 9.2 billion on 2050 (FAO 2009), more crop production is needed using alternative strategies reducing the input of chemicals and improving food quality without negatively affecting the environment (Xu et al. 2012).

The rhizosphere is often defined as the area of soil around a root where the population of microorganisms depends on inputs from the plant. The rhizosphere is a soil microenvironment where plants and microorganisms can coexist in positive, negative, or neutral interactions (Lynch and Whipps 1990; Kardol et al. 2007) (Fig. 1). It is populated with numerous organisms, including fungi, bacteria, etc. and possibly exhibiting one of the highest levels of biological biodiversity of any environment in the world (Bender et al. 2016) (Fig. 1). Given the strong interaction between plants and microbes in the rhizosphere, it can be considered an extension of the plant's genome (Berendsen et al. 2012; Verma et al. 2015). As crop yield is strongly dependent on water and nutrient uptake from the soil, the rhizosphere can be considered as a *hotspot* for interactions and therefore is a major target for improving nutrient use efficiency in crops.

Rhizosphere microbes may improve plant growth not only by making nutrients more available for uptake by the root (e.g., phosphate-solubilizing bacteria) but also through the production of phytohormones that can improve plant resistance to biotic and abiotic stresses (Berendsen et al. 2012). Plant roots secrete exudates derived from photosynthesis providing an important carbon supply for the growth of microorganisms (Brimecombe et al. 2007).

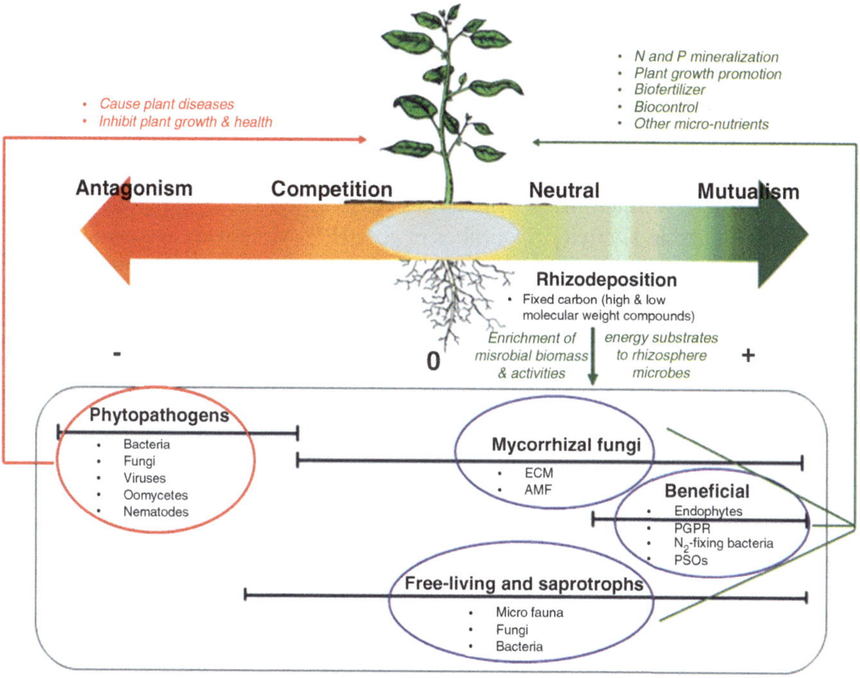

Fig. 1 Representative overview of different plant-rhizosphere microbe interactions and their activities (The figure has been adapted from Quiza et al. 2015 and modified)

Therefore, engineering the rhizosphere may be a useful target for developing more sustainable agriculture (Zhang et al. 2017). Despite the importance of the rhizosphere and the innumerable publications in this field by using new technologies related to DNA sequencing and metabolomics, to date much remains to be discovered. More information is needed to understand the mechanisms involved in rhizosphere interactions and particularly how plants control their microbiome and vice versa how microorganisms influence crops.

The present review focuses on the recent knowledge of the rhizosphere, emphasizing particularly microbe and root interactions in relation to nitrogen supply.

Finally, this chapter reports some examples of how these interactions can be used to improve crops in a background scenario of more sustainable agriculture in the context of climate change.

2 Rhizosphere as an Active Network

The "rhizosphere" term was first conceived by Hiltner (1904) to delineate the plant root-soil interface. The rhizosphere was defined as the soil microenvironment where chemical, physical, and biological properties are affected by plant roots and their metabolic activities. Thus, the rhizosphere is one of the most dynamic networks of the terrestrial ecosystem, where direct plant-soil-microbe interaction takes place. Plant roots release up to 40% of photosynthetically fixed carbon directly into soil (Bais et al. 2006) mainly in the form of high and low molecular weight compounds (Newman 1985; McNear 2013). These compounds especially low molecular weight molecules such as amino and organic acids, phenolic compounds, sugar, etc. can be used by rhizosphere microbes as energy sources, which cause enrichment of the microbial biomass and activity (Doornbos et al. 2012). Moreover, exudation significantly influences soil physical and chemical properties (Nardi et al. 2000) such as soil pH (Javed et al. 2012), soil aggregation and erosion prevention (Naveed et al. 2017; Wang et al. 2017; Meena and Meena 2017), and water holding capacity (Young 1995). On the other side, root exudation helps the plants to gain essential nutrients from the soil through acidification and oxidation/reduction processes in the rhizosphere (McNear 2013). Nitrogen (N) and phosphorus (P) are key plant growth-limiting nutrients, and rhizosphere microbes play a key role through degradation and mineralization processes to deliver these essential nutrients to the plant. Much of the soil N and P are stored in complex organic forms that plants cannot easily access often requiring breakdown and solubilization by microbes before uptake by roots can occur. Thus, the rhizosphere is the apropos niche for plant and soil microbes, where key plant-soil-microbe interactions take place giving benefits to both organisms for plant mineral nutrient acquisition and substrate for microbial energy requirements. These interactions can either be beneficial (symbiotic) for the crop, for example, plant interactions with epiphytes, such as plant growth-promoting rhizobacteria (PGPR), nitrogen-fixing bacteria, and mycorrhiza fungi, or can be negative (parasitic) with plant pathogenic microorganisms (Singh et al. 2004; Raaijmakers et al. 2009; Walkers et al. 2003; Dadhich and Meena 2014). Hence, the rhizosphere is the hotspot where the plant and soil microbial community communicate with each other through the plant roots and mediated by root exudations.

3 Root Exudates Regulating Factors

Plant roots discharge immense amounts and ranges of organic compounds into the rhizosphere soil, known as root exudates or rhizodeposition, and through these exudates, plants can directly communicate with rhizosphere microbiota. Root exudates are an important nutrient for the microbial community in rhizosphere soil and therefore encourage root-microbe colonization (Bacilio et al. 2002). Plant roots can exude many different compounds such as acids, oxygen, and ions, but most are organic molecules (Uren 2000; Bais et al. 2006). These organic-based exudates can be categorized into high and low molecular weight (HMW and LMW, respectively) compounds. The majority of exudates are LMW, including amino and organic acids, sugars, phenolic compounds, and many other secondary metabolites, while considered HMW exudates are more complicated compounds such as mucilage and cellulose (Huang et al. 2014; Rovira 1969). Although root exudates are a key driver of plant-microbe communication in rhizosphere soils, less attention has been given to the mechanisms and regulatory processes that control root exudation. Plant age and genotype, as well as other external biotic such as nutrient availability and root architecture and abiotic factors like soil properties, temperature, stress, or toxic conditions, are known to regulate the quantity and quality of root exudates (Badri and Vivanco 2009; Varma et al. 2017).

3.1 Abiotic Factors

3.1.1 Soil Properties

Soil physical and chemical properties such as pH, texture, moisture, etc. significantly regulate the root exudation process; mainly soil moisture content is one of the key drivers of the root exudation, as high soil moisture leads to hypoxia condition due to the inadequate accessibility of O_2 in the soil. Root mucilage exudation can improve the physical pathway for water and nutrient delivery to roots, and this is particularly important as soil dries (Carminati et al. 2016). Low soil moisture can cause temporary plant wilting which caused an increase in the release of amino acids from plant roots (Katznelson et al. 1954). Some bi-cropping studies suggested that high moisture content could help in the rapid transfer of maize root exudates to adjacent bean plants (Ivano 1962). Water stress, including both drought and flood conditions, can significantly impact on the quantity and quality of wheatgrass root exudates as the stress-induced exudation of organic acids such as malic, fumaric, malonic, succinic, and oxalic acids (Henry et al. 2007). Similarly, Song et al. (2012) also reported that exudation of various organic acids such as malic, lactic, acetic, succinic, citric, and maleic acid concentration was increased by osmotic stress in drought-tolerant and non-drought-tolerant maize in hydroponics. Furthermore, soil pH can have a significant effect on root exudation resulting in acidification and alkalinization. Stoltz and Greger (2002) found that the increased soil pH produced by *Eriophorum angustifolium* and *E. scheuchzeri* roots could decrease the leaching of toxic elements such as Cd, Cu, Pb, and Zn, while Wang et al. (2006) reported that a lower pH could lead to higher soil leaching of toxic elements such as Cd and Zn. Recent studies on *E. angustifolium* roots showed that changes in the rhizosphere to a more alkaline pH significantly influenced the exudation of organic acids (oxalic, succinic, and formic acids) in root mucilage under moderately toxic soil conditions (Javed et al. 2012). There are very few studies focused on how soil texture can influence the root exudation process. Soils with different textures have different chemical and physical properties which can cause direct or indirect induction or inhibition of LMW and HMW compound secretion by plant roots. Sandy substrates can induce root exudation as high secretion rates of amino acids were reported when plants were grown in quartz sand (Boulter et al. 1966). Since soil characteristics significantly affect root exudation, little attention has been given to these types of studies, which may be due to the plasticity and variability of root exudates. The study of root exudates is hindered by difficulties in the collection of representative samples, especially in the soil and under field conditions. Plant root exudates may be dissimilar in soil and hydroponics. However, recent molecular techniques combining with stable isotope probing could shed more light on the effect of soil physical characteristics on root exudation.

3.1.2 Temperature

Temperature can have major effects on root exudation. Rovira (1956) reported that higher temperature could cause the increase in the exudation of different amino acids, especially asparagine from *Solanum lycopersicum* (tomato) and *Trifolium subterraneum* (subterranean clover). Other studies showed that exudation of tannins and phenolic compounds was markedly inhibited at low temperature (4 °C) compared to the amount at 30 °C in fava beans (*Vicia faba*) (Bekkara et al. 1998). Contrariwise, Husain and Mckeen (1963) found that strawberry roots secrete higher amount of organic acids at low temperature (5–10 °C) than at higher temperature (20–30 °C), which also leads to root colonization with the pathogenic fungus *Rhizoctonia fragariae* (Husain and Mckeen 1963; Hale and Moore 1979). A generalized plant exudation response to temperature may not be found; it seems more likely that each species responds differently to temperature extremes.

3.1.3 Light Intensity

A few studies have shown that the composition and quantity of root exudates were affected by light intensity, as it is directly linked to photosynthetic C fixation, the main C source for root exudates. For example, Rovira (1956) reported that trefoils grown under full daylight released higher amounts of serine, glutamic acid, and alanine but shadow inhibited the exudation of aspartic and glutamic acid, phenylalanine, and leucine in grown plants. Many authors reported the quantity of root exudates was significantly increased by the high light intensity and prolongation of the photoperiod, for example, phosphatidylserine in Zn-deficient wheat (Cakmak et al. 1998), catechin in spotted knapweed (Tharayil and Triebwasser 2010; Yadav et al. 2018), and citrate in white tulips (Cheng et al. 2014). Although fluctuations in light intensity combined with the altered photosynthetic spectrum (longer wavelengths) affected the synthesis of secondary metabolites in the leaves of birch and woody plants, the root exudates were not measured (Lavola et al. 1998). However, Yang (2016) reported that light intensity not only influences the quantity but also significantly impacts on the composition of root exudates in sugar beet. Recently Martin et al. (2018a, b) reported that fluctuations in light exposure increased the exudation of dissolved organic carbon and protein-like and humic-like dissolved organic matter, whereas constant low and continuous light significantly inhibit the exudation of total dissolved nitrogen. These findings suggest that light intensity and exposure time are two of the key factors that regulate synthesis and secretion of root exudates, and these should get more attention for future studies on root exudates and their regulating factors.

3.1.4 Nutrient Availability in the Rhizosphere: Nitrogen as Nutrient and Sensor

Although the rhizosphere is confined in a small area (about 2 mm) between root and soil, it is an important zone affecting nutrient dynamics influenced by a plethora of microbial activities useful for improving crop uptake efficiency (Dakora and Phillips 2002). Different factors combine to determine the availability of nutrients in this zone, such as soil chemistry, plant genotype, and nutrient supply (Jones et al. 2004). The availability of nutrients depends on the activity of microorganisms, which in turn are regulated by quantity and quality of the root exudates, thus creating different local micro-ecosystem compared to the bulk soil (Neumann et al. 2009; Meena et al. 2018). Various plant mechanisms have been developed to cope nutrient-limited conditions, but the exudation of organic acids and enzymes into the soil by roots can enhance soil organic matter (SOM) decomposition thus releasing nutrients making them available for plant uptake (Charholm et al. 2015). Some plant species exhibit specific rhizosphere effects on the availability of nutrients, such as the Solanaceae influencing phosphorus mobility in the soil (Safari and Rashidi 2012). Grasses can produce phytosiderophores to increase iron availability by chelation of the metal ion, particularly under Fe-limiting conditions (Ueno et al. (2007). From the bulk soil to the root surface, the microorganism populations increase in quantity but decrease in species diversity, thereby influencing the availability of nutrients. In the rhizosphere, there is an equilibrium between the plant, soil, and microorganisms which is usually characterized by low nutrient concentrations, where a constant turnover can ensure a steady supply for roots (Shen et al. 2013). Considering phosphorus, for example, it can be present in large amounts in the soil, but it is totally unavailable for the plant as both organic and inorganic forms are insoluble (Bhattacharyya and Jha 2012; Igual et al. 2001; Gyaneshwar et al. 2002). Applications of P fertilizers may not combat this limiting condition due to the phenomena of precipitation occurring in the soil (Mckenzie and Roberts 1990). However, in the rhizosphere, some microorganisms are defined as P-solubilizing, such as *Azotobacter*, *Burkholderia*, *Erwinia*, *Flavobacterium*, and *Serratia* (Bhattacharyya and Jha 2012), and these have the ability to make this element soluble by synthesizing the low molecular weight organic acids which dissolved the inorganic phosphorus converting it to soluble forms (Zaidi et al. 2009) and therefore available for plant uptake (Mehnaz and Lazarovits 2006).

N_2 fixation is another important example of a microbe-plant interaction (discussed in Sect. 5.1), and its significance is dictated by the role that this element (N) has in plant physiology and is often limiting crop productivity. Among nutrients affecting plant growth and development, N plays a pivotal role, and nitrate, which is the main N-form in aerobic soils, can also be considered as a “signaling molecule.”

Nitrate can act as a signal to regulate plant gene expression, metabolism, physiology, growth, and development (Krouk et al. 2011; Vidal and Gutiérrez 2008). Although the capacity of the plant to assimilate N depends from on carbon turnover, an increase in biomass is principally limited by nitrogen uptake from the rhizosphere. In plants grown using nitrate-free conditions, the presence of nitrate leads to the modulation of enzymes responsible for assimilation, such as nitrate reductase (NR), nitrite reductase (NiR), and glutamine synthetase (GS), which play fundamental roles in crop production (Crawford 1995; Stitt 1999). In the last decades, genomics, bioinformatics, and systematic have described a complex regulatory network at transcriptional and posttranscriptional levels for the plant's responses to nitrate (Krouk et al. 2010). In *Arabidopsis*, nitrate and N metabolites regulate the expression of numerous genes involving a wide range of processes. Studies of the effect of nitrate supply on the quality and quantity of root exudates are missing, especially in space and time. According to Scheible et al. (2004), who provided an illustration of the signaling action of N in *Arabidopsis* grown in an N-free solution, nitrate supply determines a massive reprograming of genome expression. Within 30 min, the re-addition of nitrate induces gene expression involved in uptake, reduction of nitrate, and organic acid skeleton production. On the other hand, after longer times (about 3 h), the nitrate supply induces specific genes belonging to trehalose and hormone metabolism, protein kinases and phosphatases, receptor kinases, and transcription factors (Scheible et al. 2004), and some of these processes may be important in root-rhizosphere interactions, thereby increasing or reducing the crop nitrate uptake efficiency. As affirmed by Krouk et al. (2010), up to 10% of the transcriptome is responsive to nitrate, and many genes are considered as signals, as these genes are still nitrate-regulated in *Arabidopsis* NR-deficient mutants, where the first enzyme of the nitrate assimilation pathway is missing (Wang et al. 2004; Kumar et al. 2018).

Finally, the global nitrate signaling pathways show complex regulation either into the plants by transcriptions factor or external to the plants by nitrate availability, which in turn is strongly dependent on the rhizosphere activities. Thus, the rhizosphere is the *hotspot* where both the concentration of microorganisms is very high compared to bulk soil and the concomitant activity of roots and microorganisms allows the creation of a microenvironment favoring mineralization (Parkin et al. 2002) and denitrification (Qian et al. 1997). Finally, little is known about the mechanisms that microorganisms have on local nitrate signals in the rhizosphere.

3.2 *Biotic Factors*

3.2.1 Plant/Rhizosphere and Nutrient Use Efficiency

The current requirement in most agriculture is to reduce the inputs of fertilizers in formulating management practices for more sustainable production. New strategies for plant nutrition will require adequate consideration of rhizosphere processes for improving crop productivity (Shen et al. 2013; Zhang et al. 2017). Starting from a general point of view and focusing on specific topics, strategies to maximize the efficiency of rhizosphere processes related to plant nutrition can be assessed by manipulating root growth patterns and targeted fertilizer applications (Zhang et al. 2010; Chen et al. 2011).

Nutrient use efficiency (NUE) is determined by two components: uptake efficiency (UpE), or the root's ability to take up nutrients from the soil, and utilization efficiency (UtE), as the plant capacity to assimilate or utilize the nutrients (Good et al. 2004; Xu et al. 2012; Lupini et al. 2017; Dadhich et al. 2015). In this context, as rhizosphere is strictly in contact with the roots, UpE is the component most influenced by the rhizosphere processes and on which it is possible to identify strategies for improving NUE.

In legume species *Rhizobium* spp. accommodated in specialized nodule structures can fix a considerable amount of N to supply the plant. Other microorganisms increase in the availability of nutrients in nonlegumes (Boddey et al. 2003). There are some microorganisms, such as *Azospirillum* spp., which can supply N to plants (Assmus et al. 1995) increasing the NUE more specifically the UpE. Relationships between the roots and microorganisms in the rhizosphere favor nutrient uptake: on the one hand, there are the root exudates that provide organic compounds to microorganisms (Parkin et al. 2002), and on the other side, they provide inorganic compounds more easily taken up by the plant (Marschner 1995). Some types of bacteria form a biofilm on the roots, thereby ensuring an improvement in nutrient uptake (Beauregard et al. 2013) by providing a robust water contact between soil particles and the root. Using PCR coupled with denaturing gradient gel electrophoresis and fluorescence approaches, Briones et al. (2003) demonstrated a strong correlation among ammonia-oxidizing bacteria and NUE of some rice cultivars also underlining that the presence of these bacteria improved nitrate uptake. Therefore, nutrient uptake may be improved through the selective production of root exudates, which favor the association with specific bacteria and decreasing the abundance of others. Nutrient availability in the rhizosphere is strongly regulated by exudation, but the manipulation of high-affinity nutrient transport systems at plasma membrane level is poorly studied, and increased chelating creates nutrient-limited zones around the roots (George et al. 2005; Liu et al. 2005). Furthermore, the selectivity of root exudates could be achieved through the molecular manipulation of the transporters to improve both the mineralization processes and the numbers of beneficial microorganisms in the rhizosphere. To date, few transporters involved in plant/microbe rhizosphere cross talk were completely functionally characterized (Kretzschmar et al. 2012; Rudrappa et al. 2008), and further studies are needed in crop species. Some authors have demonstrated that plant roots secrete organic compounds to inhibit nitrification (Subbarao et al. 2013) and denitrification (Cordero and Datta 2016), which may be used in agriculture to improve NUE. Manipulating these microbial processes in the rhizosphere can be used to develop crops with higher NUE by decreasing N leaching losses.

3.2.2 Plant Root as Main Trait to Improve NUE

The major limiting factors for the twenty-first-century crop production are the scarcity of water and nutrients in the soil. The plant root system is fundamental for acquiring nutrients and water which are primarily taken up from the soil. In this context, different approaches may be used to improve NUE: (i) changing root architecture to enhance nutrient acquisition especially in nutrient-limiting conditions and (ii) managing the rhizosphere processes starting from root modifications. In fact, it is now well established that root architecture plays a fundamental role in nutrient use efficiency (Lynch 2011) and influencing microorganisms in the soil too (Wang et al. 2011; Wu et al. 2012). Improving the capacity of the root to explore the soil by modifying root architecture is a pivotal task. Thus, understanding internal and/or external factors tuning the growth and development of this organ is an important target to improve water and nutrient use efficiency. Among macroelements, N (mostly nitrate) and phosphorus are the main limiting factors for plant growth and development (Gojon et al. 2009), and plants have adopted strategies to improve nutrient uptake efficiency by modifying root morphology and architecture. Nutrient dynamics in the rhizosphere can lead to substantial and specific changes in roots. Using varying N levels, it was possible to identify specific root adaptations regulated by local or systemic signaling (Bellegarde et al. 2017; Yadav et al. 2017). At the molecular level, genes, transcription factors, proteins, and miRNAs have all been identified to have their role in root architecture and nutrient use efficiency. Mlodzínga et al. (2015) showed the role of a gene (*AHA2*), a plasma membrane H^+-ATPase in root morphology and architecture of rice under N-limiting conditions. Other molecular evidence was established in P-limiting conditions, as *OsPHR2* gene and the*OsMYB2P-1* transcription factor influenced root hair development and root architecture in rice (Wu and Wang 2008; Dai et al. 2012). QTLs were also associated with root architecture in maize (Li et al. 2015), rice (Li et al. 2009), and bean (Cichy et al. 2009). Of course, these results will be useful to use as molecular markers in breeding programs to improve root architecture and consequently nutrient use efficiency.

Root morphology is also a genetic factor modifying the rhizobiome. Diverse root types can contribute differently to microorganism selection in the rhizosphere. Recent publications point out how the microbe-community can change along roots and among root types. Edwards et al. (2015) showed how the dynamics of the root-associated microbes was strongly dependent on root niche in rice plants, whereas other authors have underlined the bacterial variation among plants, depending on age and genotypes (Lundberg et al. 2012) or different genotypes changing their *ribotype* profile (16S rRNA gene sequence) depending on host cells (Bulgarelli et al. 2012). Among different root zones, the highest number of bacteria is localized at the root tip, which may be able to select for specific microbes. In wheat seedlings, De Angelis et al. (2008), using high-density 16S rRNA microarray (PhyloChip), showed a differing number of microorganisms with the following hierarchy, bulk soil < mature root < root tips = root hairs, whereas specific colonization by *Bacillus subtilis* was observed in the root elongation zone of *Arabidopsis*, suggesting a possible protection mechanism against root pathogens (Massalha et al. 2017). The mature root zone included decomposers which may be associated with the degradation of cells from older root parts (Jones et al. 2009). Distinct communities were also confirmed in lateral roots as compared to tips and basal regions. In this respect, *Brachypodium distachyon* was employed as a monocotyledon model, and the communities assayed were dependent on root type (nodal or seminal) and root axis (Kawasaki et al. 2016a, b). Thus, as the plant genotype and root system regulates rhizosphere microbiota, more work in this field may be interesting for improving crop productivity. There is an emerging need to evaluate the differences among plant populations within the same species. Comparing wild and modern bean accessions, Pérez-Jaramillo et al. (2017) showed a difference in bacterial numbers that was highly correlated with root morphology. In conclusion, some authors affirm that root phenotype influences rhizosphere colonization by differences in cell wall structure, surface area, and exudate metabolic profiles (Pérez-Jaramillo et al. 2017; Saleem et al. 2016). Thus, it may be important to consider factors modifying root morphology such as nutrient and stresses to drive agronomic practices that optimize rhizosphere interactions.

4 Microbial Selection by Plants

The rhizosphere is where the plant and soil microbe communities are directly linked to each other based on their nutrient requirements. It is well documented that plants can shape their rhizobiome (rhizosphere microbial community), as plant genotype, growth stage, and soil type significantly affect the abundance, composition, and diversity of the rhizobiome (Berg and Smalla 2009; Philippot et al. 2013; Bulgarelli et al. 2013a, b). Selection of the rhizobiome is directly related to the plant root exudation pattern and other abiotic factors such as soil temperature, moisture, chemistry, and physical structure (Pugnaire et al. 2004; Hartmann et al. 2009; Bargett and van der Putten 2014). Recently, Burns et al. (2015) studied different predictors which can affect the rhizosphere microbial community using statistical methods, and they pointed out that plant species is the main determinant of the rhizosphere microbiome, which suggests there is an active microbial selection by plants. Plant-soil feedback strongly relies on plant species and rhizosphere microbe association (Kulmatiskietal. 2008) which implicates plant-microbe coexistence and community assembly (Bever et al. 2010). Furthermore, Burns et al. (2015) also reported that spatial location is also a key predictor which shapes the microbial community, and they also suggested that everything is not everywhere with localized hotspots on the root surface for specific populations.

Since Smalla et al. (2001) first studied the rhizosphere bacterial community using culture-independent techniques and suggested that each plant was colonized by their own bacterial community, numerous studies have focused on how different plant genotypes, plant growth period, and nutrient use efficiency have shaped or influenced abundance, composition, diversity, and functions of rhizosphere microbial communities. Due to recent advances in sequencing technology, there is a long list of studies on this topic, but here we will mainly focus on agriculturally important plants such as wheat, rice, and maize and how these plants and their genetic variation can influence the rhizosphere biome. Donn et al. (2014) studied the rhizobiome of two different wheat lines and suggested that the plant growth stage and level of adhesion of soil microbes to the root were main drivers of the bacterial community. The authors found a tenfold enrichment in the abundance of *Pseudomonas*, *Actinobacteria*, *copiotrophs*, and *oligotrophs* in the loosely bound rhizosphere soil during vegetative growth. Furthermore, *Pseudomonas* was highly abundant in the Janz wheat line, whereas *copiotrophs* were more abundant in the H45 line (Donn et al. 2014). Similar findings were observed by Kawasaki et al. (2016a, b) who showed that the binding of soil microbes to a root and root structure have a profound impact on the rhizobiome of *Brachypodium distachyon* (model plant for wheat) and its rhizosphere community was enriched by *Burkholderiales*. One recent study showed the intense effect of plant domestication and breeding on exudation which has directly influenced rhizosphere metabolites in wheat lines (Lannuci et al. 2017). Likewise, Liu et al. (2016) studied the effect of Jasmonic acid (JA) signaling on the wheat rhizosphere microbiome and showed plant organ-specific effects as JA signaling only had a significant impact on diversity and composition of root endophytes but not shoot endophytes or the rhizosphere community.

Lu et al. (2014) studied the rice rhizosphere microbiota using PLFA-based stable isotope labeling and found the rice plant had significantly influenced the microbial community through photosynthetic rhizodeposition. Breidenbach et al. (2016) studied the abundance and composition of the microbial community in the rice rhizosphere soil and showed that the abundance of archaea and bacteria was significantly higher (twofold) in the rhizosphere when compared with the bulk soil. Furthermore, these authors found profound fluctuations in *Proteobacteria*, *Gemmatimonadetes*, and *Verrucomicrobia* communities and showed their effect on functional groups such as iron reducers and fermenters. One recent study showed the autonomous behavior of rice plant in relation to the microbial community, as plants were grown in different types of soil, had the same effect on the composition of rhizosphere microbial community (Li et al. 2014). Similarly, the study of rice plants has showed that plant domestication and genotype play key roles in shaping the rhizosphere community (Lannuci et al. 2017; Shenton et al. 2016; Meena et al. 2015). Moreover, the enrichment of different bacterial phyla *Anaerolineae* and *Methanotrophs* in the wild and bred rice cultivars, respectively, were identified (Shenton et al. 2016).

Compared to other crops, there have been large numbers of studies on the maize rhizobiome. This may be due to its unique phenotypic and molecular diversity; we have discussed the more recent studies done on the maize rhizobiome. Cavaglieri et al. (2009) studied maize rhizobiome using microbial culturing-dependent techniques, and they reported that the plant growth stage has a significant influence on the rhizoplane and endo-rhizosphere community with noteworthy enrichment in some bacterial species such as *Bacillus*, *Arthrobacter*, *Azotobacter*, and *Listeria* and fungi such as *Aspergillus* and *Fusarium*. García-Salamanca et al. (2013) studied the bacterial community of the maize rhizosphere in carbonate-rich soil and reported a substantial enrichment of *Gammaproteobacteria* in rhizosphere soil which could be due to the high availability of carbonate. The rhizosphere bacterial diversity of 27 different maize inbred lines which were grown in field condition showed a remarkable influence of plant heredity on diversity and composition of rhizosphere bacterial community (Pieffer et al. 2013). Li et al. (2014) also studied the maize rhizobiome using a pyrosequencing approach, and they reported the preferential colonization of some bacterial groups such as *Proteobacteria*, *Bacteroidetes*, and *Actinobacteria* in rhizosphere. These authors also showed that the plant growth stage had a profound effect on the bacterial community composition in rhizosphere soil; fluctuations were mainly observed at the family, genus, and OTU level. Recently, few studies have focused on how maize plant N use efficiency (NUE) could shape the totals and functional diversity of rhizosphere soil microbes (Pathan et al. 2015a, b; Baraniya et al. 2016). These authors showed that the Lo5 maize line with higher NUE had stronger influence on the microbial community composition (bacteria and fungi). Both the composition and diversity of bacterial β-glucosidase and protease genes in rhizosphere soils were compared between Lo5 and T250 maize, a low NUE line. All these findings suggest the plant and its different characteristics such as plant heredity, growth stages, physiological traits, and plant nutrient requirements are key factors which drive the formation and structure of the rhizosphere microbial community. It should be noted that many of these studies have only been focused on bacteria and very few of them have focused on the fungal or archaeal communities, which suggests that future research should consider these other rhizosphere organisms in different plants.

5 Plant-Microbe Interaction

5.1 N_2-Fixing Bacteria

Nitrogen is a key nutrient for plant development, and it is usually a plant growth-limiting factor. Although more than two thirds of the global atmosphere comprise N_2, plants are incapable of using this elemental form of N. On the other side, some soil prokaryotes known as diazotrophs have a dinitrogenase enzyme which enables these prokaryotes to convert N_2 into ammonia followed by nitrite (ammonia oxidizers) and nitrate (nitrite reductase), N forms used by plants (Lam et al. 1996; Franche et al. 2009). Due to the plant's large requirement for N, some have formed beneficial symbiotic associations with these diazotrophs (N_2-fixing bacteria). This symbiotic association between plants and N_2-fixing bacteria is a kind of mutualistic symbiosis where the host plant directly consumes inorganic N such as ammonia or nitrate from the soil which is fixed by the N_2-fixing bacteria. In return the plant provides sheltered environment and fixed carbon and other nutrients to the diazotrophic prokaryotes. Legumes (Fabaceae family) are the foremost symbiotic plants that associate with *rhizobia* (gram-negative), a member of

Alphaproteobacteria (Oldroyd and Downie 2008). Legume-rhizobia interaction instigates root nodule formation, which begins with the release of flavonoids by the plant under N starvation conditions (Oldroyd et al. 2011; Mcnear 2013). Flavonoid signals are responsible for activating *nod* genes in rhizobia which encoded lipochitooligosaccharides, known as nod factors (Mcnear 2013; Mus et al. 2016). Nod factors are pivotal symbiotic signals as they induce invasion of bacteria into the host plant and nodule formation where the bacteria ultimately are housed (Mcnear 2013; Mus et al. 2016). The structure of the nod factors such as the length of backbone, size, and saturation of fatty acyl chain depend on the rhizobia species (Mcnear 2013; Mus et al. 2016) which leads to host specificity during formation of the plant-rhizobia association (Oldroyd and Downie 2008). Even though two different species of rhizobium, *etli* and *lati*, carry identical nod factors, both species have distinct host specificity (*Phaseolus* spp. and *Lotus* spp., respectively) (Cárdenas et al. 1995). Host plant selection specificity is astounding as out of millions of microbes; only a tiny number are able to create efficacious symbiosis with host plants (Mcnear 2013). Only one nonlegume plant, called *Parasponia* species (Cannabaceae family), is capable to form a symbiotic association with rhizobia (Sytsma et al. 2002). Invasion of rhizobia occurs through crack entry into the host plant *Parasponia* (Lancelle and Torrey 1985), which is different from legume plants, where the rhizobia enter through root hair curling. Furthermore, the proliferation of rhizobia and N fixation takes place after the formation of fixation threads, and these threads branch all over the nodule cells; however, they still stay in contact with the apoplast (Behm et al. 2014; Mus et al. 2016; Ashoka et al. 2017). Somehow, development of the *Parasponia-rhizobia* association is quite young and is a primitive form of nodulation. Behm et al. (2014) suggested that the *Parasponia-rhizobia* symbiosis can be used as a model system to understand the control mechanisms which emerge during the early stages of N_2-fixing mutualism evolution.

Apart from rhizobia, members of the *Actinobacteria* phylum, *Frankia* sp. (gram-positive), have been shown to nodulate (actinorhizas) with broad spectrum of woody plants, called actinorhizal plants. Actinorhizal plants are distributed in 8 different families, containing 17 genera and 150 species. Actinorhizal development processes are very similar to legume nodules, but they are much larger in size and longer lived compared to legume nodules. Actinorhizas have central vasculature and can fix the same amount of N_2 as rhizobia nodules (Mus et al. 2016). Another group of bacteria called cyanobacteria, especially filamentous cyanobacteria (mainly *Nosotc* and *Scytonema*), are able to fix N_2 with cells which are known as heterocysts. Cyanobacteria create this symbiotic association with a variety of higher and lower plants, algae, and fungi (Merks and Elhai 2002). Moreover, there are some other N-fixing bacteria, such as *Azospirillum* spp. and *Azoarcus* spp., that can colonize nonlegume plants without any nodule formation (Elmerich and Newton 2007), and these bacteria are known as N_2-fixing endophytes (Döbereiner 1992; Baldani and Baldani 2005). Consequently, it may be important to create more symbiotic associations between N_2-fixing bacteria and nonlegume plants, especially in different crops, thereby reducing the use of chemical N fertilizer and better manage greenhouse gases fluxes.

5.2 Plant Growth-Promoting Bacteria (PGPR)

Kloepper and Schroth (1978) were the first authors who coined the term "plant growth-promoting rhizobacteria" (PGPR) for bacteria inoculated on seeds that successfully colonized the plant root and promoted plant growth. Antoun and Kloepper (2001) reported that only a few bacteria (1–2%) have an aptitude to promote plant growth. PGPR are propitious free-living bacteria which can colonize the rhizosphere and rhizoplane and within the root itself (Gray and Smith 2005). Moreover, based on this colonization, PGPR can be divided into two subcategories: (i) bacteria which colonized the rhizosphere or rhizoplane are called extracellular PGPR (ePGPR) and (ii) those positioned inside root tissues are known as iPGPR (intracellular PGPR) (Viveros et al. 2010). To date, several different bacterial genera have been recognized as PGPR, many of them mainly belong to the *Proteobacteria* and *Firmicutes* phyla (Lugtenberg and Kamilova 2009; Drouge et al. 2012), though *Bacillus* spp. and *Pseudomonas* spp. are predominant (Podile and Kishore 2006).

PGPR can induce plant growth in two different ways, directly as biofertilizer or indirectly as biopesticide/biocontrol. Direct plant growth induction includes enhanced nutrient supply such as N, phosphorus, and potassium and increases in Fe through the release of siderophore, etc. or by modulation of phytohormones such as auxin and cytokinins (Arora et al. 2012; Bhardwaj et al. 2014). Indirect plant growth promotion entails control over the inhibitory effect of phytopathogens by producing antibiotics or trigger induced systemic resistance (Glick 1995; van Loon et al. 1998; Van Loon 2007). Colonization of a microbial community on the root surface was erratic, and interaction with roots occurred in various patches (McNear 2013). Nutrient availability and root surface physicochemical variations are key factors that induce changes in the abundance and structure of microbial communities (McNear 2013). Root exudation provides chemical compounds for microbes which leads to the formation of microcolonies on the root surface. Danhorn and Fuqua (2007) reported that root epidermal cell junctions, root hairs, axial grove, cap cells, etc. are common sites for the formation of bacterial colonies. Furthermore, microcolonies expand into larger bacterial biofilms which later become wrapped into an exopolymeric matrix (McNear 2013). Rudrappa et al. (2008) showed that bacterial abundance (PGPR) is one of the key factors controlling plant growth promotion. During the plant growth promotion time course, microbial biofilms act together performing quorum sensing and synchronize discharges of various compounds that directly or indirectly promote plant growth (Mcnear 2013). Each plant growth promotion mechanism and function (direct or indirect) have been discussed in detail (Gupta et al. 2015; Vecheron et al. 2013). Some recent findings also suggest that PGPR can also be used in the phytoremediation of contaminated soil (Zhuang et al. 2007; Shukla et al. 2011; Tak et al. 2013; Meena and Yadav 2015), as some PGPR enable improved plant resistance against abiotic factors, including heavy metal contamination, and in some way help plants to enhance resistance to heavy metals (Jing et al. 2007; Saharan and Nehra 2011; Tak et al. 2013).

5.3 Mycorrhizal Fungi

Among all microbial associations, fungi form symbiotic associations with plants, and between mycelial fungi and plants, the relationship is known as a mycorrhizal association. These words are derived from the Greek word *mikos* meaning fungi and *rhiza* meaning roots. In contrast to legume-rhizobia association, mycorrhizal association is pervasive and indiscriminative, resulting in colonization of nearly 80% of angiosperms and all gymnosperms. Mycorrhizal association is the earliest plant-microbe association, it first occurred approximately 450 million years ago, and this helps explain the pervasiveness of mycorrhiza overall the plant kingdom. Mainly the mycorrhizal association is a mutualistic association, where fungi provide phosphorus, water, and other micronutrient acquisitions by increasing the root surface; in return the fungi receive fixed carbon. Moreover, mycorrhizal fungi play a vital role in the fitness of natural plants (Allen et al. 1995).

Based on anatomical aspects, the mycorrhizal association is divided into two different subcategories, ectomycorrhizae and endomycorrhizae. In ectomycorrhizae (ECM), fungi grow and are colonized within root intercellular spaces by forming a *hartig* net around the root cortex (McNear 2013). In contrast, endomycorrhizae fungi are colonized within root cortical cells and form highly branched structures, called arbuscules (Harrison 2005). Moreover, endomycorrhizae fungi are subdivided into three subcategories, orchid, ericoid, and arbuscular mycorrhiza (AM), AM association being the most common occurring association. During association, fungal hyphae (hartig net or arbuscule) provide nutrients to the plant through the horizontal transfer and expanding the root surface area and exchanging photosynthetically fixed carbon from plants. While ECM prefers to associate with woody plants, AM fungi form a symbiotic association with various land plants, including many agricultural crops (Garcia et al. 2015). In stress conditions such as under extreme rainfall, the fungus can grow their hyphae outside soil nutrient-depleted zones, leading to expanded contact with the soil surface or particles which help to reach out into the soil for plant uptake (Barman et al. 2016; Datta et al. 2017). The mycorrhizal hyphal network also significantly affects soil quality by promoting soil aggregation and stability through various biochemical and biological mechanisms, which directly or indirectly increase plant productivity (Rillig and Mummey 2006).

Unlike the rhizobia symbiosis, the chemical signaling processes of mycorrhizae are less well understood. Much focus has been given to the ECM association since both partners can be grown easily, while endomycorrhizae is difficult to grow in vitro due to their obligate behavior (McNear 2013). Plant root exudates such as various flavonoids (mainly rutin) (Lagrange et al. 2001), abietic acid (Fries 1987), and strigolactones (only in AM fungus) (Akiyama et al. 2005) initiate the spore germination and hyphal growth and branching of the fungus by activation of genes in mycorrhiza fungus which produce lipochitooligosaccharides or short chitooligosaccharides, known as *Myc* factors (Maillet et al. 2011; Genre et al. 2013). There is some resemblance between *Myc* factors and nod factors (produced by rhizobia in the legume-rhizobia association) (McNear 2013), and this could explain the phenomena of "common symbiosis pathway" which originated in AM symbiosis, later adopted by legume-rhizobia association (Garcia et al. 2015). Due to these chemical dialogues, fungi interact with host plants and begin the hyphae proliferation into host plant roots, called hyphopodium (in AM), or growth between dermal cells (in ECM). Formation of arbuscules is the final step of the symbiotic process, and through these arbuscules, fungi enter into the root cell cytoplasm. Selosse et al. (2006) reported that two or more plants can share nutrients through the same mycorrhizae associations as fungal hyphae can create a "common mycorrhizal network" (CMN). Similar phenomena may occur in the ECM symbiosis, but more research is needed to understand the molecular mechanisms of ECM symbiotic associations.

5.4 Pathogenic Microorganisms

In contrast to beneficial (symbiotic/naturalistic) associations with the rhizosphere microbiota, plants can also interact with some soil microbes that inhibit plant growth and health, commonly referred as pathogenic microbes or phytopathogens. These can include fungi, bacteria, viruses, oomycetes, and nematodes. Phytopathogens are one of the major constraints to global food production and security. While many of soil-borne pathogens survive and grow into bulk soil, the rhizosphere is a key niche where pathogenic microbes form parasitic associations with plant roots. Soil-borne pathogens are more persistent when compared to pathogens associated with aboveground parts of the plant (Bruehl 1987). While information has been available on how root exudates modulate/regulate plant-symbiont association, the limited focus has been given on how rhizodeposition can attract and activate the phytopathogens. Agrios (2005) has divided plant pathogens into four main groups: virus, bacteria, fungi, and nematodes. Among these four groups, fungi are the main soil-borne pathogens compared to others. Bacteria and viruses need natural openings or wounds to enter into plant tissues. Furthermore, the soil is not a suitable habitat for nonspore-forming bacteria, while some filamentous bacteria such as *Streptomyces* spp. can grow easily in soil and able to cause infection to host plants. *Ralstonia solanacearum* (*Pseudomonas*) can colonize the xylem and cause wilt to a variety of plants, mainly tomato (Genin and Boucher 2004), pepper, and eggplants. While *Agrobacterium tumefaciens* is well known for causing crown gall (Nester et al. 2005), through the transmission of DNA (T-DNA), these two are the best-understood examples of soil-borne pathogenic bacteria. Only a few viruses can infect plant roots, as they required a vector for transmission and in soil nematodes (e.g., Nepoviruses; Brown et al. 1995), and some zoosporic fungi such as *Olpidium* (Campbell 1996) are the main vectors for viral transmission. Nematodes can be free-living worms that normally ingest other microbes such as bacteria or fungi or other nematodes. However some can form parasitic relationships with host plants. There are three main types of associations that nematodes form with host plants: (i) ectoparasitic, in which nematodes contact with only the outer body of roots; (ii) endoparasitic, in which nematodes are able to reach out the inner body of plant roots; and (iii) sedentary endoparasitic, in which nematodes persist in the inner body of roots for reproduction process (Raaijamakers et al. 2009).

Fungi and oomycetes can be soil-borne phytopathogens. While oomycetes are phylogenetically closer to blue algae, their morphological characteristics are identical to fungi, and their parasitism and diseases have similar symptoms to fungi (Raaijamakers et al. 2009). Most of the soil pathogenic fungi and oomycetes are necrotrophic, and only a few of them are biotrophic such as *Plasmodiophora brassicae* (cause of cabbage "clubroot") and *Plasmopara halstedii* ("downy mildew" in sunflower, Friskop et al. 2009). *Phytophthora sojae* is semi-biotrophic and causes powdery mildew and rusts, especially in soybean. Due to their generalist behavior, necrotrophic pathogens have a wider selection of hosts and can easily infect many plants. Fungi penetrate the host plants through the germ tubes or zoospores and infect epidermal root cells (tips, hairs, etc.) or strike at emerging shoots or seed radicles. *Pythium*, fast-growing and important soil-borne pathogen, affects tree seedling production through attacking plant seeds even before they start to emerge; the disease is called "damping off" (Vaartaja 1975; Weiland et al. 2013). Fungi produce degrading enzymes or use hydrostatic pressure to enter into host root cells (young or juvenile roots) and further colonized into the root cortex. After cortex colonization, fungal mycelia still continue to spread to other parts of the plant and sometimes grow externally and can also cause disease in adjoining plants. *Fusarium oxysporum* and *Verticillium dahlia* and other fungi can spread into the plant vascular tissues through the root endodermis and blocking water flow that causes wilting. This fungal problem can infect many different plants including potato, cotton, tomato, and eggplants and some tree species such as olive (Beckman 1987). A number of different diseases are caused by a fungus such as decay, rot, damping off, and wilt; among all of them, root rot is often the first stage. By infecting or destroying plant roots, plant water and nutrient uptake capacity is decreased causing nutrient deficiency symptoms, reduced size, and drought stress (Raaijamakers et al. 2009).

6 Concluding Remarks and Future Applications

The rhizosphere represents a plethora of interactions between plant-microbes, plant-soil, microbe-soil, and microbe-microbe, and because this network is highly complex, we can affirm that the study of this environment is still in its infancy (Zheng et al. 2017; Berendesen et al. 2012; Sihag et al. 2015). A pivotal and strong link is highlighted among soil-microorganism-plant, thereby creating a *hotspot* in which to study and improve crop health and NUE. In the present work, we have discussed the dynamic processes in the rhizosphere and especially how root exudates, as well as the rhizobiota, are influenced by biotic and abiotic factors to understand and thus improve nutrient use efficiency in the context of more sustainable agriculture. Moreover, the emphasis was given to the role of the roots both in the selection of microbes and in the efficiency of use of resources focusing particularly on nitrate as nutrient and signal.

Rhizosphere interaction studies include measurements of the microbiota, root morphology, and rarely their interactions, especially in field condition. Our limited knowledge is presented on how the plants control the microbe communities present in the soil and how different root types influence the availability of nutrients and also their absorption by rhizobiota. Furthermore, new omic technologies will contribute to identifying new signaling compounds which allow us to set new strategies for promoting plant growth and health.

References

Agrios GN (2005) Plant pathology, 5th edn. Elsevier Academic Press Inc., New York

Akiyama K, Matsuzaki K, Hayashi H (2005) Plant sesquiterpenes induce hyphal branching in arbuscular mycorrhizal fungi. Nature 435:824–827

Allen MF, Morris SJ, Edwards F, Allen EB (1995) Microbe– plant interactions in Mediterranean-type habitats: shifts in fungal symbiotic and saprophytic functioning in response to global change. In: Moreno JM, Oechel WC (eds) Global change and Mediterranean-type ecosystems. Springer, New York, pp 287–305

Antoun H, Kloepper JW (2001) Plant growth promoting rhizobacteria. In: Brenner S, Miller JH (eds) Encyclopedia of genetics. Academic, New York, pp 1477–1480

Arora NK, Tewari S, Singh S, Lal N, Maheshwari DK (2012) PGPR for protection of plant health under saline conditions. In: Maheshwari DK (ed) Bacteria in agrobiology: stress management. Springer, Heidelberg, pp 239–258

Ashoka P, Meena RS, Kumar S, Yadav GS, Layek J (2017) Green nanotechnology is a key for eco-friendly agriculture. J Clean Prod 142:4440–4441

Assmus B, Hutzler P, Kirchhof G et al (1995) In situ localization of Azospirillum brasiliense in the rhizosphere of wheat with fluorescently labeled, rRNA-targeted oligonucleotide probes and scanning confocal laser microscopy. Appl Environ Microbiol 61:1013–1019

Bacilio JM, Aguilar FS, Ventura ZE, Perez CE, Bouquelet S, Zenteno E (2002) Chemical characterization of root exudates from rice (*Oryza sativa*) and their effects on the chemotactic response of endophytic bacteria. Plant Soil 249:271–277

Badri DV, Vivanco JM (2009) Regulation and function of root exudates. Plant Cell Environ 32:666–681

Bais HP, Wier TL, Perry LG, Vivanco JM (2006) The role of root exudates in rhizosphere interaction with plants and other microorganisms. Annu Rev Plant Biol 57:233–266

Baldani JI, Baldani VLD (2005) History on the biological fixation research in graminaceous plants: special reference on the Brazilian experience. An Acad Bras Cienc 77:549–579

Baraniya D, Puglisi E, Ceccherini MT, Pietramellara G, Giagnoni L, Arenella MR, Nannipieri P, Renella G (2016) Protease encoding microbial communities and protease activity of the rhizosphere and bulk soils of two maize lines with different N uptake efficiency. Soil Biol Biochem 96:176–179

Bardgett RD, van der Putten WH (2014) Belowground biodiversity and ecosystem functioning. Nature 515:505–511

Barman J, Samanta A, Saha B, Datta S (2016) Mycorrhiza: the oldest association between plant and fungi. Resonance, General article 21: 1093–1103

Beauregard PB, Chai Y, Vlamakis H, Losick R, Kolter R (2013) Bacillus subtilis biofilm induction by plant polysaccharides. Proc Natl Acad Sci USA 110:E1621–E1630

Beckman CH (1987) The nature of wilt diseases of plants. American Phytopathological Society Press, St. Paul
Behm JE, Geurts R, Kiers ET (2014) Parasponia: a novel system for studying mutualism stability. Trends in Plant Sci 19:757–763
Bekkara F, Jay M, Viricel MR, Rome S (1998) Distribution of phenolic compounds within seed and seedlings of two *Vicia faba* varieties differing in their seed tannin content and study of their seed and root phenolic exudations. Plant Soil 203:27–36
Bellegarde F, Gojon A, Martin A (2017) Signals and players in the transcriptional regulation of root responses by local and systemic N signaling in *Arabidopsis thaliana*. J Exp Bot 68:2553–2565
Bender SF, Wagg C, van der Heijden MGA (2016) An underground revolution: biodiversity and soil ecological engineering for agricultural sustainability. Trends Ecol Evol 31:440–452
Berendsen RL, Pieterse CM, Bakker PA (2012) The rhizosphere microbiome and plant health. Trends Plant Sci 17:478–486
Berg G, Smalla K (2009) Plant species and soil type cooperatively shape the structure and function of microbial communities in the rhizosphere. FEMS Microbiol Ecol 68:1–13
Bever JD, Dickie IA, Facelli E, Facelli JM, Klironomos J, Moora M, Rillig MC, Stock WD, Tibbett M, Zobel M (2010) Rooting theories of plant community ecology in microbial interactions. Trends Ecol Evol 25:468–478
Bhardwaj D, Ansari MW, Sahoo RK, Tuteja N (2014) Biofertilizer function as key player in sustainable agriculture by improving soil fertility, plant tolerance and crop productivity. Microb Cell Fact 13:1–10
Bhattacharyya PN, Jha DK (2012) Plant growth-promoting rhizobacteria (PGPR): emergence in agriculture. World J Microbiol Biotechnol 28:1327–1350
Boddey RM, Uruquiaga S, Alves BJR et al (2003) Endophytic nitrogen fixation in sugarcane: present knowledge and future applications. Plant Soil 252:139–149
Boulter D, Jeremy JJ, Wilding M (1966) Amino acids liberated into the culture medium by pea seedling roots. Plant Soil 24:121–127
Breidenbach B, Pump J, Dumont MG (2016) Microbial community structure in the rhizosphere of rice plants. Front Microbiol 6:1537
Brimecombe MJ, De Leij FAAM, Lynch JM (2007) Rhizodeposition and microbial populations. In: Pinton R et al (eds) The rhizosphere biochemistry and organic substances at the soil–plant interface. CRC Press, Boca Raton, pp 73–109
Briones AM, Okabe S, Umemiya Y et al (2003) Ammonia – oxidizing bacteria on root biofilms and their possible contribution to N use efficiency of different rice cultivars. Plant Soil 250:335–248
Brown DJF, Robertson WM, Trudgill DL (1995) Transmission of viruses by plant nematodes. Annu Rev Phytopathol 33:223–249
Bruehl GW (1987) Soilborne Plant Pathogens. Macmillan Publishing Company, New York
Bulgarelli D, Rott M, Schlaeppi K, Ver Loren van Themaat E, Ahmadinejad N, Assenza F, Rauf P, Huettel B, Reinhardt R, Schmelzer E, Peplies J, Gloeckner FO, Amann R, Eickhorst T, Schulze-Lefert P (2012) Revealing structure and assembly cues for Arabidopsis root-inhabiting bacterial microbiota. Nature 488(7409):91–95
Bulgarelli D et al (2013a) Structure and functions of the bacterial microbiota of plants. Annu Rev Plant Biol 64:807–838
Bulgarelli D, Schlaeppi K, Spaepen S, Ver Loren van Themaat E, Schulze-Lefert P (2013b) Structure and functions of the bacterial microbiota of plants. Annu Rev Plant Biol 64:807–838
Burns JH, Anacker BL, Strauss SY, Burke DJ (2015) Soil microbial community variation correlates most strongly with plant species identity, followed by soil chemistry, spatial location and plant genus. AoB Plants 7:plv030
Cakmak I, Erenoglu B, Gülüt KY, Derici R, Römheld V (1998) Light-mediated release of phytosiderophores in wheat and barley under iron or zinc deficiency. Plant Soil 202:309–315
Campbell RN (1996) Fungal transmission of plant viruses. Annu Rev Phytopathol 34:87–108
Cárdenas L, Domínguez J, Quinto C, López-Lara IM, Lugtenberg BJJ, Spaink HP, Rademaker GJ, Haverkamp J, Thomas-Oates JE (1995) Isolation, chemical structures and biological activity of

the lipo-chitin oligosaccharide nodulation signals from Rhizobium etli. Plant Mol Biol 29 (453):464

Carminati A, Zarebanadkouki M, Kroener E, Ahmed MA, Holz M (2016) Biophysical rhizosphere processes affecting root water uptake. Ann Bot 118:561–571. https://doi.org/10.1093/aob/mcw113

Cavaglieri L, Orlando J, Etcheverry M (2009) Rhizosphere microbial community structure at different stages and root locations. Microbial Res 164:391–399

Chen XP, Cui ZL, Vitousek PM et al (2011) Integrated soil–crop system management for food security. PNAS 108:6399–6404

Cheng L, Tang X, Vance CP, White PJ, Zhang F, Shen J (2014) Interactions between light intensity and phosphorus nutrition affect the phosphate-mining capacity of white lupin (*Lupinus albus* L.). J Exp Bot 65:2995–3003

Cichy KA, Blair MW, Mendoza CHG, Snapp SS, Kelly JD (2009) QTL analysis of root architecture traits and low phosphorus tolerance in an Andean bean population. Crop Sci 49:59–68

Clarholm M, Skyllberg U, Rosling A (2015) Organic acid induced release of nutrients from metal-stabilized soil organic matter. The unbutton model. Soil Biol Biochem 84:168–176

Cordero OX, Datta MS (2016) Microbial interactions and community assembly at microscales. Curr Opin Microbiol 31:227–234

Crawford NM (1995) Nitrate: nutrient and signal for plant growth. Plant Cell 7:859–868

Dadhich RK, Meena RS (2014) Performance of Indian mustard (*Brassica juncea* L.) in Response to foliar spray of thiourea and thioglycollic acid under different irrigation levels. Indian J Ecol 41 (2):376–378

Dadhich RK, Meena RS, Reager ML, Kansotia BC (2015) Response of bio-regulators to yield and quality of Indian mustard (*Brassica juncea* L. Czernj. and Cosson) under different irrigation environments. J Appl Nat Sci 7(1):52–57

Dai X, Wang Y, Yang A, Zhang WH (2012) OsMYB2P-1, an R2R3 MYB transcription factor, is involved in the regulation of phosphatestarvation responses and root architecture in rice. Plant Physiol 159:169–183

Dakora FD, Phillips DA (2002) Root exudates as mediators of mineral acquisition in low-nutrient environments. Plant Soil 245:35–47

Datta R, Kelkar A, Baraniya D, Molaei A, Moulick A, Meena RS, Formanek P (2017) Enzymatic degradation of lignin in soil: a review. Sustain MDPI 9(7):1163; 1–18

De Angelis KM, Lindow SE, Firestone MK (2008) Bacterial quorum sensing and nitrogen cycling in rhizosphere soil. FEMS Microbial Ecol 66:197–207

Danhorn T, Fuqua C (2007) Biofilm formation by plant-associated bacteria. Annu Rev Microbiol 61(1):401–422

Döbereiner J (1992) History and new perspective of diazotrophs in association with non-leguminous plants. Symbiosis 13:1–13

Donn S, Kirkegaard JA, Perera G, Richardson AE, Watt M (2014) Evolution of bacterial communities in wheat crop rhizosphere. Environ Microbiol 17:610–621

Doornbos RF, van Loon LC, Bakker PAHM (2012) Impact of root exudates and plant defense signaling on bacterial communities in the rhizosphere. A review. Agron Sustain Dev 32:227–243

Drogue B, Dore H, Borland S, Wisniewski-Dyé F, Prigent-Combaret C (2012) Which specificity in cooperation between phytostimulating rhizobacteria and plants? Res Microbiol 163:500–510

Edwards J et al (2015) Structure, variation, and assembly of the root-associated microbiomes of rice. PNAS 112:E911–E920

Elmerich C, Newton WE (2007) Associative and endophytic Nitrogen fixing Bacteria and Cyanobacterial Associations – introduction. Springer, Dordrecht

FAO (2009) The state of food and agriculture, Rome

Franche C, Lindström K, Elmerich C (2009) Nitrogen fixing bacteria associated with leguminous and non-leguminous plants. Plant Soil 321:35–59

Fries N (1987) Ecological and evolutionary aspects of spore germination in the higher Basidiomycetes. Trans Br Mycol Soc 88:1–7
Friskop A, Marker S, Gulya T (2009) Downy mildew of sunflower. State Agricultural Experiment Station, New York
Garcia K, Delaux P, Cope KR, Ané J (2015) Molecular signals require for the establishment and maintenance of ectomycorrhizal symbiosis. New Phytol 208:79–87
Genin S, Boucher C (2004) Lessons learned from the genomeanalysis of Ralstonia solanacearum. Annu Rev Phytopathol 42:107–134
Genre A, Chabaud M, Balzergue C, Puech-Pagès V, Novero M, Rey T, Fournier J, Rochange S, Bécard G, Bonfante P, Barker DG (2013) Short-chain chitin oligomers from arbuscular mycorrhizal fungi trigger nuclear Ca^{2+} spiking in *Medicago truncatula* roots and their production is enhanced by strigolactone. New Phytol 198:190–202
George TS, Simpson RJ, Hadobas PA, Richardson AE (2005) Expression of a fungal phytase gene in *Nicotiana tabacum* improves phosphorus nutrition of plants grown in amended soils. Plant Biotechnol J 3:129–140
Giles E.D. Oldroyd, J. Allan Downie, (2008) Coordinating Nodule Morphogenesis with Rhizobial Infection in Legumes. Annual Review of Plant Biology 59 (1):519-546
Giles E.D. Oldroyd, Jeremy D. Murray, Philip S. Poole, J. Allan Downie, (2011) The Rules of Engagement in the Legume-Rhizobial Symbiosis. Annual Review of Genetics 45 (1):119-144
Glick BR (1995) The enhancement of plant growth by free-living bacteria. Can J Microbiol 41:109–117
Gojon A, Nacry P, Davidian JC (2009) Root uptake regulation: a central process for NPS homeostasis in plants. Curr Opin Plant Biol 12:328–338
Good AG, Shrawat AK, Muench DG (2004) Can less yield more? Is reducing nutrient input into the environment compatible with maintaining crop production? Trends Plant Sci 9:597–605
Gray EJ, Smith DL (2005) Intracellular and extracellular PGPR: commonalities and distinctions in the plant-bac-terium signaling processes. Soil Biol Biochem 37:395–412
García-Salamanca A, Molina-Henares MA, van Dillewijn P, Solano J, Pizarro-Tobías P, Roca A, Duque E, Ramos JL (2013) Bacterial diversity in the rhizosphere of maize and the surrounding carbonate-rich bulk soil. Microb Biotechnol 6(1):36–44
Gupta G, Parihar SS, Ahirwar NK, Snehi SK, Singh S (2015) Plant growth promoting rhizobacteria (PGPR): current and future prospective for development of sustainable agriculture. J Microbial Biochem Technol 7:2
Gyaneshwar P, Naresh Kumar G, Parekh LJ, Poole PS (2002) Role of soil microorganisms in improving P nutrition of plants. Plant Soil 245:83–93
Hale MG, Moore LD (1979) Factors affecting root exudation. Adv Agron 31:93–124
Harrison MJ (2005) Signaling in the arbuscular mycorrhizal symbiosis. Annu Rev Microbiol 59:19–24
Hartmann A, Schmid M, Van Tuinen D, Berg G (2009) Plant-driven selection of microbes. Plant Soil 321:235–257
Hennery A, Doucette W, Norton J, Bugbee B (2007) Changes in crested wheat grass root exudation caused by flood, drought and nutrient stress. J Environ Qual 36:904–912
Hiltner L (1904) Ueber neuere Erfahrungen und Probleme auf dem Gebiete der Bodenbakteriologie und unter besonderer BerUcksichtigung der Grundungung und Brache. Arb Deut Landw Gesell 98:59–78
Huang XF, Chaparro JM, Reardon KF, Zhang R, Shen Q, Vivanco JM (2014) Rhizosphere interactions: root exudates, microbes, and microbial communities. Botany 92:267–275
Husain SS, Mckeen WE (1963) Interactions between strawberry roots and Rhizoctonia fragariae. Phytopathology 53:541–545
Igual JM, Valverde A, Cervantes E, Velázquez E (2001) Phosphatesolubilizing bacteria as inoculants for agriculture: use of updated molecular techniques in their study. Agronomie 21:561–568
Ivanov VP (1962) Mutual effect through the root system of a mixed crop of corn and broad beans. Fiziol Rast 9:179–188

Javed MT, Stoltz E, Lindberg S, Greger M (2012) Changes in pH and organic acids in mucilage of *Eriophorum angustifolium* roots after exposure to elevated concentrations of toxic elements. Environ Sci Pollut Res Int. https://doi.org/10.1007/s11356-012-1413-z

Jing Y, He Z, Yang X (2007) Role of soil rhizobacteria in phytoremediation of heavy metal contaminated soils. J Zhejiang Univ Sci B 8:192–207

Jones DL, Hodge A, Kuzyakov Y (2004) Plant and mycorrhizal regulation of rhizodeposition. New Phytol 163:459–480

Jones DL et al (2009) Carbon flow in the rhizosphere: carbon trading at the soil–root interface. Plant Soil 321:5–33

Kardol P, Cornips NJ, van Kempen MML, Bakx-Schotman JMT, van der Putten WH (2007) Microbe-mediated plant–soil feedback causes historical contingency effects in plant community assembly. Ecol Monogr 77:147–162

Katznelson H, Rouatt JW, Payne TMB (1954) Liberation of amino acids by plant roots in relation to Desiccation. Nature 174:1110–1111

Kawasaki A et al (2016a) Microbiome and exudates of the root and rhizosphere of *Brachypodium distachyon*, a model for wheat. PLoS One 11:e0164533

Kawasaki A, Donn S, Ryan PR, Mathesius U, Devilla R, Jones A et al (2016b) Microbiome and exudates of the root and rhizosphere of *Brachypodium distachyon*, a model for Wheat. PLoS One 11(10):e0164533

Kloepper JW, Schroth MN (1978) In Proceedings of the 4th international conference on plant pathogenic bacteria vol. 2 (ed. Station de Pathologie Végétale et Phytobactériologie) 879–882 (Gibert-Clarey, Tours)

Kretzschmar T et al (2012) A petunia ABC protein controls strigolactone-dependent symbiotic signaling and branching. Nature 483:341–344

Krouk G, Crawford NM, Coruzzi GM, Tsay Y-F (2010) Nitrate signaling: adaptation to fluctuating environments. Curr Opin Plant Biol 13(3):265–272

Krouk G, Ruffel S, Gutierrez RA, Gojon A, Crawford NM, Coruzzi GM, Lacombe B (2011) A framework integrating plant growth with hormones and nutrients. Trends Plant Sci 16:178–182

Kulmatiski A, Beard KH, Stevens JR, Cobbold SM (2008) Plant-soil feedbacks: a meta-analytical review. Ecol Lett 11:980–992

Kumar S, Meena RS, Bohra JS (2018) Interactive effect of sowing dates and nutrient sources on dry matter accumulation of Indian mustard (*Brassica juncea* L.). J Oilseed Brassica 9(1):72–76

Lagrange H, Jay-Allgmand C, Lapeyrie F (2001) Rutin, the phenolglycoside from eucalyptus root exudates, stimulates Pisolithus hyphal growth at picomolar concentrations. New Phytol 149:349–355

Lam HM, Coschigano KT, Oliveira IC, Melo-Oliveira R, Coruzzi GM (1996) The molecular genetics of nitrogen assimilation into amino acids in high plants. Annu Rev Plant Physiol Plant Mol Biol 47:569–593

Lancelle SA, Torrey JG (1985) Early development of rhizobium induced root nodules of Parasponia rigida. I infection and early nodulation. Protoplasm 123:26–37

Lannucci A, Fragasso M, Beleggia R, Nigro F, Papa R (2017) Evolution of crop rhizosphere: impact of domestication on root exudates in Tetraploid wheat (Triticum turdigum L.). Front Plant Sci 8:2124

Lavola A, Julkunen-Tiitto R, Roininen H, Aphalo P (1998) Host-plant preference of an insect herbivore mediated by UV-B and CO2 in relation to plant secondary metabolites. Biochem Syst Ecol 26:1e12

Li JZ, Xie Y, Dai AY, Liu LF, Li ZC (2009) Root and shoot traits responses to phosphorus deficiency and QTL analysis at seedling stage using introgression lines of rice. J Genet Genomics 36:173–183

Li X, Rui J, Xiong J, Li J, He Z, Zhou J, Yannarell AC, Mackie RI (2014) Functional Potential of Soil Microbial Communities in the Maize Rhizosphere. PLoS One 9(11):e112609

Li PC, Chen FJ, Cai HG, Liu JC, Pan QC, Liu ZG, Gu RL, Mi GH, Zhang FS, Yuan LX (2015) A genetic relationship between nitrogen use efficiency and seedling root traits in maize as revealed by QTL analysis. J Exp Bot 66:3175–3188

Liu JQ, Samac DA, Bucciarelli B, Allan DL, Vance CP (2005) Signaling of phosphorus deficiency-induced gene expression in white lupin requires sugar and phloem transport. Plant J 41:257–268

Liu H, Carvalhais LC, Schenk PM, Dennis PG (2016) Effect of Jasmonic acid signalling on wheat micorbiome differ between body sites. Sci Rep 7:41766

Lu H, Murase J, Watanabe A, Sugimoto A, Kimura M (2014) Linking microbial community dynamics to rhizosphere carbon flow in a wetland rice soil. FEMS Microbiol Ecol 48:179–186

Lugtenberg B, Kamilova F (2009) Plant-growth-promoting rhizobacteria. Annu Rev Microbiol 63:541–556

Lundberg DS et al (2012) Defining the core *Arabidopsis thaliana* root microbiome. Nature 488:86–90

Lupini A, Princi MP, Araniti F, Miller AJ, Sunseri F, Abenavoli MR (2017) Physiological and molecular responses in tomato under different forms of N nutrition. J Plant Physiol 216:17–25

Lynch JP (2011) Root phenes for enhanced soil exploration and phosphorus acquisition: tools for future crops. Plant Physiol 156:1041–1049

Lynch JM, Whipps JM (1990) Substrate flow in the rhizosphere. Plant Soil 129:1–10

Maillet F, Poinsot V, Andre O, Puech-Pages V, Haouy A, Gueunier M, Cromer L, Giraudet D, Formey D, Niebel A, Andres Martinez E, Driguez H, Cecard G, Denarie J (2011) Fungal lipochitooligosaccharide symbiotic signals in arbuscular mycorrhizal. Nature 469:58–64

Marschner H (1995) Mineral Nutrition of Higher Plants. Academic, London

Martin BC, Statton J, Siebers AR, Grierson PF, Ryan MH, Kendrick GA (2018a) Colonizing tropical seagrasses increase root exudation under fluctuating and continuous low light. Limnol Oceanogr 63:S381–S391

Martin BC, Glesson D, Statton J, Siebers AR, Grierson PF, Ryan MH, Kendrick GA (2018b) Low light availability alters root exudation and reduces putative beneficial microorganisms in seagrass roots. Front Microbiol 8:2667

Massalha H et al (2017) Live imaging of root–bacteria inter-actions in a microfluidics setup. PNAS 114:4549–4554

McKenzie RH, Roberts TL (1990) Soil, and fertilizers phosphorus update. In: Proceedings of Alberta soil science workshop proceedings, February, 20–22, Edmonton, pp 84–104

McNear DH Jr (2013) The Rhizosphere- roots, soil and everything in between. Nat Educ Knowl 4:1

Meeks JC, Elhai J (2002) Regulation of cellular differentiation in filamentous cyanobacteria in free-living and plant-associated symbiotic growth states. Microbiol Mol Biol Rev 66:94–121

Meena H, Meena RS (2017) Assessment of sowing environments and bio-regulators as adaptation choice for clusterbean productivity in response to current climatic scenario. Bangladesh J Bot 46 (1):241–244

Meena RS, Yadav RS (2015) Yield and profitability of groundnut *(Arachis hypogaea* L) as influenced by sowing dates and nutrient levels with different varieties. Legum Res 38(6):791–797

Meena RS, Dhakal Y, Bohra JS, Singh SP, Singh MK, Sanodiya P (2015) Influence of bioinorganic combinations on yield, quality and economics of Mungbean. Am J Exp Agric 8(3):159–166

Meena RS, Kumar V, Yadav GS, Mitran T (2018) Response and interaction of *Bradyrhizobium japonicum* and *Arbuscular mycorrhizal* fungi in the soybean rhizosphere: a review. Plant Growth Regul 84:207–223

Mehnaz S, Lazarovits G (2006) Inoculation effects of Pseudomonas putida, Gluconacetobacter azotocaptans, and Azospirillum lipoferum on corn plant growth under greenhouse conditions. Microbial Ecol 51:326–335

Mlodzinska E, Klobus G, Christensen MD, Fuglsang AT (2015) The plasma membrane H+-ATPase AHA2 contributes to the root architecture in response to different nitrogen supply. Plant Physiol 154:270–282

Mus F, Crook MB, Garcia A, Garcia-Costas MK, Geddes BA, Kouri ED, Paramasivan P, Ryu MH, Oldroyd GE, Poole PS, Udvardi MK, Voigt CA, Ané JM, Peters JW (2016) Symbiotic nitrogen fixation and challenges to extending it to non-legumes. Appl Environ Microbiol 82:3698–3710

Nardi S, Concheri G, Pizzeghello D, Sturaro A, Rella R, Parvoli G (2000) Soil organic matter mobilization by root exudates. Chemosphere 41:653–658

Naveed M, Brown LK, Raffan AC, George TS, Bengough AG, Rosse T, Sinclair I, Koebernick N, Cooper L, Hackett CA, Hallett PD (2017) Plant exudates may stabilize or weaken soil depending on species, origin and time. Eur J Soil Sci 68:806–816

Nester EW, Gordon MP, Kerr A (2005) Agrobacterium Tumefaciens: from plant pathology to biotechnology. APS Press, St. Paul

Neumann G, George TS, Plassard C (2009) Strategies and methods for studying the rhizosphere-the plant science toolbox. Plant Soil 321:431–456

Newman EI (1985) The rhizosphere: carbon sources and microbial populations. In: Fitter AH (ed) Ecological interactions in soil. Blackwell Scientific Publications, Oxford, p 107

Oldroyd GED, Downie JA (2008) Coordinating nodule morphogenesis with Rhizobial infection in legumes. Annu Rev Plant Biol 59(1):519–546

Oldroyd GED, Murray JD, Poole PS, Downie JA (2011) The rules of engagement in the legume-Rhizobial Symbiosis. Annu Rev Genet 45(1):119–144

Parkin TB, Kaspar TC, Cambardella C (2002) Oat plant effects on net nitrogen mineralization. Plant Soil 243:187–195

Pathan SI, Ceccherini MT, Pietramellara G, Puschenreiter M, Giagnoni L, Arenella M, Varanini Z, Nannipieri P, Renella G (2015a) Enzyme activity and microbial community structure in the rhizosphere of two maize lines differing in N use efficiency. Plant Soil 387:413–424

Pathan SI, Ceccherini MT, Hansen MA, Giagnoni L, Ascher J, Arenella M, Sørensen SJ, Pietramellara G, Nannipieri P, Renella G (2015b) Maize lines with different Nitrogen Use Efficiency (NUE) also differ for molecular diversity of bacterial β-glucosidase gene and glucosidase activity in their rhizosphere. Biol Fertil Soils 51:995–1004

Peiffer JA, Spor A, Koren O, Jin Z, Tringe SG, Dangl JL et al (2013) Diversity and heritability of the maize rhizosphere microbiome under field conditions. PNAS 110:6548–6553

Pérez-Jaramillo JE, Carrión VJ, Bosse M, Ferrão LFV, de Hollander M, Garcia AAF, Ramírez CA, Mendes R, Raaijmakers JM (2017) Linking rhizosphere microbiome composition of wild and domesticated *Phaseolus vulgaris* to genotypic and root phenotypic traits. ISME J 10:2244–2257

Philippot L, Raaijmakers JM, Lemanceau P, van der Putten WH (2013) Going back to the roots: the microbial ecology of the rhizosphere. Nat Rev Microbiol 11:789–799

Podile AR, Kishore GK (2006) Plant growth-promoting rhizobacteria. In: Gnanamanickam SS (ed) Plant-associated Bacteria. Springer, Dordrecht, pp 195–230

Pugnaire FI, Armas C, Valladares F (2004) Soil as a mediator in plant-plant interactions in a semi-arid community. J Veg Sci 15:85–92

Qian JH, Doran JW, Walters DT (1997) Maize plant contributions to the root zone available carbon and microbial transformations of nitrogen. Soil Biol Biochem 29:1451–1462

Quiza L, St-Arnaud M, Yergeau E (2015) Harnessing phyto-microbiome signaling for rhizosphere microbiome engineering. Front Plant Sci 6:507

Raaijmakers JM, Paulitz TC, Steinberg C, Alabouvette C, Moënne-Loccoz Y (2009) The rhizosphere: a playground and battel filed for soil borne pathogens and beneficial microorganisms. Plant Soil 321:341–361

Rillig MC, Mummey DL (2006) Mycorrhizas and soil structure. New Phytol 171:41–53

Rovira AD (1956) Plant root excretions in relation to the rhizosphere effect. I. The nature of root exudate from oats and peas. Plant Soil 7:178–194

Rovira A (1969) Plant root exudates. Bot Rev 35:35–57

Rudrappa T et al (2008) Root-secreted malic acid recruits beneficial soil bacteria. Plant Physiol 148:1547–1556

Safari SAA, Rashidi T (2012) The relationship between available P and selected biological properties in the rhizosphere of ten crop species under glasshouse conditions. Span J Soil Sci 2:74–89

Saharan BS, Nehra V (2011) Plant growth promoting rhizobacteria: a critical review. Life Sci Med Res 21:1–30

Saleem M, Law AD, Moe LA (2016) Nicotiana roots recruit rare rhizosphere taxa as major root-inhabiting microbes. Microbial Ecol 71:469–472

Scheible WR, Morcuende R, Czechowski T, Fritz C, Osuna D, Palacios-Rojas N, Schindelasch D, Thimm O, Udvardi MK, Stitt M (2004) Genome-wide reprogramming of primary and secondary metabolism, protein synthesis, cellular growth processes, and the regulatory infrastructure of Arabidopsis in response to nitrogen. Plant Physiol 136:2483–2499

Selosse MA, Richard F, He X, Simard S (2006) Mycorrhizal networks: les liaisons dangerousness. Trends Ecol Evol 11:621–628

Shen J, Li C, Mi G, Ki L, Yuan L, Jiang R, Zhang F (2013) Maximizing root/rhizosphere efficiency to improve crop productivity and nutrient use efficiency in intensive agriculture of China. J Exp Bot 64:1181–1192

Shenton M, Iwamato C, Kureta N, Ikeo K (2016) Effect of wild and cultivated rice genotypes on rhizosphere bacterial community composition. Rice 9:42

Shukla KP, Sharma S, Singh NK, Singh V, Tiwari K, Singh S (2011) Nature and role of root exudates: efficacy in bioremediation. Afr J Biotechnol 10:9717–9724

Sihag SK, Singh MK, Meena RS, Naga S, Bahadur SR, Gaurav YRS (2015) Influences of spacing on growth and yield potential of dry direct seeded rice (*Oryza sativa* L.) cultivars. Ecoscan 9(1–2):517–519

Singh BK, Millard P, Whiteley AS, Murrell JC (2004) Unravelling rhizosphere– microbial interactions: opportunities and limitations. Trends Microbiol 12:386–393

Smalla K, Wieland G, Buchner A, Zock A, Parzy J, Kaiser S, Roskot R, Heuer H, Berg G (2001) Bulk and rhizosphere soil bacterial communities studied by denaturing gradient gel electrophoresis: plant-dependent enrichment and seasonal shifts revealed. Appl Environ Microbiol 67:4742–4751

Song F, Han X, Zhu X, Herbert S (2012) Response to water stress of soil enzymes and root exudates from drought and non-drought tolerant corn hybrids at different growth stages. Can J Soil Sci 92:501–507

Stitt M (1999) Nitrate regulation of metabolism and growth. Curr Opin Plant Biol 2:178–186

Stoltz E, Greger M (2002) Cotton grass effects of trace elements in submersed mine tailings. J Environ Qual 31:1477–1483

Subbarao GV, Sahrawat KL, Nakahara K, Rao IM, Ishitani M, Hash CT, Kishii M, Bonnett DG, Berry WL, Lata JC (2013) A paradigm shifts towards low-nitrifying production systems: the role of biological nitrification inhibition (BNI). Ann Bot 112:297–316

Sytsma KJ, Morawetz J, Pires JC, Nepokroeff M, Conti E, Zjhra M, Hall JC, Chase MW (2002) Urticalean rosids: circumscription, rosid ancestry, and phylogenetics based on rbcL, trnL-F, and ndhF sequences. Am J Bot 89:1531–1546

Tak HI, Ahmad F, Babalola OO (2013) Advances in the application of plant growth-promoting rhizobacteria in phytoremediation of heavy metals. In: Whitacre DM (ed) Reviews of environmental contamination and toxicology, vol 223. Springer, New York, pp 33–52

Thomas Danhorn, Clay Fuqua, (2007) Biofilm Formation by Plant-Associated Bacteria. Annual Review of Microbiology 61 (1):401-422

Tharayil N, Triebwasser DJ (2010) Elucidation of a diurnal pattern of catechin exudation by *Centaurea stoebe*. J Chem Ecol 36:200–204

Ueno D, Rombola AD, Iwashita T, Nomoto K, Ma JF (2007) Identification of two novel phytosiderophores secreted by perennial grasses. New Phytol 174:304–310

Uren NC (2000) Types, amounts, and possible functions of compounds released into the rhizosphere by soil-grown plants. In: Pinto R, Varanini Z, Nannipieri P (eds) The rhizosphere: biochemistry and organic substances at the soil-plant interface. Marcel Dekker, New York, pp 19–40

Vaartaja O (1975) Pythium sylvaticum in Canadian forest nurseries. Can Plant Dis Surv 55:101–102

Vacheron J, Desbrosses G, Bouffaud ML, Touraine B, Moënne-Loccoz Y, Muller D, Legendre L, Wisniewski-Dyé F, Prigent-Combaret C (2013) Plant growth-promoting rhizobacteria and root system functioning. Front Plant Sci 4:356

Van Loon LC (2007) Plant responses to plant growth-promoting rhizobacteria. Eur J Plant Pathol 119:243–254

Van Loon LC, Bakker PAHM, Pieterse CMJ (1998) Systemic resistance induced by rhizosphere bacteria. Annu Rev Phytopathol 36:453–483

Varma D, Meena RS, Kumar S (2017) Response of mungbean to fertility and lime levels under soil acidity in an alley cropping system in Vindhyan Region, India. Int J Chem Stud 5(2):384–389

Verma JP, Meena VS, Kumar A, Meena RS (2015) Issues and challenges about sustainable agriculture production for management of natural resources to sustain soil fertility and health: a book review. J Clean Prod 107:793–794

Vidal EA, Gutiérrez RA (2008) A systems view of nitrogen nutrient and metabolite responses in Arabidopsis. Curr Opin Plant Biol 11:521–529

Viveros OM, Jorquera MA, Crowley DE, Gajardo G, Mora ML (2010) Mechanism and practical considerations involved in plant growth promotion by rhizobacteria. J Soil Sci Plant Nutr 10:293–319

Walker TS, Bais HP, Grotewold E, Vivanco JM (2003) Root exudation and rhizosphere biology. Plant Physiol 132:44–51

Wang R, Tischner R, Gutierrez RA, Hoffman M, Xing X, Chen M, Coruzzi GM, Crawford NM (2004) Genomic analysis of the nitrate response using a nitrate reductase-null mutant of Arabidopsis. Plant Physiol 136:2512–2522

Wang AS, Angle JS, Chaney RL, Delorme TA, Reeves RD (2006) Soil pH effects on uptake of Cd and Zn by Thlaspi caerulescens. Plant Soil 281:325–337

Wang XR, Pan QA, Chen FX, Yan XL, Liao H (2011) Effects of coinoculation with arbuscular mycorrhizal fungi and rhizobia on soybean growth as related to root architecture and availability of N and P. Mycorrhiza 21:173–181

Wang ZH, Fang H, Chen M (2017) Effects of root exudates of woody species on the soil anti-erodibility in the rhizosphere in a karst region. China Peer J 5:e3029

Weiland J, Littke W, Haase D (2013) Forest nurseries face critical choices with the loss of methyl bromide fumigation. Calif Agric 67:153–161

Wu P, Wang X (2008) Role of OsPHR2 on phosphorus homeostasis and root hairs development in rice (*Oryza sativa* L.). Plant Signal Behav 3:674–675

Wu QS, Zou YN, Liu CY, Lu T (2012) Interacted effect of arbuscular mycorrhizal fungi and polyamines on root system architecture of citrus seedlings. J Integr Agric 11:1675–1681

Xu G, Fan X, Miller AJ (2012) Plant nitrogen assimilation and use efficiency. Annu Rev Plant Biol 63:153–182

Yadav GS, Lal R, Meena RS, Datta M, Babu S, Das LJ, Saha P (2017) Energy budgeting for designing sustainable and environmentally clean/safer cropping systems for rainfed rice fallow lands in India. J Clean Prod 158:29–37

Yadav GS, Das A, Lal R, Babu S, Meena RS, Saha P, Singh R, Datta M (2018) Energy budget and carbon footprint in a no-till and mulch based rice–mustard cropping system. J Clean Prod 191:144–157

Yang L (2016) Root exudation pattern of sugar beet (*Beta vulgaris* L.) as influenced by light intensity and P deficiency. PhD dissertation, University Göttingen

Young IM (1995) Variation in moisture contents between bulk soil and the rhizosheath of *Triticum aestivum* L. cv. New Phytol 130:135–139

Zaidi A, Khan MS, Ahemad M, Oves M (2009) Plant growth promotion by phosphate solubilizing bacteria. Acta Microbiol Immunol Hung 56:263–284

Zhang FS, Shen JB, Zhang JL, Zuo YM, Li L, Chen XP (2010) Rhizosphere processes and management for improving nutrient use efficiency and crop productivity: implications for China. In: Sparks DL (ed) Advances in agronomy, vol 107. Academic, San Diego, pp 1–32

Zhang R, Vivanco JM, Shen Q (2017) The unseen rhizosphere root-soil-microbe interaction for crop production. Curr Opin Microbiol 37:8–14

Zhuang X, Chen J, Shim H, Bai Z (2007) New advances in plant growth promoting rhizobacteria for bioremediation. Environ Int 33:406–413

Biochar and Organic Amendments for Sustainable Soil Carbon and Soil Health

G. K. M. Mustafizur Rahman, M. Mizanur Rahman, M. Saiful Alam, M. Ziauddin Kamal, H. A. Mashuk, Rahul Datta, and Ram Swaroop Meena

Abstract Organic matter is the life of soil and vital to environmental quality and sustainability. Intensive cultivation solely depending on inorganic fertilizers with lesser quantity or no organic fertilizers resulted in lower carbon content in soils of tropical and subtropical countries. This paper attempted to identify the best soil and crop management practices which ensure slower microbial decomposition of organic materials, cause a net buildup of carbon in soils, and potentially mitigate the negative effect of global warming and climate change. Biochar and other organic materials have been applied to soil as most valuable amendments for increasing carbon sequestration, soil health improvement, and reduction of greenhouse gas emission from soil. Being recalcitrant in nature, biochar is highly efficient in storing carbon in soils. Biochar possesses a larger surface area and therefore is capable of holding and exchanging cations in soils. Quantity and quality of biochar produced from different organic materials are highly variable because of various production temperature and meager oxygen control system. This review contributes to understanding details of production technologies and performance mechanisms of biochar and other organic amendments in soil. Biochar and organic materials improve soil bio-physicochemical properties, serve as a sink of atmospheric CO_2, and ensure ecological integrity and environmental sustainability.

Keywords Compost · Greenhouse gas · Mitigation · Environment · Carbon sequestration

G. K. M. M. Rahman (✉) · M. M. Rahman · M. S. Alam · M. Z. Kamal · H. A. Mashuk
Department of Soil Science, Bangabandhu Sheikh Mujibur Rahman Agricultural University (BSMRAU), Gazipur, Bangladesh
e-mail: mustafiz@bsmrau.edu.bd; mizan@bsmrau.edu.bd; saiful@bsmrau.edu.bd

R. Datta
Department of Geology and Pedology, Mendel University in Brno, Brno, Czech Republic
e-mail: rahul.datta@mendelu.cz

R. S. Meena
Department of Agronomy, Institute of Agricultural Sciences (BHU), Varanasi, UP, India
e-mail: Meenars@bhu.ac.in

R. Datta et al. (eds.), *Carbon and Nitrogen Cycling in Soil*,
https://doi.org/10.1007/978-981-13-7264-3_3

Abbreviations

BSMRAU	Bangabandhu Sheikh Mujibur Rahman Agricultural University
CEC	Cation-exchange capacity
CFU	Colony forming unit
DMBC	Dairy manure biochar
EC	Electrical conductivity
FRG	Fertilizer recommendation guide
FYM	Farmyard manure
GHG	Greenhouse gas
NETL	US National Energy Technology Laboratory
Pg C	Petagram of carbon
RHBC	Rice husk biochar
SOC	Soil organic carbon
USDE	United States Department of Education
WHC	Water holding capacity

1 Introduction

The agricultural land has been declining every year by 1% due to the new establishment of houses, industries, factories, and markets and road construction in many Asian countries including Bangladesh and India. Worldwide, it is estimated that feeding the world population will need 60% more yield by 2050 (Rosenstock et al. 2016). Such a massive increase in agricultural production needs to be accomplished without jeopardizing soil and environment. Greenhouse gas (GHG) emission like CH_4, CO_2, etc. from different agriculture practices have some notable effects on climate change, but suitable management practices could improve soil fertility and also mitigate the negative impact of climate change through sequestering the huge amount of carbon in the soil (Rahman 2013; Rahman et al. 2016; Varma et al. 2017). Appropriate use of organic materials is essential for organic amendments in the soil to address and minimize such types of challenges. The maximum organic residues remain available in the agricultural fields in stacked forms which are used as fuel by the people for cooking or get discarded in the dumping places. A remarkable nutrient loss like nitrogen, sulfur, etc. occurs as a result of burning the organic residues and may lead to air pollution by emitting different GHG gases (Tipayarom and Oanh 2007; Chen et al. 2008).

A probable novel attempt to outline the limitation of soil fertility is the recycling of different organic residues (Kamara et al. 2014). The quick decomposition of organic materials occurs in soil due to the direct application or in the form of compost. As a result, quick nutrient release and leaching loss occur from the organic sources and also emit carbon dioxide gas into the atmosphere as GHG gases. It is not a sustainable practice to conserve soil fertility because organic residues need to apply every year repeatedly through this process. Biochar can be the substitute of organic materials, and it is produced from any organic biomass. Biochar contains a huge amount of organic carbon because it has been produced by heating organic biomass through pyrolysis method at a high temperature under limited or no oxygen. Biochar can be recycled as organic amendments in the soil (Singh et al. 2010; Lehmann et al. 2011). Biochar is a stable compound, and it can stay in the soil more than hundreds of years based on its source, quality, manufacture methods, and temperature of pyrolysis (Zimmerman 2010). Biochar is the stable carbon-rich product, and the decomposition rate of biochar is very slow due to its resistant aromatic structure which retards the degradation (Baldock and Smernik 2002; Antal and Grønli 2003). Therefore, the effects of biochar on soil property improvement remain a long time than other organic residues, manures, composts, etc. The application of biochar in the soil can improve soil health (physical, chemical, and biological properties), and the duration of this effect would be endless (Amonette and Joseph 2009; Atkinson et al. 2010; Cornelissen et al. 2013; Liu et al. 2012; Meena et al. 2017a).

The agricultural soil can be amended by biochar that is the part of the crop or environmental management (Sparkes and Stoutjesdijk 2011), and it has the potential to become a new technological approach employed in agricultural systems because it has the capacity to increase nutrient availability in various soils. Additionally, biochar is the substitute of organic matter for various soils to improve soil health, quality, and crop productivity. It has also a positive impact on crop productivity when incorporated into acidic soils with low water holding capacity or low organic matter (Biederman and Harpole 2013; Jeffery et al. 2011).

Therefore, biochar is a type of pyrolyzed organic material which can play an important role in sustaining soil health and crop productivity for a long time. For this reason, recently, biochar has been considered as one of the key research materials worldwide for the academician, researchers, policy makers, and farmers.

2 Biochar

2.1 What Is Biochar?

The term "biochar" is comparatively new, but it becomes familiar day by day all around the world. The word "biochar" comes from a combination of "bio-" means "biomass" and "char" means "charcoal" (Schulz and Glaser 2012). Biochar is highly porous fine-grained charcoal, which has been produced under limited oxygen condition using organic biomass that optimizes certain special characteristics like large surface area and porosity and ability to preserve in soils for a long time with very little biological deterioration (Robertson 2014). As a result, it can be differentiated from other charcoals because it can be used as a soil amendment. Due to the huge surface area and porous characteristics of biochar, it is able to hold nutrients and water and also enhance the soil microbial activities for the improvement of soil health as compared to other soil amendments (Lehmann and Rondon 2006; Glaser et al. 2002; Warnock et al. 2007; Dadhich and Meena 2014).

One of the most fertile and productive soils has been found at Prairie of the USA (west of the Mississippi River and east of the Rocky Mountains) because the soil of that area is highly rich in naturally occurring biochar. Biochar has been using historically for soil amendment which dates back at least 2000 years (O'Neill et al. 2009). The unusual fertile soils were created by ancient time in the Amazon Basin known as Terra Preta and Terra Mulata due to extensive use of biochar through indigenous cultures (Lehmann 2007). The soils of this region still remain fertile even with centuries of leaching due to high rainfall in this tropical place. Biochar use in the soil for better and safe agriculture production also has a long history in Australia and some part of Asia particularly Japan, Korea, and China. But recently, keen interest grew in other Asian countries like Bangladesh, India, and Malaysia to use biochar as a soil amendment for sustainable crop productivity and remediation for heavy metals and other toxic organic pollutants.

2.2 *Preparation and Characterization*

2.2.1 Biochar Preparation and Production

Different types of sun-dried organic materials (rice straw, rice husk, any other crop residues, wood, grass, manure, and sewage sludge) can be used to produce biochar through pyrolysis process at 400–550 °C in the absence of oxygen or limited oxygen using biochar production stove. A modified two chambers containing pyrolysis stove have been developed by the Department of Soil Science of Bangabandhu Sheikh Mujibur Rahman Agricultural University (BSMRAU) to produce small-scale biochar for doing research (Figs. 1 and 2). The organic materials are placed in the outer chamber and put to the opposite direction to maintain limited oxygen, and then heat is gradually increased using the gas flame burner in the bottom and middle of the chamber up to 400–500 °C and held constant for 4–5 h. The produced hot biochar after pyrolysis is allowed to cool at room temperature and then powdered, weighed, and stored for the application in the agricultural field (Fig. 3). Rather than this procedure, there are many methods explained by many other scientists to produce biochar by heating plant biomass in the limited or no oxygen. The relative quality as a soil amendment of different biochar is greatly affected by the sources of organic materials and the conditions of biochar production (McClellan et al. 2007; Chun et al. 2004; Yuan et al. 2011; Ashoka et al. 2017).

Biochar only can produce through pyrolysis of biomass. The production yield of biochar depends on sources and pyrolysis temperature (Table 1). The important organic biomasses for biochar production are animal manures, crop residues, and all forestry wastes. Feedstock selection is the primary criteria for better and economical biochar production. The rice husk, rice straw, cow dung, poultry manure, and sawdust are the best for highest and economic yield and easily available biomass sources for biochar production in Bangladesh. The other researchers have also

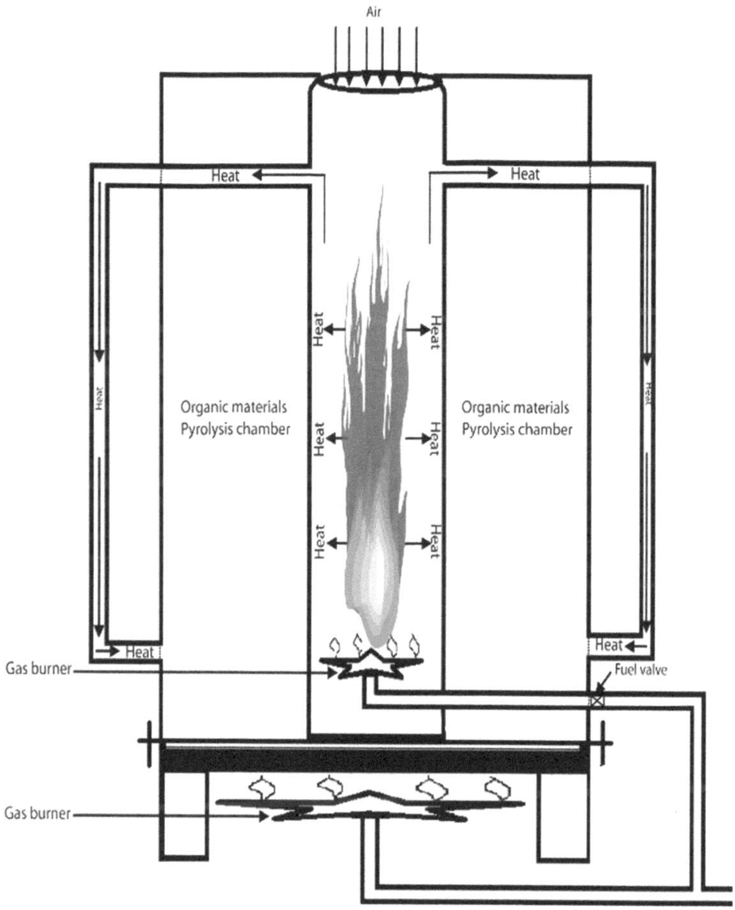

Fig. 1 Modified biochar preparation stove. (Developed by Department of Soil Science, Bangabandhu Sheikh Mujibur Rahman Agricultural University (BSMRAU), Gazipur, Bangladesh)

reported that different types of organic biomass such as plant branches and wastes, wood, magnolia leaves, other crop straw, etc. are good feedstock for biochar production (Table 1) (Wu et al. 2012; Yuan and Xu 2011; Laghari et al. 2016; Zhao et al. 2013; Kinney et al. 2012).

2.2.2 Chemical Characters of Biochar

The chemical properties of biochar derived from different organic materials are presented in Table 2. It has been reviewed and found that most of the biochar showed neutral to slightly alkaline in nature at a temperature range from 450 °C to 500 °C. But the pH of any biochar from any biomass source is independent, while it is dependent on temperature. That means biochar pH is increasing with elevating

Fig. 2 Different biochar production by modified biochar preparation stove at the Department of Soil Science, BSMRAU, Gazipur, Bangladesh

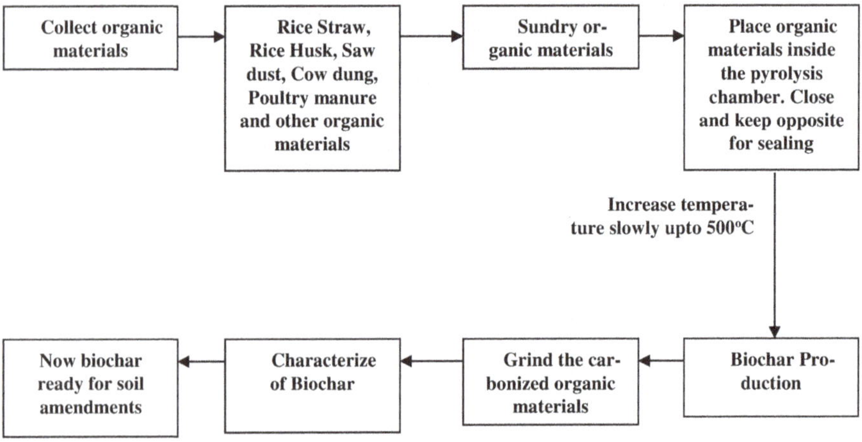

Fig. 3 Flowchart of biochar production

temperature (Table 2). On the other hand, the CEC of biochar is more dependent on the biomass used rather than the temperature for pyrolysis. The process happened due to higher pyrolysis temperature resulting in the loss of some acidic functional groups from the surface of biochar (Zhao et al. 2013).

Table 1 Effect of different biomass on biochar production yield at different pyrolysis temperature

Source of biomass	Pyrolysis temperature (°C)	Biochar production yield (%)	References
RS	450–500	35.6	Rahman (2018, Unpublished)
RH		46.1	
MSD		53.3	
PM		49.1	
CD		48.5	
RS	300	50.1	Wu et al. (2012)
	400	42.8	
	500	39.0	
	600	36.3	
	700	36.5	
CS	350	24.4	Yuan and Xu (2011)
RS		33.3	
SS		32.5	
PS		30.2	
PS	400	55.0	Laghariet al. (2016)
SD	550	28.0	Zhao et al. (2013)
ML	300	61.6	Kinney et al. (2012)
AW	600	25.0	
SGW	400	51.0	

RS rice straw, *RH* rice husk, *MSD* mixed sawdust, *PM* poultry manure, *CD* cow dung, *CS* canola straw, *SS* soybean straw, *PS* pea straw, *SD* sawdust, *ML* magnolia leaves, *AW* apple wood, *SGW* spotted gum wood

Table 2 represents the cation exchange capacity (CEC) values of different biochar. Usually, biochar encompasses high CEC, but pyrolysis temperature slightly influences the CEC of biochar (Song and Guo 2012). The biochar from crop straw demonstrates higher CEC than the biochar derived from manure (Table 2). It can be concluded that during pyrolysis, some nutrients like Ca, Mg, K, Na, and P in the organic residues promote to form O-containing functional groups on the surface of biochar which can support to produce higher CEC of biochar (Agrafioti et al. 2013).

The rice straw-derived biochar contains remarkable amount of N (0.80%), P (0.07%), S (0.005%), Ca (16.80 $Cmolkg^{-1}$), Mg (4.86 $Cmolkg^{-1}$), K (8.36 $Cmolkg^{-1}$), Na (2.10 $Cmolkg^{-1}$), and organic carbon content (44.5%). The higher pH (7.63) and CEC (32.12 Cmol kg^{-1} biochar) are also contained in rice straw biochar.

2.2.3 Biochar Potential as a Soil Amendment

Biochar can be used as soil amendments especially for acid soil correction because most of the biochar contain high pH and also a higher number of basic cations with

Table 2 Chemical properties of different biochar at different temperatures

Source of biomass	Pyrolysis temperature (°C)	pH	Organic carbon (C) (%)	CEC ($Cmol_{+}kg^{-1}$)	Total N (%)	P (%)	References
RS	450–500	8.96	46.45	32.12	0.800	0.070	Rahman (2018, Unpublished)
RH		7.75	44.50	34.67	0.673	0.065	
MSD		7.92	53.10	28.90	0.584	7.310	
PM		7.57	42.70	45.17	0.091	0.311	
CD		8.12	39.80	47.80	0.132	0.143	
RS	300	9.91	72.1	56.9	1.550	0.005	Wu et al. (2012)
	400	9.96	77.2	61.6	0.174	0.006	
	500	10.45	82.8	32.0	1.770	0.003	
	600	10.84	87.1	23.9	1.520	0.002	
	700	10.87	85.2	23.1	1.610	0.018	
CS	350	8.00	61.7	180.0	0.170	0.220	Yuan and Xu (2011)
RS		7.69	42.5	152.0	1.650	0.330	
SS		9.02	54.1	98.0	3.620	0.720	
PS		unavailable	unavailable	unavailable	unavailable	unavailable	
PSD	400	6.35	51.7	27.5	0.930	12.000	Laghari et al. (2016)
	700	9.08	73.6	24.9	0.600	21.000	
PIM	350	9.65	39.1	49.0	0.290	0.230	Zhao et al. (2013)
WS	350	8.69	58.9	87.2	0.004	0.294	
CM	500	10.20	43.7	149.0	0.064	0.100	
SD	500	10.50	75.8	41.7	0.006	unavailable	

Note: *RS* rice straw, *RH* rice husk, *MSD* mixed sawdust, *PM* poultry manure, *CD* cow dung, *CS* canola straw, *SS* soybean straw, *PS* pea straw, *PSD* pine sawdust, *PIM* pig manure, *WS* wheat straw, *CM* cow manure, *SD* sawdust, *CEC* cation-exchange capacity high CEC (Rahman 2018, Unpublished). Biochar has two parts: one is stable and the other one is easily degradable. Thus, biochar can stay a long time varying from hundreds to tens of thousands of years in the soil due to its stable properties (Yuan and Xu 2012) and as a result, enrich carbon permanently in soil and reduce GHG (CO_2) (Lehmann 2007). The aromatic ring structure of biochar might be responsible for the stability in soil. Rapid initial surface oxidation of fresh biochar occurs due to abiotic processes rather than biotic processes (Cheng et al. 2006; Yadav et al. 2018). The mineralization of biochar can be enhanced by this initial oxidation; consequently, negatively charged surface areas increase the cation exchange capacity and the cation retention which helps to improve soil fertility (Glaser et al. 2002; Cheng et al. 2006; Liang et al. 2006). Most of the research indicated that biochar application improved soil fertility and crop productivity in tropical forest soil and acid soils with low active pH. Thus, biochar can be used as a liming material for improving soil health of high aluminum toxic soil and strong acid to increase crop productivity. Therefore, the response of the biochar amendment directly involves the soil health and the crop productivity which remains dependent on the particular characteristics of the soil. As a result, biochar application may or may not bring positive effects on crop production based on the categories of soils. Chemical fertilizer plus biochar using integrated nutrient management has shown better crop yield. This might be due to the increasing soil CEC and microbial activity as a result of carbon enrichment in the soil (Dutta and Raghavan 2014).

2.2.4 The Sorption Capacity of Biochar

The sorption capacity of heavy metal like Ni in the industrial polluted soil is shown in Fig. 4. Application of different rates (1%, 5%, and 10%) of biochar shows a positive effect on the reduction of available Ni than control soil (Fig. 4). The initial amount of available Ni of the contaminated soil of the textile industry was 61.80 ppm (Fig. 4). After the end of the incubation period (90 days), the available Ni content of textile industry soil reduced to 18.20%, 25.10%, and 29.94% at the rate of 1%, 5%, and 10% biochar application, respectively (Fig. 4). Biochar contains some alkaline substances due to high pH; thus, biochar incorporation increases soil pH (Yuan et al. 2011) and accelerates the formation of heavy metal precipitation in the soil. Biochar contains many functional groups on its surface, but dominant are oxygen-containing functional groups (e.g., COOH and OH) (Lee et al. 2013), and these functional groups along with heavy metals make the surface complexes, and hence, the heavy metal adsorption increases on the surface of biochar-incorporated soils. Therefore, abovementioned, both the mechanisms involve to increase heavy metal immobilization in soil, and as a result, bioavailability and activity of heavy metal decrease in biochar-amended the soil.

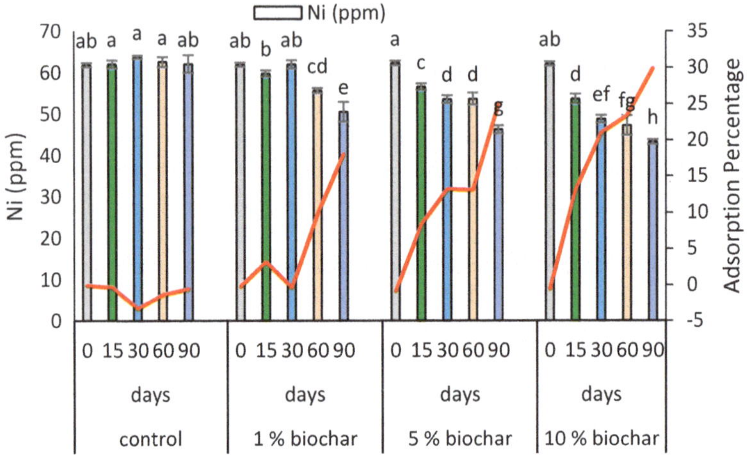

Fig. 4 Nickel (Ni) adsorption on biochar

2.2.5 Remediation of Polluted Soil for Improving Soil Fertility

Several established techniques such as physical ways, electrokinetic remediation, biological remediation, and combined remediation technologies are involved in the restoration of heavy metal pollution in the soil. But recently, biochar is popularly used as adsorbents of heavy metals from soil. Different organic biomass can be used to prepare such carbon-rich materials (biochar) through pyrolysis process at a high temperature in limited or no oxygen, which contains large surface areas with different active functional groups those are renowned as an effective tool to hold contaminants for mitigation of heavy metal contaminants. Comparatively, biochar is the best for the remediation of heavy metal contamination than any other organic absorbents because it contains all categories of adsorptive properties like large surface area, dynamic porous structure, and alkaline in nature and also contains different functional groups especially oxygen-containing functional groups (Ahmad et al. 2014; Meena and Meena 2017). Various mechanisms may play a role in controlling heavy metal removal from aqueous solutions of soil using biochar.

Besides reducing heavy metal from the soil, it is also capable of improving physical, chemical, and biological properties in soils due to its high organic carbon content (Dutta and Raghavan 2014) and high surface area (Fig. 5) (Xu et al. 2013). The maximum sorption capacity demonstrated by the dairy manure biochar (DMBC) and the rice husk biochar (RHBC) based on the findings of Langmuir modeling metals get reduced in the system of multi-metal. On the other hand, higher sorption capacity is observed in the RHBC then DMBC (Xu et al. 2013). Biochar application for soil amendment increases soil health including the soil physical, chemical, and biological properties and enhances crop production through the essential nutrient adsorption and supply to the plant (Houben et al. 2013).

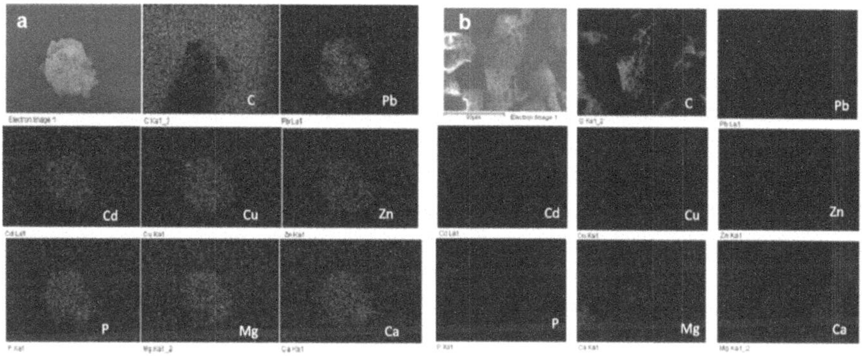

Fig. 5 SEM photographs with elemental dot maps of (**a**) dairy manure biochar (DMBC) and (**b**) rice husk biochar (RHBC) after metal sorption in the multi-metal system. (Scotti et al. 2013)

3 Organic Amendments

Soil organic matter is a key indicator of living force and soil fertility. Both anthropogenic activities and climatic condition (high temperature and humidity) lead to organic matter ruin ultimately accelerate soil degradation. Soil organic amendments provide an unpaid solution for improving degraded land by improving soil physical, chemical, and biological qualities. Any material of organic origin (plant or animal) added to the soil to improve soil properties, designated as an organic amendment. The organic amendments also become a tool for enhancing carbon stocks on low organic matter enrich soil such as degraded tropical soil.

3.1 Sources of Organic Amendments

Organic amendments irrespective of origin or decomposing state add substantial quantities of organic matter in the soil. Organic amendments usually originated from suitable plant species that may be derived directly from the nature (peat soil, peat moss, etc.) or may be obtained as by-products from food processing and agro-based industries (sawdust, oil cake, sugarcane trash, bagasse, rice husk, brans, etc.) or disposal of waste materials (different types of compost, bio-solids, processed sewage sludge, etc.), crop residues, green manuring, etc. Moreover, it is also derived from animal origin (poultry litter, cow dung, farmyard composts) and various worm origins (e.g., vermicomposting). Reclamation organic amendments like biochar, activated charcoal, gypsum, etc. are added in the degraded soil like heavy metal-polluted and salinity-affected soil to convert productive soil. A brief description of the commonly used organic amendments from various sources is given below:

3.1.1 Green Manure

A crop grown for the purpose of being plowed down in soil while green to improve the soil characteristics instead of harvested for animal and human consumption is known as green manure. To sustain the soil health, green manure is a popular practice all over the world. Generally, leguminous plants such as pea (*Pisum sativum*), cowpea (*Vigna unguiculata*), groundnut (*Arachis hypogaea*), black gram (*Vigna mungo*), lentil (*Lens culinaris*), clover (*Trifolium* spp.), soybean (*Glycine max*), mungbean (*Vigna radiata*), dhaincha (*Sesbania aculeata* and *Sesbania rostrata*), sun hemp (*Crotalaria juncea*), etc. are commonly used to prepare the green manures. Green manuring with legume plants is better than any other plants as they provide a substantial amount of nitrogen to the soil through the process of biological nitrogen fixation. Fresh leaves of the forest plants may also be used as the green manure. Farmers collect green leaves and twigs from the forest plants and incorporate into the soil (FRG 2012) to increase the soil organic matter status. This practice is termed as green leaf manure.

3.1.2 Household Waste Compost

Household wastes are organic materials that generally contain kitchen and yard wastes and wastes derived from food and wood materials. Household waste compost is prepared from the degradable waste which is generated in day to day operations at the household level. The major portion of household wastes comes in the form of food products especially the nonconsumable portion of fruits and vegetables. Proper management of household waste is necessary; otherwise, the environment may be polluted. The utilization of household waste compost for crop production will obviously reduce the dependency on synthetic chemical fertilizer application in the agriculture sector which will reduce the production cost of agricultural products as well as ensure the safe environment. Thus, the compost prepared from household waste materials might be an attractive source of soil organic matter.

3.1.3 Compost

Compost is an important type of organic manure which is produced by the decomposition of various plant and animal wastes. Good compost can be prepared from a variety of refuse materials like plant leaves, kitchen waste, banana trashes, pineapple trashes, weeds, water hyacinth, paper mill wastes, sugarcane trashes, straw, sawdust, rice husk, animal slaughter waste, etc. Compost may also be prepared from municipal waste, garbage, and leather industry waste, but it should be free from heavy metals and other toxic trace elements. At least, the harmful materials and pollutants should be below the critical level so that safe crop production can be ensured. The process of the preparation of compost through decomposition of organic materials is called composting. Compost is popularly known as black gold to the gardeners and vegetable growers as it is highly beneficial to the growing plants.

3.1.4 Farmyard Manure (FYM)

Farmyard manure is a kind of organic manure that includes the solid and liquid excreta of livestock, normally mixed with small amount of litter such as straw (mainly rice straw) used for bedding of animals. It is a by-product consisting of animal dung, urine, fodder residues, and animal bedding materials. It is one of the oldest manures used by farmers to grow different crops especially the vegetables due to its easy availability and the presence of most of the nutrients required by the crops. Cattle urine is an important part of farmyard manure as it is rich in nitrogen (FRG 2012), but the major portion of urine is lost due to the soaking by the earthen floor of the animal shed. Therefore, the animal shed should be cemented to overcome the situation. In the case of earthen floor straw, sawdust, dry weeds, rice husk, etc. might be used to reduce the urine from the cattle shed.

3.1.5 Vermicompost

Vermicompost is a kind of compost which is produced by different species of earthworms. Earthworms frequently use different types of decomposable organic wastes as their food of which only a little portion (5–10%) is absorbed by their body, while the major portion of the consumed food is released from the body as excreta in the form of pellets which is treated as vermicompost. The vermicompost contains not only excreta but also earthworm cocoons and plant growth-promoting hormones like auxins and gibberellins, enzymes, vitamins, beneficial microorganisms, etc. Vermicompost provides plant nutrients in a readily available form which is easily uptake by the plants. Moreover, the application of vermicompost in the soil improves soil chemical, physical, and biological properties.

3.1.6 Poultry Manure

Poultry manure is one of the important traditional manures that has long been used in the agricultural field for better crop production in many countries of the world. It is the organic waste derived from the feces, urine, and bedding materials of the poultry birds. To meet the increasing demand of a growing population, the number of poultry bird is increasing every year all over the world leaving huge quantities of poultry litter. It is essential to manage the poultry litter properly as early as possible. Otherwise, it may create severe environmental pollution. Though poultry manure has diverse use like biogas and electricity generation, fish feed, mushroom production, etc. but the use as organic manure in the crop field is the most important.

3.2 Preparation and Characterization

Preparation technique and nutritional composition of organic amendments vary from one to another. This chapter briefly includes the preparation techniques of some important organic amendments and their characteristics in terms of nutritional composition.

3.2.1 Green Manure

It is better to select a quick growing plant to prepare the green manure. Green manure can be prepared in two ways: First, by growing of the green manuring plants in the original land where the green manure will apply. This practice is called in situ green manuring. In this system, the crop especially the legume plant is grown in the field a few weeks before the sowing/transplanting of the main crop. The growing plants should be plowed down just before they flower or at the time of flowering. It will take 1 or 2 weeks to decompose the green manure in soil properly, and after that, the main crop can be grown in the field. Second, the green manure crop is grown in another field where the main crop will not grow. Under this practice, the green manure crop is harvested at maximum vegetative stage (just before flowering or just at the time of first flowering), cut into pieces, and incorporated into the main field just before 1–2 weeks of main crop sowing/planting. Depending on the types of the plant species, 10–15ton fresh biomass could be obtained from a hectare of land which may supply a substantial amount (60–120 kg) of nitrogen to the soil (FRG 2012). Nutrient composition of green manure greatly depends on the plant species used to prepare the manure. The data presented in Table 3 indicated that nitrogen content in *Sesbania* green manure and sun hemp green manure is similar, but P content is higher in *Sesbania*, while K content is higher in sun hemp.

3.2.2 Household Waste Compost

The composting of household waste generated from household level is a simple process. To prepare the household compost at the individual farm level, it is needed to dig an earthen pit in a suitable place nearby the homestead. The size of the individual pit largely depends on the volume of the waste material, but the ideal size might be 2 m × 1 m × 1 m. It is better to make a shed over the pit to protect the waste materials from rainfall and excessive sunlight. The daily collected household wastes like kitchen waste that includes peel of fruits, vegetable and fish wastes, yard sweepings, etc. should be gathered simultaneously and spread over to the pit. When the pit is about 30 cm deep with wastes, approximately 300 g of urea needs to apply over the surface of the waste material to accelerate the microbial decomposition of wastes, and then again, wastes should be placed into the pit, and the practice needs to continue until the pit gets full. Finally, a thin layer of cow dung or urea fertilizers is to be spread over the pit, covered with soils, and kept for about

Table 3 Nutrient composition of commonly used organic manures

Name of the manure	% N	% P	% K	Reference
Vermicompost	0.51–1.61	0.19–1.02	0.15–0.73	Nagavallemma et al. (2004)
Garden compost	0.80	0.35	0.48	Nagavallemma et al. (2004)
Compost (rural)	0.75	0.60	1.00	FRG (2012)
Compost (urban)	1.50	0.60	1.50	FRG (2012)
Green manure (*Sesbania*)	0.70	0.40	0.40	FRG (2012)
Green manure (sun hemp)	0.70	0.12	0.50	FRG (2012)
Farmyard manure	0.53	0.22	0.59	Parihar et al. (2012)
Poultry manure (decomposed)	1.25	0.70	0.95	FRG (2012)

3–4 months. Three to 4 months later, the household waste compost is ready to use in the field. The prepared compost supplies a considerable amount of essential (both macro and micro) nutrients to the plants. However, the nutrient content may vary with the variation of waste materials that are used to prepare the household waste compost. Nutrient content in household waste compost may also be influenced by the compost preparation techniques. However, research findings demonstrated that household waste compost contains 3.32% total N, 0.61% total P, and 1.59% total K (Smith and Jasim 2009).

3.2.3 Compost

The preparation of compost is almost the same as the preparation technique of household waste compost. In the composting process, it is important to maintain a suitable C:N ratio as decomposition of organic material is largely influenced by proper C:N ratio. Microbial activity can be accelerated by maintaining a good C:N ratio of the composting materials. The C:N ratio of 25:1–30:1 is suitable for better composting of the materials. Proper aeration and sufficient moisture are needed for rapid decomposition of the composting materials. The use of cellulolytic microorganisms in the composting piles will fasten the decomposition process. It takes about 3–6 months to prepare the compost. Rapid decomposition is favored by high temperature and high humidity. Addition of nitrogenous and phosphatic fertilizers to the composting materials will also accelerate the rotting of slow decomposing materials. The good-quality compost should have the following characteristics: dark brown to black in color, loose and crumbly in structure, largely insoluble in water but soluble in dilute alkali, pleasant earthy smell that doesn't attract flies and that maintain pH around neutral, and C:N ratio between 10:1 and 15:1. The nutrient composition of compost mainly depends on the nature of the composting materials. Normally, compost prepared from materials collected from urban areas contains higher amounts of nutrients as compared to the compost prepared from the materials collected from rural areas (Table 3). In general, compost contains 0.75–1.5% N, 0.35–0.6% P, and 0.48–1.5% K.

3.2.4 Farmyard Manure (FYM)

Dung obtained from domestic animals like cows, buffaloes, sheep, goats, horse, and other animals along with urine, bedding materials, and fodder residues are collected daily and placed in a pit or trench of desirable size. Depending on the number of animals, the size of the pit might be 6–7 m × 1.5–2 m × 1 m. The pit should be provided with a shed to protect the manure from rainfall and hot sun especially in the summer and rainy season. If the dung is too dry, some water might be added. The collected dung and other materials are added properly over the layer made a day before. The manuring materials should be added to the pit until the height of the added material is approximately 40–50 cm from the ground. Then the upper portion of the stack should be coated with a mixture of mud and cattle dung. Farmyard manure can also be prepared by heap method. In this method, the collected materials from the animal shed are placed uniformly in a high land until the height of the heap reached about 1 m above the ground level. The farmyard manure becomes ready to apply in the field after 3–6 months.

The quality and quantity of farmyard manure depend on the types and the ages of animals, the function of the animals, the types of animal feeds, and the care in handling and preserving the materials used to prepare the manure. Farmyard manure is used to grow different vegetables all over the world as it provides essential nutrients to the growing plants. However, a good source of farmyard manure contains 0.535 N, 0.22% P, and 0.59% K (Table 3).

3.2.5 Vermicompost

Vermicompost can be prepared using any type of biodegradable wastes including vegetable waste, crop residues, weeds, agro-industry waste, dry leaves of crops and trees, sugarcane trash, animal manures, dairy and poultry waste, etc. Though there are more than 2000 earthworm species in the world, *Eisenia fetida*, *Eisenia andrei*, and *Lumbricus rubellus* are considered as the best types of worms for vermicompost preparation. For vermicomposting, it is essential to select a shady place having cool weather and high humidity. After collection of the degradable wastes, a predigestion of the materials for a month by heaping the materials along with cattle dung slurry is recommended as the predigested material is suitable for earthworm's consumption. Vermicompost can be prepared by different methods, but it is easy to prepare the compost in the containers or tanks of different materials. Earthen bowls can also serve as containers to make the vermicompost. The predigested organic residues along with animal dung are placed in the container, and then suitable species of earthworms is released into the container. The container should be covered with thick cloth or jute bag to maintain a dark environment which is suitable for earthworm's activity. Based on a suitable condition, it takes 2–3 months for vermicompost preparation. After collection from the container, the compost should be sieved to separate fully composted material. The partially composted material, earthworms and their eggs, or cocoon should be placed in the container again for further compost preparation. Vermicompost contains a considerable amount of plant nutrients. Moreover, it provides soil vitamins, antibiotics, enzymes, and plant growth hormones (Molaei et al. 2017; Molaei et al. 2017a). Vermicompost also enhances nitrogen fixation and phosphorus solubilization as it carries certain beneficial microorganisms. The nutritional composition of vermicompost mainly depends on the type of waste material used to prepare the compost, earthworm species, preparation method, etc. Generally, vermicompost contains 0.51–1.61% N, 0.19–1.02% P, and 0.15–0.73% K (Table 3).

3.2.6 Poultry Manure

The decomposition rate of fresh manure is very high, and nutritional quality deteriorates quickly. Therefore, poultry manure needs to be processed rapidly. Fresh poultry manure should never be used in the standing crop field as it may damage the growing plants. The common practice of poultry manure preparation includes spreading the fresh poultry manure in a dry place for pile composting. But the open stockpile causes nutrient losses and environmental pollution. Therefore, it is suggested to make a shed over the manure pile or cover the stack by a plastic sheet. Due to the lower C:N ratio of fresh poultry manure, it is better to add rice husk or straw to achieve a wider C:N ratio which will ensure quality poultry manure preparation. It may take 8–10 weeks for the composting of the poultry manure by open piling method. The loss of nitrogen can be minimized by composting the fresh poultry manures under anaerobic condition. In this method, fresh poultry manure mixed with rice husk or straw is placed in a pit, sufficient water is added to maintain adequate moisture, and finally, the pit is sealed with mud plaster. Under anaerobic condition, it may take 10–12 weeks for composting of the poultry manure.

Nutritional quality of poultry manure depends on various factors including types of birds, production techniques, quality of feeds, storage and handling methods, climatic condition, age and moisture content of manures, etc. Decomposed poultry manure contains a considerable amount of plant nutrients including 1.25% N, 0.7% P, and 0.95% K (Table 3). In addition to NPK, poultry manure also contains a considerable amount of micronutrients (Amanullah et al. 2010).

3.2.7 Crop Residue

Agricultural field residue includes the materials like stem, leaves, stalk, root, etc. that are left after different management practices and harvesting activities in a field. The residue can be encompassed in soil through plowing directly into the ground or burning. Moreover, crop processing leaves several materials as an unusable product like husks, unfilled grain, seeds, and root that can also be used as soil amendment. Directly incorporation of such fibrous materials can improve soil physical properties because they act as an energy source for soil microorganisms. Crop residues from various sources contain a considerable amount of plant nutrients especially N, P, and K (FRG 2012).

3.2.8 Oil Cake

Oil cakes are milling by-products of oilseed crops (mustard, sesame, linseed, etc.) after the oil has been extracted from an oilseed. Oil cake is a greater source of N and P. Oil cake contains lower C:N ratio, but the higher mineralization rate in soil provides quicker nutrient release. Oil cake normally contains 5.0–6.2% N, 1.4–2.0% P, and 1.20% K (FRG 2012).

3.3 Organic Amendments Influence on Soil Health

Organic matter status of soil greatly influences the soil quality and soil health. But the decline of organic matter status in the soil due to intensive cropping with the utilization of huge quantities of synthetic chemical fertilizers leads to the deterioration of soil fertility, a major concern of sustainable agriculture. Though it is very difficult to recover and maintain the soil organic matter status up to a satisfactory level, research findings indicated that organic amendments from various sources improve the soil physical properties (Leroy et al. 2008), chemical properties (Scotti et al. 2013) including recovery of soil organic C stock (Zhang et al. 2015; Sihag et al. 2015), and biological fertility of soil that ultimately improves the soil health.

3.3.1 Soil Chemical Properties

3.3.1.1 Soil pH

Soil pH is one of the vital chemical properties, which regulates the nutrient availability and crop productivity. The optimum pH range of a productive soil is 6.5–7.5 because such range hastens both macro- and micronutrient availability and ensures higher microbial activity. Incorporation of organic amendment greatly influences the soil reaction (Angelova et al. 2013). Variation of soil pH by addition of organic amendment depends upon the organic residue type, application rate, inherent soil condition, and buffering capacity of a soil. Amended substances initially increase soil pH due to higher decomposition rate; later on, substantial inconsistency is observed either above or below the inherent soil pH (Wong et al. 2000). It is observed that the organic amendments lowered the buffering capacity of soil and activated the disparity of soil pH (Neilsen et al. 1998). Several mechanisms could explain the preliminary rise of soil pH, immediately after organic matter added (Angelova et al. 2013). The regulating mechanisms include organic acid anion oxidation, organic-N ammonification, organic molecule-specific absorption capacity, and anaerobic reduction activity. The organic acid anion oxidation complies with the lessening of H^+ ions or the release of OH-ions during decomposition activity, possibly playing a key role for immediate rise of soil pH (Sparling et al. 1999).

Under anaerobic conditions, reduction of higher valence Mn oxides and/or Fe oxides and hydrous oxides in soils is reduced and raised soil pH. Long-term addition of organic manure or crop residue did not show a larger increase in soil pH (Van Antwerpen and Meyer 1998). Soil pH varied with the type of organic amendment, inherent soil pH, and proton budgeting.

3.3.1.2 Soil C:N Ratio

The soil C:N ratio is one of the important indicators that explain nutrient release dynamics of added soil organic amendments. Heterotrophic microorganisms use organic matter for their body intake and release amine group and different amino acid in a fixed ratio (Datta et al. 2017c). The threshold C:N ratio for higher microbial activity is ~25–30: 1. The addition of higher C:N ratio organic matter in the soil first allows short-term N immobilization in microbial biomass and sudden drop in nutrient availability. Thus, plants suffer nutrient deficiency and impair crop yield. The lower C:N ratio in organic matter demonstrates higher mineralization rate. Several reporters explain that organic waste with below 30:1 C:N ratio indicates gross N mineralization, while above this level favor immobilization (Alexander 1977). The mineralization arises when the organic matter content in soil is more than 2%; below this, immobilization occurs (Sivapalan et al. 1985). The C:N ratio of organic matter varies according to the variety of sources. For example, the composted organic amendment such as C:N ratio of poultry manure is 6.5:1, in cereal residue 82:1, and in sawdust 664:1 (Huang et al. 2004). Moreover, different organic materials having similar C:N ratio demonstrate different mineralization behavior. The variation in decomposition rate of similar C:N ratio substance might be due to its biochemical composition, for example, different lignin, cellulose, and polyphenol content. Thus, in addition to C:N ratio, organic matter quality becomes a key factor for amendment mineralization and nutrient release. Therefore, it is essential to recognize proper quality organic amendment for ensuring sustainable carbon stock and nutrient release.

3.3.1.3 Nutrient Cycling

Sustainable management of organic amendment in soil is essential to comprehend the nutrient release pattern from organic residue. Several factors could enhance nutrient dynamics in organic-amended soil, such as organic residue matrices, residue mineralization mechanisms, sources, climate, soil properties, and management practices. Generally, organic matter contains different nutrient elements, including N, P, and S, in a relatively constant amount. One ton of organic carbon addition through organic amendment has the ability to release 8.3% N, 2.0% P and 1.4% S in soil (Kirkby et al. 2011). The nutrient flux of soil depends on the decomposition rate of the organic matter. This soil organic matter is considered as a source and sink of soil nutrient cycling. Like, cereal crops intake nutrient through uptake mechanism, of which 1/4th N &P, 1/2 of P and 3/4th of K, are accumulated in crop residue. Thus, incorporation of these residues can be an enormous source of nutrient for succeeding crops. Nitrogen cycling is interlinked with the atmosphere, soil solution, organic amendment, plant root system, and microbial activity. Organic material with high C: N ratio encourages immobilization and ingestion of N in a heterotopic organism, while a lower C:N ratio increases the mineralization rate. Submerge condition or anaerobic mineralization of organic amendments will favor the denitrification process. For a precise understanding of N budgeting and N cycling in the soil from the organic amendment, consideration of the gross N transformation is essential. Organic matter-amended 6-year experiment summarizes that farmyard composted soil had a higher nitrification rate than the mineralization rate (approximately 10.2 and 5.7 g N ton^{-1} day^{-1}, respectively); in contrast, opposite trends were seen in liquid cattle-shed waste (approximately 1.6 and 2.9 g N ton^{-1} day^{-1}, respectively). Thus, nitrification rates of dairy wastes are higher than mineralization rate including that repeated application enhances microbial activity and nitrification ability (Habteselassie et al. 2006; Meena et al. 2017b). Organic matter shows a crucial function in heightening the availability of soil phosphorus (Shen et al. 2011). In general, depending on soil type, organic amendment contributes 20–80% of total soil P. Organic amendment is also a good source of sulfur in the soil, although some soil may contain a higher amount of inherent mineral S.

3.3.1.4 Cation Exchange Capacity (CEC)

Soil exhibits CEC mainly due to the negative charge of colloidal substances (clay and humus). It is now a well-established fact that there is a strong positive correlation between organic matter and CEC of soil. The influence of organic amendments on CEC is largely interlinked with the soil texture and pH. A negative surface charge of clay is linked with the isomorphous substitution of silica by aluminum in the clay mineral permanently. In contrast, the decomposition of organic amendment releases different organic acid; the dissociation of those organic acids causes a net negative charge. The negative charge associated with fully decomposed organic matter, i.e., humus, is neutralized by cations in the soil. The CEC interrelated to the organic amendment is termed as pH-associated CEC. Actual CEC of soil is significantly influenced by soil pH. For example, the addition of the same amount of organic matter from a specific source will provide higher CEC in a neutral soil (pH ~7) than a soil with slight acidity. It is observed that the addition of organic amendment in acidic soil ($pH < 5.5$) did not show a significant effect on CEC (Murphy 2015). Higher CEC of a soil signifies the greater amount of organic matter or clay mineral present in the soil indicating its higher fertility. The relationship between the soil organic matter and CEC varies according to the soil textural class. An organic amendment can contribute 4–50 times higher CEC than a given weight of the clay. Depending on the organic residue reserve, the amendment increases CEC, first in the surface soil and subsequently in the subsurface soil. A 20-year long-term experimental result indicated that surface addition of organic residue increased the CEC of soil by 136 % (from 11 to 26 meq/100 g of soil) (Crovetto 1997).

3.3.1.5 Buffering Capacity

Globally, soil acidification is a great concern for sustainable soil management and crop productivity. At low pH, availability of macronutrients gets decreased but tends to increase micronutrient availability including several toxic elements. Buffering capacity of soil can resist the extreme variation of soil pH. At present, it is well established that organic amendment acts as a buffering agent against soil acidification (Helyar et al. 1990). Buffering capacity of organic amendment depends on several factors like source of organic residue, soil texture, clay content, CEC, and soil reaction. For example, compost can be used as a buffering agent in acidic soil, due to its higher CEC and release of nitrogen which helps to neutralize the exchangeable acidity (Latifah et al. 2018; Verma et al. 2015b). The higher buffering capacity might be associated with the higher release of exchangeable NH_4^+ ions and a humic acid fraction. In acidic condition, most of the humic fractions become protonated and provide the basis for decreasing CEC. Moreover, at strongly acidic ($pH < 5$) condition, microbial activity will slow down, thus reducing the decomposition rate of organic amendment and subsequent reduction of buffering capacity. At $pH > 5.5$, the addition of organic matter can show its inherent buffering capacity because of the enormous tendency of releasing H^+ ions from humic acid and react with hydroxyl (OH^-) for neutralization. Thus, the addition of a judicious amount of organic amendment containing inherent basic cations can actively enhance the buffering capacity of the soil.

3.3.1.6 Electrical Conductivity (EC)

Intensive application of organic amendment enhances the salt concentration and salinization. The organic amendment, more specifically composted manure, enhances soil electrical conductivity of soil and thereby increases salinity and sodicity of amended soil (Angelova et al. 2013). Organic amendment-induced soil salinity might be due to the mineralization process of releasing soluble minerals and solubilization of ions. The EC of the amended soil depends on the raw materials used for composting their biochemical composition and mineralization process (Atiyeh et al. 2002). Several reporters summarize slightly increase of EC due to organic amendment, but the value remains below the threshold level of soil salinity ($4.0\ dSm^{-1}$) (Bonanomi et al. 2014).

3.3.2 Soil Physical Properties

3.3.2.1 Aggregate Stability

Structural stability is a vital property for maintaining soil health. Aggregate stability of soil can improve other physical properties like pore space and thus accelerate gas exchange, water retention capacity, and microbial activity of a soil (Van-Camp et al. 2004). Enrichment of organic matter can ensure higher aggregate stability and lower erodibility of a stress-prone soil. Aggregate stability of soil is influenced by organic amendment in three different ways. Organic materials act as binding agents of clay particles through H-bonding and coordination with polyvalent cation. The functional group of organic matter (i.e., COO) can neutralize the negative charge of the clay surface that reduces the repulsion between clay particles. Moreover, organic matters provide cementation and encapsulation surrounding the clay particles and reduce hydrophobic nature and enhance stability. Fungal hyphen and microscopic plant roots can also hold soil particles together. Amount and type of organic amendment greatly influence the aggregate stability of the soil. The readily decomposable organic amendment has a strong and a transitory effect on the aggregate stability, while more stable lignin and cellulose-containing matter have less but consistent effects. During the decomposition process of organic matter, microorganisms release polysaccharides and raise aggregate stability by enhancing their inter-cohesion. In converse, aggregate hydrophobicity can enhance by adding the more humic compound to the clay minerals (Van-Camp et al. 2004). Mean weight and the diameter of aggregate increased with the addition of more organic carbon in the soil (Haynes 2000). However, to preserve suitable aggregate stability in soil, it is necessary to maintain a minimum of 2% SOC although the threshold level for structural ability is 3–3.5% SOC (Boix-Fayos et al. 2001). Long-term experimentation with organic amendments (crop waste and manure) in rice-wheat cropping pattern showed higher macroaggregates (Das et al. 2014). Source of the organic amendment also plays an important role in enhancing structural stability. Among the plant species, nonleguminous species exhibited greater aggregate stability, because of their higher root mass and higher rhizospheric microbial biomass. Nonleguminous plant compost-treated plot showed higher soil structural stability (28.3%) as compared with leguminous-composted one in 4-year study results (Tejada et al. 2009; Datta et al. 2017b).

3.3.2.2 Bulk Density

The bulk density is the dry weight of soil per unit bulk volume. Soil porosity, cohesiveness, and structural development of soils as well as soil health are strongly coordinated by the bulk density (Heard et al. 1988). Low values of bulk density mean a porous soil and high values a soil with low porosity. Bulk density is an integrated activity of soil solid composition (mineral and organic fraction). There is a strong negative relationship between soil organic amendment and bulk density. Improvement of soil aggregation indicates the presence of higher colloidal substances and lower bulk density. The degrees of decomposition of organic matter predominantly synchronize the bulk density of soil. A partially decomposed organic matter such as fibrous material increases the porosity of soil and decreases its bulk density. Moreover, the type of organic amendment and cropping pattern also considered as influencing factors for bulk density of soil. The application of organic amendment influences the bulk density and macroporosity of sandy loam and clay soil (Rivenshield and Bassuk 2007). In general, soil containing lower density aids easy root penetration and root growth. Lower soil bulk density confirms higher water holding capacity and nutrient use efficiency (Ikpe and Powell 2002). High bulk density accelerates surface runoff and erosion losses of soil and nutrient because of water movement through the soil restricted by compact soil. Thus, organic amendment positively affects the bulk density of all soil types because it clearly increases microporosity.

3.3.2.3 Particle Density

Particle density is used to estimate volume basis, total pore space, soil and water viscosity, soil temperature attributes, and air and water retention in soil (Hillel 1998). Several soil activities including mass flow and preservation of soil air and water, as well as nutrient transportation system within the soil, are largely dependent on the particle density-associated parameters. Organic matter is a part of the soil solid portion and has a noteworthy impact particle density. An increase of soil organic portion in soil signifies its higher particle density. Conversely, the lower ones with similar texture contain higher particle density. The addition of organic amendment decreases the particle density, mostly due to the increasing organic portion in soil, and decreases the ratio of mineral and organic matter in soil solid (Hillel 1998). There is a highly significant correlation between particle density and SOC. Depth effect of SOC on particle density displays inconsistencies. For example, variations in SOC concentration in the 0–10 cm depth explained notably larger variability in particle density (ρ_s) (~ 75% change), while for the 10–20 cm, depth variability in ρ_s was 54% and 45% for the 20–30 cm depth (Blanco-Canqui et al. 2006). Particle density increases with depth because of the simultaneous decreasing trend of organic matter content and increasing metal concentration. Soil organic matter and clay content significantly influence the particle density of a soil (Schjønning et al. 2017; Kumar et al. 2018).

3.3.2.4 Soil Water Holding Capacity

Water holding capacity (WHC) can be described as the ability of a soil to preserve water. Several soil properties like soil porosity, macro- and micropore number, the specific surface areas of soils, surface crust, slaking tendency, and absorption capacity have primarily triggered the WHC of a soil. Organic amendment and its composition significantly affect the major water retention factors, soil structure, and adsorption behavior; thereby organic amendment plays a dominant role in water retention in the soil. Higher organic matter content in soil increases aggregate stability and decreases bulk density and increases porosity and the number of micropore in soil (Haynes and Naidu 1998). Pore space provides key ways for entering soil water and gases within the soil profile. WHC of a soil strongly correlated with the soil structure and bulk density. The effect of the organic amendment on WHC tends to be greater in coarse-textured compared with fine-textured soils because WHC in heavy clay soils gets decreased with increasing SOC content. Thus, there is a strong synergistic relationship among textural components, SOC content, and WHC (Rawls et al. 2003). Plants can uptake readily available water (water held between field capacity FC, at matric suction of -10 kPa), and the unavailable ones were held between permanent wilting point PWP, at matric suction of -1500 kPa. Soil organic carbon can improve available water content in the soil (Haynes and Naidu 1998). An increase of 1% SOM can enhance 1.5% additional moisture by volume at FC (Wolf and Snyder 2003). Again, each gram of SOC can increase 50% water holding capacity of soil (Emerson and McGarry 2003). Organic amendments release humic acid substances, which can enhance water retention, available water content, and aggregate stability (Larney and Angers 2012). The organic amendment covers the soil surface and thus protects the soil from sealing and crusting by raindrop impact, thereby enhancing rainwater infiltration and reducing runoff. The short-term and long-term higher infiltration rate and cumulative infiltration were observed in the decomposed organic residue than stable residue (Mahmood-ul-Hassan et al. 2013).

3.3.2.5 Hydraulic Conductivity

Hydraulic conductivity describes the ease of movement of water through the pore space. The combined activity of hydraulic conductivity and soil porosity regulates soil water movement, soil aeration, and available water for the plant. Organic amendment plays a vital role in improving structural stability, which enhances macroporosity and then encourages hydraulic conductivity or water flow of a soil (Tisdall and Oades 1982). Macropore can regulate the soil water movement, infiltration rate, and good tilth condition of soil (Boyle et al. 1989). Several studies explained that the application of manure could raise soil structure and porosity, thereby improving the saturated hydraulic conductivity (K_{sat}) and infiltration rate (Benbi et al. 1998). It is observed that incorporation of organic amendment increased K_{sat} at all pressure heads at rooting depth (Eusufzai and Fujii 2012; Meena et al. 2015). As compared to control, field-saturated conductivity was increased by 34.4% due to the incorporation of compost while 15.9% for straw and sawdust-amended soils. Thus, organic amendment improves the hydraulic conductivity of soil through increasing the soil porosity.

3.3.3 Soil Microbial Properties

Organic amendments from different sources not only influence the soil physical and chemical properties but also greatly regulate the soil microbial properties. Soil microbes are important drivers of various nutrient cycling, nitrogen fixation, and nitrification process (Lynch and Bragg 1985; Lee and Pankhurst 1992; Tejada et al. 2009). Organic matter decomposition is one of the crucial roles of soil microorganisms. Moreover, microbial communities in the soil have significant control on the dynamics of soil carbon (Grandy et al. 2009) and ultimately influence the global carbon cycle (Doran 2002). Different types of organic amendments potentially increase soil organic matter (SOM) content (Seiter and Horwath 2004), a significant component of soil health. Improvement of SOM content in soil is important because it strongly influences the soil microbial community. The labile fractions of SOM provide the primary carbon substrate for the soil microbial communities, especially for the soil heterotrophic microorganisms. Organic amendments which increase the growth and activity of soil microbial communities indicate a strong relationship between microbial functioning and carbon content in soil (Chakraborty et al. 2011). Research studies demonstrated that organic amendments in soil ensure higher soil quality and higher biological activities including microbial populations, microbial respiration, soil microbial biomass carbon, and nitrogen content as compared to the traditional farming systems (Mäder et al. 2002; Girvan et al. 2004; Baaru et al. 2007).

3.3.3.1 Soil Biomass Carbon

Soil microbes are miracle creature in nature, while microbial biomass carbon is a responsive index of soil fertility and biotic attribute. It plays a crucial role in biogeochemical processes which are influenced by the addition of different organic and inorganic fertilizers in soils (Cerny et al. 2008). Microbial biomass in soil is a labile organic pool which unveils a quick output and acts as a regulatory dynamism of cycles of different macronutrients in crop fields. Microbes constitute about one-quarter of all living biomass on earth. The microbes perform a noble deed in the breakdown of different organic fertilizers, while microbial biomass is used as an early indicator of changing physical and chemical properties of soils because of different soils and crop management practices.

Microbial populations in soils are highly diverse, while the relation between their diversity and function influences soil structural stability and crop productivity. Soil organic matter, nutrient and water contents, physical and chemical properties, and climatic parameters influence microbial biomass in soils (Tomich et al. 2011). Soil microbial communities are influenced by land use changes and management practices. Organic farming with compost amendment has been shown to favor soil biota and provide better results in terms of biomass carbon and nitrogen compared to intensive farming with inorganic fertilizers (Santos et al. 2012; Amaral and Abelho 2016). It is found that the application of different manures and crop residues,

Table 4 Microbial biomass carbon (mg kg^{-1}) and nitrogen (mg kg^{-1}) in soil under different organic residues and chemical fertilizer management practices at different days after transplanting (DAT) of rice (Anik et al. 2017)

	Biomass carbon and nitrogen (mg kg^{-1}) at different DAT				
Treatment	0	30	60	90	120
Biomass carbon					
Rice straw	32.48	47.52	67.53	182.11	304.92
Poultry manure	100.81	114.84	121.25	300.92	431.64
Cow dung	47.30	71.28	91.58	178.34	380.12
Inorganic fertilizer	73.35	140.73	110.88	166.80	328.68
Control	16.99	35.51	49.32	118.32	199.94
Biomass nitrogen					
Rice straw	4.65	6.81	8.41	15.46	18.15
Poultry manure	5.01	13.39	13.81	18.72	31.60
Cow dung	4.70	6.99	11.55	14.37	27.18
Inorganic fertilizer	5.79	11.82	10.93	18.20	20.12
Control	3.25	6.43	6.01	11.52	9.88

adoption of crop rotation, and tillage practices alter soil organic matter dynamics by manipulating the soil environment and microbes hauling out C and N transformations (Anik et al. 2017; Zuber et al. 2018; Yadav et al. 2017a, b; Datta et al. 2014). The application of poultry manure, cow dung, and rice straw along with inorganic fertilizers contributed to higher biomass carbon compared to the sole application of inorganic fertilizers and zero input control (Table 4). Irrespective of organic materials, biomass carbon contents in soils were increased with the advancement of rice-growing periods, while the contribution of organic materials in the enrichment of biomass carbon in soils followed the order poultry manure >cow dung> rice straw (Table 4).

3.3.3.2 Soil Biomass Nitrogen

Like microbial biomass carbon in soils, biomass nitrogen is also a sensitive indicator which maintains ecological stability and strength of the environment. It reveals the soil quality in terms of soil fertility. Biomass nitrogen depends on soil physicochemical properties, microbial diversity, and soil and crop management practices including fertilizer application using organic and inorganic sources and climatic conditions. Undisturbed forest soil generally contains high amount of biomass nitrogen while lower in the grasslands and lowest in the agricultural soils (Miechówka et al. 2011). The typical value of biomass nitrogen in agricultural soils ranges from 10 to 60 g kg^{-1}; however, it may vary from <2 to >30 g kg^{-1} (Anik et al. 2017). The higher the biomass nitrogen, the higher the soil fertility. Different organic and inorganic fertilizers are applied to crop fields which determine the intensity and diversity of microbial populations in soils and thereby depict the amount of biomass nitrogen. Inorganic nitrogen fertilizer application has a significant effect on the transformation of biomass nitrogen (Table 4).

Different organic materials and their quality largely contribute in the biomass nitrogen content in soils. The C:N ratio in soils plays a strategic role in the turn over process of microbial biomass nitrogen. Research findings confirmed that the addition of a readily decomposable organic material with a low C:N ratio did not contribute greatly to the increase in the amount of soil microbial biomass N, while the addition of a material with high carbon with nitrogen ratio (C:N) prompted the immobilization of inorganic nitrogen and amplified the amount of microbial biomass N (Aoyama and Nozawa 1993). Nevertheless, the biomass forms in such a situation do not exist for a longer time period. After a certain period, immobilized N again mineralizes and supplies mineral N to the soil for crops.

3.3.3.3 Soil Microbial Population

Microbial population is considered one of the most important indicators of soil fertility which ultimately ensure the improvement of soil health. Amendment of organic materials to soil stimulates soil microbial populations which enhances the soil biological activities (Brady and Weil 1999). The increased number of microbial population in organic matter applied soils might be due to the supply of sufficient feed for the microorganisms from the organic amendments. It has also been hypothesized that application of organic manure would increase the availability of secondary nutrients as well as micronutrients in the soil, which favors the increase of soil microbial population (Krishnakumar et al. 2005). Investigation results of a long-term experiment indicated higher microbial biomass under organic management practices as compared to the traditional farming systems (Liu et al. 2007; Ram and Meena 2014). Addition of any organic manure either from plant or animal origin ensures the higher input of organic carbon in soil, which enhances the microbial population. Though it is very likely that organic amendments will enhance the microbial population, the number of microorganisms may vary from one source of organic amendments to another.

- *Fungal population*

Among the microbial communities in soil, bacteria and fungi are considered the most important constituents of soil biological characteristics. Fungi play very vital roles in organic matter build up, decomposition, mineralization, and cycling of nutrients especially nitrogen and phosphorus in soil. Fungi greatly enhance crop production through mycorrhizal associations. Fungi are also responsible for C sequestration in soil particularly in forest soils (Clemmensen et al. 2013). Organic carbon status of soil mostly enhances the soil fungal population (Girvan et al. 2004; Marfo et al. 2015; Lojkova et al. 2015). In general, the fungal population is increased in the organic manure-amended soil as compared to the control soil where no organic manure is applied. Research findings indicated a higher number of fungal population in organic manure-amended soil (13×10^4 CFU/g soil) as compared to the control

(4×10^4 CFU/g soil) soil that received no organic manure (Narasimha 2013). The higher fungal population in organic manure applied soil might be due to the favorable soil pH and optimum organic carbon supply by the organic amendments. Different sources of organic amendments greatly influence the fungal population. In a 2-year experiment, it was observed that fungal population in the first year got the significantly highest value of 25.23×10^3 CFU/g soil for vermicompost-amended plots, while the significantly lowest value of 11.37×10^3 CFU/g soil was recorded in control plot (Das and Dkhar 2012). In another experiment, farmyard manure demonstrated significantly higher fungal population among different organic amendments (Swer et al. 2011), and fungal population for different treatments follows the order as farmyard manure > plant compost > integrated compost >vermicompost> control. In most cases, it is evident that organic amendments increase fungal population, but there is also a report regarding no significant variation of the fungal population due to different types of organic amendments. Research findings revealed that the fungal population did not vary significantly due to different organic amendments in the last year of a 2-year experiment (Das and Dkhar 2012; Dhakal et al. 2016).

- *Bacterial population*

The soil is the most important habitat of bacteria. Among all microbial communities in the soil, bacteria are the most populous, even though the weight of fungi may exceed the weight of bacteria. One gram of soil can contain billions of bacteria. Organic matter status of soil is crucial for the multiplication of soil bacteria. Some bacteria may double their numbers within 30 min, especially when the soil contains adequate amounts of organic residues. Most of the soil bacteria are heterotrophic in nature and directly rely on soil organic matter as they receive their food and energy from organic substances. Research findings indicated that organic amendments commonly increase the bacterial population as compared to the no manure amendments but bacterial population may be altered by different sources of organic amendments too. Results of a long-term experiment that was managed for the last 29 years by the Department of Soil Science, BSMRAU, Gazipur, Bangladesh revealed that organic amendments significantly influence the bacterial population. Among the organic amendments, significantly the highest number of bacterial population was enumerated from rice straw-treated plots (22×10^5 CFU/g soil), and the lowest population (6×10^5 CFU/g soil) was observed in control (no organic manures) treatment. The higher microbial population in soil due to the application of organic residues might be due to higher organic carbon build up in soil and creation of suitable soil properties. Among the organic amendments, vermicompost also showed significant influence on soil bacterial population. The highest bacterial population was enumerated in vermicompost-amended plots (55.19×10^5 CFU/g dry soil) followed by the amendment of farmyard manure (54.26×10^5 CFU/g dry soil), whereas the lowest number was recorded in control treatment having a value of 30.89×10^5 CFU/g dry soil (Das and Dkhar 2011).

The rate of organic amendments is also crucial in regulating the bacterial population. Research findings clearly demonstrated that the number of spore-forming bacteria is positively correlated with the dose of applied compost (Zaccardelli et al. 2013). Moreover, some reports illustrated the decline of bacterial population due to organic amendments. Study results clearly illustrated that the addition of wheat straw in soil decreased the bacterial population (Acea and Carballas 1996). This might be due to the slow decomposition nature of the straw which resulted in lower availability of organic carbon for the bacterial community. It is also reported that high doses of organic amendments negatively influence the microbial biomass and enzyme activities (Ouni et al. 2013; Datta et al. 2017a). This negative behavior might be due to the toxic effect of the increased trace elements at a higher dose of organic amendments (Crecchio et al. 2004).

4 Biochar and Organic Amendments for Carbon Sequestration and Climate Change Mitigation

Carbon sequestration is a process of taking away of carbon from the atmosphere in a solid material through biological or physical processes and storing in the natural environment for an unlimited period. Plants naturally perform this function converting atmospheric carbon dioxide (CO_2) into organic carbon through the process of photosynthesis and incorporated into living plant matter (Rice and McVay 2002; Rahman 2010). The continuous falling off of branches and leaves from vegetation increases fresh organic materials to soils and when plants die adds more biomass which decays and becomes soil organic matter (SOM), and thus, carbon is sequestered into the soil. The sequestration of organic carbon is a requisite for soil health improvement, crop yield increment, and higher use efficiencies of plant nutrients (Rahman 2013). Soil organic carbon is a dominant factor which governs soil biological and physicochemical properties and ensures agricultural and environmental sustainability. In the tropical and subtropical climatic regions, high soil and air temperature coupled with high moisture content favors microbial decomposition of different organic fertilizers like compost, manure, crop residues, etc. added to soils, and thereby organic matter content reduces fast. Faster microbial decomposition of leftover crop residues and added organic wastes resulted in lower carbon content in soils (Rahman 2010; Agehara and Warncke 2005). Consequently, the present carbon contents and carbon stocks in soils of many countries in the tropical and subtropical climatic regions are declining at an alarming rate. It is a great challenge and global concern to increase carbon content in soils. Resource conservation strategies and best management practices are essentially required to sequester carbon in soils. The sequestration of organic carbon is the result of the long-term input-output budget, i.e., the balance between input as the addition of organic C using various organic materials and output as losses from the soil as CO_2 emission. Continual soil disturbance through plowing and other soil and crop management practices expedite microbial decomposition of organic matter, although a large amount of C is added to the soil through residues of different crops which ultimately results in either a net stockpiling or a net reduction of soil carbon in soils.

The soil has enormous potential in storing carbon more than three times of the terrestrial vegetation which depends on how we manage our soils. It is reported that the world's agricultural and degraded lands can potentially store carbon 50–66% of the historic carbon loss of 42–78 giga tons (Lal 2004). The organic carbon stocks in the world's soils are 1550 Pg, while inorganic carbons are 950 Pg (Lal 2008). Soil physical properties (texture and structure), climatic factors (rainfall and temperature), and soil and crop management practices manage the rate of soil organic carbon sequestration. Carbon stock and sequestration in soils can be increased through adoption of resource conservation strategies and best management practices like zero tillage, minimum tillage, green manuring, cover crops, balanced and optimum fertilization, manuring, sludge and compost application, and social forestry. Application of different organic amendments including biochar can play a vital role to replenish and conserve carbon in soils and ensure agricultural sustainability. Biochar is a fine-grained, highly carbonaceous material which is persistent in nature and can sequester carbon in the soil for a long time. It has also a significant contribution in developing soil structure encompassing soil particles together, and thus biochar not only sequester its own carbon but also protect inherent and applied carbon using different sources. The mineralization of biochar is much slower than any other organic materials. Research findings demonstrated a carbon mineralization rate of 1.5% in biochar-amended soils, while in the non-amended soil, it was 2.4% (Hernandez-Soriano et al. 2016). The rate of carbon reduction in biochar-mixed soils is restricted compared to cow dung and poultry manure mixed soils (Fig. 6). Biochar is known as an inert matter which contributes to the soil recalcitrant organic carbon pool which is resistant to high temperature and high moisture and to further microbial decomposition. The rates of organic carbon sequestration of different biochar in soils might be about 1 Pg C year^{-1} (Sohi et al. 2010). The half-life of carbonized materials in biochar is about 1400 years (Kuzyakov et al. 2009). However, several estimates confirmed that carbon could remain in soils at least 100 years through the application of biochar in agricultural lands if managed properly (Shackley et al. 2009). The application of rice husk and corn stover biochar significantly improves soil aggregation and structural stability, increases carbon, and improves soil health. A healthy soil containing high organic carbon is capable of exchanging different cations and holding more water in soils and decreasing CO_2 emission (Mohan et al. 2018).

Carbon added to the soil using different organic amendments and crop residues undergoes a series of transformation from labile to recalcitrant forms. The labile C upon decomposition releases CO_2 to the atmosphere which favors the globe to become warm. The labile pool of carbon in soils is vital for maintaining soil fertility, its productive capacity, and ecosystem community, while the recalcitrant carbon pool promises to sequester more carbon and inevitable for maintaining the structural integrity of the soils and long-term sustainability of the environment. The organic material that contains a higher amount of labile carbon sequesters less amount of carbon in soils while releasing more carbon to the atmosphere as shown in Table 5

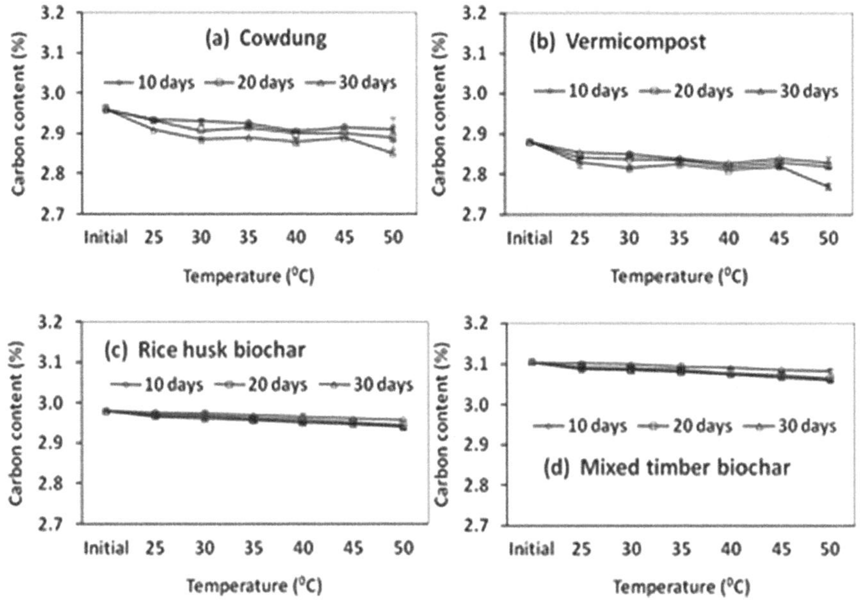

Fig. 6 Reduction in carbon contents of organic materials added soils under different temperature regimes: (**a**) cow dung, (**b**) vermicompost. (Source: Hossain et al. 2017); (**c**) rice husk biochar, and (**d**) mixed timber biochar

Table 5 Effect of organic manures, rice straw, and inorganic fertilizers on carbon sequestration and carbon loss in the soil after harvesting of five rice seasons where C was applied at 2000 kg ha^{-1} in each crop season (0–15 cm soil depth) (extracted from Rahman et al. 2016)

Treatment	Total C input (kg ha^{-1})	Labile C (g kg^{-1})	Carbon sequestration (t ha^{-1})	%C emission loss	%C sequestration	%C unaccounted
Control	–	2.92	−1.36	–	–	–
CD	10,000	5.70	3.03	36	30	34
PM	10,000	4.88	4.90	28	49	23
RS	10,000	6.98	1.04	37	10	53
STB	–	3.78	0.04	–	–	–

Note: *CD* cow dung, *PM* poultry manure, *RS* rice straw, *STB* soil test-based fertilizer, *C* carbon

(Rahman et al. 2016; Meena and Yadav 2014). The data presented in Table 5 revealed that among different organic materials, the highest amount of organic carbon sequestered in poultry manure-treated soil is followed by cow dung, rice straw, and inorganic fertilizer. Manure was found more powerful in building soil carbon than straw, possibly because of the occurrence of more humified and obstinate C forms in manure contrasted with rice straw.

Soil carbon content is influenced by tillage operations, fertility levels, cropping systems, and cropping intensity. More tillage and intensive cultivation with high

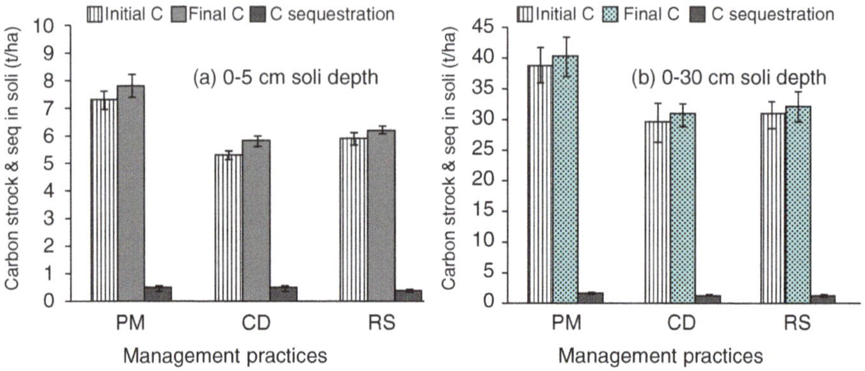

Fig. 7 Carbon stock and sequestration (t ha^{-1}) in soils under different management practices in rice-fallow-rice cropping systems in Bangladesh. (Rahman 2010), *PM* poultry manure, *CD* cow dung, *RS* rice straw

levels of inorganic fertilization enhance the decomposition of inherent as well as added organic materials in soils resulting in lower carbon in soils. The cropping systems and soil and crop management practices that provide higher carbon input might help in sustaining the carbon level and maintaining good soil health. Findings of a research study revealed that minimum tillage in an uninterrupted monoculture of cotton increased soil organic carbon by 24% compared to conventional tillage, while crop rotation increased soil carbon contents by 28% in a cotton corn (*Zeamays* L.) (Wright et al. 2008) rotation compared to continuous cotton. In another experiment, it was found that minimum tillage in rice cultivation contributed to higher carbon accumulation in soils compared to traditional tillage even though total carbon input was low under minimum tillage (Table 6). Figure 7 revealed the carbon stock and sequestration in soils of rice-fallow-rice cropping systems under different organic management practices. It was found that poultry manure contained more stable carbon and contributed to higher carbon stock as well as carbon sequestration contrasted with cow dung and rice straw (Rahman 2010). Management practices that ensure slower microbial decomposition of added residues are likely to increase carbon content in soil and thus enhance carbon store in soils. Therefore, the identification of such soil and crop management practices is the main concern for building carbon in soils and sustaining crop productivity.

The organic matter content in soil is responsible for making the agricultural system sustainable. The soil has an immense capacity to serve as an eventual sink of atmospheric CO_2. The concentration of CO_2 and other GHGs in the atmosphere is high enough to make the globe warmed and climate changed. During the last century, the planet's average surface temperature has increased by 1.1 °C which is caused by the increased emission of CO_2, CH_4, N_2O, and NO into the atmosphere. The sequestration of carbon in soils can possibly alleviate the harmful effect of global warming and climate change on crop cultivation. Different soil and crop management practices and organic materials including biochar increase soil carbon stock, and thus soil health is improved which makes it capable of acting against the negative effects of climate change (Soderstorm et al. 2014). Different management practices like diversified cropping systems, application of wastes/compost, and tillage operation coupled with balanced fertilization using organic and inorganic sources have high potential to increase carbon content in soils. Organic debris and plant residues added to soil convert into more steady humic substances and contribute in the development of various organomineral compounds and microaggregates which safeguard soil carbon from mineralization and thus help to enhance C sequestration (Lal 2016). Carbon sequestration prevents CO_2 emissions produced by human activities and remove it from the atmosphere in different ways and stores it in soils. Carbon sequestration increases with enhancing plant physiology and rate of photosynthesis which ultimately results in higher plant biomass (Lal 2004). It is found that 1 ton of carbon increases in soils can increase yields of maize, wheat, and cowpeas by 10–20, 20–40, and 0.5–1 kg ha^{-1}, respectively, while such augmentation in carbon can possibly relieve 5–15% of the worldwide emission from fossil fuel burning (Lal 2004; Verma et al. 2015).

Capturing CO_2 from large point sources such as fossil fuel power plants and enduringly storing it in different reservoirs away from the atmosphere is a process to lessen global warming (USDE & NETL 2007). It is reported that the use of organic amendments like manures, composts, biochar-compost mixtures, etc. can be the better option for the improvement of soil fertility, restoration of degraded land, and mitigation of emissions of greenhouse gases from agricultural practices and land use changes (Agegnehu et al. 2017). It is understood that the application of organic amendments to agricultural lands ensures a continuous supply of carbon pool which might reduce the emission of greenhouse gases from soil, and thus it helps in the mitigation of global warming and climate change (Rahman et al. 2016; Lehmann and Joseph 2015). Biochar releases less emission compared to fresh organic materials. Carbon dioxide emissions in chicken manure, rice straw, vermicompost, cow dung, and rice husk biochar were 19.69, 18.60, 12.16, 12.01, and 7.96%, respectively (Hossain et al. 2017). It was also reported that biochar produced from poultry manure releases more greenhouse gas compared to biochar from sugarcane straw. However, both can considerably reduce the CO_2eq emission (Novais et al. 2017).

5 Conclusion

Intensive cultivation coupled with high temperature and high moisture content favored faster decomposition of organic matter in soils. The resultant effect of faster mineralization of organic matter is declining soil fertility and increasing greenhouse gas emission. Biochar and other organic amendments potentially can sequester carbon in soil and reduce carbon dioxide emission to the atmosphere. The application of such materials improves soil health through developing soil aggregates, increasing soil nutrient and water contents, exchanging cations, easing hydraulic conductivity, and enhancing microbial activity, biomass carbon, and nitrogen while maintaining soil pH and bulk density to a favorable level for crop production. Resource conservation strategies and continuous supply of organic materials help to increase and maintain carbon content as well as the fertility of soil where biochar appeared as a promising amendment for storing carbon in the soil and mitigating global warming.

References

Acea MJ, Carballas T (1996) Microbial response to organic amendments in a forest soil. Bioresour Technol 57(2):193–199

Agegnehu G, Srivastava AK, Bird MI (2017) The role of biochar and biochar-compost in improving soil quality and crop performance: a review. Appl Soil Ecol 119:156–170

Agehara S, Warncke DD (2005) Soil moisture and temperature effects on nitrogen release from organic nitrogen sources. Soil Sci Soc Am J 69(6):1844–1855

Agrafioti E, Bouras G, Kaldis D, Diamandopoulos E (2013) Biochar production by sewage sludge pyrolysis. J Anal Appl Pyrolysis 101:72–78

Ahmad M, Rajapaksha AU, Lim JE, Zhang M, Bolan N, Mohan D, Ok YS (2014) Biochar as a sorbent for contaminant management in soil and water: a review. Chemosphere 99:19–33

Alexander M (1977) Mineralization and immobilization of nitrogen. In: Alexander M (ed) Introduction to soil microbiology, 2nd edn. Wiley, New York, pp p136–p247

Amanullah MM, Sekar S, Muthukrishnan P (2010) Prospects and potential of poultry manure. Asian J Plant Sci 9(4):172

Amaral F, Abelho M (2016) Effects of agricultural practices on soil and microbial biomass carbon, nitrogen and phosphorus content: a preliminary case study. Web Ecol 16(1):3–5

Amonette JE, Joseph S (2009) Characteristics of biochar: microchemical properties. In: Lehman J, Joseph S (eds) Biochar for environmental management: science and technology. Earthscan, London, pp 33–52

Angelova VR, Akova VI, Artinova NS, Ivanov KI (2013) The effect of organic amendments on soil chemical characteristics. Bulgarian J Agric Sci 19(5):958–971

Anik MFA, Rahman MM, Rahman GM, Alam MK, Islam MS, Khatun MF (2017) Organic amendments with chemical fertilizers improve soil fertility and microbial biomass in rice-rice-rice triple crops cropping systems. Open J Soil Sci 7:87–100

Antal MJ, Grønli M (2003) The art, science, and technology of charcoal production. Ind Eng Chem Res 42(8):1619–1640

Aoyama M, Nozawa T (1993) Microbial biomass nitrogen and mineralization-immobilization processes of nitrogen in soils incubated with various organic materials. Soil Sci Plant Nutr 39 (1):23–32

Ashoka P, Meena RS, Kumar S, Yadav GS, Layek J (2017) Green nanotechnology is a key for eco-friendly agriculture. J Clean Prod 142:4440–4441

Atiyeh RM, Lee S, Edwards CA, Arancon NQ, Metzger JD (2002) The influence of humic acids derived from earthworm-processed organic wastes on plant growth. Bioresour Technol 84(1):7–14

Atkinson CJ, Fitzgerald JD, Hipps NA (2010) Potential mechanisms for achieving agricultural benefits from biochar application to temperate soils: a review. Plant Soil 337(1–2):1–18

Baaru MW, Mungendi DN, Bationo A, Verchot L, Waceke W (2007) Soil microbial biomass carbon and nitrogen as influenced by organic and inorganic inputs at Kabete, Kenya. In: Bationo A, Waswa B, Kihara J, Kimetu J (eds) Advances in integrated soil fertility management in sub-Saharan Africa: challenges and opportunities. Springer, Dordrecht, pp 827–832

Baldock JA, Smernik RJ (2002) Chemical composition and bioavailability of thermally altered *Pinus resinosa* (Red pine) wood. Org Geochem 33(9):1093–1109

Benbi DK, Biswas CR, Bawa SS, Kumar K (1998) Influence of farmyard manure, inorganic fertilizers and weed control practices on some soil physical properties in a long-term experiment. Soil Use Manag 14(1):52–54

Biederman LA, Harpole WS (2013) Biochar and its effects on plant productivity and nutrient cycling: a meta-analysis. GCB Bioenergy 5(2):202–214

Blanco-Canqui H, Lal R, Post WM, Izaurralde RC, Shipitalo MJ (2006) Organic carbon influences on soil particle density and rheological properties. Soil Sci Soc Am J 70(4):1407–1414

Boix-Fayos C, Calvo-Cases A, Imeson AC, Soriano-Soto MD (2001) Influence of soil properties on the aggregation of some Mediterranean soils and the use of aggregate size and stability as land degradation indicators. Catena 44(1):47–67

Bonanomi G, D'Ascoli R, Scotti R, Gaglione SA, Caceres MG, Sultana S, Scelza R, Rao MA, Zoina A (2014) Soil quality recovery and crop yield enhancement by combined application of compost and wood to vegetables grown under plastic tunnels. Agric Ecosyst Environ 192:1–7

Boyle M, Frankenberger WT, Stolzy LH (1989) The influence of organic matter on soil aggregation and water infiltration. J Prod Agric 2(4):290–299

Brady NC, Weil RR (1999) Soil organic matter. In: Brady NC, Weil RR (eds) The nature and properties of soils. Prentice Hall, Upper Saddle River, pp 446–490

Cerny J, Balik J, Kulhanek M, Nedved V (2008) The changes in microbial biomass and N in long-term field experiments. Plant Soil Environ 54(5):212–218

Chakraborty A, Chakrabarti K, Chakraborty A, Ghosh S (2011) Effect of long-term fertilizers and manure application on microbial biomass and microbial activity of a tropical agricultural soil. Biol Fertil Soils 47(2):227–233

Chen KS, Wang HK, Peng YP, Wang WC, Chen CH, Lai CH (2008) Effects of open burning of rice straw on concentrations of atmospheric polycyclic aromatic hydrocarbons in Central Taiwan. J Air Waste Manage Assoc 58(10):1318–1327

Cheng CH, Lehmann J, Thies JE, Burton SD, Engelhard MH (2006) Oxidation of black carbon by biotic and abiotic processes. Org Geochem 37(11):1477–1488

Chun Y, Sheng G, Chiou CT, Xing B (2004) Compositions and sorptive properties of crop residue-derived chars. Environ Sci Technol 38(17):4649–4655

Clemmensen KE, Bahr A, Ovaskainen O (2013) Roots and associated fungi drive long-term carbon sequestration in boreal forest. Science 339(6127):1615–1618

Cornelissen G, Martinsen V, Shitumbanuma V, Alling V, Breedveld GD, Rutherford DW, Sparrevik M, Hale SE, Obia A, Mulder J (2013) Biochar effect on maize yield and soil characteristics in five conservation farming sites in Zambia. Agronomy 3(2):256–274

Crecchio C, Curci M, Pizzigallo MD, Ricciuti P, Ruggiero P (2004) Effects of municipal solid waste compost amendments on soil enzyme activities and bacterial genetic diversity. Soil Biol Biochem 36(10):1595–1605

Crovetto C (1997) Zero tillage and soil nutrition. In: Sustainable high production agriculture, now! 5th national congress of AAPRESID. Mar del Plata, Argentina, pp 73–78

Dadhich RK, Meena RS (2014) Performance of Indian mustard (*Brassica juncea* L.) in Response to foliar spray of thiourea and thioglycolic acid under different irrigation levels. Indian J Ecol 41 (2):376–378

Das BB, Dkhar MS (2011) Rhizosphere microbial populations and physico chemical properties as affected by organic and inorganic farming practices. Am-Euras J Agric Environ Sci 10:140–150

Das BB, Dkhar MS (2012) Organic amendment effects on microbial population and microbial biomass carbon in the rhizosphere soil of soybean. Commun Soil Sci Plant Anal 43(14):1938–1948

Das B, Chakraborty D, Singh VK, Aggarwal P, Singh R, Dwivedi BS (2014) Effect of organic inputs on strength and stability of soil aggregates under rice-wheat rotation. Int Agrophys 28 (2):163–168

Datta R, Vranová V, Pavelka M, Rejšek K, Formánek P (2014) Effect of soil sieving on respiration induced by low-molecular-weight substrates. Int Agrophys 28(1):119–124

Datta R, Anand S, Moulick A, Baraniya D, Pathan SI, Rejsek K, Vranova V, Sharma M, Sharma D, Kelkar A (2017a) How enzymes are adsorbed on soil solid phase and factors limiting its activity: a review. Int Agrophys 31(2):287–302
Datta R, Kelkar A, Baraniya D, Molaei A, Moulick A, Meena R, Formanek P (2017b) Enzymatic degradation of lignin in soil: a review. Sustainability 9(7):1163
Datta R, Baraniya D, Wang Y-F, Kelkar A, Meena RS, Yadav GS, Teresa Ceccherini M, Formanek P (2017c) Amino acid: its dual role as nutrient and scavenger of free radicals in soil. Sustainability 9(8):1402
Dhakal Y, Meena RS, Kumar S (2016) Effect of INM on nodulation, yield, quality and available nutrient status in soil after harvest of green gram. Legum Res 39(4):590–594
Doran JW (2002) Soil health and global sustainability: translating science into practice. Agric Ecosyst Environ 88(2):119–127
Dutta B, Raghavan V (2014) A life cycle assessment of environmental and economic balance of biochar systems in Quebec. Int J Energy Environ Eng 5(2–3):106
Emerson WW, McGarry D (2003) Organic carbon and soil porosity. Soil Res 41(1):107–118
Eusufzai MK, Fujii K (2012) Effect of organic matter amendment on hydraulic and pore characteristics of a clay loam soil. Open J Soil Sci 2(04):372
FRG (2012) Fertilizer recommendation guide, Bangladesh Agricultural Research Council (BARC) Farmgate. Dhaka 1215, 274 p
Girvan MS, Bullimore J, Ball AS, Pretty JN, Osborn AM (2004) Responses of active bacterial and fungal communities in soils under winter wheat to different fertilizer and pesticide regimens. Appl Environ Microbiol 70(5):2692–2701
Glaser B, Lehmann J, Zech W (2002) Ameliorating physical and chemical properties of highly weathered soils in the tropics with charcoal–a review. Biol Fertil Soils 35(4):219–230
Grandy AS, Strickland MS, Lauber CL, Bradford MA, Fierer N (2009) The influence of microbial communities, management, and soil texture on soil organic matter chemistry. Geoderma 150(3–4):278–286
Habteselassie MY, Miller BE, Thacker SG, Stark JM, Norton JM (2006) Soil nitrogen and nutrient dynamics after repeated application of treated dairy-waste. Soil Sci Soc Am J 70(4):1328–1337
Haynes RJ (2000) Interactions between soil organic matter status, cropping history, method of quantification and sample pretreatment and their effects on measured aggregate stability. Biol Fertil Soils 30(4):270–275
Haynes RJ, Naidu R (1998) Influence of lime, fertilizer and manure applications on soil organic matter content and soil physical conditions: a review. Nutr Cycl Agroecosyst 51(2):123–137
Heard JR, Kladivko EJ, Mannering JV (1988) Soil macroporosity, hydraulic conductivity and air permeability of silty soils under long-term conservation tillage in Indiana. Soil Tillage Res 11 (1):1–18
Helyar KR, Cregan PD, Godyn DL (1990) Soil acidity in New-South-Wales-Current pH values and estimates of acidification rates. Soil Res 28(4):523–537
Hernandez-Soriano MC, Kerré B, Kopittke PM, Horemans B, Smolders E (2016) Biochar affects carbon composition and stability in soil: a combined spectroscopy-microscopy study. Sci Rep 6:25127
Hillel D (1998) Environmental soil physics. Academic, San Diego
Hossain MB, Rahman MM, Biswas JC, Miah MMU, Akhter S, Maniruzzaman M, Choudhury AK, Ahmed F, Shiragi MHK, Kalra N (2017) Carbon mineralization and carbon dioxide emission from organic matter added soil under different temperature regimes. Int J Recycl Org Waste Agric 6(4):311–319
Houben D, Evrard L, Sonnet P (2013) Mobility, bioavailability and pH-dependent leaching of cadmium, zinc and lead in a contaminated soil amended with biochar. Chemosphere 92 (11):1450–1457
Huang GF, Wong JWC, Wu QT, Nagar BB (2004) Effect of C/N on composting of pig manure with sawdust. Waste Manag 24(8):805–813

Ikpe FN, Powell JM (2002) Nutrient cycling practices and changes in soil properties in the crop-livestock farming systems of western Niger Republic of West Africa. Nutr Cycl Agroecosyst 62 (1):37–45

Jeffery S, Verheijen FG, van der Velde M, Bastos AC (2011) A quantitative review of the effects of biochar application to soils on crop productivity using meta-analysis. Agric Ecosyst Environ 144(1):175–187

Kamara A, Mansaray MM, Kamara A, Sawyerr PA (2014) Effects of biochar derived from maize stover and rice straw on the early growth of their seedlings. Am J Agric For 2(5):232–236

Kinney TJ, Masiello CA, Dugan B, Hockaday WC, Dean MR, Zygourakis K, Barnes RT (2012) Hydrologic properties of biochars produced at different temperatures. Biomass Bioenergy 41:34–43

Kirkby CA, Kirkegaard JA, Richardson AE, Wade LJ, Blanchard C, Batten G (2011) Stable soil organic matter: a comparison of C: N: P: S ratios in Australian and other world soils. Geoderma 163(3–4):197–208

Krishnakumar S, Saravanan A, Natarajan SK, Veerabadram V, Mani S (2005) Microbial population and enzymatic activity as influenced by organic farming. Res J Agric Biol Sci 1:85–88

Kumar S, Meena RS, Bohra JS (2018) Interactive effect of sowing dates and nutrient sources on dry matter accumulation of Indian mustard (*Brassica juncea* L.). J Oilseed Brassica 9(1):72–76

Kuzyakov Y, Subbotina I, Chen H, Bogomolova I, Xu X (2009) Black carbon decomposition and incorporation into soil microbial biomass estimated by 14C labeling. Soil Biol Biochem 41 (2):210–219

Laghari M, Hu Z, Mirjat MS, Xiao B, Tagar AA, Hu M (2016) Fast pyrolysis biochar from sawdust improves the quality of desert soils and enhances plant growth. J Sci Food Agric 96(1):199–206

Lal R (2004) Soil carbon sequestration to mitigate climate change. Geoderma 123(1–2):1–22

Lal R (2008) Soil carbon stocks under present and future climate with specific reference to European ecoregions. Nutr Cycl Agroecosyst 81(2):113–127

Lal R (2016) Soil health and carbon management. Food Energy Secur 5(4):212–222

Larney FJ, Angers DA (2012) The role of organic amendments in soil reclamation: a review. Can J Soil Sci 92(1):19–38

Latifah O, Ahmed OH, Majid NMA (2018) Soil pH buffering capacity and nitrogen availability following compost application in a tropical acid soil. Compost Sci Util 26(1):1–15

Lee KE, Pankhurst CE (1992) Soil organisms and sustainable productivity. Soil Res 30(6):855–892

Lee Y, Park J, Ryu C, Gang KS, Yang W, Park YK, Hyun S (2013) Comparison of biochar properties from biomass residues produced by slow pyrolysis at 500 °C. Bioresour Technol 148:196–201

Lehmann J (2007) Bio-energy in the black. Front Ecol Environ 5(7):381–387

Lehmann J, Joseph S (eds) (2015) Biochar for environmental management: science, technology and implementation. Routledge, New York

Lehmann J, Rondon M (2006) Bio-char soil management on highly weathered soils in the humid tropics. In: Norman U et al (eds) Biological approaches to sustainable soil systems. CRC Press, Boca Raton, pp 517–530

Lehmann J, Rillig MC, Thies J, Masiello CA, Hockaday WC, Crowley D (2011) Biochar effects on soil biota–a review. Soil Biol Biochem 43(9):1812–1836

Leroy BLM, Herath HMSK, Sleutel S, De Neve S, Gabriels D, Reheul D, Moens M (2008) The quality of exogenous organic matter: short-term effects on soil physical properties and soil organic matter fractions. Soil Use Manag 24(2):139–147

Liang B, Lehmann J, Solomon D, Kinyangi J, Grossman J, O'neill B, Neves EG (2006) Black carbon increases cation exchange capacity in soils. Soil Sci Soc Am J 70(5):1719–1730

Liu B, Gumpertz ML, Hu S, Ristaino JB (2007) Long-term effects of organic and synthetic soil fertility amendments on soil microbial communities and the development of southern blight. Soil Biol Biochem 39(9):2302–2316

Liu J, Schulz H, Brandl S, Miehtke H, Huwe B, Glaser B (2012) Short-term effect of biochar and compost on soil fertility and water status of a Dystric Cambisol in NE Germany under field conditions. J Plant Nutr Soil Sci 175(5):698–707

Lojkova L, Datta R, Sajna M, Marfo TD, Janous D, Pavelka M, Formanek P (2015) Limitation of proteolysis in soils of forests and other types of ecosystems by diffusion of substrate. In: Amino acids, vol 8. Springer, Wien, pp 1690–1691

Lynch JM, Bragg E (1985) Microorganisms and soil aggregate stability. In: Stewart BA (ed) Advances in soil science, vol 2. Springer, New York, pp 133–171

Mäder P, Fliessbach A, Dubois D, Gunst L, Fried P, Niggli U (2002) Soil fertility and biodiversity in organic farming. Science 296(5573):1694–1697

Mahmood-ul-Hassan M, Rafique E, Rashid A (2013) Physical and hydraulic properties of aridisols as affected by nutrient and crop-residue management in a cotton-wheat system. Acta Sci Agron 35(1):127–137

Marfo TD, Datta R, Lojkova L, Janous D, Pavelka M, Formanek P (2015) Limitation of activity of acid phosphomonoesterase in soils. In: Amino acids, vol 8. Springer, Wien, pp 1691–1691

McClellan AT, Deenik J, Uehara G, Antal M (2007) Effects of flashed carbonized macadamia nutshell charcoal on plant growth and soil chemical properties. A century of integrating crops, soils & environment. In: International annual meeting, Nov 4–8, New Orleans, Louisiana

Meena H, Meena RS (2017) Assessment of sowing environments and bio-regulators as adaptation choice for clusterbean productivity in response to current climatic scenario. Bangladesh J Bot 46 (1):241–244

Meena RS, Yadav RS (2014) Phonological performance of groundnut varieties under sowing environments in hyper arid zone of Rajasthan, India. J Appl Nat Sci 6(2):344–348

Meena RS, Yadav RS, Meena H, Kumar S, Meena YK, Singh A (2015) Towards the current need to enhance legume productivity and soil sustainability worldwide: a book review. J Clean Prod 104:513–515

Meena RS, Kumar S, Pandey A (2017a) Response of sulfur and lime levels on productivity, nutrient content and uptake of sesame under guava (*Psidium guajava* L.) based agri-horti system in an acidic soil of eastern Uttar Pradesh, India. J Crop Weed 13(2):222–227

Meena RS, Meena PD, Yadav GS, Yadav SS (2017b) Phosphate solubilizing microorganisms, principles and application of microphos technology. J Clean Prod 145:157–158

Miechówka A, Gąsiorek M, Józefowska A, Zadrożny P (2011) Content of microbial biomass nitrogen in differently used soils of the Carpathian Foothills. Ecol Chem Eng A18(4):577–583

Mohan D, Abhishek K, Sarswat A, Patel M, Singh P, Pittman CU (2018) Biochar production and applications in soil fertility and carbon sequestration–a sustainable solution to crop-residue burning in India. RSC Adv 8(1):508–520

Molaei A, Lakzian A, Datta R, Haghnia G, Astaraei A, Rasouli-Sadaghiani M, Ceccherini MT (2017) Impact of chlortetracycline and sulfapyridine antibiotics on soil enzyme activities. Int Agrophys 31(4):499–505

Molaei A, Lakzian A, Haghnia G, Astaraei A, Rasouli-Sadaghiani M, Ceccherini MT, Datta R (2017a) Assessment of some cultural experimental methods to study the effects of antibiotics on microbial activities in a soil: an incubation study. PLoS One 12(7):e0180663

Murphy B (2015) Key soil functional properties affected by soil organic matter-evidence from published literature. IOP Conf Ser Earth Environ Sci 25:012008. https://doi.org/10.1088/1755-1315/25/1/012008

Nagavallemma KP, Wani SP, Lacroix S, Padmaja VV, Vineela C, Rao MB, Sahrawat KL (2004) Vermicomposting: recycling wastes into valuable organic fertilizer. Global theme on agroecosystems report no. 8, ICRISAT, Patancheru, Andhra Pradesh, India

Narasimha G (2013) Impact of organic manure amendment on soil physicochemical, biological and enzymatic properties. Biotechnol Indian J 7(4):154–158

Neilsen GH, Hogue EJ, Neilsen D, Zebarth BJ (1998) Evaluation of organic wastes as soil amendments for cultivation of carrot and chard on irrigated sandy soils. Can J Soil Sci 78 (1):217–225

Novais SV, Zenero MDO, Junior EFF, de Lima RP, Cerri CEP (2017) Mitigation of greenhouse gas emissions from tropical soils amended with poultry manure and sugar cane straw biochars. Agric Sci 8(09):887–903
O'Neill B, Grossman J, Tsai MT, Gomes JE, Lehmann J, Peterson J, Neves E, Thies JE (2009) Bacterial community composition in Brazilian anthrosols and adjacent soils characterized using culturing and molecular identification. Microb Ecol 58(1):23–35
Ouni Y, Lakhdar A, Scelza R, Scotti R, Abdelly C, Barhoumi Z, Rao MA (2013) Effects of two composts and two grasses on microbial biomass and biological activity in a salt-affected soil. Ecol Eng 60:363–369
Parihar CM, Rana KS, Jat ML, Jat SL, Parihar MD, Kantwa SR, Singh DK, Sharma S (2012) Carbon footprint and economic sustainability of pearl millet-mustard system under different tillage and nutrient management practices in moisture stress conditions. Afr J Microbiol Res 6 (23):5052–5061
Rahman MM (2010) Carbon sequestration options in soils under different crops and their management practices. Agriculturists 8(1):90–101
Rahman MM (2013) Nutrient-use and carbon-sequestration efficiencies in soils from different organic wastes in rice and tomato cultivation. Commun Soil Sci Plant Anal 44(9):1457–1471
Rahman F, Rahman MM, Rahman GKMM, Saleque MA, Hossain AS, Miah MG (2016) Effect of organic and inorganic fertilizers and rice straw on carbon sequestration and soil fertility under a rice–rice cropping pattern. Carbon Manage 7(1–2):41–53
Ram K, Meena RS (2014) Evaluation of pearl millet and mungbean intercropping systems in Arid Region of Rajasthan (India). Bangladesh J Bot 43(3):367–370
Rawls WJ, Pachepsky YA, Ritchie JC, Sobecki TM, Bloodworth H (2003) Effect of soil organic carbon on soil water retention. Geoderma 116(1–2):61–76
Rice C, McVay K (2002) Carbon sequestration: frequently asked question. Kansas State University. Agricultural Experiment Station and Cooperative Extension Service. MF-2564. https://www.coffey.k-state.edu/cropslivestock/crops/conservation/Carbon%20Sequestration%20Top%20Ten.pdf
Rivenshield A, Bassuk NL (2007) Using organic amendments to decrease bulk density and increase macroporosity in compacted soils. Arboricult Urban For 33(2):140
Robertson M (2014) Sustainability principles and practice. Routledge, New York
Rosenstock TS, Lamanna C, Chesterman S, Bell P, Arslan A, Richards M, Rioux J, Akinleye AO, Champalle C, Cheng Z, Corner-Dolloff C, Dohn J, English W, Eyrich AS, Girvetz EH, Kerr A, Lizarazo M, Madalinska A, McFatridge S, Morris KS, Namoi N, Poultouchidou N, Ravina da Silva M, Rayess S, Ström H, Tully KL, Zhou W (2016) The scientific basis of climate-smart agriculture: a systematic review protocol. CCAFS working paper no 138, Copenhagen, Denmark: CGIAR Research Program on Climate Change, Agriculture and Food Security (CCAFS)
Santos VB, Araújo AS, Leite L, Nunes LA, Melo WJ (2012) Soil microbial biomass and organic matter fractions during transition from conventional to organic farming systems. Geoderma 170:227–231
Schjønning P, McBride RA, Keller T, Obour PB (2017) Predicting soil particle density from clay and soil organic matter contents. Geoderma 286:83–87
Schulz H, Glaser B (2012) Effects of biochar compared to organic and inorganic fertilizers on soil quality and plant growth in a greenhouse experiment. J Plant Nutr Soil Sci 175(3):410–422
Scotti R, Conte P, Berns AE, Alonzo G, Rao MA (2013) Effect of organic amendments on the evolution of soil organic matter in soils stressed by intensive agricultural practices. Curr Org Chem 17(24):2998–3005
Seiter S, Horwath WR (2004) Strategies for managing soil organic matter to supply plant nutrients. In: Magdoff F, Weil RR (eds) Soil organic matter in sustainable agriculture. CRC Press, Boca Raton, pp 269–293
Shackley S, Sohi S, Haszeldine S, Manning D, Masek O (2009) Biochar, reducing and removing CO_2 while improving soils: a significant and sustainable response to climate change. UK Biochar Research Centre, School of Geo Sciences, University of Edinburgh

Shen J, Yuan L, Zhang J, Li H, Bai Z, Chen X, Zhang W, Zhang F (2011) Phosphorus dynamics: from soil to plant. Plant Physiol 156(3):997–1005

Sihag SK, Singh MK, Meena RS, Naga S, Bahadur SR, Gaurav YRS (2015) Influences of spacing on growth and yield potential of dry direct seeded rice (*Oryza sativa* L.) cultivars. Ecoscan 9(1–2):517–519

Singh B, Singh BP, Cowie AL (2010) Characterization and evaluation of biochar for their application as a soil amendment. Soil Res 48(7):516–525

Sivapalan K, Fernando V, Thenabadu MW (1985) N-mineralization in polyphenol-rich plant residues and their effect on nitrification of applied ammonium sulphate. Soil Biol Biochem 17 (4):547–551

Smith SR, Jasim S (2009) Small-scale home composting of biodegradable household waste: overview of key results from a 3-year research programme in West London. Waste Manag Res 27(10):941–950

Soderstorm BO, Hedlund K, Jackson LE, Kattere T, Lugato E, Thomsen IK, Jorgensen HB (2014) What are the effects of agricultural management on soil organic carbon (SOC) stock? Environ Evid 3:1–8

Sohi SP, Krull E, Lopez-Capel E, Bol R (2010) A review of biochar and its use and function in soil. In: Sparks DL (ed) Advances in agronomy. Academic, Burlington, pp 47–82

Song W, Guo M (2012) Quality variations of poultry litter biochar generated at different pyrolysis temperatures. J Anal Appl Pyrolysis 94:138–145

Sparkes J, Stoutjesdijk P (2011) Biochar: implications for agricultural productivity. Aust Bur Agric Res Econ Sci 13:23–25

Sparling GP, McLay CDA, Tang C, Raphael C (1999) Effect of short-term legume residue decomposition on soil acidity. Soil Res 37(3):561–574

Swer H, Dkhar MS, Kayang H (2011) Fungal population and diversity in organically amended agricultural soils of Meghalaya, India. J Org Syst 6(2):3–12

Tejada M, Hernandez MT, Garcia C (2009) Soil restoration using composted plant residues: effects on soil properties. Soil Till Res 102(1):109–117

Tipayarom D, Oanh NK (2007) Effects from open rice straw burning emission on air quality in the Bangkok Metropolitan region. Sci Asia 33(3):339–345

Tisdall JM, Oades J (1982) Organic matter and water-stable aggregates in soils. Eur J Soil Sci 33 (2):141–163

Tomich TP, Brodt S, Ferris H, Galt R, Horwath WR, Kebreab E, Leveau JHJ, Liptzin D, Lubell M, Merel P, Michelmore R, Rosenstock T, Scow K, Six J, Williams N, Yang L (2011) Agroecology: a review from a global-change perspective. Annu Rev Environ Resour 36:193–222

USDE & NETL (United States Department of Energy and National Energy Technology Laboratory) (2007) Carbon sequestration technology and program plan. http://www.netl.doe.gov/technologies/carbon_seq/refshelf/project%20portfolio/2007/2007Roadmap.pdf

Van Antwerpen R, Meyer JH (1998) Soil degradation II: effect of trash and inorganic fertilizer application on soil strength. Proc S Afr Sug Technol Ass 72:152–158

Van-Camp L, Bujarrabal B, Gentile AR, Jones RJ, Montanarella L, Olazabal C, Selvaradjou SK (2004) Reports of the technical working groups established under the thematic strategy for soil protection. EUR 21319 EN/1, pp 872. Office for Official Publications of the European Communities, Luxembourg

Varma D, Meena RS, Kumar S, Kumar E (2017) Response of mungbean to NPK and lime under the conditions of Vindhyan region of Uttar Pradesh. Leg Res 40(3):542–545

Verma JP, Jaiswal DK, Meena VS, Meena RS (2015) Current need of organic farming for enhancing sustainable agriculture. J Clean Prod 102:545–547

Verma SK, Singh SB, Prasad SK, Meena RN, Meena RS (2015b) Influence of irrigation regimes and weed management practices on water use and nutrient uptake in wheat (*Triticum aestivum* L. Emend. Fiori and Paol.). Bangladesh J Bot 44(3):437–442

Warnock DD, Lehmann J, Kuyper TW, Rillig MC (2007) Mycorrhizal responses to biochar in soil–concepts and mechanisms. Plant Soil 300(1–2):9–20

Wolf B, Snyder G (2003) Sustainable soils: the place of organic matter in sustaining soils and their productivity. Food Products Press, New York

Wong MTF, Gibbs P, Nortcliff S, Swift RS (2000) Measurement of the acid neutralizing capacity of agroforestry tree prunings added to tropical soils. J Agric Sci 134(3):269–276

Wright AL, Hons FM, Lemon RG, McFarland ML, Nichols RL (2008) Microbial activity and soil C sequestration for reduced and conventional tillage cotton. Appl Soil Ecol 38(2):168–173

Wu W, Yang M, Feng Q, McGrouther K, Wang H, Lu H, Chen Y (2012) Chemical characterization of rice straw-derived biochar for soil amendment. Biomass Bioenergy 47:268–276

Xu X, Cao X, Zhao L (2013) Comparison of rice husk-and dairy manure-derived biochars for simultaneously removing heavy metals from aqueous solutions: role of mineral components in biochars. Chemosphere 92(8):955–961

Yadav GS, Babu S, Meena RS, Debnath C, Saha P, Debbaram C, Datta M (2017a) Effects of godawariphosgold and single supper phosphate on groundnut (*Arachis hypogaea*) productivity, phosphorus uptake, phosphorus use efficiency and economics. Indian J Agric Sci 87(9):1165–1169

Yadav GS, Datta R, Imran Pathan S, Lal R, Meena RS, Babu S, Das A, Bhowmik S, Datta M, Saha P (2017b) Effects of conservation tillage and nutrient management practices on soil fertility and productivity of rice (*Oryza sativa* L.)–Rice system in North Eastern region of India. Sustainability 9(10):1816

Yadav GS, Das A, Lal R, Babu S, Meena RS, Saha P, Singh R, Datta M (2018) Energy budget and carbon footprint in a no-till and mulch based rice–mustard cropping system. J Clean Prod 191:144–157

Yuan JH, Xu RK (2011) The amelioration effects of low temperature biochar generated from nine crop residues on an acidic Ultisol. Soil Use Manag 27(1):110–115

Yuan JH, Xu RK (2012) Effects of biochars generated from crop residues on chemical properties of acid soils from tropical and subtropical China. Soil Res 50(7):570–578

Yuan JH, Xu RK, Zhang H (2011) The forms of alkalis in the biochar produced from crop residues at different temperatures. Bioresour Technol 102(3):3488–3497

Zaccardelli M, De Nicola F, Villecco D, Scotti R (2013) The development and suppressive activity of soil microbial communities under compost amendment. J Soil Sci Plant Nutr 13(3):730–742

Zhang H, Ding W, Yu H, He X (2015) Linking organic carbon accumulation to microbial community dynamics in a sandy loam soil: result of 20 years compost and inorganic fertilizers repeated application experiment. Biol Fertil Soils 51(2):137–150

Zhao L, Cao X, Mašek O, Zimmerman A (2013) Heterogeneity of biochar properties as a function of feedstock sources and production temperatures. J Hazard Mater 256:1–9

Zimmerman AR (2010) Abiotic and microbial oxidation of laboratory-produced black carbon (biochar). Environ Sci Technol 44(4):1295–1301

Zuber SM, Behnke GD, Nafziger ED, Villamil MB (2018) Carbon and Nitrogen content of soil organic matter and microbial biomass under long-term crop rotation and tillage in Illinois, USA. Agriculture 8(3):37

Arbuscular Mycorrhizal Fungi and Nutrient Cycling in Cropping Systems

Sergio Saia, Elisabeth Tamayo, Calogero Schillaci, and Pasquale De Vita

Abstract Soil nutrient cycling in the soil-plant system of crops relies on the effects of the agronomical practices on soil conditions, especially soil microbial population mediating soil carbon transformation (either mineralization or stabilization), nitrogen cycle including soil nitrogen transformation, uptake and return from plants, and nitrogen losses, and the fate of other elements mediating these trade-offs, including phosphorous.

Arbuscular mycorrhizal fungi (AMF) are widespread, soil-borne microbes living in an obligate symbiotic status with most of the land plants. Such symbiosis can bring considerable benefits to plant growth, uptake of nutrients, and soil conditions. However, such benefits are not frequently seen in the field since the agronomical practices, especially tillage and use of fungicides, impair its development, which can partly be overcome by inoculation. Also, the increase of CO_2 concentration with climate change and nitrogen fertilization could negatively affect the AM benefit for the plant in the short term and increase the carbon cost for the plant when aggressive AMF are used. For such a reason, there is a need for long-term studies from the agronomical point of view on the role of field inoculation with AMF on both plant

S. Saia (✉)
Council for Agricultural Research and Economics (CREA), Research Centre for Cereal and Industrial Crops (CREA-CI), Vercelli, Italy
e-mail: sergio.saia@crea.gov.it

E. Tamayo
Departamento de Microbiología del Suelo y Sistemas Simbióticos, Estación Experimental del Zaidín-CSIC, Granada, Spain

Molecular Phytopathology, Botanical Institute, Karlsruhe Institute of Technology (KIT), Karlsruhe, Germany
e-mail: eli.tamayo83@gmail.com

C. Schillaci
Department of Agricultural and Environmental Science, University of Milan, Milan, Italy
e-mail: calogero.schillaci@unimi.it

P. De Vita
Council for Agricultural Research and Economics (CREA), Research Centre for Cereal and Industrial Crops (CREA-CI), Foggia, Italy
e-mail: pasquale.devita@crea.gov.it

R. Datta et al. (eds.), *Carbon and Nitrogen Cycling in Soil*,
https://doi.org/10.1007/978-981-13-7264-3_4

yield and soil conditions. Soil could be improved by the AMF inoculation in terms of redistribution of nutrients, augmented aggregation status, increase unstable soil organic carbon content, and reduction in carbon and nitrogen losses. In the present chapter, we will show the main agronomical benefit of inoculating AMF in cropped soils, along with results from field experiment and hints on the functioning of the plant-AMF symbiosis under field conditions. Perspectives of the use of and research on the arbuscular mycorrhizal technology under field conditions are also given.

Keywords Field crops · Horticultural crops · Beneficial microbe inoculation · Fertilization · Management

Abbreviations

Al	Aluminum
AM	Arbuscular mycorrhizal
AMF	Arbuscular mycorrhizal fungi
C	Carbon
Ca	Calcium
DB	Dry biomass
EMF	Ectomycorrhizal fungi
ERM	Extra-radical mycelium
Fe	Iron
IRM	Intra-radical mycelium
K	Potassium
N	Nitrogen
Na^+	Sodium
NFB	N_2-fixing bacteria
O	Oxygen
P	Phosphorus
PGPR	Plant growth-promoting rhizobacteria
Pi	Inorganic phosphorus
Po	Organic phosphorus
S	Sulfur
SOC	Soil organic carbon
SOM	Soil organic matter

1 Introduction

Agriculture is moving toward sustainable production strategies, which include the integration of the microbial technology in cropping systems. Strategies of sustainable intensification primarily foster to increase yield and quality per unit input and long-term sustainability and to reduce energy input in agroecosystems through an optimization of the turnover of the elements and improved biological control of stresses (Godfray and Garnett 2014). Achieving such goals implies the establishment of focus at the regional, water basin, ecosystem, and plot management level (Loos et al. 2014), which should meet at one time the social expectations of food quality, security, health, and environmental protection. Among these strategies, sustainable intensification should rely on the magnification of the positive relationships among species in the stand (including plants, animals, and microbes) and components of the ecosystems (including living organisms, soil, atmosphere, etc.) (Tittonell 2014).

The importance of soil biodiversity and the potential of the soil microbial population to sustain plant yield while maintaining or reducing the environmental footprint of cultivation are known for a long time (Karlen et al. 1997; Brussaard et al. 2007; Wurzburger et al. 2017), but frequently neglected. Such a measure of increasing biodiversity, with special emphasis on soil biodiversity, is strongly needed for arable lands (Rüdisser et al. 2015; Buragohain et al. 2017). In these systems, soil-living communities are frequently threatened by long fallow and soil tillage (Roger-Estrade et al. 2010; Spurgeon et al. 2013), the latter of which also consists in a disturbance of SOC accumulation that is pivotal to foster soil biodiversity and building fertility.

In sustainable agroecosystems, soil microflora and soil fauna are of key importance to control soil-borne diseases and determine species fitness, yields and quality due to their roles in the biogeochemical cycle of elements and ecological niche (Rillig et al. 2019). For such reasons, scientific research shows a growing interest to delve into biodiversity, dynamic and significance of microbial populations in soil, and the mechanisms that rule their biological activity (Bowen and Rovira 1999).

In the soil microbial survey, mycorrhizal fungi and N_2-fixing bacteria (NFB) play a major role as symbionts of cultivated crops because of their ability to solubilize and take up nutrients, to derive nitrogen from the atmosphere, to protect the plant from abiotic and biotic stresses, and to regulate ecosystem functioning (van der Heijden et al. 1998; Barea et al. 2005; Smith and Read 2008).

Mycorrhizal fungi [Greek: μύκης (mykes)= fungus; ρίζα(rhiza) = roots] are a variegated group of fungi from various fungal clades with the common feature of the ability to interact with plants in a mutualistic relationship. This symbiosis has several implications for both the plant and the fungus, the most common of which implies the facilitation of the uptake of nutrients from the fungus to the plant and the direct or indirect delivery of photosynthates from the plant to the fungus. The mycorrhizal symbiosis appeared on earth about 460 million years ago and helped plants during the early colonization of lands (Smith and Read 2008). Given its ancestral occurrence, this association can be found in the 80% of plant species, and it is involved in major nutrient fluxes in the soil-plant system.

The ability to interact or being colonized by mycorrhizal fungi varies with the plant species. Usually, some forest trees, legumes, and C4 species are likely to strictly engage a relationship with mycorrhizas. Some plant species seem to be completely dependent on mycorrhizas for their nutrient uptake, and others do not form mycorrhizal associations. The non-mycorrhizal-host traits of some species (mainly in the families *Juncaceae*, *Cruciferae*, *Chenopodiaceae*, *Cyperaceae*, and

Table 1 Table summarizing key differences between mycorrhizal association types

Type	AMF*	EMF*	Ectendo-	Arbutoid	Monotropoid	Ericoid	Orchid
Septate hyphae	- (+)*	±	±	+	+	+	+
Hyphae in cells	+	-	+	+	+	+	+
Hyphal coils	±	-	-	-	-	+	+
Arbuscules	+	-	-	-	-	-	-
Mantle	-	+ (-)	+ (-)	+	+	-	-
Hartig net	-	+	+	+	+	-	-
Vesicles	±	-	-	-	-	-	-
Classification	Mucoromycota, Glomeromycotina (Syn. Glomeromycota)	Most basid-, some ascomycetes				Asco- (basid-)	Basid-
Symbionts of these plant taxa	Vascular plants	Gymnosperm and Angiosperm		Ericales, Arbutoid eae	'Orchidaceae; Ericales, Monotropoid eae	Ericales	Orchid aceae
Host with Chlorophyll	+	+	+	±	-	+	±**

Modified after Smith and Read (2008)

[*]AMF arbuscular mycorrhizal fungi, *EMF* ectomycorrhizal fungi; – absent; + present; values in brackets are sometimes present or absent; *Basid* basidiomycetes, *Asco* ascomycetes

**Plants in the family Orchidaceae have no chlorophyll in the early stages and sometimes in the adult stages, too

Caryophyllaceae) might be a derived trait and the outcome of specialization regarding, e.g., the plant habitat (Wang and Qiu 2006; Datta et al. 2017). Mycorrhizal symbioses can be divided into seven groups basing on histological and genetic traits (Table 1).

The most frequent groups include the arbuscular mycorrhizal fungi (AMF) and ectomycorrhizal fungi (EMF). AMF and EMF have a wide range of hosts. Other kinds of mycorrhizal symbioses have a certain specificity between some fungi *taxa* and plant families and have minor importance for crops grown in the field.

EMF is distinguished by a dense mycelium sheath around the roots and an intercellular hyphal invasion of the plant root cortex forming the *Hartig net*. From both structures, an intricate hyphal net web spreads in the surrounding soil.

AMF are polynucleate haploid organisms with an obligate symbiotic status. They are characterized by an active penetration into the root cortex and the formation of arbuscules into the root cells. Arbuscules are tree stem-like shaped organs, which set an extended surface area for nutrient exchanges between the fungus and the cell cytoplasm.

The phylogeny of AMF is under debate. Presently, all AMF species are members of the phylum *Glomeromycota* or subphylum *Glomeromycotina* (Spatafora et al. 2016). AMF can be found in various environments ranging from deserts to rainforests, aquatic environments, and areas subjected to various abiotic stresses (high salinity, sodicity, pollution, etc.). From a morphological and histological point of view, AMF can form spores, resistant reproduction structures. However, there is a lack of information about the number of AMF species not forming spores (termed "cryptic species") and their physiological traits. In addition, AMF from both an intra-radical (IRM) and an extra-radical (ERM) mycelium spread into the roots and in the soil, respectively. Some AMF species are also able to form vesicles, lipid-rich structures into the roots. AMF spores can have various flushes of germination depending on their size and species. When a spore and/or a hypha comes nearby of living host roots, development of mycelium starts with a reciprocal signaling pathway between the host plant and the AMF, and this leads in a few weeks in the penetration into the root and further formation of the IRM. When the IRM reaches the inner cortex layer, it starts growing and branching throughout the apoplast, and some hyphal branches can penetrate the inner cortical cells and, by repeated dichotomous branching, initiate the synthesis of a characteristic treelike structure, the arbuscule. The amount of arbuscules per unit length or weight of roots depends on various conditions that include the AMF, the host plant species, and environmental conditions for both the plant and the AMF. Arbuscules are responsible for direct nutrient exchange between the symbionts.

AMF spreading into the soil have a high ability to scavenge soil nutrients, especially those with low mobility such as P, thanks to the reduced size of the ERM, secretion of AM exoenzymes and other compounds efficient in or favoring the degradation of some organic compounds and the solubilizing of nutrients. Nutrients taken up by the ERM are then actively and passively transported into the IRM and the arbuscules, where part of which can be exchanged with the plant in turn of organic carbon. The amount, percentage, and rate of the exchange and form of nutrients exchanged depend on several factors that include the availability of these nutrients into the soil, their forms, the existence of stresses for the plants, the AMF and plant genotypes, etc. Also, AMF can synthesize organic compounds which have a hormone-like activity and thus modify the plant physiology irrespective of any benefit for the uptake of nutrients. Lastly, AMF can interact with soil and other microbes (including pathogens) directly by secreting allelochemical compounds, by a competition for space and indirectly by modifying nutrient availability in the soil, including the form of organic C. The activity of the AMF in the soil can also lead to soil modification that is irrespective of the benefit for the host plants. Such modification strongly regards the AM effects on soil structure and aggregate stability (Rillig and Mummey 2006) and soil-water relationship. These effects depend on the AMF direct role on aggregate size and secretion of a highly stable glycoprotein, glomalin, with a range of properties related to soil water holding capacity and sequestration of toxic elements (Rillig et al. 2002; González-Chávez et al. 2004).

2 Arbuscular Mycorrhizal Fungi in Cropped Field and Their Interaction with Bacteria

The relationship between plants and AMF has been mostly studied in controlled conditions involving the use of sterilized soil or growing media in which the AM fungus, frequently as a single species or strain, is inoculated and compared to an uninoculated control. Such setups are very far from field conditions since soil sterilization implies changes in nutrient availability, microbial activity, and soil organic structures (Thompson 1990; Trevors 1996; Berns et al. 2008). Such changes were sometimes balanced by adding a soil filtrate (discarding the AM structures, especially spores), which can reintroduce only part of the native microbial consortium. In addition, the composition of the microbial population in the filtrate can differ from those of the native soil, and lastly, microbes after filtrate addition do not easily and rapidly recolonize the same soil structures as in the native soils. Nonetheless, studies under controlled conditions have been (and still are) crucial to attribute the results to the presence of the AM symbiosis. Berns et al. (2008) also showed that soil sterilization could strongly and negatively affect soil aggregation, which trait can be strongly improved by AMF. Also, benefits of the inoculation of sterilized growing media with AMF can also indirectly depend on a plant growth depression due to the sterilization (Smith and Smitht 1981; Koide and Li 1989; Lekberg and Koide 2014), though such an effect can vary with many soils, AMF, and plant traits (Miransari et al. 2009). Lastly, the effects of AMF on plants grown in pot under controlled conditions can depend on a range of factors that depends on pot size, plant density, actual availability of nutrients per plant, and effect of pot shape and size on plant morphology (Bååth and Hayman 1984; Poorter et al. 2012; Yadav et al. 2017a), and on homogeneity of the growing media.

Under field conditions, plants and soil microbes live in patchy conditions (per aggregate size and distribution, presence and kind of the SOM fractions, etc.) and interact with several soil conditions including uneven distribution of water and nutrients and presence of a wide community of microbes, animals, and plants. In such conditions, AM symbioses are not constantly beneficial to the crop, and it is accepted that an AM benefit for the plants occurs when the cost of the AM symbiosis is lower than its potential benefit (Johnson et al. 1997). This directly implies that the AM symbiosis can benefit the plant when the activity of AMF consists in the alleviation of any stress for the plant, irrespective of the nutrient concentration in the soil. Under controlled conditions, it has been seen that AMF respond differently to the availability and ratio among nutrients (Nouri et al. 2014; Johnson et al. 2015). Nonetheless, differences in the availability and ratios among nutrients (including C for both fungi and bacteria) also occur under field conditions due to non-homogeneity of soil conditions, and AMF can have contrasting effects on plant growth, nutrient uptake, and relationship with other microbes at varying soil patchiness (Hodge et al. 2000; Saia et al. 2014b).

In cultivated soils, AMF amount and activity can be strongly depressed by the soil and crop management. Such depression depends on various management traits, the most important of which are very likely:

- the effects of soil tillage that, on the one hand, disrupts the ERM and thus the inoculum potential (Kabir 2005) and on the other hand reduces the total SOC content, which is crucial to increase the mycorrhizal benefit (Hodge 2014);
- the effects of fertilization, especially with P, that usually decrease the plant dependence on, and thus the plant feeding to, the AMF (Treseder 2004);
- and indeed, the use of fungicides and other active ingredients (a.i.) negatively affecting some aspects of the AM physiology (such as the synthesis of chitin, which is negatively affected by some insecticides and nematocides).

The abovementioned traits are responsible for the usually higher AM benefit for cultivated plants under organic cropping systems (Gosling et al. 2006), in which there is frequently less availability of nutrients in nonorganic systems, more SOM and organic residues, no use of synthetic a.i., and transient effects of the natural a.i. (Lekberg and Koide 2005).

Despite these potential benefits that were seen in many conditions, application of AMF or microbial inoculate containing AMF has led to contrasting results (Mcgonigle 1988; Lekberg and Koide 2005; Hoeksema et al. 2010; Pellegrino et al. 2015) on the AM benefit for plants, despite they were very rarely seen acting as parasitic. Such lack of constant benefit for plant growth and commercial yield is also stimulating a debate on the agronomical reward and the potential environmental risk of the introduction of exotic microbial strains in the ecosystem (Hart et al. 2018; Rillig et al. 2019), although such strains are presently considered potentially beneficial for the crops.

Inoculation with microbial consortia containing AMF to the soil in cropping systems is, however, a way to increase AMF colonization of the plant root (Lekberg and Koide 2005; Pellegrino et al. 2011, 2012, 2015, Saia et al. 2014a, 2015a, b), which is strongly related to the AM benefit for plants. Such an increase leads to direct and indirect effects on plant growth and soil behavior.

Studies on various plant species grown under field conditions yielded valuable results of the benefit of the soil inoculation with AMF (Table 2). In wheat (both bread and durum), Pellegrino et al. (2015) collected the results of 333 comparisons of inoculation compared to a non-inoculated control. They showed that hay and grain yield increased by 20% and 5% on average and that such increases were relatively constant among conditions. They also found that, on average, grain P and N uptake increased by 20% and 31% on average and that the inoculation effect on hay biomass and nutrient content was similar, despite at a lesser degree. Similar results were found in other cereals (Hamel and Smith 1991; Saia et al. 2012; Berta et al. 2014). In legumes, the effect of soil inoculation in the field on grain yield is more variable than in cereals (Kaschuk et al. 2009; Schütz et al. 2018), and positive results were seen for the biomass of forage legumes (Azcón-Aguilar et al. 1986; Barea et al. 1987; Pellegrino et al. 2011; Saia et al. 2012, 2014a, b; Shamina et al. 2018). Yield was also seen to increase after field inoculation also in a range of horticultural crops (Becagli et al. 2013; Lazzara et al. 2017; Schütz et al. 2018; Verma et al. 2015), for which pot cultivation is also applied for commercial purposes.

Table 2 Examples of studies conducted in the field about the effects of the soil inoculation with arbuscular mycorrhizal fungi (AMF) on plant yield

Group	Species: Common	Species: Linnaean nomenclature	Without inoculating	With AMF inoculation	Notes	References[*]
Cereals	Oat	*Avena sativa*	6.76	7.93	t/ha dry biomass (DB)	Saia et al. (2012) Nitrogen Workshop 90–91
	Barley	*Hordeum vulgare*	8.45	9.45	t/ha DB	
	Maize, corn	*Zea mays*	283	368	g/plant dry grain	Berta et al. (2014) Mycorrhiza 24: 161–70.
			6.40	7.60	t/ha DB	Hamel and Smith (1992) Soil Biol. Biochem24, 5: 499–501
			2.21	5.14	t/ha DB	Hamel et al. (1991) Plant and Soil 133: 177–185
			9.43	10.37	t/ha grain, mean of 3 fertilization treatments, similar results for steak and leaf biomass. Scarce effects of the AM inoculation in the nonfertilized plots	V. Cozzolino et al. (2013) J. Geoch. Expl. 129:40–44
	Bread wheat	*Triticum aestivum*	1.82	2.37	t/ha DB, average of 2 cultivars in a dry environment with irrigation	Al-Karaki et al. (2004) Mycorrhiza 14: 263–269
			0.77	0.99	t/ha DB, average of 2 cultivars in dry environment without irrigation	
			3.64	4.02	g/plant dry grain	Mohammad et al. (1998) Mycorrhiza 8: 139–144
			2.56	3.51	g/plant dry grain. Mean of various AMF inocula	Zhu et al. (2017) F Crops Res 204: 229–241
Horticultural crops	Artichoke	*Cynara scolymus*	0.83	1.07	kg/plant dry product (first-, second-, and third-order artichokes)	Ceccarelli (2010) Plant Soil 335: 311–323
	Garden asparagus	*Asparagus officinalis*	5207	5490	kg/ha in 3 years, mean of 4 areas	Stützer. In: Feldmann F, Kapulnik Y, Baar J (2008): Mycorrhiza Works, ISBN 978-3-941261-01-3; 197–208

	Watermelon	*Citrullus lanatus*	51.00	60.00	t/ha fresh product, mean of 2 experiments under various AMF species	Ortaş. In: Feldmann F, Kapulnik Y, Baar J (2008): Mycorrhiza Works, ISBN 978-3-941,261-01-3; 197–208[**]
	Cucumber	*Cucumis sativus*	21.00	22.50	t/ha fresh product, mean of 2 experiments under various AMF species	
	Eggplant	*Solanum melongena*	84.00	81.00	t/ha fresh product, mean of 2 experiments under various AMF species	
			4.32	4.43	t/ha fresh product, mean of 3 experiments under various conditions	Douds et al. (2017) Sci Hort 224: 48–52
	Hot pepper/pepper	*Capsicum annuum*	28.0	27.5	t/ha fresh product, mean of 2 experiments under various AMF species	Ortaş. In: Feldmann F, Kapulnik Y, Baar J (2008): Mycorrhiza Works, ISBN 978-3-941,261-01-3; 197–208[**]
			27.0	71.1	t/ha fresh product, mean of 3 genotypes under low P fertilization	Pivonia et al. In: Feldmann F, Kapulnik Y, Baar J (2008): Mycorrhiza Works, ISBN 978-3-941,261-01-3; 197–208[**]
			72.2	74.3	t/ha fresh product, mean of 3 genotypes under high P fertilization	
			43.7	71.7	t/ha fresh product, mean of 3 irrigation levels under low P fertilization	
			60.3	77.5	t/ha fresh product, mean of 3 irrigation levels under high P fertilization	
	Dill	*Anethum graveolens*	1.14	1.38	t/ha dry grain	Weisany et al. (2016) Europ J Agr 75: 60–71[**]
	Peppermint	*Mentha × piperita*	20.1	27.3	kg/ha DB, mean of 3 cultivars	Gupta et al. (2002) BioresTech 81: 77–79
	Tomato	*Solanum lycopersicum*	1.28	1.55	t/ha fresh product	Bona et al. (2017) Mycorrhiza 27: 1–11 13[**]
			43.9	57.1	t/ha fresh product	Ziane et al. (2017) Int J Agric Biol 19: 341–347
			55.0	61.5	t/ha fresh product, mean of 2 experiments under various AMF species	Ortaş. In: Feldmann F, Kapulnik Y, Baar J (2008): Mycorrhiza Works, ISBN 978-3-941,261-01-3; 197–208[**]

(continued)

Table 2 (continued)

Group	Species		Without inoculating	With AMF inoculation	Notes	References*
	Common	Linnaean nomenclature				
	Potato	*Solanum tuberosum*	38.3	42.2	t/ha potato yield, mean of 231 trials under authentic field conditions	Hijri (2016) Mycorrhiza 26: 209–214.
Legumes	Common bean, green bean	*Phaseolus vulgaris*	1.15	1.63	t/ha dry grain	Weisany et al. (2016) Europ J Agr 75: 60–71**
	Fenugreek	*Trigonella foenum-graecum*	6.88	7.24	t/ha DB	Saia et al. (2012) Nitrogen Workshop 90–91
	Soybean	*Glycine max*	1.52	1.61	t/ha dry product, mean of 2 fertilization levels. Hay biomass was 3.61 t/ha without and 3.78 t/ha with AMF inoculation	Ganry et al. (1982) Plant and Soil 68: 321–329
			1.41	1.37	t/ha dry product, no statistical difference between treatments	Hamel & Smith (1992) Soil Biol. Biochem 24, 5: 499–501
			0.94	2.51	t/ha DB	Hamel et al. (1991) Plant and Soil 133: 177–185
	Berseem	*Trifolium alexandrinum*	6.52	10.21	t/ha DB, sum of 2 cuts. The inoculated crop is a mean of various AMF species. AMF gave a strong improvement after the first cut regrowth	Pellegrino et al. (2011) Soil Biol Bioch 43: 367–376
			10.07	11.13	t/ha DB, sum of 3 cuts	Saia et al. (2014) PlosOne 9: e90738
	French honeysuckle	*Sulla coronaria* (syn. *Hedysarum coronarium*)	16.92	20.04	t/ha DB, sum of 4 cuts, AMF inoculated crop yielded more than the control in each cut	Barea et al. (1987) New Phytol 106: 717–725

*These references are not included in the reference list

**Data taken from pictures and computation made by the authors of this chapter

Apart from the benefits for the aboveground plant growth, and any eventual benefit for plant yield, AMF inoculation can also have positive effects on agroecosystems. These effects deal with nutrient (including C) transformation or transient stabilization in the soil, amount of residual plant (from aboveground or root) biomass returned to the soil, and relationships with soil microbes, especially bacteria. These are important components to build soil fertility. Also, it has been recently argued a direct involvement of the AM symbiosis on the greenhouse gasses released from soil. Such effect is still understudied and likely to depend on the AM uptake of inorganic nitrogen or indirectly by affecting soil water status (Lazcano et al. 2014; Storer et al. 2017) and soil pH or controlling bacterial growth (Bender et al. 2015; Bowles et al. 2016).

2.1 Interactions Between AMF and Bacteria

Bacteria are pivotal components of the C and N turnover in soil. AMF can interact with soil bacteria both directly, via a trophic and physiological dialogue, and indirectly, by modifying the environment in which bacteria live (Barea et al. 2002; Johansson et al. 2004; Toljander et al. 2007; Bonfante and Anca 2009). Also, some AMF have AM-associated bacteria living in its ERM (Bonfante and Desirò 2017; Meena et al. 2016).

Many bacteria live in the (myco)-rhizosphere and attach to the AM hyphae, which thus help them spread into the soil and into the raw organic patches. The net of AM extra-radical hyphae represents an ecological niche for bacteria. It was shown that some bacterial strains specifically respond to some AMF (Andrade et al. 1997; Artursson et al. 2005). In turn, it has been shown that plant growth-promoting rhizobacteria (PGPR) can have a stimulatory effect on growth of AMF (Azcón 1987). In addition, *Medicago truncatula* showed a common gene expression in response to a fluorescent *Pseudomonas* and an AM fungus (Sanchez et al. 2004), suggesting a potential coevolutionary trait of plant-AMF and plant-bacteria relationships.

Role of the AMF on community composition of soil bacteria is however multiple and strongly depends on indirect effects on the environment. Such effects include the modification of the aggregation status, mycorrhizosphere pH, competition for nutrients, alteration of root exudate amount and composition, and turnover of the mycorrhizal structures in the soil, which can be used as *pabulum* by the bacteria.

AMF can also directly and indirectly interact with NFB involved in the symbiotic nitrogen fixation with legumes. The symbiosis between legumes and AMF displays several similarities to the NFB-legume symbiosis by a genetic and ecophysiological point of view. Interaction among legumes, AMF and NFB is usually beneficial due to a range of factors:

- the ability of the ERM to scavenge soil for P and microelements essential for the functioning of the N_2-fixing machinery;

- the ability of the NFB symbiosis of deriving large amounts of N that is strongly needed by the AMF;
- the plant provision of photosynthates to sustain the microbial symbionts (Lum and Hirsch 2003; Chalk et al. 2006; Chang et al. 2017).

3 Transformation of Nutrients by the AMF in the Soil-Plant System: Direct and Indirect Effect

AMF and AM symbiosis can affect the nutrient amount, kind, and transformation in the soil through various direct and indirect mechanisms. At the plot level, these mechanisms include:

- the modification of the total, and especially the root, biomass of the plants;
- the modification of the above- to below-ground biomass ratio of the plants;
- the modification of the nutrient status of the plant residues;
- the direct inflow of C from photosynthesis and N from legume NFB into the soil as AMF structures;
- the direct effects of the AMF on the transformation of the SOM fractions, organic and inorganic N forms in the soil, and soil P forms;
- the indirect effects on soil microbial activity mediating C, N, and P cycles.

At the ecosystem level, AMF can also influence plant assemblage (Horn et al. 2017) and thus the traits of the plant residues that have direct implications for C and N turnover (Badagliacca et al. 2017). Such influence has a certain importance in cropped soils, where AMF can shift the competition between weeds and crop, and this particularly occurs when weeds are non-mycorrhizal species (Vatovec et al. 2005; Daisog et al. 2011), although crop residues are usually more than weed biomass. In turn, weed control can positively or negatively affect the mycorrhizal status of the plant and AMF persistence in the soil (Feldmann and Boyle 1999; Baumgartner et al. 2005). This implies that proper weed management by retaining weeds below their damage threshold and increasing their diversity can have indirect effects on soil nutrient turnover due to both the nutrient uptake from weeds and maintenance of the soil AMF community.

In general, the AM symbiosis enhances both plant above ground and root biomass (Saia et al. 2014a); thanks to the improved mineral nutrition, this implies that one main effect of the AMF on SOC content is mediated by the enhanced C fixation by the plant and likely higher amount of plant residues. Such AM benefit, especially for root biomass, is crucial for C stabilization in the soil since most of the stable SOC likely derive from root C (Rasse et al. 2005) and it is stored in aggregates (Six and Paustian 2014). However, it has been seen that AM symbiosis can increase the host shoot to root ratio (Veresoglou et al. 2011) irrespective of the effect on total plant biomass and that the effect of AMF inoculation in the field on plant root biomass can be null (Cheeke et al. 2013). Growth, especially root, depression in AM plants due to the AM symbiosis is a rare event mostly seen in pot conditions or when the plant could autonomously satisfy its nutrient needs (Fitter 1977; Buwalda and Goh 1982; Koide 1985; Koide and Li 1989; Peng et al. 1993; Graham and Abbott 2000; Li et al. 2008; Grace et al. 2009). In addition, the modification induced by the AMF to root morphology and especially the size distribution of roots can indirectly affect the root C transformation (Berta et al. 2002; Langer et al. 2010; Fan et al. 2011). Irrespective of the effect of the AMF on plant biomass and especially root growth and morphology, an indirect effect of AMF on the fate of such residual biomass toward mineralization or stabilization deals with the AM effect on plant C and N (and likely also P and other nutrients) concentration. Indeed, SOC stabilization depends on C and N availabilities and C-N ratio and thus requires given amounts of N per unit C of the residues other than the soil and that both plant microbes (and likely other organisms) satisfy their N demand. The strong need of N by AMF (but see Sect. 3.1) could thus reduce the stabilization potential of organic matter in the soil. On such aspects, an open debate is standing (Wurzburger et al. 2017; van Groenigen et al. 2017).

The role of AMF on plant C and N concentrations can be variable. As seen above, AMF can increase plant N concentration when helping to overcome stress for the plant. Nonetheless, the high N demand by the AMF has direct and strong implications for both N and C cycles in the soil-plant system (Hodge and Fitter 2010). This may lead to both increases and decreases of the aboveground and root N concentration (Hodge et al. 2010; Barrett et al. 2011; Cavagnaro et al. 2011; Saia et al. 2012, 2015a, b; Karasawa et al. 2012; Guo et al. 2013; Rivero et al. 2015; Meena and Yadav 2014).

An additional indirect effect of the AMF on C and N turnover deals with the modification of the plant root exudation. Plant exudates (in terms of amount and composition) vary with environmental conditions and can lead to a massive inflow of C into the soil (part of which can be recaptured by the plant) and modification of the bacterial activity in the rhizosphere. Root AM colonization alters the metabolism of both the leaf and the root to the formation of nonprotein amino acids and other compounds (Laparre et al. 2014; Saia et al. 2015b; Rivero et al. 2015). This is likely to alter the pattern of root exudation (Jones et al. 2004; Del-Saz et al. 2017) and directly affects the plant-bacteria relationship, as previously mentioned.

Direct effect of AMF on soil C and N turnover regards the interaction between the ERM and the soil and soil biota. AMF can drain up to a quarter of the plant photosynthates, of which, a wide part is likely ERM (Sylvia 1992; Olsson et al. 1999). Quantification of the weight of the ERM biomass in a cropped soil is hard. It has been shown that AMF can have up to some tens of meters of ERM per cm^3, which may correspond to few hundreds of external hyphal dry weights per unit soil volume (as μg hypha cm^{-3} soil) (Miller et al. 1995; Leake et al. 2004). The amount of ERM found from the latter authors in the pasture and tallgrass soils would correspond to around 700–900 kg ha^{-1} in the first 20 cm of soil. However, this amount is likely dramatically lower in cultivated soils, due to disturbance from tillage and planting. Previous reports indicated amounts of hyphae lower than 16 mg^{-1} soil or 40 μg cm^{-3} that would correspond to less than 80 kg ha^{-1} in the first 20 cm of soil (Pacovsky and Bethlenfalvay 1982; Bethlenfalvay and Ames

1987; Schubert et al. 1987). Again, 80 kg ha^{-1} in the first 20 cm of soil could still be a strong overestimation of the amount of ERM in the soil due to difficulties in the estimation of the total ERM length, the ERM specific mass per unit length, the total ERM amount depending on soil conditions, and the ERM decay rate depending on the AM and plant species (Pacovsky and Bethlenfalvay 1982; Schnepf et al. 2008; Jeske et al. 2018). Nonetheless, the tiny average diameter of the ERM hyphae (3–7 μm compared to the diameter of the finest root hair around 5–20 μm) (Datta et al. 2011) and release of AM exudates (Sato et al. 2015) imply that AMF can change the distribution in the soil of the fixed C. In addition, ERM hyphae spread into the soil toward nutrient-rich and organic patch sites (Cavagnaro et al. 2011), which implies the transportation of the bacterial community in these sites and thus the favoring of their activity.

3.1 Effects on N Transformation and Uptake

AMF play a direct role in N dynamics in the soil. It has been showed that AMF are capable of taking up large amounts of N in both inorganic and organic forms (Hawkins et al. 2000; Miller and Cramer 2005; Fitter et al. 2011; Nuccio et al. 2013; Saia et al. 2014b). Part of this N is then transferred to the plant depending on various conditions, among them the form of N taken by the AMF (Kiers et al. 2011; Fellbaum et al. 2012a, b). This consisted in contrasting N benefit for the host plant depending on the total N availability and N forms in the soil (Saia et al. 2015a). It has been shown that AMF possess at least part of the machinery for a direct N uptake from organic sources (Cappellazzo et al. 2008; Tian et al. 2010), but there is still doubt on the amount of N taken as inorganic forms by the AMF and delivered (as either organic or not) to the plant (Hodge et al. 2000; Leigh et al. 2009; Atul-Nayyar et al. 2009; Barrett et al. 2011; Miransari 2011; Belmondo et al. 2014; Saia et al. 2014b). In addition, competition between bacteria and AMF for resources has implications for the N cycle in the soil (Leigh et al. 2011; Svenningsen et al. 2018; Dhakal et al. 2016). The high ability of AMF to take up N from the soil, including the organic forms, implies that AMF can indirectly alter soil N cycling by subtracting the *pabulum* to some specific functional groups of bacteria and especially ammonia oxidizers. Also, such uptake of ammonium by the AMF and the AM induction of N transporters in the plant (Kobae et al. 2010; Pérez-Tienda et al. 2014; Drechsler et al. 2018) could imply that plants would be forced to take up more N as both ammonium and nitrate and this could reduce the activity of the denitrifiers. In various pot studies, such effects of the AMF on soil bacterial functional groups were seen, including actinomycetes and phosphate-solubilizing bacteria (Secilia and Bagyaraj 1987; Toro et al. 1997), ammonia oxidizers, and ammonifier and nitrifier bacteria (Amora-Lazcano et al. 1998).

3.2 Effects on P Transformation and Uptake

In the AM symbiosis, the plant transfers C to the AMF, and there is a reciprocal movement of mineral nutrients (mainly P, but also N, S, K, and various microelements) from fungus to plant (Casieri et al. 2013; Ferrol et al. 2016). P is essential for a correct growth and functioning of every organism, but in natural and especially cropping conditions, P supply or availability is often limited. This is partially due since the main form of soil inorganic P (Pi) is orthophosphate, which is rapidly sequestered by soil cations such as Fe, Al, and Ca, so that up to 80% of P in soil can be in unavailable forms, organic or insoluble, to plants. Mobility of those salts formed with soil cations is very low, so that the P depletion zone around roots is usually lower than 1 or 2 mm. For such reasons, plants in cropping systems usually respond strongly to the Pi fertilization. AMF are particularly efficient in taking up P from the soil and deliver it to the plant. Such benefit depends on various issues, including the low size and C cost per unit length of ERM hyphae, which can thus explore a larger volume of soil compared to roots and root hairs; the ability of AMF to take P in organic forms, which increases the actual P supply for the AM symbiosis; and the efficiency of the AMF exoenzymes and transporter system. These mechanisms and the more rapid decay of the ERM compared to the root systems can also affect soil pH in the microsites where the ERM is growing.

The sum of these aspects has direct implications for P transformation and mobilization in the soil and soil-plant system. In particular, two pathways contribute to Pi uptake from the soil in AM plants: the direct pathway by root epidermal cells and the mycorrhizal pathway, via AMF (Bucher 2007; Smith and Smith 2012; Kumar et al. 2017) (Fig. 1), the latter of which also operates for N and some micronutrients (Ferrol et al. 2016). Radiotracer experiments have made possible to assess the relative amount of Pi that enters a plant via the AMF and by the direct pathway. These studies have shown that depending on the plant and fungal species involved in the association, the mycorrhizal pathway can be responsible for 20–100% of the plant P uptake and this can occur irrespective of a plant biomass response to the inoculation (Smith et al. 2003, 2004; Facelli et al. 2010). Such uptake plays a role in the plant metabolism and potential yield. For example, there is evidence in support of a major impact of the mycorrhizal P uptake pathway on the productivity of maize under field and controlled conditions (Willmann et al. 2013).

The fungal transcripts of the Pi:H^+ transporters were shown to be predominantly expressed in ERM, and their expression was increased by low P availability (Maldonado-Mendoza et al. 2001). This implies that an indirect effect of the AMF in soil lacking P is the P redistribution after ERM decay, in addition to the P uptake enhancement for the plants. Moreover, the identification of three putative Pi:Na^+ transporters, despite still with questionable function, in the genome of the model AMF *Rhizophagus irregularis* (formerly *Glomus intraradices*) (Tisserant et al. 2013; Lin et al. 2014; Chen et al. 2018) could also imply a redistribution of the Na^+ along with Pi.

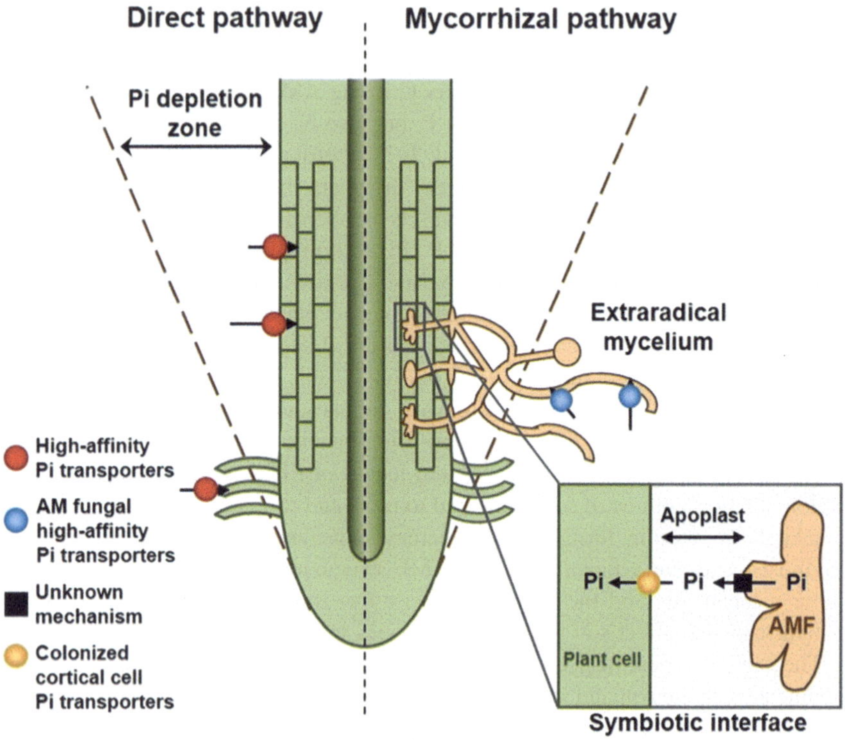

Fig. 1 Nutrient uptake pathways in arbuscular mycorrhizal plants (Modified from Smith et al. (2010))

Such as for N, after P uptake by the ERM, AMF first satisfy its own P demand, whereas the extra Pi is rapidly transported into the fungal vacuoles and condensed into polyphosphate (polyP), which is the largest P storage and mediator of long-distance P translocation in AMF (Hijikata et al. 2010). Once in the intra-radical mycelium, the polyP is hydrolyzed in Pi and transferred to the plant in the periarbuscular membrane. Several AM-induced P transporters have been identified in many plants from various families (Garcia et al. 2016), while activation of the mycorrhizal Pi uptake pathway produces downregulation of the plant Pht1 transporters located in epidermal root cells and, therefore, the direct pathway.

Besides the low level of available free Pi, soils can contain high amounts of organic P (Po), thus linked to C-containing compounds. The majority of Po is present as phosphate esters (C-O-P bonds) together with small quantities of phosphonates (C-P bonds) (Turner et al. 2005). To be used by plants and soil microorganisms, the P group of Po compounds has to be released by phosphatases, of which AMF are strong producers, along with organic acids, which catalyze the release of P from organic complexes.

In addition, it was suggested that AMF might activate part of the plant systems which respond to low P, giving rise to hydrolyzation of Po and salvaging P esters leaked from the root cell for P recycling (Li et al. 2012). Increase of P availability for the plants by the AMF has implications for both the expression of the plant yield potential and the relationship with the AMF. Garcia et al. (2016) suggested that the inhibition of the AMF by the high P content in the soil could be restored by a deficiency in other nutrients. Other authors showed that P is indeed the main soil nutrient controlling the AM symbiosis, but some microelements, and especially their ratio with P, can play an important role in the establishment and behavior of the AM symbiosis (Nouri et al. 2014; Johnson et al. 2015; Meena et al. 2017). In addition, it was seen that under high N fertilization, C cost for the plant per unit of P acquired by the AMF increased irrespective of P availability and induced a biomass decrease in the host plant (Williams et al. 2017; Treseder et al. 2018). This can explain the low AM benefit frequently found in the field when high mineral N fertilization is made. Nonetheless, benefit of the P uptake by the AMF can increase at the light of the increased CO_2 concentration in the atmosphere, which on the one hand would increase the amount of C that the plant can allocate to the AM counterpart and, on the other hand, the amount of P needed by the plants (Jakobsen et al. 2016).

4 Perspective of the Use of the Mycorrhizal Technology to Increase Plant Nutrient Uptake and Soil Nutrient Transformation in the Short and Long Term

A main aim of the present agriculture is increasing yield while preserving or enhancing the environment quality. AMF and other microbes are clearly beneficial for plants under field conditions (Schütz et al. 2018), but a few information is available about any inadvertent introduction of potentially invasive microbes in the agroecosystems (Hart et al. 2018). Nonetheless, it has been shown that exotic AMF are less persistent than native species (Piotrowski and Rillig 2008; Pellegrino et al. 2011, 2012; Yadav et al. 2018) and this reduces the chances of introducing invasive, potentially harmful microbial species. Also, proper management of cultivated plant could further reduce the chances of introducing invasive species, as shown when the relationship among plant community, soil properties, and the native AM community was studied in a mine spoil (Krüger et al. 2017). In addition, the structuring of the AM communities in cultivated soils should also consider the fertilization strategy (Yang et al. 2018) and genetic aspects of the plant species (De Vita et al. 2018).

Soil inoculation with symbiotic microbes is rarely made with single species and strains and usually includes microbes from various reigns including bacteria and fungi. Such inocula show a range of species, strains, concentration of the propagules, and sometimes the addition of organic compounds and inorganic elements. Such traits imply that the AM effect on the crop depends not solely on the addition of the

AMF or the whole inoculum per se but also on the agronomical practices affecting the relationship among these microbes, among those in the inoculum and soil, and lastly between the microbes in the inoculum and soil conditions. In particular, it was shown that AMF could strongly affect the bacterial composition during organic matter decay (Gui et al. 2017) and that organic matter is pivotal to foster the AM benefit for the plant (Thirkell et al. 2016) and enhances AM activity and growth in the soil (Yang et al. 2018).

A clear understanding of the use of beneficial inocula under field conditions is still needed (Berruti et al. 2016). Nonetheless, AM inocula showed promising effects on plant yield and quality and had been evaluated in several conditions that include both field and horticultural crops (Pellegrino et al. 2015; Schütz et al. 2018). However, the majority of studies about the effects of AM inoculation on plant performances or environmental conditions under field conditions comes from short-time studies, and there is a scarcity of information about the long-term effect of the one-time or repeated inoculation.

With regard to short-term studies, inoculation of AMF is likely to increase plant yield, especially when aiding the crop in overcoming stresses, either as nutrient lack or unavailability or other abiotic or biotic stresses. Also, studies on the relationship between SOC or carbon from plant residues and AMF were mostly conducted in pot, where plants can easily colonize the soil volume and roots can suffer from high-temperature stress or by preventing the access to the nutrient to the roots, which is not a condition standing in the authentic field environment. Indeed, this kind of studies yields invaluable information about the AMF role in plant nutrient uptake, but they are hardly transferable to the common cropping conditions, even for horticultural crops grown in pots.

Studies on the persistence of the introduced AMF are rare and mostly point to a short-term persistence (Harinikumar and Bagyaraj 1996; Hart et al. 2001; Farmer et al. 2007; Pellegrino et al. 2012; Verbruggen et al. 2013; Nicolás et al. 2015; Imperiali et al. 2017; Ashoka et al. 2017), which does not imply that persistent AMF also have persistent effects on plant yield and soil quality. In addition, the potentially positive but transient effects of soil inoculation must be managed according to the other agronomical management strategies applied and their effects on crop yield (Verzeaux et al. 2017; Rillig and Lehmann 2019). Among these strategies, tuning the actual amount of inoculum per unit area or weight of soil could be important to achieve significant yield responses (Imperiali et al. 2017), since most of the field experiments use a high amount of inoculum (Al-Karaki et al. 2004; Saia et al. 2015a) that is presently far from being used by farmers.

It is very likely that the positive effects of AMF inoculation in the field seen in the short time correspond to long-term benefits for the crops, the soil, and the whole environment (Wilson et al. 2009). Such environmental benefits could be enhanced by the augmented plant tolerance to stresses after AM inoculation, since high stress for the plants limits at one time the net primary production, SOM accumulation, and sequestration, which is the primary component of soil fertility and positively interacts with AMF to sustain plant yield. However, future climate change scenarios could turn the AM benefit for the environment over if C and N losses from the soil will exceed the accumulation of plant biomass, SOM, and stabilization of nutrients (Cheng et al. 2012; Lazcano et al. 2014; Jakobsen et al. 2016; Storer et al. 2017). Such balancing, especially if considering that reduction of N losses could strongly overcome the damage of the potential increase of C losses, calls for life-cycle assessments scaled in the long term for the use of the microbial inocula, including an estimation of their effects to the native soil microbes and conditions.

5 Conclusions

In conclusion, field experiments aimed at evaluating the role and effects of the inoculation of AMF are needed and should be conducted under common agronomical conditions. Furthermore, long-term field experiments of AMF inoculation (either once or repeated) in agricultural soils should also be conducted to achieve information on the long-term benefits of the AMF.

Apart from some agronomical management practices that can impair the AM activity (soil tillage, fertilization, and use of active ingredients) (Yadav et al. 2017b), while indeed increasing plant yield by removing other limitation to the expression of the yield potential, practices that should be considered and investigated to increase the AM benefit of the plants include:

- the concentration of the inoculum in the soil and the intimacy between the inoculum and plant roots, since a limit of the establishment of the inoculation in the short-cycle crops is that the complete establishment of the AM symbiosis can take a wide part of the early growth of the plants, when the root system is not well developed and relying on AMF;
- the relationship between the inoculum and soil conditions and especially the need of sustaining the early growth of the AMF (e.g., with the addition of organic residues) which in turn would increase the AM benefit for the plant;
- the availability of AMF strains resistant to the fungicides or with high growth rate after germination and, at the same time, an acceptable benefit for the plant;
- the relationship between plant genotypes and AMF and the breeding for high AM benefits (Singh et al. 2012; Ellouze et al. 2016; De Vita et al. 2018);
- the availability in the market of pure inocula of single taxa and strains (e.g., AMF, or bacteria, or other fungi) to be mixed according to the main limitation for the crop growth (e.g., uptake of P in the early phases, uptake of microelements during flowering, etc.). Such a perspective is very far to be reached and would require the knowledge of the effects and/or behavior of every inoculum or strain in the field at varying environmental conditions (Treseder et al. 2018), the availability in the market of each of such inocula, and the expertise of the farmer in their use.

References

Al-Karaki G, McMichael B, Zak J (2004) Field response of wheat to arbuscular mycorrhizal fungi and drought stress. Mycorrhiza 14:263–269. https://doi.org/10.1007/s00572-003-0265-2

Amora-Lazcano E, Vázquez MM, Azcón R (1998) Response of nitrogen-transforming microorganisms to arbuscular mycorrhizal fungi. Biol Fertil Soils 27:65–70. https://doi.org/10.1007/s003740050401

Andrade G, Mihara KL, Linderman RG, Bethlenfalvay GJ (1997) Bacteria from rhizosphere and hyphosphere soils of different arbuscular-mycorrhizal fungi. Plant Soil 192:71–79. https://doi.org/10.1023/A:1004249629643

Artursson V, Finlay RD, Jansson JK (2005) Combined bromodeoxyuridine immunocapture and terminal-restriction fragment length polymorphism analysis highlights differences in the active soil bacterial metagenome due to Glomus mosseae inoculation or plant species. Environ Microbiol 7:1952–1966. https://doi.org/10.1111/j.1462-2920.2005.00868.x

Ashoka P, Meena RS, Kumar S, Yadav GS, Layek J (2017) Green nanotechnology is a key for eco-friendly agriculture. J Clean Prod 142:4440–4441

Atul-Nayyar A, Hamel C, Hanson K, Germida J (2009) The arbuscular mycorrhizal symbiosis links N mineralization to plant demand. Mycorrhiza 19:239–246. https://doi.org/10.1007/s00572-008-0215-0

Azcón R (1987) Germination and hyphal growth of Glomus mosseae in vitro: effects of rhizosphere bacteria and cell-free culture media. Soil Biol Biochem 19:417–419. https://doi.org/10.1016/0038-0717(87)90032-0

Azcón-Aguilar C, Barea JM, Azcón R, Díaz-Rodriguez RM (1986) Assessment of field situations for the feasibility of vesicular-arbuscular mycorrhizal inoculation, using a forage legume as test plant. Agric Ecosyst Environ 15:241–252. https://doi.org/10.1016/0167-8809(86)90123-4

Bååth E, Hayman DS (1984) Effect of soil volume and plant density on mycorrhizal infection and growth response. Plant Soil 77:373–376. https://doi.org/10.1007/BF02182939

Badagliacca G, Ruisi P, Rees RM, Saia S (2017) An assessment of factors controlling N2O and CO2 emissions from crop residues using different measurement approaches. Biol Fertil Soils 53:547–561. https://doi.org/10.1007/s00374-017-1195-z

Barea JM, Azcon-Aguilar C, Azcon R (1987) Vesicular-arbuscular mycorrhiza improve both symbiotic N2 fixation and N uptake from soil as assessed with a 15N technique under field conditions. New Phytol 106:717–725. https://doi.org/10.1111/j.1469-8137.1987.tb00172.x

Barea JM, Azcón R, Azcón-Aguilar C (2002) Mycorrhizosphere interactions to improve plant fitness and soil quality. Antonie van Leeuwenhoek Int J Gen Mol Microbiol 81:343–351. https://doi.org/10.1023/A:1020588701325

Barea JM, Werner D, Azcón-Aguilar C, Azcón R (2005) Interactions of arbuscular mycorrhiza and nitrogen-fixing symbiosis in sustainable agriculture. In: Werner D, Newton WE (eds) Nitrogen fixation in agriculture, forestry, ecology, and the environment. Springer, Dordrecht, pp 199–222

Barrett G, Campbell CD, Fitter AH, Hodge A (2011) The arbuscular mycorrhizal fungus Glomus hoi can capture and transfer nitrogen from organic patches to its associated host plant at low temperature. Appl Soil Ecol 48:102–105. https://doi.org/10.1016/j.apsoil.2011.02.002

Baumgartner K, Smith RF, Bettiga L (2005) Weed control and cover crop management affect mycorrhizal colonization of grapevine roots and arbuscular mycorrhizal fungal spore populations in a California vineyard. Mycorrhiza 15:111–119. https://doi.org/10.1007/s00572-004-0309-2

Becagli S, Lazzara L, Fani F et al (2013) Relationship between methanesulfonate (MS-) in atmospheric particulate and remotely sensed phytoplankton activity in oligo-mesotrophic central Mediterranean Sea. Atmos Environ 79:681–688. https://doi.org/10.1016/j.atmosenv.2013.07.032

Belmondo S, Fiorilli V, Pérez-tienda J et al (2014) A dipeptide transporter from the arbuscular mycorrhizal fungus Rhizophagus irregularis is upregulated in the intraradical phase. Front Plant Sci 5:1–11. https://doi.org/10.3389/fpls.2014.00436

Bender SF, Conen F, Van der Heijden MGA (2015) Mycorrhizal effects on nutrient cycling, nutrient leaching and N2O production in experimental grassland. Soil Biol Biochem 80:283–292. https://doi.org/10.1016/j.soilbio.2014.10.016
Berns AE, Philipp H, Narres H-D et al (2008) Effect of gamma-sterilization and autoclaving on soil organic matter structure as studied by solid state NMR, UV and fluorescence spectroscopy. Eur J Soil Sci 59:540–550. https://doi.org/10.1111/j.1365-2389.2008.01016.x
Berruti A, Lumini E, Balestrini R, Bianciotto V (2016) Arbuscular mycorrhizal fungi as natural biofertilizers: let's benefit from past successes. Front Microbiol 6:1559. https://doi.org/10.3389/fmicb.2015.01559
Berta G, Fusconi A, Hooker JE (2002) Arbuscular mycorrhizal modifications to plant root systems: scale, mechanisms and consequences. In: Gianinazzi S, Schüepp H, Barea JM, Haselwandter K (eds) Mycorrhizal technology in agriculture. Birkhäuser Basel, Basel, pp 71–85
Berta G, Copetta A, Gamalero E et al (2014) Maize development and grain quality are differentially affected by mycorrhizal fungi and a growth-promoting pseudomonad in the field. Mycorrhiza 24:161–170. https://doi.org/10.1007/s00572-013-0523-x
Bethlenfalvay GJ, Ames RN (1987) Comparison of two methods for quantifying extraradical mycelium of vesicular-arbuscular mycorrhizal fungi1. Soil Sci Soc Am J 51:834. https://doi.org/10.2136/sssaj1987.03615995005100030049x
Bonfante P, Anca I-A (2009) Plants, mycorrhizal fungi, and bacteria: a network of interactions. Annu Rev Microbiol 63:363–383. https://doi.org/10.1146/annurev.micro.091208.073504
Bonfante P, Desirò A (2017) Who lives in a fungus? The diversity, origins and functions of fungal endobacteria living in Mucoromycota. ISME J 11:1727–1735. https://doi.org/10.1038/ismej.2017.21
Bowen GD, Rovira AD (1999) The rhizosphere and its management to improve plant growth. Adv Agron 66:1–102. https://doi.org/10.1016/S0065-2113(08)60425-3
Bowles TM, Barrios-Masias FH, Carlisle EA et al (2016) Effects of arbuscular mycorrhizae on tomato yield, nutrient uptake, water relations, and soil carbon dynamics under deficit irrigation in field conditions. Sci Total Environ 566–567:1223–1234. https://doi.org/10.1016/j.scitotenv.2016.05.178
Brussaard L, de Ruiter PC, Brown GG (2007) Soil biodiversity for agricultural sustainability. Agric Ecosyst Environ 121:233–244. https://doi.org/10.1016/j.agee.2006.12.013
Bucher M (2007) Functional biology of plant phosphate uptake at root and mycorrhiza interfaces. New Phytol 173:11–26. https://doi.org/10.1111/j.1469-8137.2006.01935.x
Buragohain S, Sharma B, Nath JD, Gogaoi N, Meena RS, Lal R (2017) Impact of ten years of bio-fertilizer use on soil quality and rice yield on an inceptisol in Assam, India. Soil Res. https://doi.org/10.1071/SR17001
Buwalda JG, Goh KM (1982) Host-fungus competition for carbon as a cause of growth depressions in vesicular-arbuscular mycorrhizal ryegrass. Soil Biol Biochem 14:103–106. https://doi.org/10.1016/0038-0717(82)90052-9
Cappellazzo G, Lanfranco L, Fitz M et al (2008) Characterization of an amino acid permease from the endomycorrhizal fungus Glomus mosseae. Plant Physiol 147:429–437. https://doi.org/10.1104/pp.108.117820
Casieri L, Ait Lahmidi N, Doidy J et al (2013) Biotrophic transportome in mutualistic plant–fungal interactions. Mycorrhiza 23:597–625. https://doi.org/10.1007/s00572-013-0496-9
Cavagnaro TR, Barrios-Masias FH, Jackson LE (2011) Arbuscular mycorrhizas and their role in plant growth, nitrogen interception and soil gas efflux in an organic production system. Plant Soil 353:181–194. https://doi.org/10.1007/s11104-011-1021-6
Chalk P, Souza R, Urquiaga S et al (2006) The role of arbuscular mycorrhiza in legume symbiotic performance. Soil Biol Biochem 38:2944–2951. https://doi.org/10.1016/j.soilbio.2006.05.005
Chang C, Nasir F, Ma L, Tian C (2017) Molecular communication and nutrient transfer of arbuscular mycorrhizal fungi, symbiotic nitrogen-fixing bacteria, and host plant in tripartite symbiosis. In: Legume nitrogen fixation in soils with low phosphorus availability. Springer International Publishing, Cham, pp 169–183

Cheeke TE, Cruzan MB, Rosenstiel TN (2013) Field evaluation of arbuscular mycorrhizal fungal colonization in *Bacillus thuringiensis* toxin-expressing (Bt) and non-Bt maize. Appl Environ Microbiol 79:4078–4086. https://doi.org/10.1128/AEM.00702-13

Chen ECH, Morin E, Beaudet D et al (2018) High intraspecific genome diversity in the model arbuscular mycorrhizal symbiont Rhizophagus irregularis. New Phytol. https://doi.org/10.1111/nph.14989

Cheng L, Booker FL, Tu C et al (2012) Arbuscular mycorrhizal fungi increase organic carbon decomposition under elevated CO2. Science 337:1084–1087. https://doi.org/10.1126/science.1224304

Daisog H, Sbrana C, Cristani C et al (2011) Arbuscular mycorrhizal fungi shift competitive relationships among crop and weed species. Plant Soil 353:395–408. https://doi.org/10.1007/s11104-011-1040-3

Datta S, Kim CM, Pernas M et al (2011) Root hairs: development, growth and evolution at the plant-soil interface. Plant Soil 346:1–14. https://doi.org/10.1007/s11104-011-0845-4

Datta R, Baraniya D, Wang Y-F, Kelkar A, Meena RS, Yadav GS, Teresa Ceccherini M, Formanek P (2017) Amino acid: its dual role as nutrient and scavenger of free radicals in soil. Sustainability 9(8):1402

De Vita P, Avio L, Sbrana C et al (2018) Genetic markers associated to arbuscular mycorrhizal colonization in durum wheat. Sci Rep. https://doi.org/10.1038/s41598-018-29020-6

Del-Saz NF, Romero-Munar A, Cawthray GR et al (2017) Arbuscular mycorrhizal fungus colonization in *Nicotiana tabacum* decreases the rate of both carboxylate exudation and root respiration and increases plant growth under phosphorus limitation. Plant Soil 416:97–106. https://doi.org/10.1007/s11104-017-3188-y

Dhakal Y, Meena RS, Kumar S (2016) Effect of INM on nodulation, yield, quality and available nutrient status in soil after harvest of green gram. Legum Res 39(4):590–594

Drechsler N, Courty PE, Brulé D, Kunze R (2018) Identification of arbuscular mycorrhiza-inducible nitrate transporter 1/peptide transporter family (NPF) genes in rice. Mycorrhiza 28:93–100. https://doi.org/10.1007/s00572-017-0802-z

Ellouze W, Hamel C, DePauw RM et al (2016) Potential to breed for mycorrhizal association in durum wheat. Can J Microbiol 62:263–271. https://doi.org/10.1139/cjm-2014-0598

Facelli E, Smith SE, Facelli JM et al (2010) Underground friends or enemies: model plants help to unravel direct and indirect effects of arbuscular mycorrhizal fungi on plant competition. New Phytol 185:1050–1061. https://doi.org/10.1111/j.1469-8137.2009.03162.x

Fan L, Dalpé Y, Fang C et al (2011) Influence of arbuscular mycorrhizae on biomass and root morphology of selected strawberry cultivars under salt stress. Botany 89:397–403. https://doi.org/10.1139/b11-028

Farmer MJ, Li X, Feng G et al (2007) Molecular monitoring of field-inoculated AMF to evaluate persistence in sweet potato crops in China. Appl Soil Ecol 35:599–609. https://doi.org/10.1016/j.apsoil.2006.09.012

Feldmann F, Boyle C (1999) Weed-mediated stability of arbuscular mycorrhizal effectiveness in maize monocultures. Angew Bot 73:1–5

Fellbaum CR, Gachomo EW, Beesetty Y et al (2012a) Carbon availability triggers fungal nitrogen uptake and transport in arbuscular mycorrhizal symbiosis. Proc Natl Acad Sci U S A 109:2666–2671. https://doi.org/10.1073/pnas.1118650109

Fellbaum CR, Mensah JA, Pfeffer PE et al (2012b) The role of carbon in fungal nutrient uptake and transport: implications for resource exchange in the arbuscular mycorrhizal symbiosis. Plant Signal Behav 7:1509–1512. https://doi.org/10.4161/psb.22015

Ferrol N, Tamayo E, Vargas P (2016) The heavy metal paradox in arbuscular mycorrhizas: from mechanisms to biotechnological applications. J Exp Bot 67:6253–6265. https://doi.org/10.1093/jxb/erw403

Fitter AH (1977) Influence of mycorrhizal infection on competition for phosphorus and potassium by two grasses. New Phytol 79:119–125. https://doi.org/10.1111/j.1469-8137.1977.tb02187.x

Fitter HA, Helgason T, Hodge A (2011) Nutritional exchanges in the arbuscular mycorrhizal symbiosis: implications for sustainable agriculture. Fungal Biol Rev 25:68–72. https://doi.org/10.1016/j.fbr.2011.01.002

Garcia K, Doidy J, Zimmermann SD et al (2016) Take a trip through the plant and fungal transportome of mycorrhiza. Trends Plant Sci 21:937–950. https://doi.org/10.1016/j.tplants.2016.07.010

Godfray HCJ, Garnett T (2014) Food security and sustainable intensification. Philos Trans R Soc B Biol Sci 369:20120273–20120273. https://doi.org/10.1098/rstb.2012.0273

González-Chávez MC, Carrillo-González R, Wright SF, Nichols KA (2004) The role of glomalin, a protein produced by arbuscular mycorrhizal fungi, in sequestering potentially toxic elements. Environ Pollut 130:317–323. https://doi.org/10.1016/j.envpol.2004.01.004

Gosling P, Hodge A, Goodlass G, Bending GDD (2006) Arbuscular mycorrhizal fungi and organic farming. Agric Ecosyst Environ 113:17–35. https://doi.org/10.1016/j.agee.2005.09.009

Grace EJ, Cotsaftis O, Tester M et al (2009) Arbuscular mycorrhizal inhibition of growth in barley cannot be attributed to extent of colonization, fungal phosphorus uptake or effects on expression of plant phosphate transporter genes. New Phytol 181:938–949. https://doi.org/10.1111/j.1469-8137.2008.02720.x

Graham JH, Abbott LK (2000) Wheat responses to aggressive and non-aggressive arbuscular mycorrhizal fungi. Plant Soil 220:207–218. https://doi.org/10.1023/a:1004709209009

Gui H, Purahong W, Hyde KD et al (2017) The arbuscular mycorrhizal fungus Funneliformis mosseae alters bacterial communities in subtropical forest soils during litter decomposition. Front Microbiol 8:1120. https://doi.org/10.3389/fmicb.2017.01120

Guo W, Zhao R, Zhao W et al (2013) Effects of arbuscular mycorrhizal fungi on maize (Zea mays L.) and sorghum (*Sorghum bicolor* L. Moench) grown in rare earth elements of mine tailings. Appl Soil Ecol 72:85–92. https://doi.org/10.1016/j.apsoil.2013.06.001

Hamel C, Smith DL (1991) Plant development in a mycorrhizal field-grown mixture. Soil Biol Biochem 23:661–665. https://doi.org/10.1016/0038-0717(91)90080-4

Harinikumar KM, Bagyaraj DJ (1996) Persistence of introduced Glomus intraradices in the field as influenced by repeated inoculation and cropping system. Biol Fertil Soils 21:184–188. https://doi.org/10.1007/BF00335932

Hart MM, Reader RJ, Klironomos JN (2001) Life-history strategies of arbuscular mycorrhizal fungi in relation to their successional dynamics. Mycologia 93:1186–1194. https://doi.org/10.2307/3761678

Hart MM, Antunes PM, Chaudhary VB, Abbott LK (2018) Fungal inoculants in the field: is the reward greater than the risk? Funct Ecol 32:126–135. https://doi.org/10.1111/1365-2435.12976

Hawkins H, Johansen A, George E (2000) Uptake and transport of organic and inorganic nitrogen by arbuscular mycorrhizal fungi. Plant Soil 226:275–285. https://doi.org/10.1023/a:1026500810385

Hijikata N, Murase M, Tani C et al (2010) Polyphosphate has a central role in the rapid and massive accumulation of phosphorus in extraradical mycelium of an arbuscular mycorrhizal fungus. New Phytol 186:285–289. https://doi.org/10.1111/j.1469-8137.2009.03168.x

Hodge A (2014) Interactions between arbuscular mycorrhizal fungi and organic material substrates. Adv Appl Microbiol 89:47–99. https://doi.org/10.1016/B978-0-12-800259-9.00002-0

Hodge A, Fitter AH (2010) Substantial nitrogen acquisition by arbuscular mycorrhizal fungi from organic material has implications for N cycling. Proc Natl Acad Sci U S A 107:13754–13759. https://doi.org/10.1073/pnas.1005874107

Hodge A, Robinson D, Fitter AH (2000) An arbuscular mycorrhizal inoculum enhances root proliferation in, but not nitrogen capture from, nutrient-rich patches in soil. New Phytol 145:575–584. https://doi.org/10.1046/j.1469-8137.2000.00602.x

Hodge A, Helgason T, Fitter AH (2010) Nutritional ecology of arbuscular mycorrhizal fungi. Fungal Ecol 3:267–273. https://doi.org/10.1016/j.funeco.2010.02.002

Hoeksema JD, Chaudhary VB, Gehring CA et al (2010) A meta-analysis of context-dependency in plant response to inoculation with mycorrhizal fungi. Ecol Lett 13:394–407. https://doi.org/10.1111/j.1461-0248.2009.01430.x

Horn S, Hempel S, Verbruggen E et al (2017) Linking the community structure of arbuscular mycorrhizal fungi and plants: a story of interdependence? ISME J 11:1400–1411. https://doi.org/10.1038/ismej.2017.5

Imperiali N, Chiriboga X, Schlaeppi K et al (2017) Combined field inoculations of pseudomonas bacteria, arbuscular mycorrhizal fungi, and entomopathogenic nematodes and their effects on wheat performance. Front Plant Sci 8:1809. https://doi.org/10.3389/fpls.2017.01809

Jakobsen I, Smith SE, Smith FA et al (2016) Plant growth responses to elevated atmospheric CO 2 are increased by phosphorus sufficiency but not by arbuscular mycorrhizas. J Exp Bot 67:6173–6186. https://doi.org/10.1093/jxb/erw383

Jeske ES, Tian H, Hanford K et al (2018) Long-term nitrogen fertilization reduces extraradical biomass of arbuscular mycorrhizae in a maize (Zea mays L.) cropping system. Agric Ecosyst Environ 255:111–118. https://doi.org/10.1016/j.agee.2017.11.014

Johansson JF, Paul LR, Finlay RD (2004) Microbial interactions in the mycorrhizosphere and their significance for sustainable agriculture. FEMS Microbiol Ecol 48:1–13. https://doi.org/10.1016/j.femsec.2003.11.012

Johnson NC, Graham JH, Smith FA (1997) Functioning of mycorrhizal associations along the mutualism-parasitism continuum. New Phytol 135:575–585. https://doi.org/10.1046/j.1469-8137.1997.00729.x

Johnson NC, Wilson GWT, Wilson JA et al (2015) Mycorrhizal phenotypes and the law of the minimum. New Phytol 205:1473–1484. https://doi.org/10.1111/nph.13172

Jones DL, Hodge A, Kuzyakov Y (2004) Plant and mycorrhizal regulation of rhizodeposition. New Phytol 163:459–480. https://doi.org/10.1111/j.1469-8137.2004.01130.x

Kabir Z (2005) Tillage or no-tillage: impact on mycorrhizae. Can J Plant Sci 85:23–29. https://doi.org/10.4141/P03-160

Karasawa T, Hodge A, Fitter AH (2012) Growth, respiration and nutrient acquisition by the arbuscular mycorrhizal fungus Glomus mosseae and its host plant *Plantago lanceolata* in cooled soil. Plant Cell Environ 35:819–828. https://doi.org/10.1111/j.1365-3040.2011.02455.x

Karlen DL, Mausbach MJ, Doran JW et al (1997) Soil quality: a concept, definition, and framework for evaluation (A guest editorial). Soil Sci Soc Am J 61:4. https://doi.org/10.2136/sssaj1997.03615995006100010001x

Kaschuk G, Kuyper TW, Leffelaar PA et al (2009) Are the rates of photosynthesis stimulated by the carbon sink strength of rhizobial and arbuscular mycorrhizal symbioses? Soil Biol Biochem 41:1233–1244. https://doi.org/10.1016/j.soilbio.2009.03.005

Kiers ET, Duhamel M, Beesetty Y et al (2011) Reciprocal rewards stabilize cooperation in the mycorrhizal symbiosis. Science 333:880–882. https://doi.org/10.1126/science.1208473

Kobae Y, Tamura Y, Takai S et al (2010) Localized expression of arbuscular mycorrhiza-inducible ammonium transporters in soybean. Plant Cell Physiol 51:1411–1415. https://doi.org/10.1093/pcp/pcq099

Koide R (1985) The nature of growth depressions in sunflower caused by vesicular-arbuscular mycorrhizal infection. New Phytol 99:449–462. https://doi.org/10.1111/j.1469-8137.1985.tb03672.x

Koide RT, Li M (1989) Appropriate controls for vesicular-arbuscular mycorrhiza research. New Phytol 111:35–44. https://doi.org/10.1111/j.1469-8137.1989.tb04215.x

Krüger C, Kohout P, Janoušková M et al (2017) Plant communities rather than soil properties structure arbuscular mycorrhizal fungal communities along primary succession on a mine spoil. Front Microbiol 8. https://doi.org/10.3389/fmicb.2017.00719

Kumar S, Meena RS, Yadav GS, Pandey A (2017) Response of sesame (*Sesamum indicum* L.) to sulphur and lime application under soil acidity. Int J Plant Soil Sci 14(4):1–9

Langer I, Syafruddin S, Steinkellner S et al (2010) Plant growth and root morphology of Phaseolus vulgaris L. grown in a split-root system is affected by heterogeneity of crude oil pollution and mycorrhizal colonization. Plant Soil 332:339–355. https://doi.org/10.1007/s11104-010-0300-y

Laparre J, Malbreil M, Letisse F et al (2014) Combining metabolomics and gene expression analysis reveals that propionyl- and butyryl-carnitines are involved in late stages of arbuscular mycorrhizal symbiosis. Mol Plant 7:554–566. https://doi.org/10.1093/mp/sst136

Lazcano C, Barrios-Masias FH, Jackson LE (2014) Arbuscular mycorrhizal effects on plant water relations and soil greenhouse gas emissions under changing moisture regimes. Soil Biol Biochem 74:184–192. https://doi.org/10.1016/j.soilbio.2014.03.010

Lazzara S, Militello M, Carrubba A et al (2017) Arbuscular mycorrhizal fungi altered the hypericin, pseudohypericin, and hyperforin content in flowers of *Hypericum perforatum* grown under contrasting P availability in a highly organic substrate. Mycorrhiza 27:345–354. https://doi.org/10.1007/s00572-016-0756-6

Leake J, Johnson D, Donnelly D, Muckle G, Boddy L, Read D (2004) Networks of power and influence: the role of mycorrhizal mycelium in controlling plant communities and agroecosystem functioning. Can J Bot 82:1016–1045

Leigh J, Hodge A, Fitter AH (2009) Arbuscular mycorrhizal fungi can transfer substantial amounts of nitrogen to their host plant from organic material. New Phytol 181:199–207. https://doi.org/10.1111/j.1469-8137.2008.02630.x

Leigh J, Fitter AH, Hodge A (2011) Growth and symbiotic effectiveness of an arbuscular mycorrhizal fungus in organic matter in competition with soil bacteria. Fems Micriobiol Ecol 76:428–438

Lekberg Y, Koide RT (2005) Is plant performance limited by abundance of arbuscular mycorrhizal fungi? A meta-analysis of studies published between 1988 and 2003. New Phytol 168:189–204. https://doi.org/10.1111/j.1469-8137.2005.01490.x

Lekberg Y, Koide RT (2014) Integrating physiological, community, and evolutionary perspectives on the arbuscular mycorrhizal symbiosis 1. Botany 92:241–251. https://doi.org/10.1139/cjb-2013-0182

Li H, Smith FA, Dickson S et al (2008) Plant growth depressions in arbuscular mycorrhizal symbioses: not just caused by carbon drain? New Phytol 178:852–862. https://doi.org/10.1111/j.1469-8137.2008.02410.x

Li C, Gui S, Yang T et al (2012) Identification of soybean purple acid phosphatase genes and their expression responses to phosphorus availability and symbiosis. Ann Bot 109:275–285. https://doi.org/10.1093/aob/mcr246

Lin K, Limpens E, Zhang Z et al (2014) Single nucleus genome sequencing reveals high similarity among nuclei of an endomycorrhizal fungus. PLoS Genet 10:e1004078. https://doi.org/10.1371/journal.pgen.1004078

Loos J, Abson DJ, Chappell MJ et al (2014) Putting meaning back into "sustainable intensification". Front Ecol Environ 12:356–361. https://doi.org/10.1890/130157

Lum MR, Hirsch AM (2003) Roots and their symbiotic microbes: strategies to obtain nitrogen and phosphorus in a nutrient-limiting environment. J Plant Growth Regul 21:368–382. https://doi.org/10.1007/s00344-003-0003-1

Maldonado-Mendoza IE, Dewbre GR, Harrison MJ (2001) A phosphate transporter gene from the extra-radical mycelium of an arbuscular mycorrhizal fungus Glomus intraradices is regulated in response to phosphate in the environment. Mol Plant-Microbe Interact 14:1140–1148. https://doi.org/10.1094/MPMI.2001.14.10.1140

Mcgonigle TP (1988) A numerical analysis of published field trials with vesicular-arbuscular mycorrhizal fungi. Funct Ecol 2:473–478. https://doi.org/10.2307/2389390

Meena RS, Yadav RS (2014) Phonological performance of groundnut varieties under sowing environments in hyper arid zone of Rajasthan, India. J Appl Nat Sci 6(2):344–348

Meena RS, Bohra JS, Singh SP, Meena VS, Verma JP, Verma SK, Shiiag SK (2016) Towards the prime response of manure to enhance nutrient use efficiency and soil sustainability a current need: a book review. J Clean Prod 112:1258–1260

Meena RS, Gogaoi N, Kumar S (2017) Alarming issues on agricultural crop production and environmental stresses. J Clean Prod 142:3357–3359

Miller AJ, Cramer MD (2005) Root nitrogen acquisition and assimilation. Plant Soil 274:1–36. https://doi.org/10.1007/s11104-004-0965-1

Miller RM, Jastrow JD, Reinhardt DR (1995) External hyphal production of vesicular-arbuscular mycorrhizal fungi in pasture and tallgrass prairie communities. Oecologia 103:17–23. https://doi.org/10.1007/BF00328420

Miransari M (2011) Arbuscular mycorrhizal fungi and nitrogen uptake. Arch Microbiol 193:77–81. https://doi.org/10.1007/s00203-010-0657-6

Miransari M, Bahrami HA, Rejali F, Malakouti MJ (2009) Effects of arbuscular mycorrhiza, soil sterilization, and soil compaction on wheat (Triticum aestivum L.) nutrients uptake. Soil Tillage Res 104:48–55. https://doi.org/10.1016/j.still.2008.11.006

Nicolás E, Maestre-Valero JF, Alarcón JJ et al (2015) Effectiveness and persistence of arbuscular mycorrhizal fungi on the physiology, nutrient uptake and yield of Crimson seedless grapevine. J Agric Sci 153:1084–1096. https://doi.org/10.1017/S002185961400080X

Nouri E, Breuillin-Sessoms F, Feller U, Reinhardt D (2014) Phosphorus and nitrogen regulate arbuscular mycorrhizal symbiosis in *Petunia hybrida*. PLoS One 9:e90841. https://doi.org/10.1371/journal.pone.0090841

Nuccio EE, Hodge A, Pett-Ridge J et al (2013) An arbuscular mycorrhizal fungus significantly modifies the soil bacterial community and nitrogen cycling during litter decomposition. Environ Microbiol 15:1870–1881. https://doi.org/10.1111/1462-2920.12081

Olsson PA, Thingstrup I, Jakobsen I, Bååth E (1999) Estimation of the biomass of arbuscular mycorrhizal fungi in a linseed field. Soil Biol Biochem 31:1879–1887. https://doi.org/10.1016/S0038-0717(99)00119-4

Pacovsky RS, Bethlenfalvay GJ (1982) Measurement of the extraradical mycelium of a vesicular-arbuscular mycorrhizal fungus in soil by chitin determination. Plant Soil 68:143–147. https://doi.org/10.1007/BF02374736

Pellegrino E, Bedini S, Avio L et al (2011) Field inoculation effectiveness of native and exotic arbuscular mycorrhizal fungi in a Mediterranean agricultural soil. Soil Biol Biochem 43:367–376. https://doi.org/10.1016/j.soilbio.2010.11.002

Pellegrino E, Turrini A, Gamper HA et al (2012) Establishment, persistence and effectiveness of arbuscular mycorrhizal fungal inoculants in the field revealed using molecular genetic tracing and measurement of yield components. New Phytol 194:810–822. https://doi.org/10.1111/j.1469-8137.2012.04090.x

Pellegrino E, Öpik M, Bonari E, Ercoli L (2015) Responses of wheat to arbuscular mycorrhizal fungi: a meta-analysis of field studies from 1975 to 2013. Soil Biol Biochem 84:210–217. https://doi.org/10.1016/j.soilbio.2015.02.020

Peng S, Eissenstat DM, Graham JH et al (1993) Growth depression in mycorrhizal citrus at high-phosphorus supply (analysis of carbon costs). Plant Physiol 101:1063–1071. https://doi.org/10.1104/pp.101.3.1063

Pérez-Tienda J, Corrêa A, Azcón-Aguilar C, Ferrol N (2014) Transcriptional regulation of host NH4+ transporters and GS/GOGAT pathway in arbuscular mycorrhizal rice roots. Plant Physiol Biochem 75:1–8. https://doi.org/10.1016/j.plaphy.2013.11.029

Piotrowski JS, Rillig MC (2008) Succession of arbuscular mycorrhizal fungi: patterns, causes, and considerations for organic agriculture. Advances 97:111–130. https://doi.org/10.1016/S0065-2113(07)00003-X

Poorter H, Bühler J, van Dusschoten D et al (2012) Pot size matters: a meta-analysis of the effects of rooting volume on plant growth. Funct Plant Biol 39:839. https://doi.org/10.1071/FP12049

Rasse DP, Rumpel C, Dignac M-F (2005) Is soil carbon mostly root carbon? Mechanisms for a specific stabilisation. Plant Soil 269:341–356. https://doi.org/10.1007/s11104-004-0907-y

Rillig MC, Lehmann A (2019) Exploring the agricultural parameter space for crop yield and sustainability. New Phytol. https://doi.org/10.1111/nph.15744

Rillig MC, Mummey DL (2006) Mycorrhizas and soil structure. New Phytol 171:41–53. https://doi.org/10.1111/j.1469-8137.2006.01750.x

Rillig MC, Wright SF, Eviner VT (2002) The role of arbuscular mycorrhizal fungi and glomalin in soil aggregation: comparing effects of five plant species. Plant Soil 238:325–333. https://doi.org/10.1023/A:1014483303813

Rillig MC, Aguilar-Trigueros CA, Camenzind T et al (2019) Why farmers should manage the arbuscular mycorrhizal symbiosis. New Phytol. https://doi.org/10.1111/nph.15602

Rivero J, Gamir J, Aroca R et al (2015) Metabolic transition in mycorrhizal tomato roots. Front Microbiol 6:598. https://doi.org/10.3389/fmicb.2015.00598

Roger-Estrade J, Anger C, Bertrand M, Richard G (2010) Tillage and soil ecology: partners for sustainable agriculture. Soil Tillage Res 111:33–40. https://doi.org/10.1016/j.still.2010.08.010

Rüdisser J, Tasser E, Peham T et al (2015) The dark side of biodiversity: spatial application of the biological soil quality indicator (BSQ). Ecol Indic 53:240–246. https://doi.org/10.1016/j.ecolind.2015.02.006

Saia S, Ruisi P, Amato G et al (2012) Effects of soil inoculation with arbuscular mycorrhizal fungi on plant growth and nutrient uptake of some Mediterranean species grown under rainfed field conditions. In: Richards K, Fenton O, Watson CJ (eds) 17th International Nitrogen Workshop, Wexford, pp 90–91

Saia S, Amato G, Frenda AS et al (2014a) Influence of arbuscular mycorrhizae on biomass production and nitrogen fixation of berseem clover plants subjected to water stress. PLoS One 9:e90738. https://doi.org/10.1371/journal.pone.0090738

Saia S, Benítez E, García-Garrido JM et al (2014b) The effect of arbuscular mycorrhizal fungi on total plant nitrogen uptake and nitrogen recovery from soil organic material. J Agric Sci 152:370–378. https://doi.org/10.1017/S002185961300004X

Saia S, Rappa V, Ruisi P et al (2015a) Soil inoculation with symbiotic microorganisms promotes plant growth and nutrient transporter genes expression in durum wheat. Front Plant Sci 6:815

Saia S, Ruisi P, Fileccia V et al (2015b) Metabolomics suggests that soil inoculation with arbuscular mycorrhizal fungi decreased free amino acid content in roots of durum wheat grown under N-limited, P-rich field conditions. PLoS One 10:e0129591. https://doi.org/10.1371/journal.pone.0129591

Sanchez L, Weidmann S, Brechenmacher L et al (2004) Common gene expression in *Medicago truncatula* roots in response to *Pseudomonas fluorescens* colonization, mycorrhiza development and nodulation. New Phytol 161:855–863. https://doi.org/10.1046/j.1469-8137.2004.00997.x

Sato T, Ezawa T, Cheng W, Tawaraya K (2015) Release of acid phosphatase from extraradical hyphae of arbuscular mycorrhizal fungus Rhizophagus clarus. Soil Sci Plant Nutr 61:269–274. https://doi.org/10.1080/00380768.2014.993298

Schnepf A, Roose T, Schweiger P (2008) Growth model for arbuscular mycorrhizal fungi. J R Soc Interface 5:773–784. https://doi.org/10.1098/rsif.2007.1250

Schubert A, Marzachí C, Mazzitelli M et al (1987) Development of total and viable extraradical mycelium in the vesicular–arbuscular mycorrhizal fungus Glomus clarum Nicol. & Schenck. New Phytol 107:183–190. https://doi.org/10.1111/j.1469-8137.1987.tb04892.x

Schütz L, Gattinger A, Meier M et al (2018) Improving crop yield and nutrient use efficiency via biofertilization—a global meta-analysis. Front Plant Sci 8. https://doi.org/10.3389/fpls.2017.02204

Secilia J, Bagyaraj DJ (1987) Bacteria and actinomycetes associated with pot cultures of vesicular–arbuscular mycorrhizas. Can J Microbiol 33:1069–1073. https://doi.org/10.1139/m87-187

Shamina Imran Pathan TV, Giagnoni L, Datta R, Baldrian P, Nannipieri P, Renella G (2018) Microbial expression profiles in the rhizosphere of two maize lines differing in N use efficiency. Plant Soil:401–413

Singh AK, Hamel C, DePauw RM, Knox RE (2012) Genetic variability in arbuscular mycorrhizal fungi compatibility supports the selection of durum wheat genotypes for enhancing soil ecological services and cropping systems in Canada. Can J Microbiol 58:293–302. https://doi.org/10.1139/w11-140

Six J, Paustian K (2014) Aggregate-associated soil organic matter as an ecosystem property and a measurement tool. Soil Biol Biochem 68:A4–A9

Smith SE, Read D (2008) Mycorrhizas in agriculture, horticulture and forestry. In: Mycorrhizal symbiosis. Elsevier, Amsterdam, pp 611–XVIII

Smith SE, Smith FA (2012) Fresh perspectives on the roles of arbuscular mycorrhizal fungi in plant nutrition and growth. Mycologia 104:1–13. https://doi.org/10.3852/11-229

Smith FA, Smitht SE (1981) Mycorrhizal infection and growth of *Trifolium subterraneum*: use of sterilized soil as a control treatment. New Phytol 88:299–309

Smith SE, Smith FA, Jakobsen I (2003) Mycorrhizal fungi can dominate phosphate supply to plants irrespective of growth responses. Plant Physiol 133:16–20

Smith SE, Smith FA, Jakobsen I (2004) Functional diversity in arbuscular mycorrhizal (AM) symbioses: the contribution of the mycorrhizal P uptake pathway is not correlated with mycorrhizal responses in growth or total P uptake. New Phytol 162:511–524. https://doi.org/10.1111/j.1469-8137.2004.01039.x

Smith SE, Facelli E, Pope S, Andrew Smith F (2010) Plant performance in stressful environments: interpreting new and established knowledge of the roles of arbuscular mycorrhizas. Plant Soil 326:3–20. https://doi.org/10.1007/s11104-009-9981-5

Spatafora JW, Chang Y, Benny GL et al (2016) A phylum-level phylogenetic classification of zygomycete fungi based on genome-scale data. Mycologia 108:1028–1046. https://doi.org/10.3852/16-042

Spurgeon DJ, Keith AM, Schmidt O et al (2013) Land-use and land-management change: relationships with earthworm and fungi communities and soil structural properties. BMC Ecol 13:46. https://doi.org/10.1186/1472-6785-13-46

Storer K, Coggan A, Ineson P, Hodge A (2017) Arbuscular mycorrhizal fungi reduce nitrous oxide emissions from N 2 O hotspots. New Phytol. https://doi.org/10.1111/nph.14931

Svenningsen NB, Watts-Williams SJ, Joner EJ et al (2018) Suppression of the activity of arbuscular mycorrhizal fungi by the soil microbiota. ISME J. https://doi.org/10.1038/s41396-018-0059-3

Sylvia DM (1992) Quantification of external hyphae of vesicular-arbuscular mycorrhizal fungi. Methods Microbiol 24:53–65. https://doi.org/10.1016/S0580-9517(08)70086-2

Thirkell TJ, Cameron DD, Hodge A (2016) Resolving the 'nitrogen paradox' of arbuscular mycorrhizas: fertilization with organic matter brings considerable benefits for plant nutrition and growth. Plant Cell Environ 39:1683–1690. https://doi.org/10.1111/pce.12667

Thompson JP (1990) Soil sterilization methods to show VA-mycorrhizae aid P and Zn nutrition of wheat in vertisols. Soil Biol Biochem 22:229–240. https://doi.org/10.1016/0038-0717(90)90092-E

Tian C, Kasiborski B, Koul R et al (2010) Regulation of the nitrogen transfer pathway in the arbuscular mycorrhizal symbiosis: gene characterization and the coordination of expression with nitrogen flux. Plant Physiol 153:1175–1187. https://doi.org/10.1104/pp.110.156430

Tisserant E, Malbreil M, Kuo A et al (2013) Genome of an arbuscular mycorrhizal fungus provides insight into the oldest plant symbiosis. Proc Natl Acad Sci 110:20117–20122. https://doi.org/10.1073/pnas.1313452110

Tittonell P (2014) Ecological intensification of agriculture—sustainable by nature. Curr Opin Environ Sustain 8:53–61. https://doi.org/10.1016/j.cosust.2014.08.006

Toljander JF, Lindahl BD, Paul LR et al (2007) Influence of arbuscular mycorrhizal mycelial exudates on soil bacterial growth and community structure. FEMS Microbiol Ecol 61:295–304. https://doi.org/10.1111/j.1574-6941.2007.00337.x

Toro M, Azcon R, Barea J (1997) Improvement of arbuscular mycorrhiza development by inoculation of soil with phosphate-solubilizing rhizobacteria to improve rock phosphate bioavailability ((sup32)P) and nutrient cycling. Appl Environ Microbiol 63:4408–4412

Treseder KK (2004) A meta-analysis of mycorrhizal responses to nitrogen, phosphorus, and atmospheric CO2 in field studies. New Phytol 164:347–355. https://doi.org/10.1111/j.1469-8137.2004.01159.x

Treseder KK, Allen EB, Egerton-Warburton LM et al (2018) Arbuscular mycorrhizal fungi as mediators of ecosystem responses to nitrogen deposition: a trait-based predictive framework. J Ecol 106:480–489. https://doi.org/10.1111/1365-2745.12919

Trevors JT (1996) Sterilization and inhibition of microbial activity in soil. J Microbiol Methods 26:53–59. https://doi.org/10.1016/0167-7012(96)00843-3

Turner BL, Frossard E, Baldwin DS (eds) (2005) Organic phosphorus in the environment. CABI, Wallingford

van der Heijden MGA, Klironomos JN, Ursic M et al (1998) Mycorrhizal fungal diversity determines plant biodiversity, ecosystem variability and productivity. Nature 396:69–72. https://doi.org/10.1038/23932

van Groenigen JW, van Kessel C, Hungate BA et al (2017) Sequestering soil organic carbon: a nitrogen dilemma. Environ Sci Technol 51:4738–4739. https://doi.org/10.1021/acs.est.7b01427

Vatovec C, Jordan N, Huerd S (2005) Responsiveness of certain agronomic weed species to arbuscular mycorrhizal fungi. Renew Agric Food Syst 20:181–189. https://doi.org/10.1079/RAF2005115

Verbruggen E, van der Heijden MG, Rillig MC, Kiers ET (2013) Mycorrhizal fungal establishment in agricultural soils: factors determining inoculation success. New Phytol 197:1104–1109. https://doi.org/10.1111/j.1469-8137.2012.04348.x

Veresoglou SD, Menexes G, Rillig MC (2011) Do arbuscular mycorrhizal fungi affect the allometric partition of host plant biomass to shoots and roots? A meta-analysis of studies from 1990 to 2010. Mycorrhiza 22:227–235

Verma JP, Jaiswal DK, Meena VS, Meena RS (2015) Current need of organic farming for enhancing sustainable agriculture. J Clean Prod 102:545–547

Verzeaux J, Hirel B, Dubois F et al (2017) Agricultural practices to improve nitrogen use efficiency through the use of arbuscular mycorrhizae: basic and agronomic aspects. Plant Sci 264:48–56. https://doi.org/10.1016/j.plantsci.2017.08.004

Wang B, Qiu Y-L (2006) Phylogenetic distribution and evolution of mycorrhizas in land plants. Mycorrhiza 16:299–363

Williams A, Manoharan L, Rosenstock NP et al (2017) Long-term agricultural fertilization alters arbuscular mycorrhizal fungal community composition and barley (Hordeum vulgare) mycorrhizal carbon and phosphorus exchange. New Phytol 213:874–885. https://doi.org/10.1111/nph.14196

Willmann M, Gerlach N, Buer B et al (2013) Mycorrhizal phosphate uptake pathway in maize: vital for growth and cob development on nutrient poor agricultural and greenhouse soils. Front Plant Sci 4:533

Wilson GWT, Rice CW, Rillig MC et al (2009) Soil aggregation and carbon sequestration are tightly correlated with the abundance of arbuscular mycorrhizal fungi: results from long-term field experiments. Ecol Lett 12:452–461. https://doi.org/10.1111/j.1461-0248.2009.01303.x

Wurzburger N, Brookshire ENJ, McCormack ML, Lankau RA (2017) Mycorrhizal fungi as drivers and modulators of terrestrial ecosystem processes. New Phytol 213:996–999. https://doi.org/10.1111/nph.14409

Yadav GS, Lal R, Meena RS, Babu S, Das A, Bhomik SN, Datta M, Layak J, Saha P (2017a) Conservation tillage and nutrient management effects on productivity and soil carbon sequestration under double cropping of rice in North Eastern Region of India. Ecol Indic. https://doi.org/10.1016/j.ecolind.2017.08.071

Yadav GS, Datta R, Imran Pathan S, Lal R, Meena RS, Babu S, Das A, Bhowmik S, Datta M, Saha P (2017b) Effects of conservation tillage and nutrient management practices on soil fertility and productivity of rice (*Oryza sativa* L.)–rice system in North Eastern Region of India. Sustainability 9 (10):1816

Yadav GS, Das A, Lal R, Babu S, Meena RS, Saha P, Singh R, Datta M (2018) Energy budget and carbon footprint in a no-till and mulch based rice–mustard cropping system. J Clean Prod 191:144–157

Yang W, Gu S, Xin Y et al (2018) Compost addition enhanced hyphal growth and sporulation of arbuscular mycorrhizal fungi without affecting their community composition in the soil. Front Microbiol 9. https://doi.org/10.3389/fmicb.2018.00169

Using Traditional and Simulation Methods for C and N Cycling Studies with Additional Periods of Human Civilisation: Replicating the Procedures at Regional Levels Advocate

Paul Ola Igboji

Abstract Carbon and nitrogen cycling are two essential processes in soil health and ecosystem stability. Many researchers and scholars have supported these processes as immensely vital for continued life on the planet earth. Not just their cycling but carbon and nitrogen sink are needed in the era of greenhouse gases emissions, associated global warming and climate change. The sinking of nitrogen naturally and through the activities of nitrogen fixers is also a panacea to nitrogen mobility, accumulated losses and constant replenishment with costly and environment-damaging inorganic fertilisers. An English experience may give clue to global action plan on enhancing soil carbon and nitrogen sink for various periods of human civilisation. These periods were grouped into prehistoric, historical and agricultural revolution, post-agricultural revolution, Green Planet and Post Green Planet. Besides, several types of land management were compared with reference to effective and efficient carbon sequestration, namely, grassland under permanent pasture on 5-year ley prior to grazing, grassland under permanent pasture sown with red clover prior to grazing, arable land under barley and deciduous woodland. By the aid of modelling, simulated data was generated for over 8500 years of English agriculture and compared with field data. The study showed that all these land management practices sequester little or no carbon but required integrated approaches. Nevertheless, the practices if continued were found to be sustainable, as serious changes that may require other sustainable options were forecasted over subsequent 25–30 years.

Keywords Inorganic fertilisers · Ecosystem stability · Nitrogen cycling · Carbon cycling

---

P. O. Igboji (✉)
Department of Soil Science and Environmental Management, Faculty of Agriculture and Natural Resources Management, Ebonyi State University, Abakaliki, Nigeria
e-mail: p.igboji@ebsu.edu.ng

R. Datta et al. (eds.), *Carbon and Nitrogen Cycling in Soil*,
https://doi.org/10.1007/978-981-13-7264-3_5

Abbreviations

ANOVA	Analysis of Variance
AR	Agricultural revolution
ATP	Adenosine triphosphate
BD	Bulk density
CEC	Cation exchange capacity
$CHCl_3$	Chloroform
CMC	Carboxyl methyl cellulose
COOH	Carboxylic
CPMAS	Cross polarisation magic angle spinning
Cult	Cultivation
Cv	Coefficient of variation
DMSO	Dimethylsulphoxide
DOC	Dissolved organic carbon
DOY	Day of the year
E(df)	Error degree of freedom
ECL	Chemiluminescence
EF	Electrostatic factor
EMBRACE	Earth Model Bias Reduction and Assessing Abrupt Climate Change
ESR	Electronic spin resonance
Fa	Aromaticity fraction
FA	Fulvic acid
FACE	Free-air carbon dioxide enrichment
FLSD	Fisher's least significant difference
FTIR	Fourier transform infrared spectroscopy
FYM	Farmyard manure
GP	Green Planet
GPP	Gross primary productivity
H^+	Hydrogen ion
HÁ	Humic acid
HA	Prehistoric agriculture
HÁg	Historic agriculture
HS	Humic substances
$K_2S_2O_8$	Potassium silicate
K_2SO_4	Potassium sulphate
KCl	Potassium chloride
KD	Kilo Dalton
MBC	Microbial carbon
MRT	Mean residence time
MUB	Modified universal buffer
MW	Molecular weight
N_2H_2	Hydrazine
NMR	Nuclear magnetic resonance

OD	Above Datum
OH^-	Hydroxyl
OM	Organic matter
PAR	Post-agricultural revolution
PD	Particle density
PGP	Post Green Planet
PHAg	Prehistoric agriculture
PNP	p-Nitrophenyl Phosphate
POM	Particulate organic matter
R^2	Coefficient of determination
RPM	Resistant plant material
Rpm	Revolution per minute
RQ	Respiratory quotient
RT	Research time
Se	Standard error
SIR	Substrate-induced respiration
SOC	Soil organic carbon
SOM	Soil organic matter
SOMNET	Soil Organic Matter Network
SON	Soil organic nitrogen
SWC	Soil water content
WFPS	Water-filled pore space
WHC	Water holding capacity
XAD-8	Polymethylmethacrylate resin

1 C-Cycle

Micro-organisms drive C and nutrient cycling in the soils. Soils play a dual role of source and sink for greenhouse gases, particularly CO_2, CH_4 and N_2O. Figure 1 shows the global C cycle (Pidwimy 2005; RCEP 1996; Houghton et al. 1995). Globally, soils are the main reservoir of C. EE (1993) gave the amount of C in decomposing plant litter, and soil organic matter (SOM) is likely to have exceeded the quantity in living biomass by more than a factor of 2 or 3. Besides, the scientists estimated 9322 Mt C in English soils, peat and litter and 114 Mt C in English vegetation (with 80% of this value in forests and woodlands).

The release of CO_2 is a natural process by the oxidation of SOM and organic litter by the microbial populations. According to RCEP (1996) the rate of CO_2 evolved is dependent on soil management, with quick losses when forests or grasslands are turned into arable land. Similarly, C dominate in soil changed from arable land to grassland or forests. But, the building of soil C following upgrading to pasture takes over ten times longer than when the pasture land has been ploughed (RCEP 1996).

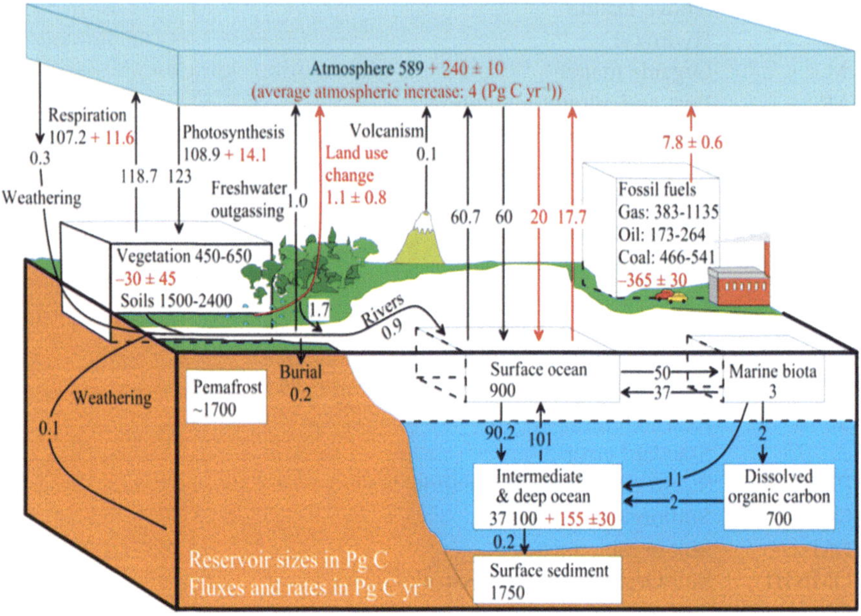

Fig. 1 Simplified diagram of the global carbon cycle. Numbers denote reservoir mass, also called "carbon stocks", in Pg C (1 Pg C = 10 15 g C) and annual carbon exchange fluxes (in Pg C year^{-1}) between the atmosphere and its two major sinks, the land and ocean. Black numbers and arrows indicate reservoir mass and exchange fluxes estimated for the time before the Industrial Era, about 1750. Red arrows and numbers indicate annual "anthropogenic" fluxes averaged over the 2000–2009 time period. These fluxes are a perturbation of the carbon cycle during the Industrial Era post 1750. Red numbers in the reservoirs represent cumulative changes of anthropogenic carbon over the Industrial Period 1750–2011. (The diagram model is taken from Sarmiento and Gruber (2002), and all numbers of stocks and fluxes are reproduced from Ciais et al. (2014))

It has been observed that C builds much more slowly when an arable piece of land converts into pasture. Often, 49 t C ha^{-1} may be added over 275 years, and half of this in the first 38 years (RCEP 1996; Cannell et al. 1994). The activities of man on natural C cycle releases CO_2 to the atmosphere, especially those deposited as chalk, limestone and fossil fuels (RCEP 1996).

Some scientist like Durán et al. (2017) worked on climate effects of the northern hardwood forests of North America. In their findings, they projected warmer environment in the few decades arising from increases in soil temperature, decreases in the availability of water and the rapid changes in winter snow pack and soil frost.

These events, according to the authors are bound to affect C and N cycling. For them, more attention of the researchers has been on the tides of climate change on soil properties, primarily on the upper-organic part of the soil profile, e.g. forest floor, with little known about the deeper mineral soil horizons. They worked extensively on the landscape gradient of the Hubbard Brook Experimental Forest in New Hampshire, USA, on the effects of climate change and expected scenarios for the next 50–100 years, with attendant C and N cycling. In their study, lower elevation, soil temperature, soil freeze and thaw reduced soil inorganic N and lower N mineralization and vice versa. According to Durán et al. (2017), the difference in N pools and fluxes were consistent for all soil horizons investigated, while sensitivity to climate change varied with soil depth. Hence, for accuracy of findings, the authors recommended inclusion of climate change on soil profile investigations. The authors observed that N cycling was more reactive to climate change than N cycling processes, which implies decoupling of C and N cycles in coming decades, with attendant consequences for ecosystem function. A further search by the authors revealed that climate sensitivity was more in summer than spring and during the warmer and less snowy years, which is clear evidence of climate change arising from increases in temperatures and drop in water availability over the coming decades.

1.1 C-Sequestration

C-sequestration is simply the ability of the ecosystem to trap and safely store C in soil or water medium, instead of allowing it to remain or escape to the atmosphere (FAO 2000; Pretty and Ball 2001). The whole scientific community believes in/relay on the impacts of the greenhouse gases such as CH_4, N oxides and water vapour, with considerable opinion and debates on international policies for mitigating greenhouse gases and controversial climate change (Körner 2003; Ogunseitan 2005).

Globally, C sink goes with the reduction of CO_2 primary sources like the use of fossil fuel with cleaner technologies like solar energy, wind energy, tidal power and hydropower. Another suggested method is the role of micro-organisms in C sink for terrestrial and marine ecosystems that includes ocean fertilisation (Ogunseitan 2005). The process of ocean fertilisation involves augmenting the natural ocean levels of iron and other trace elements such as manganese to trigger phytoplankton productivity. For experts, it is believed that fertilising the Antarctic Ocean with iron can permit the phytoplankton to convert all the available nutrients into new organic matter (up to 1.7–2.8 billion t of C), with more than 5% of that C sequestered in deep waters (Ogunseitan 2005; Lackner 2003).

Generally, other scientists (Post and Kwon 2000; Paul et al. 2003) have supported the accumulation and turnover of SOM as a major factor in soil fertility studies and proper functioning of the ecosystem. This helps to establish whether soils are the real asset for C sink or sources in the global C cycle. For Sollins et al. (1999) and Paul et al. (2001), the dynamic complexes of SOM involve a wide array of organic constituents with the various mean residence time (MRT) that vary over space and time.

Johnson et al. (2003) also reported the role of soil in balancing fossil fuel CO_2 emissions. Nevertheless, the potential of sols for C-sequestration is highly controversial. While some scientists believe it is possible under a proper management, others are sceptical (Lal et al. 1998; Johson et al. 2003; Post and Kwon 2000; Schlesinger 1977, 1990). These are based on field evidences with conflicting reports, like conversion of forest or grassland to agriculture that is known to cause massive losses in soil C (Mann 1986; Post and Kwon 2000), and returning the land to native vegetation reaccumulates C in soils (Jenkinson 1991; Johnson et al. 2003). Other authors like Compton and Boone (2000), Compton et al. (1998) and Richter et al. (1999) found little or no reaccumulations, while Post and Kwon (2000) found large differences in other agricultural practices which included both positive and negative changes. They gave average values of 0.33–0.34 Mg C ha^{-1} $year^{-1}$ C accumulation of C, which is equivalent to rates noted by Schlesinger (1990); he contended that the potential for C accumulation on a global scale is low. Other scientists such as Pretty and Ball (2001) assessed the loss of SOM and C during intensive land cultivation covering long experiments in both Europe and North America where higher equilibria rose with sustainable management practices. The most significant gain comes from converting an arable piece of land into agroforestry, with increased woody biomass above the ground from increased SOM. Pretty and Ball (2001) assessed grassland within rotations, zero tillage or no-till farming, green manures, amendments of straw and manures to sustain agricultural systems up to 0.3–0.6 t C ha^{-1}, increasing to several tonnes per hectare under tree intercrop and grazing systems.

Johnson and Curtis (2001) observed that forest harvesting followed by reforestation caused little or no change in soil C, irrespective of the intensity of harvest. Other authors like Turner and Lambert (2000) recorded large net to large net gains (Johnson and Todd 1988), even over a relatively short period of time (Knoepp and Swank 1997).

Several studies in tropical dryland ecosystem (Farage et al. 2005) used data from India, Kenya, Nigeria and Argentina to model changes in soil C based on farm practices and native technologies. References were made to pre-cultivation C stocks and the amounts lost in the course of continuous cultivation. The authors discovered that additions of OM to such soils by using different quantities of farmyard manures, green manures, legumes in rotations and vermicompost and the use of fallows in rotations increased soil C and agricultural yields. Similarly, when trees were used as part of agroforestry, C stocks were increased, while addition of inorganic fertiliser alone leads to a decrease in the soil C stock in all systems assessed but increased a little when used with zero tillage. On the other hand, zero tillage alone increased soil C, especially when OM was added to the soil.

2 N Cycle and N-Fixation

Nitrogen is essential for growth and reproduction of all organisms. Figure 2 has presented the principal processes within the terrestrial ecosystem. Globally, N is estimated to be distributed as follows: atmosphere (3,800,000 billion tonnes), plant biomass on land (12 billion tonnes), plant biomass in oceans (0.3 billion tonnes) and dead OM in oceans (550 billion tonnes).

Major factors govern fixation of gaseous N, including lightning and high-temperature combustion which produces oxides of N or steps in the manufacture of nitrogenous fertilisers. Globally, over 276 billion N are fixed per year, out of

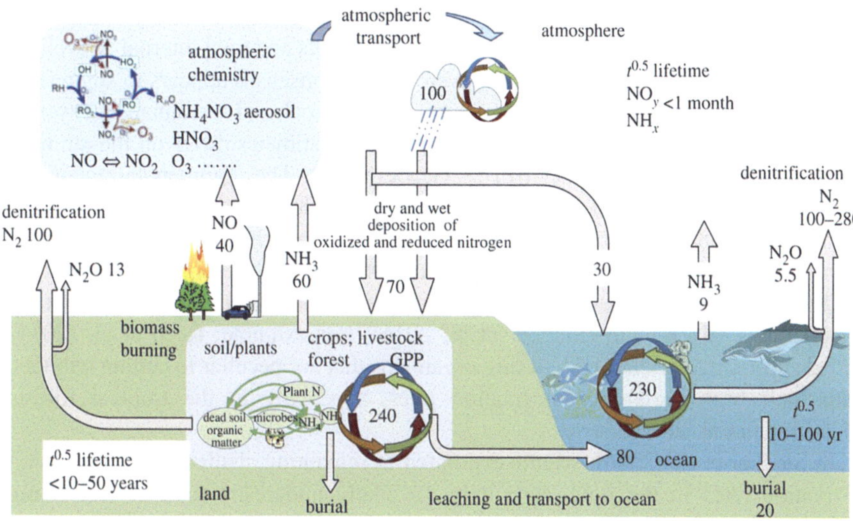

Fig. 2 The global nitrogen cycle in the twenty-first century. (Adapted from Fowler et al. 2013)

which over 85% is fixed biologically (with over 140 million tonnes fixed on terrestrial ecosystem and 100 million tonnes in the aquatic ecosystem), while around 15% comes from industrial processes (RCEP 1996). The N cycle is, however, modified by inputs of N compounds (fixed N) from human activities principally the application to soil of fertilisers containing NO_3, NH_4 and NH_4-forming substances such as urea and the wet and dry deposition on soil of airborne N compounds from industrial and other combustion sources (principally as NO_3, N oxides and HNO_3) and from livestock production (as NH_3^+ and NH_4) (Ogunseitan 2005; Levia and Frost 2003; Newman and Banfield 2002; Weinberg 1997; Imbert and Blondeau 1998; Sellers et al. 1989; Falkowski 1994; Falkowski et al. 1998, 2000; Lovelock 1995; RCEP 1996; Benjamin and Honeyman 1992; Schlesinger 1991).

The imbalances in biogeochemical cycles lead to various environmental problems such as NO_3 leaching from soils. This leads to eutrophication of aquatic ecosystem and subsequent pollution of water bodies. Eutrophication leads to N-limitation in marine ecosystem and P-limitation in freshwater ecosystem. Similarly, acidification of soils and water has attributed to industrial deposition of N and S compounds, especially in the industrialised world.

2.1 *Openness of the Nitrogen Cycle*

A new subject of attention is the "openness of the nitrogen cycle". This topic is important for scientists exploring how N cycling varies at large scale. This is what

Stocker et al. (2016) described as "openness of the nitrogen cycle". They described it as the process of quantifying the ratio between N inputs and total internal N cycling.

According to Stocker et al. (2016), even though consensus appears to suggest an increase in N cycle openness from boreal to tropical and savannah ecosystem (Cleveland et al. 1999), any models and empirical estimates differ on the strength of the gradient (Sonke Zaehle MPI-BCG, Germany). The argument/augment lies partly on the estimates of biological N-fixation, with recent statistics giving the global rate of 58 Tg N $year^{-1}$ (Vitousek et al. 2013), which is much lower than the quote provided by Cleveland et al. (1999) of 100–290 Tg N $year^{-1}$. Further results by Sullivan et al. (2014) confirm this downward correction. Nevertheless, huge uncertainties still remain (Stocker et al. 2016). For example, Reed et al. (2011) added surprising varieties of N-fixing organisms that are peculiar to certain habitats, with their high variability in fixation rates, especially in the tropical forests (Batterman et al. 2013).

In Stocker et al. (2016), scientists pointed to remaining challenges in measuring N-fixation rates in the field, especially the likely underestimation of free-living N-fixation in the recent estimates. For example, while Elbert et al. (2012) gave N-fixation in cryptogamic covers as 49 Tg N $year^{-1}$, many others differ, with Stocker et al. (2016) recommending model approaches to the actual quantification of the magnitude of N-fixation flux, in addition to studies geared towards understanding its controls and energetic C cost. Indeed, according to the researchers, mechanisms have been identified that imply a limited plant control over N-fixation rates (Menge and Hedin 2009), including labile C exports by plants to free-living N-fixing heterotrophs (Reed et al. 2011). There are also debates on physiological processes of N-fixation. While some scientists hold atmospheric deposition of N to be underestimated, others are of the view that deposition of organic N has been ignored but can be substantial in remote areas due to their longer atmospheric lifetimes and transport ranges compared to NO_x (Neff et al. 2002). Many scientists are of the view that the reassessment of N cycle openness in the boreal system is long overdue (Stocker et al. 2016).

One important scenario is the global mineralisation rates compared with fractions of total gross primary productivity (GPP) allocated to below ground. The derived pattern reveals a high C cost of P acquisition but a low apparent C cost of N acquisition in the tropics, consistent with open N cycling (Stocker et al. 2016). By contrast, low biomass production *efficiency* (ratio of biomass production to GPP in tropical forests tends to reflect C costs of nutrient acquisition (Stocker et al. 2016)). The spatial scale of the studies varies substantially, and various possibilities exist to explain this discrepancy (higher rates of autotrophic respiration among others). Even as additional research has been recommended to reconcile many of these unanswered questions (Stocker et al. 2016), their findings suggest a large-scale gradient of N cycle openness, total belowground allocation and biomass production efficiency as prime benchmarks for a new generation of C-N models. The summary of these findings is presented in (Fig. 3).

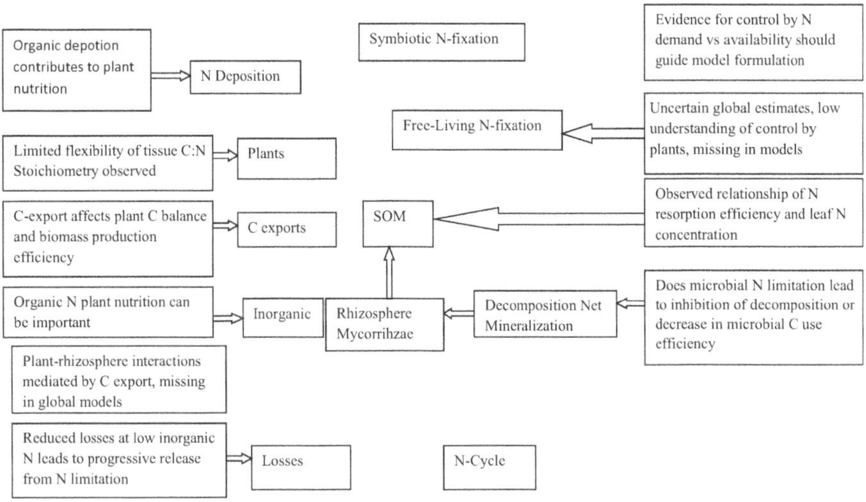

Fig. 3 Summary of findings and research priorities emerging from the workshop "Terrestrial nitrogen cycling in Earth system models". The thickness of arrows qualitatively illustrates the relative magnitude of fluxes. N cycle openness is illustrated by the relative flux magnitudes of losses or inputs (arrows for N deposition, symbiotic N-fixation and free-living N-fixation vs internal cycling (circles), soil organic matter. (Source: Stocker et al. 2016))

2.2 *Nitrogen Constraints on Plant CO_2 Responses*

Many pieces/types of researches on *elevated* (high) CO_2 effect on leaf level photosynthesis have been attributed to variations in the balance of supply and demand of C and N (Stocker et al. 2016). For Zaehle et al. (2014), most available C-N models have concentrated on measuring free-air CO_2 enrichment (FACE) experiments, with positions of enhanced net primary productivity (NPP) by the increase of C-N at the tissue level. These scientists noted the failure of the models to capture the measured movement towards below ground C accumulation, with subsequent N-uptake. Again, the leaf N-concentration across noted European forests has given more information on their phylogeny and wide differences within plant species which even though is sometimes large is yet temporal with regard to changes in environmental variables (Zaehle et al. 2014).

These differences call for monitoring of their changes in different plant tissues and their response to atmospheric CO_2 vis-a-vis soil nutrient availability. This way monitoring of environmental changes in different plant tissues will guide model developers in adding the role of mycorrhizas (Phillips et al. 2013) and rhizosphere plant-interface factors (Finzi et al. 2015). Thus happens because mycorrihizal associations limit N-availabilty under low N availability due to immobilization factors, that affect negatively C-fertility (Franklin et al. 2014).

Other pieces of researches have shown low N and positive biomass accumulation under *elevated* high CO_2 in ectomycorrhizal fungi, absolving arbuscular mycorrhizas plants (Stocker et al. 2016), suggesting similar activities for N-fixing species, while Larmola et al. (2014) described sustained C buildup with low N deposition in peatlands.

Similarly, Meyerholt et al. (2016) and Stocker et al. (2016) reveal N-fixation and implications for rising CO_2 stressing the need to ratify N-fixation controls over the global land C balance, as this may be related to C-N models that have differentiated the impacts of soil inorganic N pool and the way N pool and N loss can be simulated based on pool size with the absence of N-fixation extrapolation (Walker et al. 2015). Hence, the C costs for N fixation remains a subsiding rhizosphere activity and Therefore, mycorrihizas, it has been suggested can successfully model C allocation and guide the estimation of environmental changes (Stocker et al. 2016). Based on these findings, it has been suggested (Franklin et al. 2012) that more simple models that can confidently optimise principles and practices of C-N cycling be developed as such data will help in extrapolating individual processes and set off constraints in resource availability and trade-offs arising from physiological limitations (Stocker et al. 2016).

2.3 The Representation of Soil Nitrogen Cycling in Models

The CENTURY modelling approaches to transformations of C and N are very basic in the understanding of terrestrial nitrogen and carbon cycling in the earth systems. These modelling approaches have stood the test of time, especially for monitoring effects of land use changes on soil C nutrient dynamics. Many researchers (Janssens et al. 2010) have observed responses to N deposition and subsequent reduction of heterotrophic respiration relative to NPP, opposite to what CENTURY-type models predict and consistent with reduced respiration, with increasing N availability.

Attention has also been drawn to the C-use efficiency of microbes that tend to decrease with subsequent increase in organic matter C-N. Hence, the use of microbial functions has been suggested in understanding the activities of these soil microbes. This will buttress plant-soil associations. This phenomenon can be captured by models, making data required for evaluation to be more simple and unique. This will increase understanding of such ecosystem processes (Stocker et al. 2016).

3 Biological Properties That Govern C and N Cycling

3.1 Whole Ecosystem and Soil Respiration

All living cells require a regular source of energy. For heterotrophic microflora, energy is got in the course of transformation of organic materials such as cellulose, proteins, nucleotides and humic compounds. Energy is derived through cell redox reactions when electrons are transferred from the donor to the recipient. During aerobic respiration, organic matter is oxidised by micro-organisms. In this scenario, oxygen acts as a terminal acceptor of the electrons. Hence, the index for measuring CO_2 production (Alef 1995; Nannipieri et al. 1990) is proxy to the evaluation of O_2 consumption.

Soil respiration is one of the traditional and most often used indexes for measuring microbial activities in soils (Kieft and Rosacker 1991). Basal respiration is done in absence of organic substrate to the soil, while substrate-induced respiration (SIR) involves the addition of organic substrates to the soil such as glucose. Respiration is not only restricted to micro-organisms, but it is also carried out by other organisms inhabiting the soils. Like other metabolic activities, respiration depends on the physiological state of the cells and is governed by other factors (Alef 1995). Such factors include soil moisture, temperature, the availability of nutrients, soil structure and tillage. It has been observed that air-drying drastically reduces soil respiration, while remoistened soils, however, show very high initial activities, probably as a result of the high concentrations of easily degradable organic compounds such as amino and organic acids caused by chemical and physical processes at the moistening of dry soils (Clark and Kemper 1967; Anderson and Slinger 1975; Wilson and Griffin 1975, Kowalenko and Ivarson 1978; Kroeckel and Stolp 1986; Kieft 1987; Datta et al. 2017b).

When air-dried soil containing carbonate is remoistened, there is a release of CO_2. In this case, it is recommended that O_2 consumption is to be measured (Anderson 1982; Kieft 1987). Researchers have also reported decreases in soil respiration with the depth of soil and increases with soil organic matter and microbial biomass (Stotzky 1965; Thalmann 1968; Parkinson et al. 1971; Alexander 1977; Gray and Williams 1977; Anderson and Domsch 1978a, b; Domsch et al. 1979; Sparling 1981a, b; Alef and Kleiner 1987; Alef et al. 1988; Suttner and Alef 1988; Alef 1990).

Researchers have shown effects of cultivation methods, agrochemicals, pesticides and heavy metals on soil respiration (Jaggi 1976; Anderson et al. 1984; Malkomes 1985; Carlisle and Trevors 1986; Domsch and Schroder 1986; Wilke 1986; Somerville and Greaves 1987; Schlosser 1988; Schuster 1988).

Traditionally, basal soil respiration can be followed for long periods of time. The changes in the composition of aerobic microflora occur during long-term incubation. In the case of the SIR method, a change in population is expected when the incubation period is longer than 4–6 h (Anderson and Domsch 1978b). The incubation temperature used to vary between 20 and 30 °C along with the water holding capacity ranging between 50 and 70%. The pH value of the measurement is usually that of the soil in water (Alef 1995). A summary of methods used in estimating whole ecosystem and soil respiration is presented in Table 1.

3.2 *Microbial Biomass*

Microbial biomass is the portion of the organic matter that consists of living micro-organisms that are smaller than 5–10 μm^3. These living micro-organisms are

Table 1 Methods of estimating whole ecosystem and soil respiration

Condition of measurement	Methods of determination	Remarks	References
Soil respiration in the laboratory	1. Incubation of soils in jars, closed petri dishes or different types of flask	CO_2 is usually adsorbed in NaOH and determined by HCl titration	Isermeyer (1952), Pochon and Tardieux (1962), Jaggi (1976), and Rowell (1994)
	2. Electrical conductivity of the NaOH solution following incubation of soil samples		Anderson and DOMSCH (1978a) and Cheng and Coleman (1989)
	3. Use of gas chromatography		Brookes and Paul (1987) and Trevors (1985)
	4. Infrared spectroscopy		Heinemeyer et al. (1989)
	5. Use of labelled CO_2 ($^{14}CO_2$)	This is following the decomposition of specific organic compounds in the soil	Nakas and Klein (1981)
	6. Use of Warburg apparatus	Measures the O_2 consumption in incubated soils	Domsch (1962) and Stotzky (1965)
	7. Use of electrorespirometer		Birch and Melville (1969) and Kroeckel and Stolp (1986)
Soil respiration in the field	1. Placing of NaOH solution in an open glass jar above the soil surface and covering the area to be measured with a metal cylinder closed at the upper end	Used for determining CO_2 evolved from undisturbed soils	Anderson (1982)
	2. Use of gas chromatography	Measures CO_2 and O_2 concentrations at various soil depths	Richter (1972) and Anderson (1982)
	3. Automated monitoring of biological trace gas production and consumption	Estimates gas concentration under field conditions in a covered soil	Brumme and Beese (1995)
Whole ecosystem and soil respiration in the field	1. Micrometeorological technique		Franzluebbers et al. (2002), Verma (1990), and Norman et al. (1992)
	2. Static chamber with alkali absorption method	Nocturnal trial	Franzluebbers et al. (2002), and Zibilske (1994)
	3. Portable environmental gas monitor linked to soil respiration chamber		Sowerby et al. (2000), and Ball et al. (2000)

generally expressed in mg C kg^{-1} soil or μg C g^{-1} dry weight, ranging from 1% to 5% of SOM (Smith et al. 1990). Table 2 summarises the various methods used in its determination.

3.3 *Enzyme Activities*

Enzymes are substances produced by living organisms which act as catalyst to bring about a specific biochemical reaction without being used up. They show extreme specificity in catalysing biological reactions. There is a systematic classification of enzymes based on the recommendation of the International Enzyme Commission. The new system has divided enzymes into six principal classes further classifying each class into four categories according to the type of reaction each group is capable of catalysing. For example, the recommended name, systematic name and classification name, for phosphodiesterase is phosphoric diester hydrolase while its classification number is EC 3.1.4 (Alef and Nannipieri 1998).

Researches into soil enzymes are traditional, with many activities linked with biotic and abiotic factors that contribute to the overall efficiency and effectiveness of soil enzymes. According to Burns (1982), an enzyme may be involved physically with proliferating animal and microbial and plant cells, and it can also be found in the cytoplasm (within the periplasm of gram-negative bacteria or sometimes attached to the outer surface of cells). They have also been reported in non-proliferating cells (e.g. microbial spores or protozoan cysts), in entirely dead cells or cell debris. Many enzymes are found as an extracellular soluble molecule. They are sometimes temporarily associated in enzyme-substrate complexes, adsorbed to clay minerals or associated with humic colloids (Datta et al. 2017). Some of these categories according to Alef and Nannipieri (1998) may represent various stages in the life of an enzyme. An intracellular enzyme may still function after the cell dies and thus it becomes associated with cell debris. It may be released in the aqueous phase and may eventually be adsorbed in an active form by the soil colloids. In enzyme-clay and enzyme-organic polymer molecules, there are remarkable resistance to proteolytic and thermal denaturation (Sarkar et al. 1980; Burns 1982, 1986; Trasar-Cepeda and Gil-Sotres 1987, 1988; Nannipieri et al. 1988).

Several methods exist for the measurement of enzyme activities. These are summarised in Table 3 (Burns 1982; Nannipieri et al. 1990; Nannipieri 1994).

Alef and Nannipieri (1995) cautioned about the interpretation of results arising from the measurement of soil enzyme activities. According to these authors, these measurements represent the maximum potential because the incubation conditions for enzyme assays are chosen to ensure optimum rates of catalysis. Hence, the concentration of substrate is sometimes is in excess, with the optimum values of pH and temperature selected to ensure the highest rate of enzyme activity. Similarly, the volume of the reaction mix is important to allow free diffusion of substrate. Hence, Nannipieri (1994) warned researchers to avoid concluding the fact that enzyme assays have no significance in ecological and agricultural matters.

Table 2 Methods used in determining soil microbial biomass

Parameter	Method of determination	Remarks	References
Soil microbial biomass	1. Staining and counting of microbial biomass		Babiuk and Paul (1970), Trolldenier (1973), Anderson and Slinger (1975), Paul and Johnson (1977), Söderström (1977), Torsvik and Goksoyr (1978), Lundgren (1981), Sparling and West (1990), Jenkinson (1988), Kaiser et al. (1992) and van de Werf and Verstraete (1987)
	2. Use of physiological parameters such as ATP content, respiration and heat output		
	3. The fumigation -incubation technique	1. Soil fumigation only kills the microbial biomass and does not affect nonliving OM	Shen et al. (1984), Brookes et al. (1985), Vance et al. (1987a, b, c), Joergensen et al. (1990), Martens (1985) and Mueller et al. (1992)
		2. The flush in respiration exclusively derives from the microbial biomass	
		3. The number of organisms killed in the unfumigated soil is negligible compared with that in fumigated soil	
		4. The fraction of dead microbial biomass carbon mineralised over a given period does not differ in different soils	
		5.The method is not recommended for acidic soils (H_2O < 4.5) because soil inoculation can be difficult	
		6. The method is unsuitable for soils recently treated with OM, because the large microflora of the unfumigated soil decomposes the substrate more effectively than the smaller microflora of the fumigated soil	
		7. This method cannot be used when fresh roots are present in soil because cell membranes of young	

(continued)

Table 2 (continued)

Parameter	Method of determination	Remarks	References
		living roots are affected by $CHCl_3$	
	4.Modified fumigation incubation procedure	1. Suitable for determining microbial biomass N for waterlogged soil	Inubushi et al. (1984)
		2. In calcareous soils, low in OM, errors can occur due to the decomposition of HCO_3^-. This is reduced by placing beakers with soda lime in desiccators holding fumigated and unfumigated soils	
	5.Colourimetric		Chaussod et al. (1986)
	6.Gas chromatography		Martens (1985) and Chaussod and Nicolardot (1982)
	7.Fumigation-extraction		Vance et al. (1987c) and Kaiser et al. (1992)
	8.Dichromate oxidation	OM is oxidised and Cr (+VI) reduced to Cr(+III). Amount of dichromate left is back titrated	Vance et al. (1987c)
	9.Ultraviolet persulphate oxidation	Extracted OC is oxidised by UV light to CO_2, which can be measured by IR or photospectrometry	Wu et al. (1990)
	10.Ninhydrin nitrogen reaction	Ninhydrin forms a purple complex with molecules containing AA, peptides and proteins	Moore and Stein (1948) and Lamothe and McCormick (1973)
	11. SIR	O_2 uptake or CO_2 evolution immediately after the amendment with quantities of glucose	Sparling and West (1990)
	12. Respiration-simulation method	Based on continuous monitoring of O_2 uptake by soil supplied with readily degradable OM in a respirator	Van de Werf et al. (1995) and van de Werf and Verstraete (1987)

Scientists have established that enzyme measurements do answer qualitative questions about specific metabolic processes and in combination with other measurements (ATP, AEC, CO_2 evolution) may increase the understanding of the effect of agrochemicals, cultivation practices and environmental and climatic factors on the microbiological activity of the soil (Skujiņš 1978; Nannipieri 1994; Alef and

Table 3 Major soil enzymes and methods of determination

Enzyme	Activity	Method of assay	Remarks	References
Urease	Catalyses the hydrolysis of urea to CO_2 and NH_3, with the formation of carbamate as an intermediate product	1.Determination of NH_3 liberated on incubation of toluene-treated soil with buffered urea solution	1. Widely distributed in nature	Bremner and Mulvaney (1978), Gosewinkel and Broadbent (1984), Kandeler and Gerber (1988), McCarty et al. (1989), Skujiņš and McLaren (1969), Douglas and Bremner (1970), Bremner and Mulvaney (1978), Mulvaney and Bremner (1979), Hoffman and Schmidt (1953), Tabatabai and Bremner (1972), Zantua and Bremner (1975), Frankenberger and Johanson (1986), Kandeler and Gerber (1988), Kissel and Cabrera (1988), Moyo et al. (1989), Nannipieri et al. (1978), Nannipieri et al. (1974) and Burns et al. (1972)
		2. Estimation of the rate of urea hydrolysis in soils by residual urea or $^{14}CO_2$ liberated after incubation	2.Catalyses the hydrolysis of hydroxyurea, dihydroxyurea and semicarbazide	
		3. Use of buffer to control pH, or addition of toluene to inhibit microbial proliferation	3.Denatured at 70 °C	
			4. Incubation temperature of assay ranges from 15 to 42 °C	
			5. Urease extracted from soil is resistant to thermal and proteolytic denaturation	
Phosphatase	Catalyses the hydrolysis of phosphate esters		1. Enzymes with relatively broad specificity, capable of acting on a number of different structurally related substrates, but at widely different rates	Florkin and Stotz (1964), Beck (1973), Burns (1978), Chhonkar and Tarafdar (1981), Dick et al. (1983), Tarafdar and Jungk (1987), Alef et al. (1995), Speir and Ross (1978), Eivazi and Tabatabai (1977) and Nannipieri et al. (1988)
			2. Acid phosphatase is predominant in acid soils, alkaline phosphatase in alkaline soils	

(continued)

Table 3 (continued)

Enzyme	Activity	Method of assay	Remarks	References
			3. Assays carried out at neutral pH (6.5–7)	
			4. Optimum temperature for assays (40–60 °C)	
			5. Activity affected also by OM content, soil moisture and anaerobiosis	
Cellulase	Catalyses the hydrolysis of cellulose to Ŋ-glucose	Based on the determination of either released sugars or evolved $^{14}CO_2$, using cotton strips, radioisotope-labelled cellulose and carboxyl methyl cellulose (CMC)	Enzymatic hydrolysis of cellulose depends on degree of crystallinity, the nature of associated substances and surface area	Lee and Fan (1980), Eriksson and Wood (1985), Sinsabaugh and Linkins (1989), Alef and Nannipieri (1995), Hayano (1986), Hunt (1977), Schröder and Gewehr (1977), Schröder and Urban (1985), Sinsabaugh and Linkins (1988), Tateno (1988), Kshattriya et al. (1992), Speir and Ross (1981), Benefield (1971), Hope and Burns (1987), Kiss et al. (1978), Clarke and Stone (1965), Yamane et al. (1970), Hayano (1986), Rhee et al. (1987), Stutzenberger (1972), Gottschalk et al. (1981), Joliff et al. (1989), Benefield (1971), Isbister et al. (1980), Sato (1981) and Schinner and vonMersi (1990)

Nannipieri 1995). For example, Curci et al. (1997) discovered that enzyme activity was higher in the uppermost 20 cm of soil in plots tilled by shallow ploughing, as against the situation in soils tilled by deep ploughing.

3.4 Humic Matter in Soils: Influence of Environment and Agriculture

Organic matter in soils is commonly grouped into humic and non-humic substances. Humic substances are organic materials derived principally from the decaying plant remains but with the (*average*) plant components considerably altered by the soil animal and microbial populations. Abiological chemical reactions also alter the entire process giving rise to a complex mixture of macromolecules whose composition is dependent on the site, more especially vegetation and pedogenic processes (Anderson et al. 1984).

Humic substances have been tested with different isolation, purification and fractionation techniques. It was in 1982 that the International Humic Substances Society published methodologies for their extraction, purification and fractionation (from both soils and water) (Anderson et al. 1984). These scientists gave details of the levels of humic fluvic and hydrophilic acids in soil and water. Another reliable method is the use of cross polarisation magic angle spinning (CPMAS) ^{13}C nuclear magnetic resonance (NMR) spectroscopy. From sugar and amino acid analysis of humic substances, it has been found that each humic component in each environment possesses an individuality that distinguishes it from other components in the same environment and from the same humic components in different environments (Hayes 1991).

3.4.1 Origins, Compositions and Structures of Humic Substances

There is no evidence on genetic or biological control of the synthesis of humic substances. Hence, there is a lack of regularity in the manner in which the molecules that aggregate the macromolecules are sequenced. There is evidence of intramolecular and intermolecular interactions that give unique secondary and tertiary structure (Hayes 1991). However, there are no rigid regularities between such interactions and associations. Even though interactions and associations can form between humic substances and other organic and inorganic molecules, they are seldom random and do not constitute part of any biological functions attributed to humic substances. Hence humic substances do not meet any of the criteria for a structure as they apply for proteins (Hayes 1991).

There are two biological processes that give rise to humic substances. One is "degradative" which involves biological transformations of the organic macromolecules that are refractory in nature, namely, lignins, paraffinic substances, cutins, melanins and suberins (Nip et al. 1986; Hayes 1991).

The other are polysaccharides and proteins which form readily available substrates for micro-organisms, which eventually give rise to humic substances, even though the origins of such substances are more likely to be from the micro-organisms which proliferate on the labile substrates, rather than from the substrates themselves (Hayes 1991). Another possibility is that of amino acids and peptides released from proteins and sugars, as well as oligosaccharides released from polysaccharides which undergo "browning" reactions to give rise to humic substances. Another account is from quinones (from oxidised phenols) which are also known to give rise to humic-type substances. Such pathways are part of the second or synthetic process, for the genesis of humic substances (Hayes 1991).

To understand this, Hayes (1985) clarified the "browning" reaction or the "melanoidin" theory, when he observed that monomeric reducing sugars, such as glucose, could condense with amino acids such as glycine to form brown macromolecule substances. Similarly, polycondensation reactions between glycine and 2-oxopropanal (methyl glyoxal) were considered by Enders et al. (1948) to provide plausible processes for the formation of humic substances. By regulating the ratios of the reactants, the macromolecular substances produced can have elemental contents and charge and other characteristics similar to soil humic acids (Schuffelen and Bolt 1962). More recently there has been an emphasis on the role of quinines from di- and polyhydroxybenzene structures with –OH groups in the 1,2- and 1,4- ring positions on the synthesis of humic substances (Flaig 1988). Researchers have also discovered that lignins can give rise to the appropriate phenols, and, again, fungi are known also to synthesise phenols (many of which are components of melanins), the coloured secondary metabolites formed during fungal degradation of saccharides (Martin et al. 1980; Datta et al. 2017a). Likewise, Aiken et al. (1985), considered humic substances to be a "general category of naturally occurring biogenic heterogeneous organic substances that can generally be characterised as being yellow to black in colour, of high molecular weight and refractory". Similarly, Hayes and Swift (1978) based on proposals by Kononova and Somne (1966) considered humic substances to be the amorphous, macromolecular, brown-coloured components of SOM which bear no morphological resemblances to plant or animal tissues from which they were derived and which can be differentiated into broad general classes on the basis of solubility differences in aqueous acids and bases. According to these authors, humic acids are the components of humic substances precipitated when extracts in dilute aqueous alkali are acidified to pH 1. Fulvic acids are the components which are soluble in aqueous acids and base; humin is the term applied to the components which are insoluble in aqueous acids and bases. However, these three categories (humic, fulvic and humin acids) referring to gross mixtures and the elemental composition of these mixtures and their chemical and physicochemical properties can vary with their origins and with the environments in which they are formed (Hayes 1991).

Table 4 Methodologies for successive extraction of humic substances from a soil [Yield (% of total OM)]

Extractant	Humic acid	Fulvic acid	Cumulative total	pH value of extractant
Water	0.0	2.8	2.8	–
DMF	15.0	2.2	20.0	6.8
Sulfolane	4.1	1.0	25.1	3.7
DMSO	0.7	0.2	26.0	5.9
Pyridine	14.8	0.6	41.4	11.6
EDA	23.2	6.3	70.9	13.0

3.4.2 Isolation of Humic Substances from Soil

The properties used for the isolation of humic substances from the soil and the processes of the dissolution of the macromolecules in various solvent systems were described by Hayes (1985). The author's findings reveal that humic substances are mostly ionised at pH of most agricultural soils, leading to negative charges of the conjugated bases in the acid groups being balanced by divalent and polyvalent cations. Table 4 shows the data relating to the successive extraction of humic substances from the soil (Hayes et al. 1975; Swift 1985).

3.4.3 Purification and Fractionation of Extracts

Humic acid is extracted using organic and aqueous solvents. These give rise to mixtures containing humic and non-humic molecules. To facilitate the extraction, purification and fractionation of extracts, water scientists have developed a resin treatment process that separates the hydrophilic substances from the humic substances in waters. This is known as the poly (methylmethacrylate) resin, XAD-8. It binds the H^+-exchanged humic substances, but it allows salts and small molecules and macromolecules organic substances (such as polysaccharides) to pass through the resin column. The humic substances are recovered by raising the pH causing the acidic groups to ionise and the macromolecules to desorb from the resins (Hayes 1991). This procedure is followed for both the fulvic acid fraction that is in the supernatant, whenever any aqueous alkaline extracts are precipitated at pH 1. Thus, the substances retained, when this fraction is applied under acidic conditions to XAD-8 resins and subsequently eluted when the pH of the solvent is raised, are true fulvic acids (Hayes 1991).

Nevertheless, according to Hayes (1991), humic acid precipitates cannot be applied directly to resins. Rather, these acids can be dissolved in DMSO and passed into XAD-8 resin columns, followed by subsequent elution with acidified water (pH 1–2) which removes the DMSO and polar substances. Then, the humic acids are subsequently eluted as the pH is raised. According to the author, the procedure allows the fractionation of the humic acids using a pH gradient system. It also allows humic substances to be recovered from the acidified DMSO extracts of soils. While these extracts are applied to XAD-8 columns, the DMSO is eventually removed, followed by the humic substances recovery (Hayes 1991).

Furthermore, fractionations of the humic substances on the basis of molecular size differences are most frequently used to analyse the humic substances further. The procedures generally involve uses of gel permeation chromatography (Swift 1989; DeNobilli et al. 1989), ultrafiltration or centrifugation (Hayes 1991). For gel permeation and ultrafiltration procedures to be effective, it is recommended that the humic substances should not interact (either adsorb to or be rejected by the gel or membrane) with the media used (Hayes 1991).

The methods used by Appelqvist (1990) in fractionating sodium humate preparations were remarkable. The author used ultrafiltration (Sartorius membranes of pore sizes and nominal molecular size exclusion values of 5000, 20,000 and 100,000 daltons) and gel chromatography (Sephacryl S-200 gel, with cross-linked dextran from Pharmacia). The scientist observed differences between the samples excluded by the gel (MW > 150,000) and those retained by the membranes which nominally kept the materials of 100,000 MW and above (>100,000), 20,000 and above (100,000–20,000 MW fraction) and 5000 and above (20,000–5000). As the molecular sizes decreased, the carbon contents also decreased (52.5%, 51.6%, 49.2%, 47.4%), and the E_4/E_6 ratios increased. The E_4 (solvent mixture absorbance at 465_{nm}) and E_6 (solvent mixture absorbance at 665_{nm}) are indicators of differences in solution conformations. It aids comparison of humic substances in different solvent systems. Hence, such data support the concept of greater aromaticity, for the higher molecular weight materials (Chen et al. 1977). This is in keeping with the concept of greater numbers of aromatic carboxylic acids, in the higher molecular weight substances. Further use of Fourier transform infrared spectroscopy (FTIR) indicated increased aliphatic characteristics for the lower molecular weight components, while Nip et al. (1986) proposed a procedure that generated H-NMR spectra for the high molecular weight materials and confirmed that phenols were present in the aromatic structures. This substantiated the data from most potentiometric titrations (Nip et al. 1986).

There were also differences between the >150,000 MW (obtained by gel filtration) and the >100,000 MW (obtained by ultrafiltration) fractions; no major differences from further examinations of the titration data were found. Other differences, the workers attributed to the amino acid content of the high molecular weight materials isolated by ultrafiltration and gel filtration (43.24 and 38.25 nmol mg^{-1}, respectively). The scientists gave the corresponding values for the total sugars as 20.6 and 25.7 μg mg^{-1}. Prior to all analyses, the humic acids (which were isolated in 0.1 M NaOH from a sapric histosol) were dissolved in DMSO-HCl (1% v/v) and adsorbed on and recovered (by back elution with 0.1 M NaOH) from XAD-8 resin. The resin treatment had separated the humic acid fraction sugar and the amino acid containing residues which were not covalently bonded to the humic acid "core" or "backbone" structures.

Therefore, the most useful of the physicochemical procedures for studying the structures of humic substances has focused on molecular size and shape and charge characteristics (Hayes 1991). For such studies it has been desirable to work with molecules that are relatively homogeneous with respect to sizes and shapes.

Details of these procedures in the study of humic substances of liquid and solid-state NMR have been detailed by Wershaw (1985), Malcolm (1989), Wilson (1989) and Steelink et al. (1989). Hence, advances in the use of liquid and solid-state NMR analyses now make it possible to obtain reasonably quantitative and well-resolved spectra for humic substances. The access to cross polarisation magic angle spinning (CPMAS) ^{13}C NMR has done much to improve resolution and applications of this procedure to the study of humic substances in the solid state (Hayes 1991).

3.4.4 Impact on Agriculture and the Environment

The role of SOM on the soil structure and fertility is very important. Many scientists have used humic substances to access the bioavailability of elements and chemical compounds in the environment. Their findings are diverse. Some believe that the slow degradation of humic substances may support the fact of not being the major source of nutrients (C, N, P and S) for soil biota, except for some highly specialised organisms, mainly fungi, which can utilise humic substances as a source of energy. However, the release of the above nutrients has been recognised to be more important in aquatic systems where humic substances are likely to be transformed more rapidly due to processes such as photoalteration (Hayes 1991).

4 Simulation of C and N Cycling

Modelling helps to understand the principal mechanisms affecting ecosystem functioning and the causes of disturbances to them. They are essential for long-term predictions and in making recommendations aimed at reducing harmful effects and preventing environmental disturbances. Many authors have demonstrated the benefits of using computer models in agriculture (Farage et al. 2005; Paul et al. 2003; Webb et al. 2003; Pumpanen et al. 2003; Qian et al. 2003; Smith et al. 1996, 1997).

Although mechanistic soil-crop models are increasingly accepted as valuable tools in analysing agronomical or environmental issues, potential users are faced with an equally increasing number of available models (Gabrielle et al. 2002; Plentinger and Penning de Vries 1995; Farage et al. 2005; Paul et al. 2003; Webb et al. 2003; Pumpanen et al. 2003; Qian et al. 2003; Smith et al. 1996, 1997). Besides little information is given as to the validity, limits and potential applications of these models, which would provide some guidance in the selection of the most appropriate package. As a result, models are chosen based on practical criteria such as code accessibility or the existence of user-friendly interfaces rather than the scientific assessment of their expected performance (Gabrielle et al. 2002). Again, although testing the models over long timescales may not be possible in all cases, the description of processes in the model can be validated. The most important data for such research are long-term experiments with a duration of more than 20 years, with information available about SOM pool dynamics during the experimental period.

Among the models that have been tested in various ecosystems and cropping systems are the CENTURY 4.0, RothC, CERES, NCSOIL, SNDIAL and STICS. These have been evaluated in the domain of carbon and nitrogen cycling and sequestration (Gabrielle et al. 2002; Plentinger and Penning de Vries 1995; Farage et al. 2005; Paul et al. 2003; Webb et al. 2003; Pumpanen et al. 2003; Qian et al. 2003; Smith et al. 1996, 1997). Century 5 is a research version of the CENTURY Model that has not been extensively tested or validated. For that reason, it has not released for public download except on special request (Ojima 2006).

RothC 26.3 model was developed with data from the long-term Broadbalk experiment at Rothamsted for monitoring the turnover of carbon in non-waterlogged topsoils that allows for the effects of soil type, temperature, moisture content and plant cover on the turnover process. It needs few inputs, and those it needs are easily obtainable (it needs a few inputs which are easily obtainable) (Coleman and Jenkinson 2005). It uses a monthly time step to calculate total organic carbon, microbial carbon and $\Delta\ ^{14}C$ (from which the equivalent radiocarbon age of the soil can be calculated) on year to century timescale. It is an extension of the earlier model described by Hart (1984). A version replacing the monthly time steps by continuous processes has been published by Parshotam (1996), while King et al. (1997) incorporated Roth C into a much larger model for the global C cycling.

The CENTURY 4.0 Agroecosystem Model manual by Parton et al. (1993) describes the best understanding to date of the biogeochemistry of C, N, P and S (Parton et al. 1993). The primary purposes of the model are to provide a tool for ecosystem analysis, to test the consistency of data and to evaluate the effects of changes in management and climate on ecosystems. It was developed to deal with a wide range of cropping system rotations and tillage practices and for the systematic analysis of the effects of management, global change on productivity and sustainability of agroecosystems (grasslands, forest, crops and savannahs). This version 4.0 integrates the effects of climate- and soil-driven variables including agricultural management in the simulation of C, N and H_2O dynamics in the soil-plant system. Simulation of complex agricultural management systems, including crop rotations, tillage practices, fertilisation, irrigation, grazing and harvest methodologies, is now possible in this enhanced release of the model (Parton et al. 1993; Metherell 1992; Parton et al. 1983), Parton et al. 1987). CENTURY operates on a monthly time step and is adequate for simulation of medium- to long-term (10 to >10,000 y) changes in soil total C and other ecosystem parameters in response to the changes in climate, land use and management. It also has the advantage of additional sub-models (Parton et al. 1993).

4.1 CENTURY Model Overall Structure

The soil organic matter sub-model simulates the flow of C, N, P and S through plant litter and the different inorganic and organic pools in the soil. The model runs using the following major inputs:

- Monthly average maximum and minimum air temperature
- Monthly precipitation
- Lignin content of plant material
- Plant N, P and S content
- Soil texture
- Atmospheric and soil N inputs and
- Initial soil C, N, P and S levels

The input variables are available for most natural and agricultural ecosystems and can be generally estimated from existing literature (Parton et al. 1993). The following description is mainly taken from Parton et al. (1993).

4.2 CENTURY Key Model Processes and Assumptions

In CENTURY Model, there are three types of organic matter, namely, active, slow and passive, with various rates of decomposition. The belowground litter accumulation also differs with the various microbial communities that facilitate the decomposition of the surface litter. The pools and flows of C are illustrated in Fig. 4. The belowground and organic animal excreta belowground biomass consist of two types, namely, structural pool and metabolic pool, based on the lignin to N ratio in the residue. There are also slow decompositions of structural residue pool. While the structural pool residue has slower decomposition rates, the metabolic ones have higher rates of decomposition. On the other hand, the structural pool contains all the plant lignin. Again, the plant residues and SOM are broken down by soil microbes during respiration leading to loss of CO_2, which increases with soil and sand level, with three products of SOM pools with defined characteristics and decomposition rates.

The overall decomposition rate is dependent on soil moisture and temperature. The overall decomposition rate rises with different cultivation activities. The model uses soil temperature at the surface layer, and soil moisture is worked out from the soil water at the 0–30 cm soil depth, with the addition of current month evapotranspiration. The level of decomposition of structural materials is dependent on the fraction of the structural material lignin concentration; while the lignin fraction of the plant material is assumed to directly identify the slow C pools as the structural plant material decomposes. The soil microbes and microbial products are the active pool with a turnover time of months to a few years based on environmental condition and sand level. The soil texture also influences the turnover rate of the soil SOM. This is

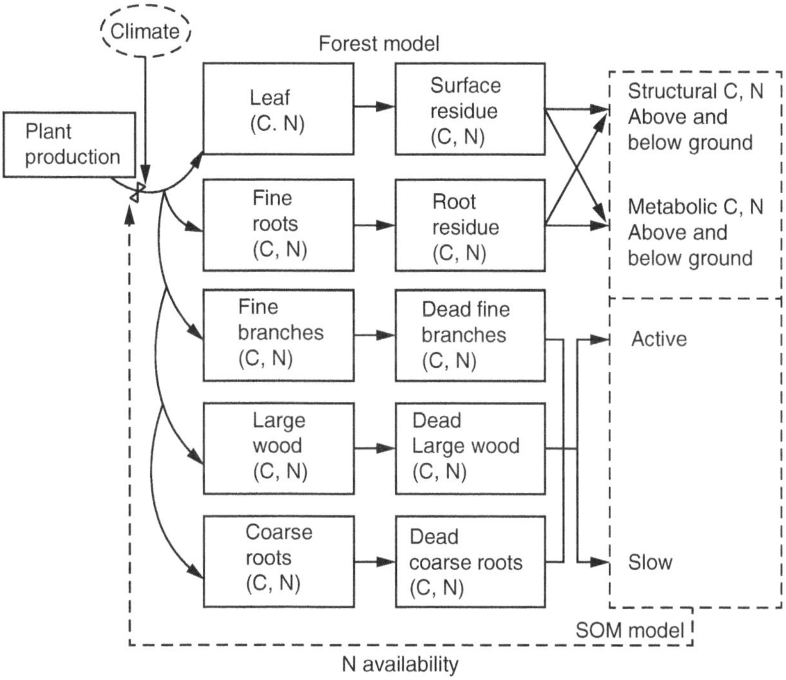

Fig. 4 Compartment diagram for the forest CENTURY Model (soil organic matter component). (Adapted from Ryan 1996)

higher for sandy soils, while clay soils have higher stabilisation rates and higher efficiency in restructuring active SOM.

On the other hand, soil surface microbial pool is not dependent on soil texture because it transfers materials openly into the slow SOM pool, while the structural pool is composed of resistant materials from the active and surface microbial pools with the average turnover time of 20–50 years. The other passive pools which are resistant to breakdown cover the physical and chemical SOM with the turnover time of 400–2000 years. The model runs on the assumption of different levels of decomposition products which pass through the passive pool and which rise with clay content, while other fraction of the active pool is lost via leaching as organic matter. Leaching is also based on the rate of the decomposition of soil SOM, as well as the clay level of the soil. Hence, less loss for clay soils occurs especially where drainage of water falls below the 30 cm soil depth.

The CENTURY Model has N, P and S pools which symbolise the C pools, with each pool having different levels of C to various element ratios that are calculated based on the method described by McGill and Cole (1981) model. Again, N is assumed to be linked with C, as C-N ratios are limitedly small, as ester bonds of P and S permit C-P and C-S ratios of wide variations. While the ratios of the structural pool are fixed at higher rates, that of metabolic pool is configured to permit nutrient dynamics in line with plant debris.

Each material that is moving into SOM phase is directly related to the rate of each element in the mobile inorganic mineral pool surrounding the surface soil layers. The low nutrient rates in the mobile pools result into high C ratios in the different SOM pools. There is a flow of N, P and S between SOM pools and various C movements, with the quantity of each element flowing out of a given pool being the same with the product of C flow plus the element to C ratio of the pool. The process of mineralisation and immobilisation of N, P and S continues to take place in order to balance the nutrient ratios. In the English case studies, soil total carbon and soil microbial respiration associated with decomposing surface litter (0–20 cm) depth were simulated.

The model simulates a wide variety of crops and grasslands by altering a number of crop- specific parameters. In the English case study, the model was configured using arable land under barley, grassland under permanent pasture (one on 5-year ley before grazing and the second sown with red clover before grazing) and deciduous woodland. Other cropping practices from historical records such as wheat (*Triticum aestivum* L.), maize (*Zea mays* L.), potato (*Solanum tuberosum* L.) and soybean (*Glycine max* L.) were also included. Hence in amending the crop/tree file using File.100 CENTURY updating facility, the Geescroft_Wilderness_Rothamsted_UK (succession) for woodland vegetation, Grass_Clover_Pasture Geescroft for the clover plot, Rothamsted_Grass for the 5-year ley plot and Brly_ V1 for the barley plot were used. The parameterization values for these crops/trees were obtained from Rothamsted data archive as described by Farage et al. (2005). The values are replicated in Tables 5 and 6.

The CENTURY Model makes certain assumptions, namely, that plant available soil N will be preferred by the crop, taking into account the limitations by P and S, before determining the symbiotic N_2 fixation. This has to do with the permanent pasture sown with red clover in the English scenario. On the other hand, fertiliser addition is normally by fixed amounts or calculated automatically according to the crop requirements. There is automatic option in the model that considers the quantity required to sustain crop growth given a given productivity index (with minimum and maximum levels). The OM is specified in a special file (omad.100 file) which is part of CENTURY file.100 updating facility (Table 7). The fertiliser file (Table 8) was amended using the fert.100 file. The entire fertiliser type and rates used at Writtle English scenario except potassium fertiliser that is not supported by the model were included in this amendment.

In CENTURY 4.0 Agroecosystem Model configuration is done based on several yardsticks. When grains are harvested, the model takes into consideration the removal of grain and notes the live shoots that are transferable to what is called standing dead or surface residue. The genetic maximum and moisture limitations during the months of flowering and filling of grains are established by harvest index. It also includes the limited moisture derived from normal potential transpiration for defined months. The proportion of aboveground N, P and S in a given grain remains crop-specific constants. This is calculated using the square root of the moisture limitation term, hence defining the higher grain nutrient level, particularly when harvest index has been reduced as a result of moisture limitations. It is also assumed

Table 5 Crop parameters values used in updating crop/tree file using file. 100 CENTURY updating utility in line with Writtle English agroecosystem

Crop parameter	Description	Wheat medium harvest index	Potato	Standard maize	Soybean
Prdx (1)	Potential aboveground monthly production for crops – g C m^2	300	150.0	360.0	300
ppdf(1)	Optimum temperature for production for parameterization of a Poisson density function curve to simulate temperature effect on growth	18	17.0	30.0	27
ppdf(2)	Maximum temperature for production for parameterization of a Poisson density function curve to simulate temperature effect on growth	35	5.0	45.0	40
ppdf(3)	Left curve shape for parameterization of a Poisson density function curve to simulate temperature effect on growth	0.7	1.2	1.0	1.0
Ppdf(4)	Right curve shape for parameterization of a Poisson density function curve to simulate temperature on growth	5.0	5.0	2.5	2.5
Bioflg	Flag indicating whether production should be reduced by physical obstruction = 0 production should be reduced = 1 production should be reduced	0	0.0	0.0	0.0
biok5	Level of aboveground standing dead +10% struck (1) at which production is reduced to half maximum due to physical obstruction by dead material (g m^2)	1800	1800.0	1800.0	1800
Pltmrf	Planting month reduction factor to limit seeding growth; set to 1.0 for grass	0.4	0.4	0.5	0.5
Fulcan	Value of *aglivc* at full canopy cover, above which potential production is not reduced	150	150.0	150.0	150
frtc(1)	Initial fraction of C allocated to roots; for Great Plains equation based on precipitation, set to 0	0.6	0.6	0.5	0.5
frtc(2)	Final fraction of C allocated to roots	0.1	0.8	0.1	0.1
frtc(3)	Time after planting (months with soil temperature greater than *rtdtmp*) at which the final value is reached	3.0	2.0	3.0	3
biomax	Biomass level (g biomass m^{-2}) above which the minimum and maximum C/E ratios of new shoot	600	600.0	650.0	800

(continued)

Table 5 (continued)

Crop parameter	Description	Wheat medium harvest index	Potato	Standard maize	Soybean
	increments equal *pramn (*,2)* and *pramx(*,2)*, respectively				
pramn (3,1)	Minimum C/E ratio with zero biomass; (1,1) = N; (2,1) = P and (3,1) = S	100	100.0	190.0	150
pramn (3,2)	Minimum C/E ratio with biomass greater than or equal to *biomax*; (1,2) = N; (2,2) = P and (3,2) = S	200	200.0	150.0	150
pramx (3,1)	Maximum C/E ratio with zero biomass; (1,1) = N; (2,1) = P; (3,1) = S	230	230.0	230.0	230
pramx (3,2)	Maximum C/E ratio with biomass greater than or equal to *biomax*; (1,2) = 1; (2,2) = P and (3,2) = S	270	270.0	230.0	230
prbmn (3,2)	Parameters for computing minimum C-N ratio for belowground matter as a linear function of annual precipitation; (1,1) = N, intercept; (2,1) = P, intercept; (3,1) = S, intercept; (1,2) = N, slope; (2,2) = P, slope; (3,2) = S, slope	0.0	0.0	0.0	0.0
prbmx (3,2)	Parameters for computing maximum C-N ratio for belowground matter as a linear of annual precipitation; (1,1) = N, intercept; (2,1) = P, intercept; (3,1) = S, slope; (1,2) = N, slope; (2,2) = P, slope; (3,2) = S, slope	0.0	0.0	0.0	0.0
fligni (1,1)	Intercept for equation to predict lignin content fraction based on annual rainfall for aboveground material	0.15	0.2	0.1	0.1
fligni (2,1)	Slope for equation to predict lignin content fraction based on annual rainfall for aboveground material. For crops, set to 0	0.0	0.0	0.0	0
fligni (1,2)	Intercept for equation to predict lignin content fraction based on annual rainfall for aboveground material	0.06	0.1	0.1	0.1
fligni (2,2)	Slope for equation to predict lignin content fraction based on annual rainfall for belowground material. For crops, set to 0	0.0	0.0	0.0	0
Himax	Harvest index maximum (fraction of aboveground live C in grain)	0.35	0.4	0.38	0.3
Hiwsf		0.5	0.0	0.0	0.0

(continued)

Table 5 (continued)

Crop parameter	Description	Wheat medium harvest index	Potato	Standard maize	Soybean
	Harvest index water stress factor = 0 no effect of water stress; = 1 no grain yield with maximum water stress				
himon(1)	Number of months prior to harvest to which to begin accumulating water stress effect on harvest index	1	1.0	3.0	2
himon(2)	Number of months prior to harvest in which to stop accumulating water stress effect on harvest index	0	1.0	2.0	1
Efrgrn(3)	Fraction of the aboveground E which goes to grain; (1) = N (2) = P (3) = S	0.6	0.6	0.6	0.6
Vlossp	Fraction of aboveground plant N which is volatilised (occurs only at harvest)	0.04	0.1	0.1	0.1
fsdeth(1)	Maximum shoot death rate at very dry soil conditions (fraction/month); for getting the monthly shoot death rate, this fraction is multiplied times a reduction factor depending on the soil water status	0.0	0.0	0.0	0
fsdeth(2)	Fraction of shoots which die during senescence month; must be greater than or equal to 0.4	0.0	0.0	0.0	0
fsdeth(3)	Additional fraction of shoots which die when aboveground live C is greater than *fsdeth(4)*	0.0	0.0	0.0	0
fsdeth(4)	The level of aboveground C above which shading occurs and shoot senescence increases	200	200.0	500.0	500
Fallrt	Fall rate (fraction of standing dead which falls each month)	0.12	0.1	0.1	0.1
Rdr	Maximum root death rate at very dry soil conditions (fraction/month); for getting the monthly root death rate, this fraction is multiplied times a reduction factor depending on the soil water status	0.05	0.0	0.1	0.1
Rtdtmp	Physiological shutdown temperature for root death and change in shoot/root ratio	2.0	2.0	2.0	2.0
crprtf(3)	Fraction of E translocated from grass/crop at death (1) = N (2) = P (3) = S	0.0	0.0	0.0	0
snfxmx (1)	Symbiotic N-fixation maximum for grass/crop (Gn fixed/Gc new growth)	0.0	0.0	0.0	0

(continued)

Table 5 (continued)

Crop parameter	Description	Wheat medium harvest index	Potato	Standard maize	Soybean
del ^{13}C	Delta ^{13}C value for stable isotope labelling	−27	−27.0	−15.0	−27.0
CO_2ipr (1)	In grass/crop system, the effect on plant production ratio of doubling the atmospheric CO_2 concentration from 350 ppm to 700 ppm	1.2	1.3	1.2	1.2
CO_2itr(1)	In grass/crop system, the effect on transpiration rate of doubling the atmospheric CO_2 concentration from 350 ppm to 700 ppm	0.8	0.8	0.8	0.8
CO_2ice (1,2,3)	In grass/crop system, the effect on C/E ratios of doubling the atmospheric CO_2 concentration from 350 ppm to 700 ppm; (1,1,1) = minimum C-N; (1,2,1) = maximum C-N; (1,1,2) = minimum C-P; (1,2,2) = maximum C-P; (1,1,3) = minimum C-S; (1,2,3) = maximum C-S	0	1.0	1.0	1.0
CO_2irs(1)	In grass/crop system, the effect on root-shoot ratio of doubling the atmospheric CO_2 concentration from 350 ppm to 700 ppm	1.0	1.0	1.0	1.0

Adapted from Farage et al. (2005)

that certain level of the aboveground nitrogen is lost to volatilisation at harvest. The crop harvest index allows for the harvest of roots and hay crops or straw removal after grain harvest. The crop may be killed at harvest as for cereal crops, with a fraction of roots and shoots deemed to be unaffected by harvest operations, and this can stimulate the growth of stubbles.

The model also gives provisions for the death of shoots and roots during the growing season, where shoot and roots are directly related to available water in the entire soil profile and root zone, respectively. These are subject to multiplicative effects of crop-specific maximum death rates. The shoot death rates are increased through shading effects of live biomass, especially at critical levels, while the root death is provided for to cover physiologically active roots, which are subject to soil temperature being greater than 2 Celsius. In the months of senescence, the shoot death rate is set to fixed fraction of the live biomass, withstanding the dead material transferred to the surface litter based on crop-specific fall rate. In the Writtle English scenario, grain harvest index was calculated based on 50% straw removal and 50% standing on the field.

The effects of grazing on plant production were fully described by Holland et al. (1992) and Ojima et al. (1990). These scientists also specified the effects of grazing

Table 6 Other categories of crop and vegetation parameter whose values were used in updating crop/tree file by aid of the file. 100 CENTURY updating utility in line with Writtle English agroecosystem. Details of parameter symbols are in Parton et al. (1993)

Crop/tree parameter	Geescroft Wilderness Rothamsted UK succession (woodland vegetation)	Successional understory (Geescroft)		Grass clover pasture (Geescroft)	Rothamsted grass	Barley (Brly V1).
DECID	1	PRDX (1)	200	350	270	300
PRDX(2)	800	PPDF (1)	27	22	18	17
PRDX(3)	500	PPDF (2)	45	35	35	35
PPDF(1)	25	PPDF (3)	1	0.8	1.2	0.5
PPDF(2)	45	PPDF (4)	3	3.5	3	5.0
PPDF(3)	1	BIOFLG	1	1	1	0
PPDF(4)	3	BIOK5	1800	200	60	1800
CERFOR (1,1,1)	25	PLTMRF	0.2	0.5	1	0.4
CERFOR (1,1,2)	150	FULCAN	100	150	100	150
CERFOR (1,1,3)	300	FRTC (1)	0.2	0.5	0	0.6
CERFOR (1,2,1)	40	FRTC (2)	0.1	0.5	0	0.1
CERFOR (1,2,2)	200	FRTC (3)	1	1	0	3.0
CERFOR (1,2,3)	250	BIOMAX	400	400	400	600
CERFOR (1,3,1)	100	PRAMN (1,1)	13	8.5	10	12.0
CERFOR (1,3,2)	400	PRAMN (2,1)	390	100	390	100
CERFOR (1,3,3)	1100	PRAMN (3,1)	340	125	340	100
CERFOR (1,4,1)	150	PRAMN (1,2)	15	8.5	15	57
CERFOR (1,4,3)	4000	PRAMN (2,2)	390	100	390	160
CERFOR (1,5,1)	150	PRAMN (3,2)	340	125	340	200
CERFOR (1,5,2)	500	PRAMX (1,1)	20	11	20	25
CERFOR (1,5,3)	4000	PRAMX (2,1)	440	133	440	200
CERFOR (2,1,1)	30	PRAMX (3,1)	440	160	440	230
CERFOR (2,1,2)	500	PRAMX (1,2)	25	11	40	125

(continued)

Table 6 (continued)

	Geescroft Wilderness Rothamsted					
Crop/tree parameter	UK succession (woodland vegetation)	Successional understory (Geescroft)		Grass clover pasture (Geescroft)	Rothamsted grass	Barley (Brly V1).
CERFOR (2,1,3)	300	PRAMX (2,2)	440	133	440	260
CERFOR (2,2,1)	60	PRAMX (3,2)	440	160	440	270
CERFOR (2,2,2)	600	PRBMN (1,1)	20	17	30	45
CERFOR (2,2,3)	250	PRBMN (2,1)	390	100	390	390
CERFOR (2,3,1)	200	PRBMN (3,1)	340	125	340	340
CERFOR (2,3,2)	2000	PRBMN (1,2)	0	0	0	0
CERFOR (2,3,3)	1100	PRBMN (2,2)	0	0	0	0.0
CERFOR (2,4,1)	500	PRBMN (3,2)	0	0	0	0.0
CERFOR (2,4,2)	2500	PRBMX (1,1)	30	22	40	60.0
CERFOR (2,4,3)	4000	PRBMX (2,1)	420	133	420	420.0
CERFOR (2,5,1)	500	PRBMX (3,1)	420	160	420	420.0
CERFOR (2,5,2)	2000	PRBMX (1,2)	0	0	0	0.0
CERFOR (2,5,3)	4000	PRBMX (2,2)	0	0	0	0.0
CERFOR (3,1,1)	25	PRBMX (3,2)	0	0	0	0.0
CERFOR (3,1,2)	150	FLIGNI (1,1)	0.05	0.04	0.02	0.15
CERFOR (3,1,3)	300	FLIGNI (2,1)	0	0	0.01	0.0
CERFOR (3,2,1)	46.9	FLIGNI (1,2)	0.06	0.12	0.26	0.06
CERFOR (3,2,2)	250	FLIGNI (2,2)	0.1	0	−0.0015	0.0
CERFOR (3,2,3)	250	HIMAX	0	0	0	0.35
CERFOR (3,3,1)	130	HIWSF	0	0	0	0.25
CERFOR (3,3,2)	1100	HIMON (1)	2	2	2	1

(continued)

Table 6 (continued)

Crop/tree parameter	Geescroft Wilderness Rothamsted UK succession (woodland vegetation)	Successional understory (Geescroft)		Grass clover pasture (Geescroft)	Rothamsted grass	Barley (Brly V1).
CERFOR (3,3,3)	1100	HIMON (2)	1	1	1	1
CERFOR (3,4,1)	557	EFRGRN (1)	0	0	0	0.6
CERFOR (3,4,2)	4000	EFRGRN (2)	0	0	0	0.6
CERFOR (3,4,3)	4000	EFRGRN (3)	0	0	0	0.6
CERFOR (3,5,1)	450	VLOSSP	0.15	0.02	0.15	0.04
CERFOR (3,5,2)	4000	FSDETH (1)	0.2	0.3	0.2	0.0
CERFOR (3,5,3)	4000	FSDETH (2)	0.95	0.4	0.95	0.0
DECW(1)	0.2	FSDETH (3)	0.2	0.1	0.2	0.0
DECW(2)	0.01	FSDETH (4)	150	500	150	200
DECW(3)	0.04	FALLRT	0.18	0.5	0.15	0.12
FCFRAC (1,1)	0.23	RDR	0.05	0.6	0.25	0.05
FCFRAC (2,1)	0.27	RTDTMP	2	2	2	2.0
FCFRAC (3,1)	0.15	CRPRTF (1)	0	0	0	0.0
FCFRAC (4,1)	0.25	CRPRTF (2)	0	0	0	0.0
FCFRAC (5,1)	0.1	CRPRTF (3)	0	0	0	0.0
FCFRAC (1,2)	0.23	SNFXMX (1)	0	0.04	0	0.0
FCFRAC (2,2)	0.27	DEL^{13}C	−18	−27	−24	−27
FCFRAC (3,2)	0.15	CO_2IPR	0.25	0	0	1.20
FCFRAC (4,2)	0.25	CO_2ITR	0.77	0	0	0.80
FCFRAC (5,2)	0.1	CO_2ICE (1,1,1)	1	0	0	1
LEAFDR (1)	0	CO_2ICE (1,1,2)	1	0	0	1

(continued)

Table 6 (continued)

	Geescroft Wilderness Rothamsted					
Crop/tree parameter	UK succession (woodland vegetation)	Successional understory (Geescroft)		Grass clover pasture (Geescroft)	Rothamsted grass	Barley (Brly V1).
LEAFDR (2)	0	CO_2ICE (1,1,3)	1	0	0	1
LEAFDR (3)	0	CO_2ICE (1,2,1)	0.75	0	0	1.3
LEAFDR (4)	0	CO_2ICE (1,2,2)	1	0	0	1
LEAFDR (5)	0	CO_2ICE (1,2,3)	1	0	0	1
LEAFDR (6)	0	CO_2IRS	1	0	0	1.00
LEAFDR (7)	0					
LEAFDR (8)	0					
LEAFDR (9)	0.5					
LEAFDR (10)	0.8					
LEAFDR (11)	1					
LEAFDR (12)	0					
BTOLAI	0.01					
KLAI	1000					
LAITOP	−0.47					
MAXLAI	6					
MAXLDR	0					
FORRTF (1)	0.5					
FORTF (2)	0.5					
FORRTF (3)	0					
SAPK	1500					
SWOLD	6					
WDLIG (1)	0.15					
WDLIG (2)	0.2					

(continued)

Table 6 (continued)

Crop/tree parameter	Geescroft Wilderness Rothamsted UK succession (woodland vegetation)	Successional understory (Geescroft)	Grass clover pasture (Geescroft)	Rothamsted grass	Barley (Brly V1).
WDLIG (3)	0.22				
WDLIG (4)	0.22				
WDLIG (5)	0.22				
WOODDR (1)	0.95				
WOODDR (2)	0.4				
WOODDR (3)	0.07				
WOODDR (4)	0.01				
WOODDR (5)	0.01				
SNFXMX (2)	0				
$DEL^{13}C$	0				
CO_2IPR	0				
CO_2ITR	0				
CO_2ICE (1,1,1)	0				
CO_2ICE (1,1,2)	0				
CO_2ICE (1,1,3)	0				
CO_2ICE (1,2,1)	0				
CO_2ICE (1,2,2)	0				
CO_2ICE (1,2,3)	0				
CO_2IRS	0				
BASFC2	2				
BASFCT	100				
SITPOT	2600				

Adapted from Farage et al. (2005)

Table 7 Details of CENTURY environment files

Name of file	Description
Century.bat	Batch file used to run CENTURY and view
Centurym.exe	The CENTURY executable model
Centurym.tab	Table file generated by the TIME-ZERO™ to handle I/O
Centurym.dat	Master list of all variables used in CENTURY, not to be modified by the user
Temp.sav	File required by VIEW
Centuryx.exe	The stand-alone CENTURY executable model
Fix.100	File with fixed parameters primarily relating to organic matter decomposition and not normally adjusted between runs
<site>.100	Site-specific parameters such as precipitation and soil texture and the initial conditions for soil organic matter; the name of the file is provided by the user
Crop.100	Crop option file
Cult.100	Cultivation option file
Fert.100	Fertilisation option file
Fire.100	Fire option file
Graz.100	Grazing option file
Harv.100	Harvest option file
Irrig.100	Irrigation option file
Omad.100	Organic matter addition option file
Tree.100	Tree option file
Trem.100	Tree removal option file
*Def	For each *.100 file, there is a corresponding ".def" file which contains the definitions of each parameter needed for each option; the format of these ASCII files should not be modified by the user
Sample.wth	Sample weather file
C14data	Sample ^{14}C data file

on the removal of vegetation. This returns mineral nutrients to the soil and changes the root-shoot ratio while increasing the nitrogen content of the live shoots and roots. The model depends on three options for describing the effect of grazing on plant productivity (except the removal of vegetation or the return of plant nutrients by the livestock). In option 2, there is a light grazing effect (Holland et al. 1992), and this includes a constant root to shoot ratio which does not change with grazing and which establishes a linear decrease in potential plant productivity with increased grazing capacity. Option 3 is heavy grazing (Holland et al. 1992), which comprises of complicated grazing optimum where the aboveground plant productivity increases for moderate grazing and decreases for heavy grazing (the assumption is that over 40% are removed each month). Another thing that is made constant is the root to shoot ratio especially for low to moderate grazing. This decreases for heavy grazing.

Table 8 A typical Fert.100 file as amended using CENTURY file.100 updating utility

Symbol	Description
A	Automatic_maintain_production_at_minimum_concentrations
A90	Automatic_maintain_production_at_90%_of_maximum
A80	Automatic_maintain_production_at_80%_of_maximum
A75	Automatic_maintain_production_at_75%_of_maximum
MAX	Automatic_fertilizer_to_achieve_maximum_plant_nutrient_concentration
MED	Automatic_medium_nutrient_concentrations
N5	5 g N m^{-2}
N100	10 g N m^{-2}
N85	8.5 g N m^{-2}
N63	6.3 g N m^{-2}
N10	10 g N m^{-2} 7 g P m^{-2}
N5P25	5 g N_2.5 g P m^{-2}
N3P8	3.3 g N_8 g P m^{-2}
N0P35	0 g N_3.5 g P m^{-2}
N12	12 g N m^{-2}
N45	4.5_g N m^{-2}
N3	3_g N m^{-2}
N1	1_g N m^{-2}
PS1	Superphosphate_125 kg ha^{-1}
PS2	Superphosphate_250_kg ha^{-1}
PS3	Superphosphate_188_kg ha^{-1}
P12	1.2 g P m^{-2}
P35	3.5 g P m^{-2}
PS4	Superphosphate_376 kg ha^{-1}
PS5	Superphosphate_564 kg ha^{-1}

Regarding the three options, it is assumed that nutrient content of new shoot will increase with reference to the residual biomass.

In the Writtle English scenario, option 2 (light grazing) was in practice during the research period, and this was assumed up to 2055, while other options (medium to high grazing) were featured in most sites in the past. This information was included in the events schedule for most of these sites (Tables 11, 12, 13, 14, 15, 16 and 17).

The CENTURY Model uses different cultivation practices based on the well-established levels of shoots, roots, standing dead and surface litter that is standing dead or surface and soil litter pools (whichever is more important). The second scenario is where cultivation practices like ploughing, sweep tillage, thinning activities and herbicide are used. For this option, each cultivation practice considers parameters of multiple soil effects based on the disturbance as a result of cultivation or organic matter decomposition rates especially for structural, active, slow and passive pools. The values for these parameters range from 1.0 to about 1.6 with the actual value being dependent on the degree of soil stirring and disruption caused by each implement. In the Writtle English scenario, only one aspect of cultivation (ploughing or drilling) per time was chosen during events scheduling (Tables 11, 12, 13, 14, 15, 16 and 17) as supported by the model even when more than one cultivation activity took place on the same land per time (Parton et al. 1987).

4.3 CENTURY Inputs Requirement, Weather and Management Information

The weather information required for the configuration of the CENTURY 4.0 Agroecosystem Model includes monthly minimum and maximum temperature. By the aid of EVENT 100 tool, the user selects four options for weather information, namely, an average value for each month of the year in each block of simulation, while the second chooses the mean monthly temperature for each year and then generates precipitation from skewed distribution (Nicks 1974). In the absence of skewness, the user chooses a normal distribution (even though this is likely to increase the mean precipitation, especially when the coefficient of variation for the precipitation is high.) In the third option, the monthly values for precipitation and minimum and maximum temperatures are chosen from the start of the weather file, to enable the fourth option to read from the same file without going back. In the Writtle English scenario, the geographical, weather and soil information used in updating the site.100 file was based on Writtle record for 35 years (Writtle 2004). This is presented in Table 9. Other inputs made use of model default values.

4.4 The CENTURY Environment

With the aid of VIEW output programme and two utilities, we are able to run the CENTURY 4.0 Agroecosystem Model as fully described by Parton et al. (1993). FILE 100 programme works by assisting the user to create and update the 12 files that make up the CENTURY. There is another programme known as EVENT 100 that helps the users to establish schedules for various events surrounding agricultural crops and activities that will take place during the simulation. The 12 data files supply the CENTURY Model the various input values required, with each file containing sub-number of variables, namely, Cult.100 file, which includes the values related to cultivation practices.

Again, within each file, there are several options which define the variables for several variations of the same given event or scenario, e.g. ploughing or rod-weeder. Each given option defines variables to be used in the simulation of that same option, with each data input file bearing the extension 100 to confirm it as a CENTURY file. All these files are updated, with the addition of new options using the FILE 100 programme. With reference to time variables and schedules of events during

Table 9 A typical geographical and soil information used in configuring weather and site.100 files through the CENTURY file.100 updating facility

Period	Parameter	Jan	Feb	Mar	Apr	May	Jun	Jul	Aug	Sep	Oct	Nov	Dec
1971–2000	Maximum air temp. (°C)	7.1	7.4	10.2	12.7	16.6	19.5	22.2	22.3	18.9	14.7	10.1	7.9
2001		6.5	8.1	9.0	12.3	17.8	19.9	23.2	23.1	17.5	17.6	10.7	6.6
2002		8.9	11.0	11.7	14.6	16.6	20.1	21.9	23.0	19.7	14.5	11.8	8.2
2003		7.3	7.4	12.5	14.8	17.8	22.6	23.6	25.6	21.2	13.8	11.9	8.1
Mean		7.5	8.5	10.9	13.6	17.2	20.5	22.7	23.5	19.3	15.2	11.1	7.7
1971–2000	Minimum air temp. (°C)	1.3	0.9	2.5	3.6	6.3	9.3	11.5	11.3	9.4	6.6	3.5	2.2
2001		0.3	1.4	2.9	4.1	7.6	9.2	12.7	13.0	9.4	10.5	3.1	−0.1
2002		2.1	4.1	3.2	4.1	8.0	10.1	11.8	12.7	8.5	6.2	5.5	4.1
2003		1.6	0.3	1.7	3.6	6.7	11.1	12.6	12.7	6.3	4.9	5.3	1.9
Mean		1.3	1.7	2.6	3.9	7.2	9.9	12.2	12.4	8.4	7.1	4.4	2.0
1971–2000	Precipitation (mm)	54.4	35.2	42.5	42.3	45.4	52.8	39.3	49.4	51.3	58.2	54.4	52.8
2001		57.7	103.5	90.4	63.6	42.1	56.6	63.0	53.3	105.7	106.2	47.0	15.5
2002		50.8	67.5	34.5	42.0	67.4	44.9	49.8	38.0	29.9	63.4	125.6	101.8
2003		59.6	22.9	18.2	28.1	36.6	42.7	49.0	5.2	19.7	33.2	95.3	54.3
Mean		55.6	57.3	46.4	44.0	47.9	49.3	50.3	36.5	51.7	65.3	80.6	56.1

SITLAT (Site latitude) – (51° 44'); SITLNG (Site longitude) – (0° 26'); Sand (0.35%), Silt (0.38%), Clay (0.27%), Bulk density (1.10 Kg l^{-1}), pH (6.33)

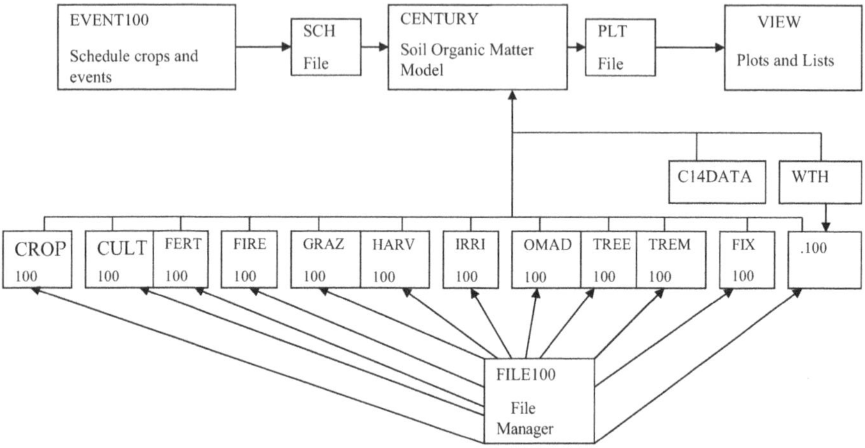

Fig. 5 The CENTURY Model environment showing the relationship between programmes and the file structure

the simulation, the extension file used is the one marked with the extension name "sch". That file is created and updated using the EVENT 100 programme.

To run the CENTURY Model, the user has to instal the programme on the computer, followed by using the FILE 100 programme to update values or create as many options as possible using the various .100 data files, and then the EVENT 100 programme stimulates the time and schedules events which will occur. Then the user runs the model based on the procedure described by Parton et al. (1993). The model environment showing the relationship between programmes and the file structure is shown in Fig. 5, while the details of the century environment are shown in Table 7. The units of major parameters are:

- Time step (1 month) or 1/12 years or 0.083333
- Minimum time (year)
- Soil organic matter (grammes C, N, P or S m^{-2})
- Plant material (grammes C, N, P or S m^{-2})
- Mineral pools (grammes N, P or S m^{-2})
- Temperature (degree centigrade)
- Precipitation (centimetres per month)

The above files and procedure were used in amending necessary files and creating events scenarios in the Writtle English scenario.

4.5 CENTURY Parameterisation and Events Scheduling

There is a procedure for parameterisation and events scheduling (schedule). This is based on the procedure described by Parton et al. (1993). The model has provisions for estimating the steady state of soil C and N levels in the grassland systems as developed for the US Great Plains. The parameterisation of the model simulates SOM dynamics in the top 20 cm of the soil. Hence, any depth or parameterisation of depth above that will be of no effect. The scientists have made provision for filling the teething part of the model parameterisation in terms of calculating C, N, P and S levels for varying soil fractions.

The active soil fraction comprises of the live soil microbes and microbial products, which can only be determined by using the microbial fumigation methods as described by Jenkinson and Rayner (1977). After the estimation of the microbial biomass, its doubling takes care of the microbial products. For Parton et al. (1993), soil-active organic matter is up to two or three times the live microbial biomass, while in most soils, the fraction is composed of two to four percent of soil total C.

Scientists describe the part known as slow SOM fraction built up in the model as part of lignin-derived plant materials which stabilise the microbial products. Hence, these fractions are about 55% of total SOM and based on their recent findings (Elliott and Cambardella 1991) they have discovered that 40% of soil total SOM is lignin-derived plant material. The authors also compared the different sizes of the pool of the C simulated against measured values of SOM, which show that such pool is 1.6 times the level of particulate organic matter (POM) (Metherell et al. 1993).

In the Writtle English scenario, four sites under the following land management, arable land under barley and grassland under permanent pasture, were sown with red clover a year to stocking and grassland under permanent pasture on 5-year ley to stocking, and deciduous woodlands were scheduled according to tillage, cropping, fertilisation, pasture and general management as detailed for seven periods of English agriculture and as derived from the following authors: Chambers and Mingay (1966), Thirsk (1991, 2000), Cantor (1987), Holderness and Turner (1991), Perry (1973), Pretty (2002) and Wormel (1999; Table 10). The sites before simulation were forest, with Geescroft_Wilderness Rothamsted_UK (succession) for woodland vegetation used to initialize the lignin values of the forest system (Farage et al. 2005). The output period was monthly (Jan–Dec) for all periods, while the output interval was 100 years between 6000–5000 BC and 4998 BC to 964 AD. For the periods 965–1514 AD and 1515–1734 AD, the output interval was 20 years each, and for other periods up to 2054 AD, it was a year each.

4.6 *Criteria Used in Selecting or Grouping Periods of Human Civilisation in the English Experience*

Prehistoric agriculture was the time when predominantly all English land was regarded as pristine woodland. Historic agriculture was the period dominated by traditional shepherds and husbandmen whose activities included returning of wheat straw to the land during seedbed preparation and trying of certain innovations like the use of red clover seeds imported from Holland to enrich the soil fertility.

Table 10 Criteria used in selecting or grouping periods of human civilisation in the English experience

Period	Time	Years	Features
Prehistoric agriculture	6000 BC to 1200 AD	7200	The time when predominantly all English land was regarded as pristine woodland
Historic agriculture	1201–1699	498	This period was dominated by traditional shepherds and husbandmen whose activities included returning of wheat straw to the land during seedbed preparation and trying of certain innovations like the use of red clover seeds imported from Holland to enrich soil fertility
Agricultural revolution	1700–1904	204	This time was marked by the use of agricultural equipment like plough, harrow and thresher, chemical fertiliser and soil drainage technology, which improved seeds in intensive and extensive arable and pastoral farming
Post-agricultural revolution	1905–1986	81	This period was marked by the enclosure of open fields, common lands, meadows, and wastes, conversion of land to more profitable uses, example laying of old arable land to permanent pasture or long leys, ploughing and putting under suitable rotation common and rough hills, overgrown with weeds, or that was bare through overgrazing
Green Planet	1987–2025	38	This time is marked by environmental movements (Earth First, Greenpeace, Green Party) and national and international treaties, as well as code of conduct for the protection and preservation of air, water and soil resources
Post Green Planet	2026–2055	29	This is a period where some environmental problems are envisaged, ranging from higher greenhouse gases scenario and climatic change
Research time	2002–2004	2	This is the time when the permanent pasture evaluated had low stocking density and when some practices like the use of legumes to enrich pasture were practised. It is also a period which saw the use of arable land for winter barley, together with inorganic fertiliser and agrochemicals

Agricultural revolution was the time marked by the use of agricultural equipment like plough, harrow and thresher, chemical fertiliser and soil drainage technology, which improved seed in intensive and extensive arable and pastoral farming. Post-agricultural revolution was the period marked by the enclosure of open fields, common lands, meadows, and wastes, conversion of land to more profitable uses, example laying of old arable land to permanent pasture or long leys, ploughing and putting under suitable rotation common and rough hills, overgrown with weeds, or that was bare through overgrazing. Green Planet is the time marked by environmental movements (Earth First, Greenpeace, Green Party) and national and international treaties as well as the code of conduct for the protection and preservation of air, water and soil resources. Post Green Planet is the period where some environmental

Table 11 Events schedule for site 1 (arable land under barley) using the CENTURY 4.0 model

Block	Start Year	End Year	Rept.	Output Year	Output Month	Output Interval	Weather Type	Field Comment
1	−6000	−5000	1	−6000	1	100	M	GWRS forest
2	−4999	−4999	1	−4999	1	1	M	GWRS forest
3	−4998	964	1	−4998	1	100	M	GWRS forest
4	965	1514	1	965	1	20	M	Pasture
5	1515	1734	1	1515	1	20	M	Wheat
6	1735	1954	1	1735	1	1	M	Wheat
7	1955	1964	10	1955	1	1	M	Barley
8	1965	1974	10	1965	1	1	M	Barley
9	1975	1984	10	1975	1	1	M	Wheat
10	1985	1994	10	1985	1	1	M	Wheat
11	1995	2004	10	1995	1	1	M	Barley
12	2005	2054	1	2005	1	1	M	Barley

Rept. (Number of years), GWRS (Geescroft Wilderness Rothamsted Succession for woodland vegetation, M (mean values), Block header information (Arable land under barley)

problems are envisaged, ranging from higher greenhouse gases scenario and climatic change. Research time was the time when the permanent pasture evaluated had low stocking density and when some practices like the use of legumes to enrich pasture were practised. It is also a period which saw the use of arable land for winter barley, together with inorganic fertiliser and agrochemicals. The years of coverage of this grouping is presented in Table 10 (Igboji 2006; Igboji et al. 2015; RCEP 1996; Briggs and Courtney 1985; Davis et al. 1992; Fowler 1983; Dent and Russell 1966; Simmons and Tooley 1981; Chambers and Mingay 1966; Thirsk 1991, 2000; Cantor 1987 Holderness and Turner 1991; Perry 1973; Pretty 2002; Holderness and Turner 1991; Wormel 1999; Grenville and Grant 1988 Writtle 2004).

4.7 A Typical Scheduling of Arable Land Under Barley for English Experience

The arable land under barley was assumed under GWRS (Geescroft Wilderness Rothamsted Succession for woodland vegetation) from 6000 to 964. Cropping and management activities (historic and prehistoric) started from 965 to 1954 and 1955–2054, respectively. Altogether 12 blocks with various types of cropping and soil management were scheduled as below (Table 11). The details are presented in Table 12.

Table 12 Details of scheduled scenario in blocks 4–12 of arable land under barley

Scheduled block	Period	Month	Activities
Block 4	965–1514	April–October	Moderate rough grazing
Block 5–6	1515–1954	1. April	1. Spring wheat with OM (wheat straw) addition during seedbed preparation (1515–1734)
		2. April	2. Spring wheat with 125 kg ha^{-1} superphosphate fertiliser top-dressing (1735–1954)
		3. September	3. Harvesting of grain with 50% straw removal (1515–1954)
Block 7–8	1955–1974	1. April	1. Spring barley with 125 kg ha^{-1} superphosphate fertiliser top-dressing
		2. September	2. Harvesting of grain with 50% straw removal
Block 9–10	1975–1994	1. April	1. Spring wheat with 125 kg ha^{-1} superphosphate fertiliser top-dressing
		2. September	2. Harvesting of grain with 50% straw removal
Blocks 11–12	1995–2054	1. October	1. Winter barley with 125 kg ha^{-1} superphosphate seedbed fertiliser
		2. April	2. Top-dressing with 3 g N m^{-2} fertiliser
		3. September	3. Harvesting of grain with 50% straw removal

4.8 *Scheduling of Grassland Under Permanent Pasture Sown with Red Clover a Year Before Stocking for English Experience*

This site was assumed under GWRS (Geescroft_Wilderness_Rothamsted_Succession for woodland vegetation) from 6000–964. Cropping and management activities (historic and prehistoric) started from 965 to 1954 and 1955–2054, respectively. Altogether 12 blocks with various types of cropping and soil management were scheduled as below (Table 13). The details are presented in Table 14.

4.9 *Scheduling of Grassland Under Permanent Pasture on 5 Year Ley Before Stocking for English Experience*

This site was assumed under GWRS (Geescroft_Wilderness_Rothamsted_Succession for woodland vegetation) from 6000–964. Cropping and management activities (historic and prehistoric) started from 965 to 1954 and 1955–2054, respectively. Altogether 12 blocks with various types of cropping and soil management were scheduled as below (Table 15). The details are presented in Table 16.

Table 13 Events schedule for grassland under permanent pasture that was sown with red clover a year before stocking using the CENTURY 4.0 model

Block header information (grassland under permanent pasture sown with red clover)								
Block	Start year	End year	Rept.	Output year	Output month	Output interval	Weather type	Field comment
1	−6000	−5000	1	−6000	1	100	M	GWRS forest
2	−4999	−4999	1	−4999	1	1	M	GWRS forest
3	−4998	964	1	−4998	1	20	M	GWRS forest
4	965	1514	1	965	1	20	M	Pasture
5	1515	1734	1	1515	1	20	M	Wheat
6	1735	1954	1	1735	1	20	M	Wheat
7	1955	1964	10	1955	1	1	M	Pasture
8	1965	1974	10	1965	1	1	M	Ley/pasture
9	1975	1984	10	1975	1	1	M	Ley/pasture
10	1985	1994	10	1985	1	1	M	Ley/pasture
11	1995	2004	10	1995	1	1	M	Clover/pasture
12	2005	2054	1	2005	1	1	M	Clover/pasture

Rept. (Number of years), GWRS (Geescroft_Wilderness_Rothamsted_Succession for woodland vegetation, M (mean values)

4.10 Scheduling of Deciduous Woodland for English Experience

This site was assumed under GWRS (Geescroft_Wilderness_Rothamsted_Succession for woodland vegetation) from 8000 BC to 2054 AD. Only three blocks were scheduled (Table 17).

5 Measured Versus Simulated Soil Total Carbon

A comparison was made between measured and simulated soil total C for the various land management. The simulated results were in agreement with the field measurements similar to the results found in the deciduous woodland, while the differences in the other sites were based on land management (Fig. 6a). Further relationship was established between measured and simulated soil total carbon (MSTC = 20.2 + 0.6SSTC g kg^{-1}, $R^2 = 0.68$ and MSTC = 18.9 + 0.6SSTC g kg^{-1}, $R^2 = 0.72$)- Fig. 5d (Igboji et al. 2015; Igboji 2006).

Table 14 Details of scheduled scenario in blocks 4–12 of grassland under permanent pasture sown with red clover a year before stocking

Scheduled block	Period	Month	Activities
Block 4	965–1514	April–October	Moderate rough grazing
Blocks 5–6	1515–1954	1. April	1. Spring wheat with OM (wheat straw) addition during seedbed preparation (1515–1734)
		2. April	2. Spring wheat with 125 kg ha^{-1} superphosphate fertiliser top-dressing (1735–1954)
		3. September	3. Harvesting of grain with 50% straw removal
Block 7	1955–1964	1. September (1st year)	1. Seeding of rye grass seeds
		2. April (2nd year)	2. Top-dressing with 1.2 g m^{-2} of phosphorus
		3. June (2nd year)	3. Harvesting of hay
		4. April (3rd year)	4. Top-dressing of grass established last y with 1.2 g m^{-2} of phosphorus
		5. March–October (4th to 5th year)	5. Low grazing of grass
		6. March–October (6th year)	6. Rejuvenation of grass with rye seeds
		7. March–October (7th–8th year)	7. Low grazing of grass
Block 8	1965–1974	1. March–October (1st–4th year)	1. Rejuvenation of grass with rye grass seeds, with fertiliser top-dressing at the rate of 0 g N_3.5 g P m^{-2}; 3.5 g P m^{-2}; 1.2 g P m^{-2}; and 3.3 g N_8 g P m^{-2} in years 5, 6, 7 and 9, as well as 3.5 g P m^{-2}; 0 g N_3.5 g P m^{-2}; 1.2 g P m^{-2} in years 6, 7 and 8, respectively
		2. March–October (5th–10 year)	2. Low grazing of grass
Block 9–10	1975–1994	1. September (1st year)	1. Seeding of rye grass seeds
		2. April (2nd year)	2. Top-dressing with 1.2 g P m^{-2} fertiliser
		3. March–October (3rd–8th year)	3. Low grazing of grass
		4. September (9th year)	4. Rejuvenation of grass with red clover
		5. April (10th year)	5. Top-dressing with 1.2 g P m^{-2} fertiliser
		6. June (10th year)	6. Harvesting of hay

(continued)

Table 14 (continued)

Scheduled block	Period	Month	Activities
Block 11	1995–2004	1. March–October (1s –2nd year)	1. Low grazing of grass
		2. March–October (3rd–8th year)	2. Rejuvenation of rye grass with red clover
		3. March–October (9th–10th year)	3. Low grazing of grass
Block 12	2005–2054	March–October (1st–10th year)	Low grazed grassland under permanent pasture was assumed

Table 15 Events schedule for grassland under permanent pasture that was on 5-year ley before stocking using the CENTURY 4.0 model

Block	Start year	End year	Rept.	Output year	Output month	Output interval	Weather type	Field comment
1	−6000	−5000	1	−6000	1	100	M	GWRS forest
2	−4999	−4999	1	−4999	1	1	M	GWRS forest
3	−4998	964	1	−4998	1	20	M	GWRS forest
4	965	1514	1	965	1	20	M	Pasture
5	1515	1734	1	1515	1	20	M	Wheat
6	1735	1954	1	1735	1	20	M	Wheat
7	1955	1964	10	1955	1	1	M	Pasture
8	1965	1974	10	1965	1	1	M	Ley/pasture
9	1975	1984	10	1975	1	1	M	Ley/pasture
10	1985	1994	10	1985	1	1	M	Ley/pasture
11	1995	2004	10	1995	1	1	M	Ley/pasture
12	2005	2054	1	2005	1	1	M	Ley/pasture

Rept. (Number of years), GWRS (Geescroft Wilderness Rothamsted Succession for woodland vegetation), M (mean values), Block header information (Grassland under permanent pasture, 5-year ley)

Table 16 Details of scheduled scenario in blocks 4–12 of grassland under permanent pasture that was on 5-year ley before stocking

Scheduled block	Period	Month	Activities
Block 4	965–1514	April to October	Moderate rough grazing
Blocks 5–6	1515–1954	1. April	1. Spring wheat, with OM (wheat straw) addition during seedbed preparation (1515–1734)
		2. May	2. Spring wheat, with 125 kg ha^{-1} superphosphate fertiliser top-dressing (1735–1954)
		3. September	3. Harvesting of grain with 50% straw removal
Block 7	1955 to 1964	1.September (1st year)	1. Seeding of rye grass seeds
		2. April (2nd year)	2. Top-dressing with 1.2 g m^{-2} of phosphorus
		3. June (2nd year)	3. Harvesting of hay
		4. April (3rd year)	4. Top-dressing of grass established last y with 1.2 g m^{-2} of phosphorus
		5. March–October (4th to 5th year)	5. Low grazing of grass
		6. March–October (6th year)	6. Rejuvenation of grass with rye seeds
		7. March–October (7th–8th year)	7. Low grazing of grass
Block 8	1965–1974	1. March–October (1st–4th year)	1. Rejuvenation of grass with rye grass seeds, with fertiliser top-dressing at the rate of 0 g N_3.5 g P m^{-2}; 3.5 g P m^{-2}; 1.2 g P m^{-2}; and 3.3 g N_8 g P m^{-2} in years 5, 6, 7 and 9, as well as 3.5 g P m^{-2}; 0 g N_3.5 g P m^{-2}; and 1.2 g P m^{-2} in years 6, 7 and 8, respectively
		2. March–October (5th–10 year)	2. Low grazing of grass
Block 9	1975–1984	1. September (1st year)	1. Seeding of rye grass seeds
		2. April (2nd year)	2. Top-dressing with 1.2 g P m^{-2} fertiliser
		3. March–October (3rd–8th year)	3. Low grazing of grass
		4. September (9th year)	4. Rejuvenation of grass with rye grass seeds
		5. April (10th year)	5. Top-dressing with 1.2 g P m^{-2} fertiliser
		6. June (10th year)	6. Harvesting of hay

(continued)

Table 16 (continued)

Scheduled block	Period	Month	Activities
Block 11	1995–2004	1. March–October (1st–2nd year)	1. Low grazing of grass
		2. March–October (3rd–8th year)	2. Rejuvenation of grass with various types and levels of seedbed and top-dressing fertilisers, namely, 5 g N_2.5 g P m^{-2}; 0 g N_3.5 g P m^{-2}; and 3.3 g N_8 g P m^{-2} seedbed fertilisers in years 4, 5 and 7; 3.5 g P m^{-2}; 1.2 g P m^{-2}; 6.3 g N_2.3 g P m^{-2}; and 5 g N_2.5 g P m^{-2} top-dressing fertilisers in years 4–7
		3. March–October (9th–10th year)	3. Low grazing of grass
Block 12	2005–2054	March–October (1st–10th year)	Low grazed grassland under permanent pasture was assumed

Table 17 Events schedule for deciduous woodland using the CENTURY 4.0 model

Block	Start year	End year	Rept.	Output year	Output month	Output interval	Weather type	Field comment
1	−6000	−5000	1	−6000	1	100	M	GWRS forest
2	−4999	−4999	1	−4999	1	1	M	GWRS forest
3	−4998	2054	1	−4998	1	1	M	GWRS forest

Rept. (Number of years), GWRS (Geescroft Wilderness Rothamsted Succession for woodland vegetation, M (mean values), Block header information (deciduous woodland)

5.1 Simulated Soil Total Carbon from Historic Agriculture to 2055 AD

5.1.1 Arable Land Under Barley

This site was under barley during the model period, but previously it was under different cropping and nutrient management system for the past 40 years and was assumed to be in model time management up to 2055 as indicated in the schedule (Tables 11 and 12). The changes in simulated soil total carbon from 1942 to 2055 on 39–40-year basis are shown in (Fig. 6b). The decline in soil total carbon was rapid between 1942 and 1948 (from 21.5 t ha^{-1} to 9.5 t ha^{-1}) arising from conventional tillage practices and crop rotations (where some crops depleted soil nutrients faster than others) during the post-agricultural revolution; after which it remained more stable up to the end of 1981 (a period that witnessed conversion of some arable land to pasture or long leys). The rate of turnover in simulated soil total carbon between 1982 and 2021 was not >10 t ha^{-1} and between 2022 and 2055 remained at nearly 7 t ha^{-1} (Igboji et al. 2015; Igboji 2006).

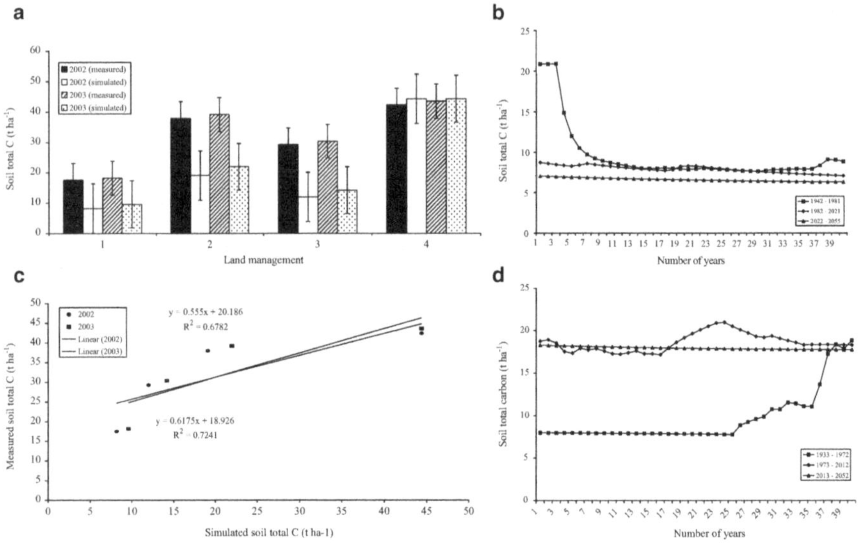

Fig. 6 (**a**) Measured versus simulated SOC; (**b**) arable land under barley; (**c**) regression of measured versus simulated SOC; (**d**) grassland under permanent pasture sown with red clover a year before stocking: 1, arable land under barley; 2, grassland under permanent pasture sown with red clover before grazing; 3, grassland under permanent pasture on 5-year ley before grazing; 4, deciduous woodland

5.1.2 Grassland Under Permanent Pasture Sown with Red Clover a Year Before Stocking

This site was sown with red clover as a part of rye grass rejuvenation before stocking during the model period, but previously it was under various cropping/nutrient and pasture management for the past 40 years and was assumed to be in model time management up to 2055 as indicated in the schedule (Tables 13 and 14). Figure 6d shows changes in simulated soil total carbon in this site from 1933 to 2052 on 39–40 years basis. The rate in decline in soil carbon was nearly uniform between 1933 and 1958 (< 10 t ha^{-1}) based on the conventional tillage and fertilisation practices at the agricultural revolution period. However, the amount of soil total carbon recorded between 1973 and 2012 was >15 t ha^{-1} due to long leys, and the greatest stability in soil total carbon at higher levels compared to other periods was between 2013 and 2052 based on the red clover introduced as a part of rye grass rejuvenation coupled with low grazing at the Green and Post Green Planet time (Igboji et al. 2015; Igboji 2006).

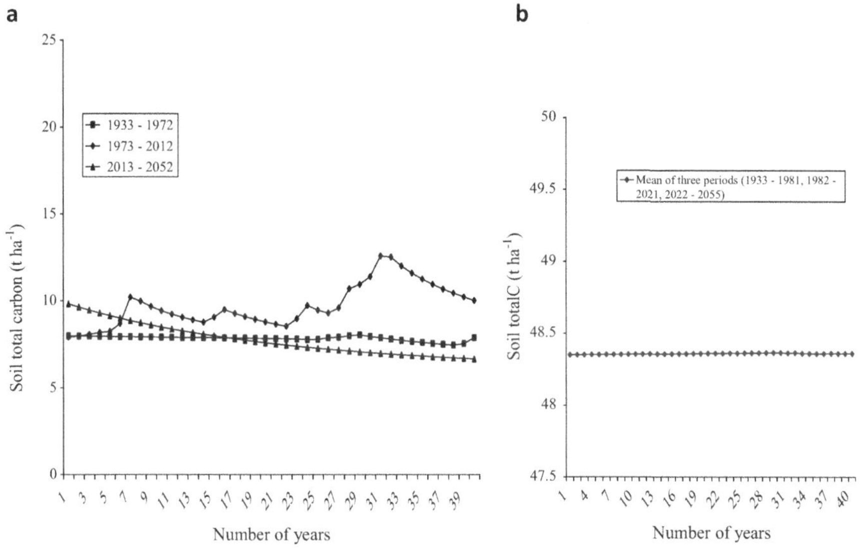

Fig. 7 (**a**) Grassland under permanent pasture on 5 year ley before grazing; (**b**) Deciduous woodland; 1 = Arable land under barley; 2 = Grassland under permanent pasture sown with red clover before grazing; 3 = Grassland under permanent pasture on 5 year ley before grazing; 4 = Deciduous woodland

5.1.3 Grassland Under Permanent Pasture on 5-Year Ley Before Stocking

This site was on 5-year ley in the course of rye grass rejuvenation before stocking during the model period but previously under various cropping/nutrient and pasture management for the past 40 years and was assumed to be in model time management up to 2055 as indicated in the schedule (Tables 15 and 16). The changes in the simulated soil total carbon in this site from 1933 to 2052 were analysed on 39–40-year basis (Fig. 7a). The trend reveals higher carbon level between 1973 and 2012 (5–7.5 t ha^{-1}) based on low grazing and intermittent ley in rye grass rejuvenation (Igboji et al. 2015; Igboji 2006).

5.1.4 Deciduous Woodland

This site was deciduous woodland during the model period. It had been the same for the past 40 years and assumed to remain so up to 2055 (Table 17). The changes in the simulated soil total C on 39–40-year basis are shown in Fig. 7b. This site did not vary to any significant proportion in simulated soil total carbon from 1933 to 2055 (from 48.2 t ha^{-1} to 48.4 t ha^{-1}) based on no-tillage or nutrient management (Igboji et al. 2015; Igboji 2006).

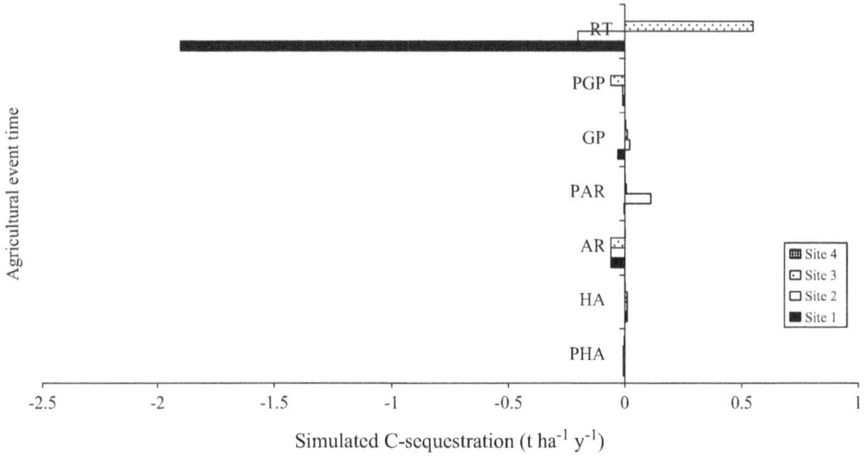

Fig. 8 Simulated C-sequestration (t ha^{-1} $year^{-1}$) of a soil under four types of management: arable land under barley (site 1), grassland under permanent pasture sown with red clover a year before stocking (site 2), grassland under permanent pasture on 5 year ley before stocking (site 3), deciduous woodland (site 4). Prehistoric agriculture (7200 years), historic agriculture (498 years), agricultural revolution (204 years), post-agricultural revolution (81 years), Green Planet (38 years), Post Green Planet (29 years), research time (2 years)

5.2 *The Effects of Time and Land Management on C-Sequestration*

The simulated soil total carbon in all the sites at the inception of prehistoric agriculture (6000 BC) was 61.5 t ha^{-1}. By the end of this period (1200 AD), it stood at 15 t ha^{-1} in the three sites (arable land under barley, grassland under permanent pasture sown with red clover a year before stocking and grassland under permanent pasture on 5-year ley before stocking). The change in soil C within this period (7200 years) across the three sites was −46.5 t ha^{-1} with no C sequestered. It is only in the site 4 (deciduous woodland) that soil total C ended with 47.1 t ha^{-1} by 1200 AD, which is a change of −14.4 t ha^{-1} with no C sequestered (Fig. 8). These are effects of moderate rough grazing practised in the three out of four sites at this period (Tables 11, 12, 13, 14, 15 and 16), while SOM decomposition might have influenced the deciduous woodland results (Igboji et al. 2015; Igboji 2006).

During the historic agriculture (1201–1699), simulated soil total C started with 15 t ha^{-1} in three sites apart from the deciduous woodland, increasing to 20.9 t ha^{-1} in the site 1 (arable land under barley), 20.5 t ha^{-1} in site 2 (clover pasture) and 20.6 t ha^{-1} in the site 3 (ley pasture). These correspond to changes of 5.9, 5.5 and 5.6 t ha^{-1}, respectively, for these 498 years and C-sequestration of 0.01 t ha^{-1} $year^{-1}$ in the three sites. In the site 4 (deciduous woodland), simulated soil total C started with 47 t ha^{-1} in 1201 AD ending with 47.3 t ha^{-1} by 1699 AD which is a change of 0.3 t ha^{-1} and C-sequestration of 0.001 t ha^{-1} $year^{-1}$ (Fig. 8). This is the aftermath of soil organic matter restoration (like the inclusion of wheat straw during seedbed preparation (Igboji et al. 2015; Igboji 2006).

During the agricultural revolution (1700–1904), simulated soil total C started with 20.9, 20.5, 20.6 and 47.3 t ha^{-1} in site 1 (arable land under barley), site 2 (clover pasture), site 3 (ley pasture) and site 4 (deciduous woodland), respectively. By the end of this period (1904), simulated soil total C fell to 8.5 t ha^{-1} in site 1, 8.3 t ha^{-1} in site 2 and 8.2 t ha^{-1} in site 3 and increased to 47.5 t ha^{-1} in site 4. This represents soil total C of −12.4, −12.2, −12.4 and 0.1 t ha^{-1} and no C-sequestration in three sites but 0.001 t ha^{-1} $year^{-1}$ in site 4 for these 204 years, respectively (Fig. 8). All the sites, except the deciduous woodland (site 4), were under different cropping and nutrient management as scheduled (Tables 11, 12, 13, 14, 15 and 16). These cropping and nutrient management coupled with tillage practices depleted soil C in three sites except the deciduous woodland (Igboji et al. 2015; Igboji 2006).

During the post-agricultural revolution (1905–1986), simulated soil total C started with 8.5, 8.3, 8.2 and 47.4 t ha^{-1} in sites 1–4, respectively, decreasing to 8.2 t ha^{-1} in site 1 (arable land under barley) but increasing to 17.3, 8.7 and 47.5 t ha^{-1} in site 2 (clover pasture), site 3 (ley pasture) and site 4 (deciduous woodland), respectively. These are changes of −0.3, 9, 0.5 and 0.1 t ha^{-1} and C-sequestration of 0.11, 0.01 and 0.001 t ha^{-1} $year^{-1}$ in sites 2, 3 and 4 for these 81 years, respectively (Fig. 8). The introduction of red clover in site 2 and ley in site 3 as part of rye grass rejuvenation (Tables 11, 12, 13, 14, 15 and 16) led to increments in soil C. Soil total C was further depleted in site 1 (arable land under barley) due to constant pulverisation and crop uptake, while it increased a little in deciduous woodland (Igboji et al. 2015; Igboji 2006).

Within the Green Planet (1987–2025), simulated soil total C started with 8.2, 17.3, 8.7 and 47.4 t ha^{-1} in the sites 1–4, started decreasing to 6.9 and 8.2 t ha^{-1} in the sites 1 and 3 but kept increasing to 18 and 47.5 t ha^{-1} in sites 2 and 4, respectively. The changes in the simulated soil total C over these 38 years were − 1.3, 0.7, −0.5 and 0.1 t ha^{-1} in sites 1–4, respectively, while the C-sequestration in sites 2 and 4 were 0.02 and 0.003 t ha^{-1} $year^{-1}$, respectively. The order of cropping, nutrient management and tillage practices as detailed in Tables 11, 12, 13, 14, 15 and 16 influenced these results. Nevertheless, apart from the arable and ley plots where there were depletion of soil total C arising from continuous cropping and grazing, the clover pasture and woodland retained soil C (Igboji et al. 2015; Igboji 2006).

During the Post Green Planet (2026–2055), simulated soil total C started with 6.9, 18, 8.2 and 47.5 t ha^{-1} in sites 1–4, respectively and kept decreasing to 6.5, 17.7, 6.6 t ha^{-1} in sites 1–3, respectively but retained at 47.5 t ha^{-1} in site 4 (deciduous woodland). For these 29 years the soil C turnovers were −0.4, −0.3, −1.6 and 0 t ha^{-1} and no C-sequestration. This will be the aftermath of retaining current land management practices without returning the arable and pasture lands to occasional leys or without the use of legumes like red clover in rye grass rejuvenation (Tables 11, 12, 13, 14, 15 and 16) (Igboji et al. 2015, Igboji 2006).

At the research time (2002–2004), simulated soil total C started with 8.3, 19.4, 11.4 and 47.4 t ha^{-1} in sites 1–4, respectively, and decreased to 4.6 and 18.8 t ha^{-1} in sites 1 and 2 but increased to 12.5 t ha^{-1} in site 3 and remained at 47.4 t ha^{-1} in site 4 (deciduous woodland). The soil C turnover over these 2 years was − 3.7, −0.6, 1.1 and 0 t ha^{-1} and C-sequestration of 0.55 t ha^{-1} $year^{-1}$ in site 3 only (Fig. 8). The long ley (Tables 15 and 16) enhanced C turnover in the grassland under permanent pasture on 5-year ley before stocking, just as the deciduous woodland (Table 17) remained at steady state, while arable land under barley (Tables 11 and 12) depleted soil C. The loss in soil C over this period in the red clover plot (Tables 13 and 14) regained over the Green Planet time but lost again just as in the long ley as from 2026 to 2055 suggesting when the review of the current management options will be necessary (Igboji et al. 2015; Igboji 2006).

5.2.1 The Effects of Land Management on Measured and Simulated Soil Total C

The data fit of measured to simulated soil total C (Fig. 6a) and that of measured to simulated soil microbial respiration (Fig. 6c) to no significant difference especially in the deciduous woodland shows the reliability of this prediction. Similarly, the measured and simulated soil total C and soil microbial respiration in the remaining sites are in line with the changes in management. For example, the incorporation of organic matter (wheat straw) during seedbed preparation before the agricultural revolution increased simulated soil total C (as in arable land under barley) (Fig. 6b) and simulated soil microbial respiration (as in grassland under permanent pasture sown with red clover a year before stocking) during post-agricultural revolution (Fig. 6a), while the monocrop of barley, straw removal during harvest and constant tillage as seen during the research time in the arable piece of land depleted the soil organic carbon. In this same plot, constant pulverisation, together with the incorporation of straw left behind during harvesting, raised the soil total porosity and enhanced both the measured and simulated soil microbial respiration. It was in the deciduous woodland that the greatest stability in measured and simulated soil total C and microbial respiration was achieved due to lack of management (Fig. 6b) (Igboji et al. 2015; Igboji 2006).

In Foereid et al. (2004), simulated soil respiration were in fairly good agreement with these observed values (18–70 μg CO_2 g^{-1} soil day^{-1} that is 2.6×10^{-10} μmol m^{-2} s^{-1} or 9.2×10^{-8} μmol m^{-2} h^{-1}) using CENTURY 4.0 model, in grass soil with and without roots and *Miscanthus* soil of 11 and 18 years (all with and without added nitrogen). In their simulation only the treated soil overestimated respiration. According to these workers, the model accounted sufficiently for the changes in CO_2 production.

Therefore, one of the principal factors that affect simulated soil total C and microbial respiration is land management. Others include model parameterization and configuration. For example, the current work was based on the available data (such as minimum and maximum monthly air temperature, monthly precipitation, soil texture, bulk density). Other pieces of information made use of default values. Secondly, the simulation made use of model fitted moisture function which can vary over ecosystems. For example, Pumpanen et al. (2003) investigated the importance of soil moisture on predicted CO_2 efflux by running a model with and without a moisture function. When the model was run without the moisture factor, the predicted CO_2 effluxes were significantly overestimated. The addition of the moisture function improved the accuracy of the model prediction. However, when applied with the moisture function, the model tended to slightly underestimate high CO_2 effluxes and overestimated low effluxes. The underestimation of high effluxes was mainly caused by the parameterization. In their model respiration was restricted to the O-horizon which was already at a volumetric water content of 0.35 m^3 m^3. Similarly, the effects of fine root respiration, moisture and temperature on the soil respiration have been studied. For example, Widén and Majdi (2001) found soil water content as low as 0.01 m^3 m^3 not to affect fine root respiration, while Pumpanen et al. (2003) attributed some of the differences between measured and predicted CO_2 effluxes during autumn and spring to seasonal variation in the proportion of root respiration as well as temperature response. In Igboji et al. (2015) and Igboji (2006), the assumption is that soil microbial respiration is a combination of root and soil respiration and that soil temperature and moisture influence measured and simulated results.

In Durán et al. (2017), studies of climate of the northern hardwood forests of North America projected it will become significantly warmer in the coming decades. (It was projected that climate will become significantly warmer in the coming decades.) The reasons included associated increases in the soil temperature, decreases in water availability and changes in the winter snow pack and the soil frost. These are likely to affect soil carbon (C) and nitrogen (N) cycling. According to these scientists, most studies of the effects of climate change on soil function have focused on the upper-organic part of the soil profile (e.g. forest floor), and little is known about the effects on deeper mineral soil horizons. They exploited an elevation/orientation gradient at the Hubbard Brook Experimental Forest (New Hampshire, USA) to evaluate how variation in climate, similar to that projected to occur over the next 50–100 years, affects soil C and N pools and transformation rates in the different soil horizons of northern hardwood forests. Lower elevation, south-facing plots with higher soil temperature, less soil moisture and snow and increased frequency of soil freeze/thaw events had less soil inorganic N content and lower potential net N mineralization rates compared to higher elevation, north-facing plots. These differences in N pools and fluxes were consistent for all soil horizons, but sensitivity to climate variation increased with soil depth, confirming that assessments of climate change effects that do not consider variation throughout the soil profile are likely to be incomplete and potentially inaccurate. The researchers noted that nitrogen cycling processes were more sensitive to climate variation than C cycling processes, suggesting a decoupling of C and N cycles in coming decades, with important implications for ecosystem function. Since soil processes showed greater sensitivity to climate variation in summer than in spring, and in the warmer and less snowy year of sampling, the effects of climate change might become more pronounced as temperatures increase and snow fall and water availability decrease in the coming decades (Durán et al. 2017).

Similarly, understanding the degree to which N availability limits C uptake under global environmental change has been reported as an unresolved challenge (Stocker et al. 2016). First, according to these authors, generation C only vegetation models, lacking explicit representation of N cycling, projected a substantial and increasing land C sink under rising atmospheric CO_2 concentrations. This prediction is currently being questioned for not taking into account the potentially limiting effect of N availability, which is necessary for the plant growth (Hungate et al. 2003). More recent global models include coupled C and N cycles in land ecosystems (C-N models) and are widely assumed to be more realistic (Stocker et al. 2016). However, the inclusion of more processes has not consistently improved their performance in capturing the observed responses of the global C cycle (Wenzel et al. 2014). With the advent of a new generation of global models, including coupled C, N and P cycling, model complexity is sure to increase; but reliability may not, unless greater attention is paid to the correspondence of model process representations and empirical evidence (Stocker et al. 2016). It was in this context that the Nitrogen Cycle Workshop at Dartington Hall, Devon, UK, was held on 1–5 February, 2016 (Stocker et al. 2016). Organised by I. Colin Prentice and Benjamin D Stocker (Imperial College, London, UK), the workshop was funded by the European Research Council, Project Earth Model Bias Reduction and Assessing Abrupt Climate Change (EMBRACE). These scientists gathered ecologists and ecosystem modellers to identify key uncertainties in terrestrial C-N cycling and to discuss processes that are missing or poorly represented in current models. They recommended that future experiments should be specifically designed to identify constraints set by resource availability and trade-offs caused by intrinsic physiological limitations.

6 Conclusion

If the English experience is replicated in all the regions of the world, based on recommendations of Stocker et al. (2016), there will be enough database to anchor on past, present and future C and N cycling. In tropical Nigeria, Igboji is attempting collating data and information of 8500–10,000 years to replicate the English experience and the recommendations of Stocker et al. (2016), but the limitation of the author has been a lack of data and record keeping as available in civilised nations of the world. But the current book *Sustainable C and N Cycling* is an awakening call for collective action all over the world including Nigeria.

Many scientists have contributed immensely to the study of C and N cycling as mentioned in this chapter. These great men and women have shaped the face of world agriculture and the environment. They are continuously modifying and developing new skills and methods (including potable instrumentation) that make researches and sciences much easier. Most of these technologies trickle down to developing world when the methods and ideas are nearly obsolete in the developed world. But, it still pays to use any available methods and techniques (including those ones chronicled in this chapter) to address continuously world problems (including the controversial issues of C and N cycling) and associated greenhouse gas emissions, global warming and climate change. Civilisation has come with problems that demand constant attention and solution. Hence, the periods of human civilisation (so named periods of English agriculture) call for more investigations and debate and replication all over the world (according to their various periods of civilisation), in view of the fact that civilisation to political, social, economic, religious and cultural scientists are too broad to be covered in environmental and agricultural contexts.

References

Aiken GR, McKnight DM, Wershaw RL, MacCarthy P (1985) In: Aiken GR, McKnight DM, Wershaw RL, MacCarthy P (eds) Humic substances in soil, sediment, and water: geochemistry, isolation, and characterization. John Wiley, New York

Alef K (1990) Bestimmung mikrobieller Aktivität und Biomasse in Boden und Kompost. Umweltwiss Schadst Forsch 2(2):76–78

Alef K (1995) Soil respiration. In: Alef K, Nannipieri P (eds) Methods in applied soil microbiology and biochemistry, 2nd edn. Academic, London, pp 214–219

Alef K, Kleiner D (1987) Applicability of arginine ammonification as indicator of microbial activity in different soils. Biol Fertil Soils 5(2):148–151

Alef K, Nannipieri P (1995) Cellulase activity. In: Alef K, Nannipieri P (eds) Methods in applied soil microbiology and biochemistry. Academic, London, p 345

Alef K, Nannipieri P (1998) Methods in applied soil microbiology and biochemistry, 2nd edn. Academic, London

Alef K, Beck T, Zelles L, Kleiner D (1988) A comparison of methods to estimate microbial biomass and N-mineralization in agricultural and grassland soils. Soil Biol Biochem 20(4):561–565

Alef K, Nannipieri P, Trazar-Cepeda C (1995) Phosphatase activity. In: Alef K, Nannipieri P (eds) Methods in applied soil microbiology and biochemistry. Academic, London, pp 335–336

Alexander M (1977) Soil microbiology. Wiley, New York

Anderson JPE (1982) Soil respiration. In: Page AL, Miller RH, Keeney DR (eds) Methods of soil analysis, part 2, chemical and microbiological properties, 2nd edn. American Society of Agronomy, Madison

Anderson J, Domsch K (1978a) Mineralization of bacteria and fungi in chloroform-fumigated soils. Soil Biol Biochem 10(3):207–213

Anderson J, Domsch K (1978b) A physiological method for the quantitative measurement of microbial biomass in soils. Soil Biol Biochem 10(3):215–221

Anderson JR, Slinger JM (1975) Europium chelate and fluorescent brightener staining of soil propagules and their photomicrographic counting—I. Methods Soil Biol Biochem 7 (3):205–209

Anderson HA, Stewart M, Miller JD, Hepburn A (1984) Organic nitrogen in soils and associated surface waters. In: Wilson WS (ed) Advances in soil organic matter research: the impact on agriculture and environment. The Royal Society of Chemistry, Cambridge

Appelqvist IAM (1990) The binding of Cu [Superior] 2 [plus] and AI [superior] 3 [plus] ions by a high molecular weight humic acid fraction using continuous flow stirred cell-flow injection analysis. University of Birmingham, Birmingham

Babiuk L, Paul EA (1970) The use of fluorescein isothiocyanate in the determination of the bacterial biomass of grassland soil. Can J Microbiol 16(2):57–62

Ball A, Milne E, Drake B (2000) Elevated atmospheric-carbon dioxide concentration increases soil respiration in a mid-successional lowland forest. Soil Biol Biochem 32(5):721–723

Batterman SA, Hedin LO, Van Breugel M, Ransijn J, Craven DJ, Hall JS (2013) Key role of symbiotic dinitrogen fixation in tropical forest secondary succession. Nature 502(7470):224
Beck T (1973) Uber die Eignung von Modellversuchen bei der Messung der biologischen Aktivitat von Boden. Bayer Landw Jb 50:270–288
Benefield C (1971) A rapid method for measuring cellulase activity in soils. Soil Biol Biochem 3 (4):325–329
Benjamin MM, Honeyman BD (1992) Trace metals. In: Butcher SS, Charlson RJ, Orans G, Wolfe GV (eds) Global biogeochemical cycles. Academic Press, New York, pp 318–352
Birch J, Melville M (1969) An electrolytic respirometer for measuring oxygen uptake in soils. Eur J Soil Sci 20(1):101–110
Bremner JM, Mulvaney RL (1978) Urease activity in soils. In: Burns RG (ed) Soil enzymes. Academic, New York, pp 149–196
Briggs DJ, Courtney FM (1985) Agriculture and environment: the physical geography of temperate agriculture systems. Longman Group Ltd., London
Brookes PD, Paul EA (1987) A new automated technique for measuring respiration in soil samples. Plant Soil 101(2):183–187
Brookes P, Landman A, Pruden G, Jenkinson D (1985) Chloroform fumigation and the release of soil nitrogen: a rapid direct extraction method to measure microbial biomass nitrogen in soil. Soil Biol Biochem 17(6):837–842
Brumme R, Beese F (1995) Automated monitoring of biological trace gas production and consumption. In: Alef K, Nannipieri P (eds) Methods in applied soil microbiology and biochemistry. Academic, London, pp 468–472
Burns RG (1978) Enzyme activity in soil: some theoretical and practical considerations. In: Burn RG (ed) Soil enzymes. Academic, London, pp 73–75
Burns RG (1982) Enzyme activity in soil: location and a possible role in microbial ecology. Soil Biol Biochem 14(5):423–427
Burns RJ (1986) Interaction of enzymes with soil mineral and organic colloids. In: Huang PM, Schnitzer M (eds) Interactions of soil minerals with natural organics and microbes, SSSA Special Publication 17. SSSA, Madison, pp 429–452
Burns RG, Pukite A, McLaren A (1972) Concerning the location and persistence of soil urease 1. Soil Sci Soc Am J 36(2):308–311
Cannell MGR, Milne R, Dewar RC, Howard PJA (1994) Carbon pools and sinks in British vegetation and soils. NERL Annual Report
Cantor L (1987) The changing English countryside (1400–1700). Routledge and Kegan Paul Ltd, London
Carlisle SM, Trevors JT (1986) Effect of the herbicide glyphosate on respiration and hydrogen consumption in soil. Water Air Soil Pollut 27(3–4):391–401
Chambers JD, Mingay GE (1966) The agricultural revolution 1750–1880. Batsford, London
Chaussod R, Nicolardot (1982) Mesure de la biomasse microbienne dans les sols cultivés. I Approche cinétique et estimation simplifiée du carbone facilement minéralisable. Rev Ecol Biol Sol 19:501–512
Chen Y, Senesi N, Schnitzer M (1977) Information provided on humic substances by E4/E6 ratios 1. Soil Sci Soc Am J 41(2):352–358
Cheng W, Coleman DC (1989) A simple method for measuring CO2 in a continuous air-flow system: modifications to the substrate-induced respiration technique. Soil Biol Biochem 21 (3):385–388
Chhonkar P, Tarafdar J (1981) Characteristics and location of phosphatases in soil-plant system. J Indian Soc Soil Sci 29(2):215–219
Ciais P, Sabine C, Bala G, Bopp L, Brovkin V, Canadell J, Chhabra A, DeFries R, Galloway J, Heimann M (2014) Carbon and other biogeochemical cycles. In: Climate change 2013: the physical science basis. Contribution of working group I to the fifth assessment report of the intergovernmental panel on climate change. Cambridge University Press, Cambridge, pp 465–570

Clark FE, Kemper W (1967) Microbial activity in relation to soil water and soil aeration. Irrig Agric Lands 11:472–480
Clarke A, Stone BJ (1965) Properties of a β-(1→ 4)-glucan hydrolase from Aspergillus niger. Biochem J 96(3):802
Cleveland CC, Townsend AR, Schimel DS, Fisher H, Howarth RW, Hedin LO, Perakis SS, Latty EF, Von Fischer JC, Elseroad AJ (1999) Global patterns of terrestrial biological nitrogen (N2) fixation in natural ecosystems. Glob Biogeochem Cycles 13(2):623–645
Coleman K, Jenkinson DS (2005) ROTHC-26.3 A model for the turnover of carbon in soil-Model description and windows users guide, November 1999 issue (modified April 2005). Rothamsted Research, Harpenden
Compton JE, Boone RD (2000) Long-term impacts of agriculture on soil carbon and nitrogen in New England forests. Ecology 81(8):2314–2330
Compton JE, Boone RD, Motzkin G, Foster DR (1998) Soil carbon and nitrogen in a pine-oak sand plain in central Massachusetts: role of vegetation and land-use history. Oecologia 116 (4):536–542
Curci M, Pizzigallo M, Crecchio C, Mininni R, Ruggiero P (1997) Effects of conventional tillage on biochemical properties of soils. Biol Fertil Soils 25(1):1–6
Datta R, Anand S, Moulick A, Baraniya D, Pathan SI, Rejsek K, Vranova V, Sharma M, Sharma D, Kelkar A (2017) How enzymes are adsorbed on soil solid phase and factors limiting its activity: a review. Int Agrophys 31(2):287–302
Datta R, Kelkar A, Baraniya D, Molaei A, Moulick A, Meena R, Formanek P (2017a) Enzymatic degradation of lignin in soil: a review. Sustainability 9(7):1163
Datta R, Baraniya D, Wang Y-F, Kelkar A, Meena RS, Yadav GS, Teresa Ceccherini M, Formanek P (2017b) Amino acid: its dual role as nutrient and scavenger of free radicals in soil. Sustainability 9(8):1402
Davis B, Walter B, Ball N, Fitter A (1992) The soil: the new naturalistic survey of British natural history. Harper Collins, London
DeNobilli M, Gjessing E, Sequi P (1989) In: Hayes MHB, MacCarthy P, Malcolm RL, Swift RS (eds) Humic substances II. In search of structure. Wiley, Chichester, p 561
Dent JM, Russell EJ (1966) A history of agricultural science in Great Britain. George Allen and Unwin, London
Dick W, Juma N, Tabatabai M (1983) Effects of soils on acid phosphatase and inorganic pyrophosphatase of corn roots. Soil Sci 136(1):19–25
Domsch KH (1962) Bodenatmung, Sammelbericht über Methoden und Ergebnisse. Zbl Bakt Abt II 116:33–78
Domsch KH, Schroder M (1986) Einfluβ einiger Herbizide auf den Mikrobiellen Biomasse-Kohlenstoff und den Mineralstickstoffgehalt des Bodens. In: DFG-Forschungsbericht Herbizide 11. VCH verlagsgesellschaft mbH, Weinheim
Domsch K, Beck T, Anderson J, Söderström B, Parkinson D, Trolldenier GJ (1979) A comparison of methods for soil microbial population and biomass studies. Z Pflanzenernähr Bodenkd 142 (3):520–533
Douglas L, Bremner J (1970) Extraction and colorimetric determination of urea in soils 1. Proc Soil Sci Soc Am J 34(6):859–862
Durán J, Morse JL, Rodríguez A, Campbell JL, Christenson LM, Driscoll CT, Fahey TJ, Fisk MC, Mitchell MJ, Templer PH (2017) Differential sensitivity to climate change of C and N cycling processes across soil horizons in a northern hardwood forest. Soil Biol Biochem 107:77–84
Eivazi F, Tabatabai MJ (1977) Phosphatases in soils. Soil Biol Biochem 9(3):167–172
Elbert W, Weber B, Burrows S, Steinkamp J, Büdel B, Andreae MO, Pöschl U (2012) Contribution of cryptogamic covers to the global cycles of carbon and nitrogen. Nat Geosci 5(7):459
Elliott E, Cambardella CA (1991) Physical separation of soil organic matter. Agric Ecosyst Environ 34(1–4):407–419

Enders C, Tschapek M, Glane R, Kolloid Z (1948) Concepts of the origins, composition and structure of humic substances. In: Wilson WS (ed) Advances in soil organic matter: the impact on agriculture and environment. Royal Society of Chemistry, Cambridge, pp 3–5
Eriksson KE, Wood TM (1985) Biodegradation of cellulose. In: Higuchi T (ed) Biosynthesis and biodegradation of wood components. Academic, London, pp 469–503
Europe's Environment (EE) (1993) The Dobris assessment. Prepared by the European Environmental Agency Force. HMSO, London
Falkowski PG (1994) The role of phytoplankton photosynthesis in global biogeochemical cycles. Photosynth Res 39(3):235–258
Falkowski PG, Barber RT, Smetacek VJ (1998) Biogeochemical controls and feedbacks on ocean primary production. Science 281(5374):200–206
Falkowski P, Scholes R, Boyle E, Canadell J, Canfield D, Elser J, Gruber N, Hibbard K, Högberg P, Linder SJ (2000) The global carbon cycle: a test of our knowledge of earth as a system. Science 290(5490):291–296
FAO (2000) Carbon sequestration options under the clean development mechanism address land degradation. World resources reports. Rome, FAO
Farage P, Pretty J, Ball A (2005) Carbon sequestration in tropical dryland agroecosystems. Carbon sequestration in dryland tropics modelling report. Centre for Environment and Society/Department of Biological Sciences, University of Essex, Essex
Finzi AC, Abramoff RZ, Spiller KS, Brzostek ER, Darby BA, Kramer MA, Phillips RP (2015) Rhizosphere processes are quantitatively important components of terrestrial carbon and nutrient cycles. Glob Chang Biol 21(5):2082–2094
Flaig W (1988) Humic substances and their role in the environment. Wiley, Chichester
Florkin M, Stotz E (1964) Report of the Enzyme Commission of the International Union of Biochemistry. International Union of Biochemistry, Amsterdam
Foereid B, de Neergaard A, Høgh-Jensen H (2004) Turnover of organic matter in a Miscanthus field: effect of time in Miscanthus cultivation and inorganic nitrogen supply. Soil Biol Biochem 36(7):1075–1085
Fowler PJ (1983) The farming of prehistoric Britain. Cambridge University Press, Cambridge
Fowler D, Coyle M, Skiba U, Sutton MA, Cape JN, Reis S, Sheppard LJ, Jenkins A, Grizzetti B, Galloway JN (2013) The global nitrogen cycle in the twenty-first century. Philos Trans R Soc Lond Ser B Biol Sci 368(1621):20130164
Frankenberger W Jr, Johanson J (1986) Use of plasmolytic agents and antiseptics in soil enzyme assays. Soil Biol Biochem 18(2):209–213
Franklin O, Johansson J, Dewar RC, Dieckmann U, McMurtrie RE, Brännström Å, Dybzinski R (2012) Modeling carbon allocation in trees: a search for principles. Tree Physiol 32(6):648–666
Franklin O, Näsholm T, Högberg P, Högberg MN (2014) Forests trapped in nitrogen limitation–an ecological market perspective on ectomycorrhizal symbiosis. New Phytol 203(2):657–666
Franzluebbers K, Franzluebbers A, Jawson M (2002) Environmental controls on soil and whole-ecosystem respiration from a tallgrass prairie. Soil Sci Soc Am J 66(1):254–262
Gabrielle B, Mary B, Roche R, Smith P, Gosse G (2002) Simulation of carbon and nitrogen dynamics in arable soils: a comparison of approaches. Eur J Agron 18(1–2):107–120
Gosewinkel U, Broadbent FE (1984) Conductimetric determination of soil urease activity. Commun Soil Sci Plant Anal 15(11):1377–1389
Gottschalk G, Andressen JR, Hippe H (1981) In: Starr MP, Stolp H, Truper HG, Balows A, Schlegel HG (eds) The prokaryote, vol II. Springer, Berlin, pp 1767–1803
Gray TRG, Williams ST (1977) Soil micro-organisms. Longman, London
Grenville A, Grant A (1988) The countryside of medieval England. Basil Blackwell Ltd, Oxford
Hart P (1984) Effects of soil type and past cropping on the nitrogen supplying ability of arable soils. University of Reading, Reading
Hayano K (1986) Cellulase complex in a tomato field soil: induction, localization and some properties. Soil Biol Biochem 18(2):215–219

Hayes MHB (1985) In: Aiken GR, McKnight DM, Wershaw RL, MacCarthy P (eds) Humic substances in soil, sediment and water. Wiley, New York, p 329
Hayes MHB (1991) Concepts of the origins, composition and structure of humic substances. In: Hayes MHB, MacCarthy P, Malcolm RL, Swift RS (eds) Humic substances II. In search of structure. Wiley, New York, p 3
Hayes MH, Swift RJ (1978) The chemistry of soil organic colloids. Wiley, New York, pp 179–320
Hayes M, Swift R, Wardle R, Brown J (1975) Humic materials from an organic soil: a comparison of extractants and of properties of extracts. Geoderma 13(3):231–245
Heinemeyer O, Insam H, Kaiser E, Walenzik G (1989) Soil microbial biomass and respiration measurements: an automated technique based on infra-red gas analysis. Plant Soil 116 (2):191–195
Hoffman E, Schmidt WJBZ (1953) Uber das Enzymsystem unserer Kulturboden. 2 Urease. Biochem Z 324(2):125–127
Holderness BA, Turner M (1991) Land, labour and agriculture (1700–1920). Hambledon Press, London
Holland EA, Parton WJ, Detling JK, Coppock DLJT (1992) Physiological responses of plant populations to herbivory and their consequences for ecosystem nutrient flow. Am Nat 140 (4):685–706
Hope C, Burns RG (1987) Activity, origins and location of cellulases in a silt loam soil. Biol Fertil Soils 5(2):164–170
Houghton RA, Hackler JL, Daniels RC (1995) Continental scale estimates of the biotic carbon flux from land cover change: 1850–1980, Rep. ORNL/. CDIAC-79 NDP-050, R. C. Daniels. Oak Ridge Natl. Lab, Oak Ridge
Hungate BA, Dukes JS, Shaw MR, Luo Y, Field CB (2003) Nitrogen and climate change. Science 302(5650):1512–1513
Hunt HW (1977) A simulation model for decomposition in grasslands. Ecology 58(3):469–484
Igboji PO (2006) The effects of land management on the biological properties of East Anglian agricultural soils. University of Essex, Essex
Igboji PO, Pretty JN, Ball AS (2015) Effect of park grassland management on C-fluxes in temperate ecosystem. Elixir Int J Elixir Agric 82:2415–2434
Imbert M, Blondeau RJ (1998) On the iron requirement of lactobacilli grown in chemically defined medium. Curr Microbiol 37(1):64–66
Inubushi K, Wada H, Takai YJ (1984) Determination of microbial biomass-nitrogen in submerged soil. Soil Sci Plant Nutr 30(3):455–459
Isbister J, Shippen R, Caplan JJ (1980) A new method for monitoring cellulose and starch degradation in soils. Bull Environ Contam Toxicol 24(1):570–574
Isermeyer H (1952) Eine einfache Methode zur Bestimmung der Bodenatmung und der Karbonate im Boden. Zeitschrift für Pflanzenernährung, Düngung, Bodenkunde 56(1–3):26–38
Jaggi W (1976) Die Bestimmung der CO_2-Bildung als Maβ der bodenbiologischen Aktivitat. Schweiz Landwirtch Forsch 15(314):317–380
Janssens I, Dieleman W, Luyssaert S, Subke J-A, Reichstein M, Ceulemans R, Ciais P, Dolman AJ, Grace J, Matteucci GJ (2010) Reduction of forest soil respiration in response to nitrogen deposition. Nat Geosci 3(5):315
Jenkinson D (1988) Determination of microbial biomass carbon and nitrogen in soil. CAB International, Wallingford, pp 368–386
Jenkinson D (1991) The Rothamsted long-term experiments: are they still of use? Agron J 83 (1):2–10
Jenkinson D, Rayner JJ (1977) The turnover of soil organic matter in some of the Rothamsted classical experiments. Soil Sci 123(5):298–305
Joergensen R, Brookes P, Jenkinson D (1990) Survival of the soil microbial biomass at elevated temperatures. Soil Biol Biochem 22(8):1129–1136
Johnson DW, Curtis PS (2001) Effects of forest management on soil C and N storage: meta analysis. For Ecol Manag 140(2–3):227–238

Johnson DW, Todd D (1988) Nitrogen fertilization of young yellow poplar and loblolly pine plantations at differing frequencies. Soil Sci Soc Am J 52(5):1468–1477

Johnson DW, Todd D, Tolbert VR (2003) Changes in ecosystem carbon and nitrogen in a loblolly pine plantation over the first 18 years. Soil Sci Soc Am J 67(5):1594–1601

Joliff G, Edelman A, Klier A, Rapoport GJ (1989) Inducible secretion of a cellulase from Clostridium thermocellum in Bacillus subtilis. Appl Environ Microbiol 55(11):2739–2744

Kaiser E, Mueller T, Joergensen R, Insam H, Heinemeyer O (1992) Evaluation of methods to estimate the soil microbial biomass and the relationship with soil texture and organic matter. Soil Biol Biochem 24(7):675–683

Kandeler E, Gerber H (1988) Short-term assay of soil urease activity using colorimetric determination of ammonium. Biol Fertil Soils 6(1):68–72

Kieft TL (1987) Microbial biomass response to a rapid increase in water potential when dry soil is wetted. Soil Biol Biochem 19(2):119–126

Kieft TL, Rosacker LL (1991) Application of respiration-and adenylate-based soil microbiological assays to deep subsurface terrestrial sediments. Soil Biol Biochem 23(6):563–568

King AW, Post WM, Wullschleger SD (1997) The potential response of terrestrial carbon storage to changes in climate and atmospheric CO2. Climate Change 35(2):199–227

Kiss S, Dragan-Bularda M, Radulescu D (1978) Soil polysaccharides: activity and agricultural importance. In: Burns RG (ed) Soil enzymes. Academic, New York, pp 117–147

Kissel DE, Cabrera ML (1988) Factors affecting urease activity. In: Bock BR, Kissel DE (eds) Ammonia volatilization from urea fertilizers. TVA, National Fertilizer, Muscle Shoals

Knoepp JD, Swank WT (1997) Forest management effects on surface soil carbon and nitrogen. Soil Sci Soc Am J 61(3):928–935

Kononova MJ, Somne PP (1966) Soil organic matter. Pergamon Press, Oxford/New York

Körner C (2003) Slow in, rapid out--carbon flux studies and Kyoto targets. Science 300 (5623):1242–1243

Kowalenko C, Ivarson K (1978) Effect of moisture content, temperature and nitrogen fertilization on carbon dioxide evolution from field soils. Soil Biol Biochem 10(5):417–423

Kroeckel L, Stolp H (1986) Influence of the water regime on denitrification and aerobic respiration in soil. Biol Fertil Soils 2(1):15–21

Kshattriya S, Sharma G, Mishra R (1992) Enzyme activities related to litter decomposition in forests of different age and altitude in North East India. Soil Biol Biochem 24(3):265–270

Lackner KS (2003) A guide to CO2 sequestration. Science 300(5626):1677–1678

Lal R, Kimble J, Follett R (1998) Management of carbon sequestration in soil. In: Lal R (ed) Land use and soil C pools in terrestrial ecosystems. CRC Press, Boca Raton, pp 1–10

Lamothe PJ, McCormick PG (1973) Role of hydrindantin in the determination of amino acids using ninhydrin. Anal Chem 45(11):1906–1911

Larmola T, Leppänen SM, Tuittila E-S, Aarva M, Merilä P, Fritze H, Tiirola M (2014) Methanotrophy induces nitrogen fixation during peatland development. Proc Natl Acad Sci U S A 111(2):734–739

Lee Y-H, Fan L (1980) Properties and mode of action of cellulase. In: Advances in biochemical engineering, vol 17. Springer, New York, pp 101–129

Levia DF Jr, Frost E (2003) A review and evaluation of stemflow literature in the hydrologic and biogeochemical cycles of forested and agricultural ecosystems. J Hydrol 274(1–4):1–29

Lovelock J (1995) New statements on the Gaia theory. Microbiologia 11(3):295–304

Lundgren B (1981) Fluorescein diacetate as a stain of metabolically active bacteria in soil. Oikos 36:17–22

Malcolm RL (1989) In: Hayes MHB, MacCarthy P, Malcolm RL, Swift RS (eds) Humic substances II. In search of structure. Wiley, Chichester, p 339

Malkomes H-P (1985) Einfluß des Herbizids Dinoseb-acetat und dessen Kombination mit einem Phospholipid auf bodenmikrobiologische Aktivitäten unter Labor-und Gewächshausbedingungen/Influence of the herbicide dinoseb acetate and its combination with a phospholipid on microbial activities in soil under laboratory and field conditions. Z Pflanzenkrankh Pflanz/J Plant Dis Protect 92:489–501

Mann L (1986) Changes in soil carbon storage after cultivation. Soil Sci 142(5):279–288

Martens R (1985) Limitations in the application of the fumigation technique for biomass estimations in amended soils. Soil Biol Biochem 17(1):57–63

Martin J, Haider K, Kassim G (1980) Biodegradation and stabilization after 2 years of specific crop, lignin, and polysaccharide carbons in soils. Congr Int Soc Soil Sci 44(6):1250–1255

McCarty G, Bremner J, Chai H (1989) Effect of N-(n-butyl) thiophosphoric triamide on hydrolysis of urea by plant, microbial, and soil urease. Biol Fertil Soils 8(2):123–127

McGill W, Cole C (1981) Comparative aspects of cycling of organic C, N, S and P through soil organic matter. Geoderma 26(4):267–286

Menge DN, Hedin LO (2009) Nitrogen fixation in different biogeochemical niches along a 120,000-year chronosequence in New Zealand. Ecology 90(8):2190–2201

Metherell AK (1992) Simulation of soil organic matter dynamics and nutrient cycling in microbial assimilation of nitrogen. Biol Fertil Soils 12:261–264

Metherell AK, Parton WJ, Cambardella CA, Peterson GA, Harding LA, Cole CV (1993) Simulation of soil organic matter dynamics in dryland wheat-fallow cropping systems. In: Proceedings of the international symposium on soil processes and management systems: greenhouse gas emissions and carbon sequestration, Ohio State University, April 5–9

Meyerholt J, Zaehle S, Smith MJ (2016) Variability of projected terrestrial biosphere responses to elevated levels of atmospheric CO2 due to uncertainty in biological nitrogen fixation. Biogeosciences 13(5):1491–1518

Moore S, Stein WH (1948) Photometric nin-hydrin method for use in the chromatography of amino acids. J Biol Chem 176:367–388

Moyo C, Kissel D, Cabrera M (1989) Temperature effects on soil urease activity. Soil Biol Biochem 21(7):935–938

Mueller T, Joergensen R, Meyer B (1992) Estimation of soil microbial biomass C in the presence of living roots by fumigation-extraction. Soil Biol Biochem 24(2):179–181

Mulvaney R, Bremner J (1979) A modified diacetyl monoxime method for colorimetric determination of urea in soil extracts. Commun Soil Sci Plant Anal 10(8):1163–1170

Nakas JP, Klein DA (1981) Use of an amino acid mixture to estimate the mineralization capacity of grassland soils. Soil Biol Biochem 13:427–428

Nannipieri P (1994) Productivity, sustainability and pollution. In: Parkhurst CE, Double BM, Gupta VV, Grace PR (eds) Soil biota: management in sustainable farming systems. CSIRO, Melbourne, pp 238–244

Nannipieri P, Ceccanti B, Cervelli S, Sequi P (1974) Use of 0· 1 m pyrophosphate to extract urease from a podzol. Soil Biol Biochem 6(6):359–362

Nannipieri P, Ceccanti B, Cervelli S (1978) Stability and kinetic properties of humus-urease complexes. Soil Biol Biochem 10(2):143–147

Nannipieri P, Ceccanti B, Bianchi D (1988) Characterization of humus-phosphatase complexes extracted from soil. Soil Biol Biochem 20(5):683–691

Nannipieri P, Grego S, Ceccanti B, Bollag J, Stotzky G (1990) Ecological significance of the biological activity in soil. In: Bollag JM, Stotzky G (eds) Soil biochemistry, vol 6. Marcel Dekker, New York/Basel, pp 293–355

Neff JC, Holland EA, Dentener FJ, McDowell WH, Russell KM (2002) The origin, composition and rates of organic nitrogen deposition: a missing piece of the nitrogen cycle? Biogeochemistry 57(1):99–136

Newman DK, Banfield JF (2002) Geomicrobiology: how molecular-scale interactions underpin biogeochemical systems. Science 296(5570):1071–1077

Nicks AD (1974) Stochastic generation of the occurrence, pattern, location of maximum amount of daily rainfall. In: Proceedings symposium on statistical hydrology, United States Department of Agriculture Misc, vol 1275, pp 154–171

Nip M, Tegelaar E, De Leeuw J, Schenck P, Holloway P (1986) A new non-saponifiable highly aliphatic and resistant biopolymer in plant cuticles. Naturwissenschaften 73(10):579–585
Norman J, Garcia R, Verma SJ (1992) Soil surface CO2 fluxes and the carbon budget of a grassland. J Geophys Res Atmos 97(D17):18845–18853
Ogunseitan O (2005) Microbial diversity. Blackwell Publishing Company, Malden
Ojima D (2006) Century soil organic matter model version 5. Natural Resource Ecology Laboratory 5, Fort Collins
Ojima DS, Parton W, Schimel D, Owensby C (1990) Simulated impacts of annual burning on prairie ecosystems. In: Collins SL, Wallance L (eds) Fire in North American prairies. University of Oklahoma, Norman
Parkinson D, Gray TR, Williams ST (1971) Methods for study-ing the ecology of soil micro-organisms. Blackwells, Oxford
Parton WJ, Anderson DW, Cole CV, Stewart JWB (1983) Simulation of soil organic matter formation and mineralization in semi-arid agroecosystems. In: Lowrance RR, Todd RL, Asmussen LE, Leonard RA (eds) Nutrient cycling in agricultural ecosystem, Special Publication No 23. The University of Georgia, College of Agriculture Experiment Stations, Athens, pp 553–550
Parton W, Schimel DS, Cole C, Ojima D (1987) Analysis of factors controlling soil organic matter levels in Great Plains Grasslands. Soil Sci Soc Am J 51(5):1173–1179
Parton WJ, Metherell AK, Harding LA, Cole CV (1993) CENTURY: Soil organic matter model environment. Technical documentation, agroecosystem version 4.0. Great Plains System Research. USDA-ARS, Fort Collins
Paul E, Johnson R (1977) Microscopic counting and adenosine 5′-triphosphate measurement in determining microbial growth in soils. Appl Environ Microbiol 34(3):263–269
Paul E, Collins H, Leavitt S (2001) Dynamics of resistant soil carbon of Midwestern agricultural soils measured by naturally occurring 14C abundance. Geoderma 104(3–4):239–256
Paul E, Morris S, Six J, Paustian K, Gregorich E (2003) Interpretation of soil carbon and nitrogen dynamics in agricultural and afforested soils. Soil Sci Soc Am J 67(5):1620–1628
Perry PJ (1973) British agriculture (1875–1914). Methuen and Co. Ltd, London
Phillips RP, Brzostek E, Midgley MG (2013) The mycorrhizal associated nutrient economy: a new framework for predicting carbon–nutrient couplings in temperate forests. New Phytol 199 (1):41–51
Pidwimy M (2005) Introduction to the biosphere. In: Fundamaentals of physical geography. University of British Columbia, Okanagan
Plentinger M, Penning de Vries F (1995) CAMASE, register of agroecosystems models, version 1. Centre for Agricultural Publishing and Documentation, Wageningen
Pochon J, Tardieux P (1962) Techniques d'analyse en microbiologie du sol. Editions de la Tourelle, St Mande
Post WM, Kwon KC (2000) Soil carbon sequestration and land-use change: processes and potential. Glob Chang Biol 6(3):317–327
Pretty J (2002) Reconnecting people, land and nature. Earthscan, London
Pretty J, Ball A (2001) Agricultural influences on carbon emissions and sequestration: a review of evidence and the emerging trading options. Centre Environ Soc Occ Pap 3:31
Pumpanen J, Ilvesniemi H, Hari P (2003) A process-based model for predicting soil carbon dioxide efflux and concentration. Soil Sci Soc Am J 67(2):402–413
Qian Y, Bandaranayake W, Parton W, Mecham B, Harivandi M, Mosier AJ (2003) Long-term effects of clipping and nitrogen management in turfgrass on soil organic carbon and nitrogen dynamics. J Environ Qual 32(5):1694–1700
RCEP (1996) Commission on Environmental Pollution, 1996. Sustainable use of soil. HMSO, Royal Commission on Environmental Pollution (RCEP), London
Reed SC, Cleveland CC, Townsend AR (2011) Functional ecology of free-living nitrogen fixation: a contemporary perspective. Annu Rev Ecol Evol Syst 42:489–512

Rhee Y, Hah Y, Hong S (1987) Relative contributions of fungi and bacteria to soil carboxymethylcellulase activity. Soil Biol Biochem 19(4):479–481
Richter J (1972) Zur Methodik des Bodengashaushaltes. II. Ergebnisse und Diskussion. Z Pflanzenern Bodenkde 132(3):220–239
Richter DD, Markewitz D, Trumbore SE, Wells CG (1999) Rapid accumulation and turnover of soil carbon in a re-establishing forest. Nature 400(6739):56
Rowell DL (1994) Soil science: methods and applications, vol 1. Longman Scientific Ltd, London
Ryan MJ (1996) Comparing models of ecosystem function for temperate conifer forests. I. Model description and validation. In: Global change: effects on coniferous forests and grasslands. Wiley, Chichester
Sarkar J, Batistic L, Mayaudon J (1980) Les hydrolases du sol et leur association avec les hydrates de carbone. Soil Biol Biochem 12(4):325–328
Sarmiento JL, Gruber N (2002) Sinks for anthropogenic carbon. Phys Today 55(8):30–36
Sato K (1981) Relations between soil microflora and CO 2 evolution upon decomposition of cellulose. Plant Soil 61(1–2):251–258
Schinner F, Von Mersi W (1990) Xylanase-, CM-cellulase-and invertase activity in soil: an improved method. Soil Biol Biochem 22(4):511–515
Schlesinger WH (1977) Carbon balance in terrestrial detritus. Annu Rev Ecol Syst 8(1):51–81
Schlesinger WH (1990) Evidence from chronosequence studies for a low carbon-storage potential of soils. Nature 348(6298):232
Schlesinger W (1991) The global carbon cycle. Wiley, New York
Schlosser HJ (1988) Auswertung ökotoxikologischer Forschung zur Belastung von Ökosystemen durch Chemikalien: Bundesministerium für Forschung u. Technologie Förderkennzeichen 03 7393 4. Projektleitung Biologie, Ökologie, Energie (PBE) d. Kernforschungsanlage Jülich
Schröder D, Gewehr H (1977) Stroh-und Zelluloseabbau in verschiedenen Bodentypen. Z Pflanzenernähr Bodenk 140(3–4):273–284
Schröder D, Urban B (1985) Bodenatmung, Celluloseabbau und Dehydrogenaseaktivität in verschiedenen Böden und ihre Beziehungen zur organischen Substanz sowie Bodeneigenschaften. Forschung 38:166–172
Schuffelen AC, Bolt GH (1962) Landbouwk tidjschr. ste Jaargang No 4/5
Schuster E (1988) EinfluB von Pflanzenschutzmittel-Spritzfolgen und-Kombinationen auf die mikro-biologische Aktivitat des Bodens. Freiland-und Laborversuche. Dissertations arbeit, Fach Boden-Kunde Universitat Trier
Sellers PJ, Dickinson RE, Randall DA, Betts AK, Hall F, Berry G, Senesi N, Steelink C (1989) Humic substances. In: Hessen et al (eds) Humic substances II. In search of structure. Wiley, Chichester, p 372
Shen S, Pruden G, Jenkinson D (1984) Mineralization and immobilization of nitrogen in fumigated soil and the measurement of microbial biomass nitrogen. Soil Biol Biochem 16(5):437–444
Simmons IG, Tooley MJ (1981) The environment in British prehistory. Duckworth, London
Sinsabaugh R, Linkins A (1988) Adsorption of cellulase components by leaf litter. Soil Biol Biochem 20(6):927–931
Sinsabaugh RL, Linkins A (1989) Ellulase mobility in decomposing leaf litter. Soil Biol Biochem 21(2):205–209
Skujiņš J (1978) History of abiotic soil enzyme research. In: Burns RG (ed) Soil enzymes. Academic, London, pp 1–49
Skujiņš J, McLaren A (1969) Assay of urease activity using 14C-urea in stored, geologically preserved, and in irradiated soils. Soil Biol Biochem 1(1):89–99
Smith J, Paul E, Bollag J, Stotzky G (1990) The significance of soil microbial biomass estimations. In: Bollag J-M, Stotzky G (eds) Soil biochemistry, vol 6. Marcel Dekker, New York, pp 357–396
Smith P, Smith JU, Powlson DS (1996) Soil organic matter network (SOMNET): model and experimental metadata. GCTE Task 3.3.1. Global Change and Terrestrial Ecosystems Rep. No 7, Wallingford

Smith P, Smith J, Powlson D, McGill W, Arah J, Chertov O, Coleman K, Franko U, Frolking S, Jenkinson D (1997) A comparison of the performance of nine soil organic matter models using datasets from seven long-term experiments. Geoderma 81(1–2):153–225

Söderström B (1977) Vital staining of fungi in pure cultures and in soil with fluorescein diacetate. Soil Biol Biochem 9(1):59–63

Sollins P, Glassman C, Paul E, Swanston C, Lajtha K, Heil J (1999) Soil carbon and nitrogen. Elliott Management Corporation, New York, pp 89–105

Somerville L, Greaves MP (1987) In: Somerville L, Greaves MP (eds) Pesticide effects on soil microflora. Taylor and Francis, London/New York

Sowerby A, Blum H, Gray TR, Ball AS (2000) The decomposition of *Lolium perenne* in soils exposed to elevated CO2: comparisons of mass loss of litter with soil respiration and soil microbial biomass. Soil Biol Biochem 32(10):1359–1366

Sparling G (1981a) Heat output of the soil biomass. Soil Biol Biochem 13(5):373–376

Sparling G (1981b) Microcalorimetry and other methods to assess biomass and activity in soil. Soil Biol Biochem 13(2):93–98

Sparling G, West A (1990) A comparison of gas chromatography and differential respirometer methods to measure soil respiration and to estimate the soil microbial biomass. Pedobiologia 34 (2):103–112

Speir T, Ross D (1978) Soil phosphatase and sulphatase. Soil Enzymes 203:197–250

Speir T, Ross D (1981) A comparison of the effects of air-drying and acetone dehydration on soil enzyme activities. Soil Biol Biochem 13(3):225–229

Steelink C, Wershaw RL, Thorn KA, Wilson MA (1989) In: Hayes MHB, MacCarthy P, Malcolm RL, Swift RS (eds) Humic substances II: In search of structure. Wiley, Chichester, p 281

Stocker BD, Prentice IC, Cornell SE, Davies-Barnard T, Finzi AC, Franklin O, Janssens I, Larmola T, Manzoni S, Näsholm T (2016) Terrestrial nitrogen cycling in Earth system models revisited. New Phytol 210(4):1165–1168

Stotzky G (1965) Microbial respiration. In: Methods of soil analysis. Part 2, Agronomy Monograph 9. UFRGS, Porto Alegre, pp 1550–1572

Stutzenberger FJ (1972) Cellulolytic activity of Thermomonospora curvata: optimal assay conditions, partial purification, and product of the cellulase. Appl Microbiol 24(1):83–90

Sullivan BW, Smith WK, Townsend AR, Nasto MK, Reed SC, Chazdon RL, Cleveland CC (2014) Spatially robust estimates of biological nitrogen (N) fixation imply substantial human alteration of the tropical N cycle. Proc Natl Acad Sci 111:8101

Suttner T, Alef K (1988) Correlation between the arginine ammonification, enzyme activities, microbial biomass, physical and chemical properties of different soils. Zentralbl Mikrobiol 143(8):569–573

Swift RS (1985) Fractionation of soil humic substances. In: Aiken GR, McKnight DM, Wershaw RL, MacCarthy P (eds) Humic substances in soil, sediment and water: geochemistry, isolation and characterisation. Wiley, New York, pp 387–408

Swift RS (1989) In: Hayes MHB, MacCarthy P, Malcolm RL, Swift RS (eds) Humic substances II. In search of structure. Wiley, Chichester, p 467

Tabatabai M, Bremner J (1972) Assay of urease activity in soils. Soil Biol Biochem 4(4):479–487

Tarafdar J, Jungk A (1987) Phosphatase activity in the rhizosphere and its relation to the depletion of soil organic phosphorus. Biol Fertil Soils 3(4):199–204

Tateno M (1988) Limitations of available substrates for the expression of cellulase and protease activities in soil. Soil Biol Biochem 20:117–118

Thalmann A (1968) Zur Methodik der Bestimmung der DehydrogenaseaktivitAt im Boden mittels triphenytetrazoliumchlorid (TTC). Landwirtsch Forsch 21:249–258

Thirsk J (1991) The agrarian history of England and Wales, vol 111. Cambridge University Press, New York

Thirsk J (2000) The agrarian history of England and Wales (1850–1914), vol VII. Cambridge University Press, New York

Torsvik VL, Goksoyr J (1978) Determination of bacterial DNA in soil. Soil Biol Biochem 10 (1):7–12

Trasar-Cepeda MC, Gil-Sotres F (1987) Phosphatase activity in acid high organic matter soils in Galicia (NW Spain). Soil Biol Biochem 19(3):281–287

Trasar-Cepeda MC, Gil-Sotres F (1988) Kinetics of acid phosphatase activity in various soils of Galicia (NW Spain). Soil Biol Biochem 20(3):275–280

Trevors JT (1985) Oxygen consumption in soil: effect of assay volume. Soil Biol Biochem 17:385–386

Trolldenier G (1973) The use of fluorescence microscopy for counting soil microorganisms. Bull Ecol Res Commun (Stockholm) 17:53–59

Turner J, Lambert M (2000) Change in organic carbon in forest plantation soils in eastern Australia. For Ecol Manag 133(3):231–247

Van de Werf H, Verstraete W (1987) Estimation of active soil microbial biomass by mathematical analysis of respiration curves: development and verification of the model. Soil Biol Biochem 19 (3):253–260

Van de Werf H, Genouw G, Van Vooren L, Verstraete W (1995) The determination of active microbial biomass by the respiration simulation method of Chapter 8 'Microbial biomass'. In: Alef K, Nannipieri P (eds) Methods in applied soil and biochemistry. Academic Press Ltd., London, pp 405–408

Vance E, Brookes P, Jenkinson D (1987a) Microbial biomass measurements in forest soils: determination of kc values and tests of hypotheses to explain the failure of the chloroform fumigation-incubation method in acid soils. Soil Biol Biochem 19(6):689–696

Vance E, Brookes P, Jenkinson D (1987b) Microbial biomass measurements in forest soils: the use of the chloroform fumigation-incubation method in strongly acid soils. Soil Biol Biochem 19 (6):697–702

Vance ED, Brookes PC, Jenkinson DS (1987c) An extraction method for measuring soil microbial biomass C. Soil Biol Biochem 19(6):703–707

Verma SB (1990) Micrometeorological methods for measuring surface fluxes of mass and energy. Remote Sens Rev 5(1):99–115

Vitousek PM, Menge DN, Reed SC, Cleveland CC (2013) Biological nitrogen fixation: rates, patterns and ecological controls in terrestrial ecosystems. Philos Trans R Soc Lond B Biol Sci: Biol Sci 368(1621):20130119

Walker AP, Zaehle S, Medlyn BE, De Kauwe MG, Asao S, Hickler T, Parton W, Ricciuto DM, Wang YP, Wårlind D (2015) Predicting long-term carbon sequestration in response to CO2 enrichment: how and why do current ecosystem models differ? Glob Biogeochem Cycles 29 (4):476–495

Webb J, Bellamy P, Loveland P, Goodlass G (2003) Crop residue returns and equilibrium soil organic carbon in England and Wales. Soil Sci Soc Am J 67(3):928–936

Weinberg ED (1997) The Lactobacillus anomaly: total iron abstinence. Perspect Biol Med 40 (4):578–583

Wenzel S, Cox PM, Eyring V, Friedlingstein PJ (2014) Emergent constraints on climate carbon cycle feedbacks in the CMIP5 Earth system models. J Geophys Res Biogeo 119(5):794–807

Wershaw RL (1985) In: Aiken GR, McKnight DM, Wershaw RL, MacCarthy P (eds) Humic substances in soil, sediment and water. Wiley, New York, p 561

Widén B, Majdi H (2001) Soil CO2 efflux and root respiration at three sites in a mixed pine and spruce forest: seasonal and diurnal variation. Can J For Res 31(5):786–796

Wilke B-M (1986) Einfluß verschiedener potentieller anorganischer Schadstoffe auf die mikrobielle Aktivität von Waldhumusformen unterschiedlicher Pufferkapazität. Dr.-Haus Bayreuth Verlag-Ges. Springer Verlag, Berlin/Heidelberg/New York

Wilson MJHS (1989) Solid-state nuclear magnetic resonance spectroscopy of humic substances: basic concepts and techniques. Wiley, New York, pp 310–338

Wilson JM, Griffin D (1975) Water potential and the respiration of microorganisms in the soil. Soil Biol Biochem 7(3):199–204

Wormel P (1999) Essex Farming (1900–2000). Alberton Books, Colchester

Writtle College (2004) Crop production handbook: data and general information (2003–2004). Writtle College, Chelmsford Press, Essex

Wu J, Joergensen R, Pommerening B, Chaussod R, Brookes P (1990) Measurement of soil microbial biomass C by fumigation-extraction-an automated procedure. Soil Biol Biochem 22 (8):1167–1169

Yamane K, Suzuki H, Nisizawa K (1970) Purification and properties of extracellular and cell-bound cellulase components of *Pseudomonas fluorescens* var. cellulosa. J Biochem 67(1):19–35

Zaehle S, Medlyn BE, De Kauwe MG, Walker AP, Dietze MC, Hickler T, Luo Y, Wang YP, El-Masri B, Thornton P (2014) Evaluation of 11 terrestrial carbon–nitrogen cycle models against observations from two temperate Free-Air CO 2 Enrichment studies. New Phytol 202 (3):803–822

Zantua M, Bremner J (1975) Comparison of methods of assaying urease activity in soils. Soil Biol Biochem 7(4–5):291–295

Zibilske LJ (1994) Carbon mineralization . . . methods of soil analysis: Part 2—Microbiological and biochemical properties, vol 5.2, pp 835–863. https://doi.org/10.2136/sssabookser5.2.frontmatter

Small Farmers and Sustainable N and P Management: Implications and Potential Under Changing Climate

Mehmood Ali Noor, Muhammad Mohsin Nawaz, Mahmood ul Hassan, Alam Sher, Tariq Shah, Muhammad Mohsin Abrar, Umair Ashraf, Sajid Fiaz, Mohammed A. Basahi, Waqas Ahmed, and Wei Ma

Abstract The use of inorganic fertilizers has been an efficient way to increase agricultural production. Ever-increasing global population threatening food security, risks of changing climate affecting plant productivity, and the need for environment-friendly agriculture are all requiring the rational use of fertilizers to improve their use efficiencies. The increase in agricultural production in the recent past and in the future is associated with intensified (many folds) use of nitrogen (N) fertilizers,

M. A. Noor (✉) · M. M. Nawaz · W. Ma (✉)
Institute of Crop Sciences, Chinese Academy of Agricultural Sciences, Key Laboratory of Crop Physiology and Ecology, Ministry of Agriculture, Beijing, China
e-mail: mnoor@caas.cn; mcheema@caas.cn; mawei02@caas.cn

M. u. Hassan
Department of Plant Nutrition, College of Resources and Environmental Sciences, China Agricultural University, Beijing, China

A. Sher
School of Agronomy, Anhui Agricultural University, Hefei, Anhui, China

T. Shah
Key Laboratory of Oil Crop Biology and Genetic Improvement, Oil Crops Research Institute of the Chinese Academy of Agricultural Sciences, Wuhan, Hubei, China
e-mail: tariqshah@aup.edu.pk

M. M. Abrar · W. Ahmed
National Engineering Laboratory for Improving Quality of Arable Land, Institute of Agricultural Resources and Regional Planning, Chinese Academy of Agricultural Sciences, Beijing, China
e-mail: mohsin_abrar@caas.cn

U. Ashraf
Department of Botany, University of Education, Lahore, Punjab, Pakistan
e-mail: umair.ashraf@ue.edu.pk

S. Fiaz
State Key Laboratory of Rice Biology, China National Rice Research Institute, Hangzhou, China

M. A. Basahi
Department of Biology, College of Science & Arts, Sajir, Shaqra University, Shaqra, Saudi Arabia

R. Datta et al. (eds.), *Carbon and Nitrogen Cycling in Soil*,
https://doi.org/10.1007/978-981-13-7264-3_6

whose irrational use threatens the neighboring micro- and macro-environments by polluting them (e.g., eutrophication) and thus affecting the functioning of nearby animal and plant ecosystems. Therefore, this chapter critically focuses on these challenges faced by the small farmers with persistence farming in managing the precise use of nitrogenous and phosphorus (P) fertilizers. A highlight is given on the socioeconomic features of this persistence farming, which are the key drivers for decision-making in all the agricultural activities of this type of farming. In addition, the interaction of fertilizer management and crop production is provided with respect to small farms, and the review on management strategies for rational use of fertilizers proposed by esteemed international organizations and agencies is described and analyzed with respect to scientific achievements. Best management practices for N and P fertilizers and their significance in agricultural production and plant functioning are proposed also. In last, the management of these two agriculturally important nutrients (N and P) is concluded for sustainable productivity on small farms, and the major players involved in this regard are highlighted.

Keywords Sustainable productivity · Small farmers · Subsistence farming · Nitrogen · Phosphorus · Best management practices

Abbreviations

AN	Ammonium nitrate
BMPs	Best management practices
CAN	Calcium ammonium nitrate
ENSET	Efficient Nutrient Supply in East Africa
FAO	Food and Agriculture Organization
H_2S	Hydrogen sulfide
HHV	High heat value
HLPE	High-Level Panel of Experts of the Committee on Food Security
IFA	International Fertilizer Association
IFAD	International Fund for Agricultural Development
MOB	Methane-oxidizing bacteria
NH_4^+	Ammonium N
NMP	Nutrient management plan
NO_3^-	Nitrate N
NUE	Nitrogen use efficiency
PGPRs	Plant growth-promoting rhizobacteria
SPAD	Soil plant analysis development
UAN	Urea ammonium nitrate
UNEP	United Nations Environment Programme
UNIDO	United Nations Industrial Development Organization
WT	Wild type

1 Smallholding Farms in a Bigger Picture

Agriculture is facing multifaceted pressures from the ever-increasing world population. While on one side, the production of food to meet the needs of the increasing population is one challenge – even if the current resources remain as such – the resource allocation to agriculture is on the downfall. The nature of challenges might differ across regions, but the intensity of risks associated is everywhere. For example, the availability of land resources under increased urbanization can be the most challenging scenario in developed and rapidly developing countries in contrast to land-abundant or least developed countries. However, least developed countries have their own problems such as low productivity and low efficiencies in resource utilization. Then come the sustainability and climate change issues. Agriculture doesn't only play a major role in anthropogenic climate change but is also the most vulnerable sector affected by the changing climate. And the gravest issue is that the impacts of climate change on agriculture are more serious in the least developed countries, putting more pressure on such nations.

Agriculture contributes to greenhouse gas emissions directly from agricultural activities as well as indirectly through the manufacturing of synthetic fertilizers. Green Revolution, characterized by input-responsive cultivars, resulted in more and more use of fertilizers and other inputs. Scientists are continuously working to improve fertilizer use efficiencies either by improving crop productivity potentials or soil productivity and fertilizer management through bettering and innovating agricultural practices. Although much efforts have been made to increase the efficiency of inputs used, the task is still continuing. Another aspect is the dissemination of these improved practices and their adoption by farmers. The success of any new technology or practice – whether linked to sustainability, climate change, or resource use efficiency – depends on how it is perceived by farmers in context to their targets. Here come the prevailing socioeconomic conditions in which farmers have to make these decisions. These include the economic conditions of farmers, the purpose of farming, the scale of farming, and much more. In short, a sustainable fertilizer management practice may not be as much welcomed by a subsistence smallholder farmer in South Asia as it could be by a commercial farm in Australia.

The scale of farming and the purpose of farming are two important drivers in the decision-making process at the farm. Small farmers are generally less interested in those practices which do not produce immediate benefits in terms of production. As most of the small farmers are practicing subsistence farming, they are least interested in environmental costs associated with agricultural practices. Nevertheless, they are exposed to the same risks associated with climate change and may be at a high-risk intensity. Scale and distribution of these smallholder farms are not the same all over the world, yet the importance of such farms is significant in facing challenges of food security, climate change, and sustainability.

1.1 Farm Size and Distribution

Though there has always been a consensus on the importance of small farms in agriculture, clear picture of landholdings and area operated by smallholders is missing. There are many sources which provide such information, but most of the times they either lack documentation or are outdated. One major constraint in this regard is lack of clarity in defining the terms such as small farms and family farms which are most of the times used interchangeably or in combination. Other terms with which small-scale farming is often confused are resource-poor, subsistence, low-input, low-income, or low-technology farming (Abele and Frohberg 2003). However, even if these terms can be used interchangeably in some contexts, they don't mean the same all the time. Also, what is a small farm in one country may not fall in the same category in another country. A report by High-Level Panel of Experts of the Committee on Food Security (HLPE), while surveying definitions of smallholders, acknowledged the relativity of the concept of small and its dependence on agroecological as well as socioeconomic considerations (Bosc et al. 2013). The most common approach adopted in many sources is defining small farms on the basis of the size of landholding (or livestock number). Having many pros such as availability of internationally comparable empirical data, this approach, however, fails to consider the quality of resources, cropping systems, disparities across regions, and many other factors that have significant impacts on the state of farm's productivity and efficiency (Nagayets 2005).

Going through literature, we find that average farm sizes and trends in changes in average farm sizes over time, land distribution, and distribution of small farms vary across countries and regions. Some of the findings of different authors are given in Table 1. Although Africa is considered as a land-abundant region (Deininger and Byerlee 2011), the distribution of this arable but uncultivated land is not even across the region but concentrated in a few countries. And the decrease in average farm size was found in all land-constrained African countries in contrast to land-abundant countries from 1980 to 2000 (Jayne et al. (2014)). In general, it can be said that developing countries saw a decrease in average farm size compared to developed countries where average farm size has increased. And high-income countries have unequal farmland distribution than that in developing regions.

Recently, Lowder et al. (2016) reported comprehensive estimates for a total number of farms worldwide, as well as the trends of changes in average farm size and distribution of different farms by using agricultural censuses data from 167 countries. They also analyzed the data at global and regional scale and compared countries by income. The 167 countries of which they used data comprise 90% of world farmland, and 97% of the population is active in agriculture which is 96% of total world population, so their estimates were derived from well-representative data. The reported lower-bound estimate for a total number of farms worldwide is 570 million farms, and they expected that the actual figure is more considering various limitations during data analyses. Region-wise, Asia is ranked first by accounting for 74% of these 570 million farms, followed by sub-Saharan Africa

Table 1 Land distribution and distribution of small farms vary across countries

Author, year	Data used	Geographical coverage	Time	Findings
Eastwood et al. (2010)	Agricultural censuses	Global	1930–1990	Log of average farm size increased in North America and Europe and decreased in Asia and Africa
FAO (2013)	Agricultural censuses	Global	1930–2000	Decrease in average farm size for most regions except Europe. Decrease from 1960 to 1980 followed by a slight increase in Africa. Increase in Australia, New Zealand, the United States, Canada, Argentina, and Uruguay
Bosc et al. (2013)	Agricultural censuses	Global	1930–2000	Decrease in average farm size for Africa as a whole as well as in China. Farmland distribution: as per 2000 round, 73% of farms are smaller than 1 ha, and 85% are smaller than 2 ha. On average for 14 African countries, 80% of holdings are smaller than 2 ha in size; they operate about 25% of the agricultural land. In the European Union, 50% of farms are smaller than 2 ha in size and operate only about 2% of the agricultural land
Masters et al. (2013)	Rural and urban population data	Africa and Asia	1950–2050	Since 1950 average farm sizes have been decreasing for Africa and Asia as a whole, but in recent years they have begun increasing for Asia as a whole, while they will continue to decrease in Africa for quite some time
Adamopoulos and Restuccia (2014)	Agricultural censuses	Global	1990	Average farm size is larger in countries with higher average per capita GDP. Farmland distribution: in countries with high average incomes, farms larger than 20 ha operate 70% of the land, while in the poorest countries, 70% of the land is operated by farms smaller than 5 ha

(9%), Europe and Central Asia (7%), Latin America and the Caribbean (4%), and the Middle East and North Africa (Teeling et al. 2005). Among countries, China (35%) and India (24%) are leading the tables for share in total farms worldwide (Table 2). These two countries contribute largely in the share of upper-middle- and lower-middle-income countries, i.e., respectively, 47% and 36% of total farms worldwide

Table 2 World's top share of agricultural and farmland holdings

Agricultural holdings (in millions)		Farmland (in a million ha)	
Top 5 countries	Share of the world's total (%)	Top 5 countries	The share of the world's total (%)
China	35	China	11
India	24	Australia	9
Indonesia	4	United States	8
Russian Federation	4	Brazil	5
Bangladesh	3	Russian Federation	4

(China in former and India in the latter group). Low-income and high-income countries have, respectively, 13% and 4% of total farms worldwide.

It is generally perceived that average farm size has increased in developed countries in contrast to developing countries. However, categorizing countries on an income basis, i.e., low-income, lower-middle-income, upper-middle-income, and high-income, provides some varying trends in these categories. Average farm size has decreased in most of the countries belonging to the first two categories, whereas an increase in most countries belonging to third and a clear increase in the fourth category are found.

1.2 Significance in Agriculture

Agriculture is considered as a potential sector to invest in to cope with poverty and hunger especially in regions where a larger share of the population is directly or indirectly linked to agriculture (World Bank 2007; FAO 2012). However, it is debated that what scale and type of farming are appropriate to promote in achieving these targets (Larson et al. 2013). Many claims are being made about the importance of small farmers in agriculture production and food security. These claims range from the importance in national or regional scenarios to a global level. Some proponents claim that small farmers contribute a major share in global food production (International 2013), while others are more moderate by claiming that most of the food consumed in Africa and Asia comes from small farms (Bosc et al. 2013; IFAD 2013). Looking at global farmland distribution among farm size classes, Lowder et al. (2016) conclude that its implausible that small farms (<2 ha) operating 12% of global farmland are responsible for producing major share in global food production; however, the study acknowledges their share in regional and/or national food production.

Wiggins et al. (2010) highlighted three important aspects while discussing the significance of small farms, i.e., the productivity and efficiency of these farms, their potential in reducing poverty, and the fate of small farms in changing agrarian structures and development. Until now there has been a considerable consensus that small farms produce more yield per hectare than their larger counterparts (Barrett et al. 2010; Cornia 1985; Eastwood et al. 2010; FAO 2014b; Heltberg 1998; Larson et al. 2013; Meena et al. 2018); however, it still remains unclear how the productivity of small farms will shape up under changing technologies and access to inputs as well as output supply chains. Some factors for this higher productivity at small farms are family labor, motivation to work, self-supervision, and flexibility in farm operation management because most of the small farms also practice subsistence agriculture and economy of household heavily depends on farming. Recently, agriculture has come under limelight after policy-makers realized the potential of agriculture in coping poverty and the fact that it has been neglected previously in favor of industrialization.

While efficiency and equity can be advocated as grounds to bring small farms at the center of agricultural development (Hazell et al. 2010), some experts argue that case of small farmers is worsening in prevailing contemporary challenges (Ellis 2005; Maxwell 2005). Many of these challenges identified by different researches may not come under the scope of this chapter. However, few important challenges impacting small farms are changing production methods, environmental degradation, climate change, and changes in agricultural research. Small farms are suffering more by issues like climate change and environmental or soil degradation. On the other hand, apparently less research is focused on improving production technologies specifically for small farms. And most of the practices recommended to improve production, sustainability, and resource use efficiencies and mitigation strategies either lack of applicability at small farms or are not adopted by small farms due to neglected focus on dissemination of these technologies and/or lack of conducive socioeconomic environment. In short, considering the share of small farms in total global farms and the share of world population directly or indirectly involved with these farms, it is of immense importance that small farms must not be neglected in the face of challenges such as climate change. Small farms should be focused on innovative and sustainable agricultural practices under these challenging scenarios, and proper policies should be developed to ensure the adoption of these practices at these farms.

1.3 Soil Microbes: Role in Soil Health and Plant Productivity

Fertile soils with high organic matter are usually rich in soil microbes which include bacteria, actinomycetes, fungi, soil algae, and cyanobacteria. These microbes are not only present in thousands to millions in numbers per unit of soil but also diverse in their structures and functions (Alexander 1977). Microbial communities in the soils excrete diverse types of gluey compounds, i.e., mucilages, polysaccharides, etc., which help in cementing the soil aggregates and maintaining soil structures. Except for excretions, hairlike structures called "hyphae" spread like net below the soil surface and aggregate the soil particles like a hairnet, thus helping in stabilizing soil structures (Alexander 1977; Schadt and Classen 2007). Generally, the activity and functionality of soil microbes, as well as microbial population dynamics, normally depend on soil moisture, temperature, pH, aeration, minerals/inorganic nutrients, and the available substrate. The favorable soil conditions would definitely lead to increased activities of soil microbes (Stotzky and Norman 1961). Besides the role of soil microbes in improving soil structures, these microbes are also involved in the cycling of various nutrients like carbon (C), nitrogen (N), phosphorus (P), and sulfur (S) actively. Microbes are specialized enough in their functions that they not only involved the transformation processes of one element to another (e.g., ammonium N (NH_4^+) to nitrate N (NO_3^-)) but also control the amounts of these nutrients available to the plants. Hence, conversions of the nonavailable form of plant nutrients to available form, recycling of nutrients, and degradation/decomposition processes are the paramount contributions of soil microbes to the agroecosystem (Paul 2014; Sylvia et al. 2005). Furthermore, biological nitrogen fixation is exclusively governed by soil microbes/N-fixers that transform the inert N_2 into available N to the plants. Soil microbes, e.g., *Rhizobium* spp., make symbiotic relationships with some plants (esp. Leguminosae family), while some are free-living, e.g., *Azotobacter* spp. are promising microbes with significant contribution in nitrogen fixation and its availability to the plants (Rodelas et al. 1999).

Soil microbes are among the factors that affect nutrient availability to the plants. The soil-plant interrelationships are potentially governed by microorganisms present in the rhizosphere, while their presence as an individual and/or in communities may have far-reaching effects on soil health and plant productivities directly or indirectly (Chaparro et al. 2012). Soil microbes often have mutual relationships with plants which enable them to coexist in the intimate vicinity of the plant roots (Nihorimbere et al. 2011). Plant interactions with the communities of microbes present in the rhizosphere are complex enough that the individual plant-microbe interaction is difficult to probe (Mendes et al. 2011). The soil microbes can be pathogenic and beneficial as well. However, plants change the composition of root exudates according to the external stimuli with the root proximity (De-la-Peña et al. 2010; Flores 1999). The beneficial microbial community can also be artificially developed in the root zone by adding externally into the soil to improve soil fertility, nutrient availability, and plant growth (Guiñazú et al. 2009; Kirankumar et al. 2010). Once applied, these microbes develop communities in the soil and sustain themselves with developing symbiotic and/or nonsymbiotic relationships with the plants, hence avoiding the attack of other pathogens (Lucas 2010). The presence of beneficial soil microbes helps in the mineralization process within the soil and the nutrients available for plants. For example, application of plant growth-promoting rhizobacteria (PGPRs) in the soil improved the microbial biomass and hence the soil fertility status (Fließbach et al. 2009).

Seed inoculation with PGPRs is the best way to develop beneficial microbial communities within the rhizosphere to achieve better yields. For instance, Adesemoye et al. (2009) obtained better yields of tomato in plants applied with PGPRs with lower doses of N compared to no PGPRs with full N dose. Hence, PGPRs colonize themselves rapidly with enhanced mineralization processes at lower initial N, P, and K levels than higher (that could affect the efficiency of microbes) (Shaharoona et al. 2008). Seed inoculation of alfalfa with different strains of phosphate-solubilizing bacteria was studied by Guiñazú et al. (2009) and found that *Bacillus* and *Pseudomonas* spp. have promising effects on the root/shoot length, dry biomass, and symbiotic characteristics of alfalfa plants. The morphological growth and grain yield of lentil as well as P use efficiency were higher when the seed was inoculated with phosphate-solubilizing bacteria than no-inoculation (Smith et al. 2007). Similarly, mycorrhizal fungi have a substantial role in improving growth and nutrient use especially in nutrient-depleted soils (Zaidi et al. 2003). Application of arbuscular mycorrhizal fungi in the growing medium could also promote the growth and yield of plants along with the nutrient use efficiencies (Heidari and Karami 2014). The increased nutrient uptake, i.e., N, P, K, Ca, and Mg, was reported in maize and soybean grown under mycorrhizal-supplemented medium than non-mycorrhizal conditions (Khalil et al. 1994). Agronomic and/or crop management practices also largely affect soil microbiome; hence the use of organic fertilizers/manures and/or plant-derived decomposed materials and less intensive farming with minimum disturbance of soil could improve the soil structures and microbial populations in the soil (Crowder et al. 2010; Liu et al. 2007; Reeve et al. 2010).

2 Fertilizer Management and Crop Production

The environment, soils, and crop systems which are frequently in use by farmers of small farms of Asia and sub-Saharan Africa encounter different challenges for fertilizer management and application (Masso et al. 2017). Management plans should not only produce agronomic crops economically but also protect our water resources (Vanlauwe et al. 2015). Best management practices must include achievable management options and must be research-proven. Management plans are depending on past and current soil management, climate, specific site, operator expertise, and crops grown. Modern apprehensions about plant nutrients in the landscape have amplified interest in environmentally sustainable and economically sound fertilizer management. For fertilizer management, the fertilizer industry mainly focuses on four Rs: right rate, right fertilizer source, right place, and right time (Njoroge and Zingore 2015).

Fertilizer management has three prime goals:

(a) Match crop nutrient requirements to fertilizer nutrients.
(b) Cope fertilizer applications intelligently.
(c) Abate the nutrient transport from fields to water bodies.

All these three prime goals are discussed below:

2.1 *Match Fertilizer Nutrients to Crop Nutrient Requirements*

Soil testing: Fields should be analyzed for nutrients including N, P, K, and pH at least every 3 years and preferably more often. A soil having pH ranging from 6.5 to 7.0 is ideal for the majority of agronomic crops, but pH of soil less or greater than

Table 3 pH ranges for common crops

Sl. no	Crop	pH range
1.	Alfalfa	6.5–8.5
2.	Barley	5.5–8.5
3.	Blueberry	4.0–6.0
4.	Corn	6.0–7.5
5.	Cotton	5.5–8.5
6.	Clovers	6.0–7.5
7.	Potato	5.5–6.5
8.	Soybeans	6.0–7.5
9.	Sugar beet	6.5–8.5

this not only affects nutrient availability but also leads to toxicity (Table 3). Proper management of the soil pH exploits the effectiveness of the nutrients including N, P, K, etc. in the soil under practice (Hochmuth et al. 2014). Few examples include:

- Availability of potassium, phosphorus, and nitrogen is optimum at 6.5–7.5.
- At pH levels, less than 6.0, availability of phosphorus decreases due to iron and aluminum tie-up.
- Phosphorus availability declines again at pH levels greater than 7.5, due to calcium tie-up.
- Manganese and aluminum become toxic to plants at pH values of soil less than 5.0.
- At pH values of soil greater than 7.5, iron, manganese, copper, and zinc become unavailable and insoluble; this may lead to micronutrient deficiencies.

Nutrient Budgeting Manure analysis, soil testing, nutrient removal, and nutrient uptake data account for sources and outflows of all nutrients (Matheyarasu et al. 2017). Information in this regard makes it possible to calculate fertilizer application rates, specifically if animal manures are applied, and allows "what-if" analysis of different scenarios of fertilizer application rate (Matheyarasu et al. 2017).

Plant Nutrient Analysis Soil testing along with chemical analysis of different concentrations of plant nutrient in tissue may appraise the nutrient availability and soil fertility program (Marles 2017; Roe et al. 2015). Thus, it is valuable when "bad" and "good" sections of a field under evaluation can be contrasted.

2.2 Manage Fertilizer Applications Wisely

Soil Test-Based Recommendations Result of each soil test relevant to potash and phosphate is rated with index or category. Africa uses five soil tests which includes very low, low, medium, high, and very high (Tamene et al. 2017). The category compares the fertilizer amount present in the soil to the amount required by the plants (Solanki et al. 2017).

Using the Right Fertilizer for the Situation Usually, different fertilizer sources for the same nutrient seem to work differently; it is due to inherent differences between the fertilizer materials which are usually not taken into consideration. Plants cannot differentiate between sources of the specific nutrient. Nutrient ions, for instance, nitrate or phosphate, when present in the soil in the form of the solution are the same, no matter what their source is. Nitrogen (N) fertilizer efficiency particularly depends on the product along with managing practice (Lu and Tian 2017).

Proper Placement of Fertilizers Correct fertilizer placement is critical for efficiency. UAN solutions in the form of broadcast sprays must be avoided on hot or dry days unless the plant material is cultivated in irrigated condition or rain is imminent. Incorporation of fertilizers in the form of animal manure helps to prevent their immediate movement in the landscape.

Proper Application Timing Fertilizer application timing is crucial because the availability of nutrient to the plants is reduced over time. Nitrogen should be applied at the time of uptake by crop (Wu et al. 2017). Too early application of N fertilizer results in its leaching, and also the application of fertilizer as a supplement at the late growing stage is crucial. Application of inorganic phosphorus fertilizers in the fall before a spring-sown crop, as P is not mobile in the soil. Though phosphorus fixation in soils is common. Thus phosphorus is not very efficient because it mainly does not depend on application time (Khabarov and Obersteiner 2017).

Equipment Maintenance and Calibration Efficient nutrient applications can be achieved through equipment calibration and maintenance. Know the correct application width for the equipment and the material being applied; avoid overlaps within the field and onto field borders.

Precision Technology Precision technologies may allow more efficient fertilizer management of nutrient-deficient, acidic, or more responsive soil areas.

2.3 Minimize the Potential Transport of Nutrients from Fields to Water Bodies

Conservation Tillage Some nutrients, such as P ions, are closely bound to soil particles, so soil management that minimizes erosion also minimizes movement of those nutrients (Carr 2017). These management practices include strip-tillage, mulch tillage, no-tillage, or ridge-tillage (Carr 2017).

Proper Storage of Animal By-products Proper storage of poultry litter is important. Many poultry growers have dry stack sheds to store litter, but farmers are acquiring litter may need to store it temporarily.

Control Water Flow on and off Fields Controlling water flow with surface and subsurface drainage management reduces nutrient, pathogen, and pesticide runoff into downstream waters.

Maintain Buffers Sediment delivery to nearby water bodies along with rill and sheet erosion can be reduced by planting buffers between nutrient applications. Nutrients that move from planted areas are absorbed by planted buffer thus preventing their entry into surface waters (Datta et al. 2017a).

Use Cover Crops as a Nutrient Scavenger and Erosion Control Erosion can be reduced through crop residue and cover crops. Cover crops play a vital role in N movement reduction in the landscape by "N scavenging" left in the soil by the preceding crop.

2.4 Food for Food (Significance of Fertilizers in Crop Production)

Crop productivity mainly depends on phosphorous (P) and nitrogen (N) fertilization. Inefficient use of nutrients by plants results in environmental pollution, as greenhouse gases are released into the atmosphere and soluble forms of nitrogen (N), phosphorus (P), and carbon (C) leached into watercourses. Improvement in nutrient use efficiencies in agriculture is necessary for the development of sustainable agriculture, more efficient use of mineral fertilizers, nutrient management strategies, increased recovery and recycling of waste nutrients and better exploitation of the substantial organic and inorganic reserves of nutrients (Zahoor et al. 2014).

2.4.1 Nitrogen

Tisdale et al. (1993) comprehensively reviewed the nitrogen requirement of crops. Nitrogen is one of the indispensably important plant nutrients (Bandaogo et al. 2014; Liu et al. 2015; Wu et al. 2017). Nitrogen is a major yield-determining factor in maize production; its availability in adequate amount is essential throughout the growing season (Noor 2017; Sun et al. 2017). Nitrogen present in the decomposing organic matter is converted into N_4H^+ by soil microorganisms including fungi and bacteria through the mineralization process (Pidwirny 2002).

When nitrogen is supplied in adequate amount, then it will lead to a dark green color with vigorous vegetative growth. But an imbalance of nitrogen or its excess in relation to other nutrients including P, K, Mn, Zn, and S not only prolongs the growing period but also delays crop maturity (Fig. 1) (Gojon 2017). Nitrogen deficiency in plants leads to stunted growth and yellowing of leaves. This chlorosis or yellowing is usually spotted on the lower leaves first, while upper leaves remain green. Severe shortage of N will turn leaves brown (Gojon 2017).

Fig. 1 The maize ear phenotypes in GS1-deficient mutants and overexpressing lines are from N conditions which are suboptimal (N^+) or limiting (N^-): (**a**) wild-type (WT), gln1.4, gln1.3, and gln1.3/gln1.4 mutants; (**b**) WT null segregants and T4 transgenic lines 1 and 9 overexpressing Gln1-3 cDNA (cited from Hirel et al. 2007)

2.4.2 Phosphorous

Phosphorus being a second important nutrient after nitrogen limits the agricultural production in various regions of the globe (Kogbe and Adediran 2003; Souza et al. 2014). Phosphorus is the structural component of RNA and DNA, the two genetic entities which are crucial in a living organism for growth and reproduction. Both humans and plants derive their internal energy from P-based compounds, i.e., ADP (adenosine diphosphate) and ATP (adenosine triphosphate). This implies that an inadequate phosphorus supply will result in reduced RNA synthesis (the protein maker) leading to abnormal growth (Freitas et al. 2013). Phosphorus (P) deficiency in plants results in stunted growth, thin stems, and a limited root system. In some plants, older leaves turn to purple due to anthocyanins or purple pigment accumulation. Some crops produce small ear containing smaller and fewer kernels than usual due to phosphorus deficiency. In addition, grain yield is also severely affected (Jones et al. 2003). Phosphorus is usually concentrated in the seed in most plants, which is ultimately harvested (Fig. 2).

Fig. 2 Cob size severely reduced due to P deficiency (Source: http://bit.ly/Yara_UK)

2.5 Why Food (Fertilizer) Must Not Be Wasted

Fertilizer wastage hypothetically impacts various environmental components such as soil landscape, land resources, air, water, human health, and biodiversity (Savci 2012; Meena and Yadav 2014). Land resources are impacted which include changes in land use, for instance, grasslands, croplands, forests' habitat, marginal croplands, and water resources are changed (Savci 2012).

2.5.1 Environmental Cost for Manufacturing

Different processes of fertilizer production which are abating emissions up to some extent do not release harmful quantities of wastes to the atmosphere. The following emissions have a negative impact on the environment:

- Acidic fumes and ammonia maybe created in the surrounding areas of plants ultimately damaging the vegetation.
- Fluoride emissions not only damage the vegetation but are also harmful for livestock that consumes affected vegetation.
- Excessive release of dust in the atmosphere may cause damage to vegetation, particularly to pine forests.
- The release of excessive plant nutrients to the aquatic environment from the runoff water or scrubber system may result in eutrophication of water resources, particularly in closed reservoirs and lakes.

2.5.1.1 Global Warming

In different developing countries, including Pakistan, India, Bangladesh, etc., grain production mainly depends on fertilizer use, primarily nitrogenous fertilizers which increase up to many folds due to industrialization to fulfill increased food production

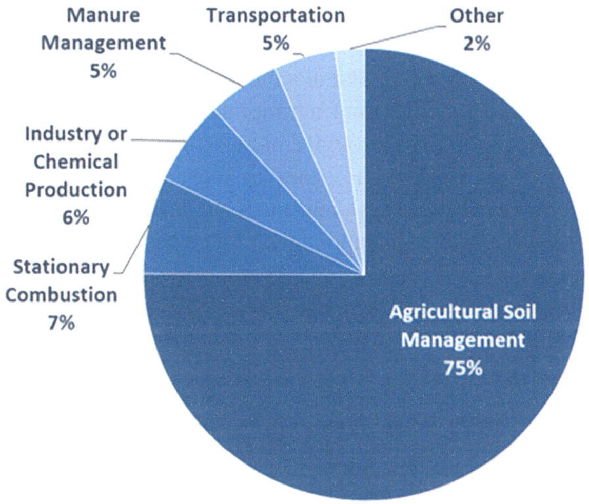

Fig. 3 Greenhouse gas emission from different sources

demand. Excessive N fertilizer use alters the global N cycle and thus results in the decline of total factor productivity ultimately decreasing NUE (nitrogen use efficiency). The N cycle is the transformation or conversion of N_2 (molecular nitrogen) into other reactive forms of N and then back to the original state of N. The nitrogen cycle mainly consists of five different steps including nitrogen fixation, nitrification, assimilation, ammonification, and denitrification (Pathak et al. 2016). As N_2O is a potent greenhouse gas, it has a direct contribution to global warming. Nitrogen fertilizer contributes 70% of the total N_2O emission and 77% of the total NO^{3-}(nitrous oxide) emission from agricultural soils of Africa (Fig. 3) (Pathak et al. 2010).

Nitrogen application in the form of fertilizer to agricultural soils not only affects nitrogenous compound emission, but it also affects the uptake and emission of methane (CH_4) and carbon dioxide (CO_2) from soils. Aerobic soils serve as a major sink of atmospheric CH_4 (methane), and CH_4 oxidation by the MOB (methane-oxidizing bacteria) is one of the significant methane removal processes.

2.5.1.2 Deteriorating Soil Health

According to different studies and researches, the chemical fertilizers' effects on the soil are not instantaneously obvious, as soils have a strong power of buffering due to their constituents. Over time, the pollution, soil degradation reactions occurring in the soil, and deterioration of soil fertility lead to deterioration of the current element balance. Besides this, toxic substances which accumulate within different vegetables ultimately affect animals and humans in a negative way.

2.5.1.3 Pollution

When fertilizer is applied in an inadequate amount, it may cause a loss in production and quality of important agronomic crops. But too much application of fertilizer causes air pollution through emission of nitrogen oxides such N_2O, NO, and NO_2. At the present time, some gases in the atmosphere such as carbon dioxide (CO_2), hydrogen sulfide (H_2S), and methane (CH_4), with chlorofluorohydrocarbons which are associated with halogen compounds, are present in the tropospheric ozone layer and contribute to the greenhouse effect. At the global level, atmospheric nitrogen oxide (N_2O) increases from 0.2 to 0.3% every year (Atılgan et al. 2007).

2.5.2 Economical Costs

The production of inorganic fertilizer hinges on the mining of two essential elements: K (potassium) and P (phosphorus). Chemical/inorganic fertilizers have been artificially manufactured that contain nutrients' indefinite quantities. Natural fertilizers are less expensive and are almost free compared to commercial chemical fertilizers. The latter contains components that may be toxic to the respiratory system and skin. Organic sources, rocks, and petroleum products are utilized during the manufacturing process of these fertilizers. Chemical fertilizers reduce soil fertility with the passage of time, while organic fertilizers not only improve but also sustain the soil fertility.

2.5.2.1 More Food (Fertilizer Application) for Less Food (Crop)

Farmers of developing countries especially in Africa and Asia do not have knowledge of optimum doses, fertilizer placement methods, and time of fertilizer application due to which they misleadingly use high fertilizer doses to maximize yield. Environmental, soil, and economic problems are created due to excessive supply of fertilizers. Some of the disadvantages of the excessive use of fertilizer are mentioned below.

- Their cost is relatively higher compared to organic fertilizers.
- Salts and other compounds are also constituents of chemical fertilizer besides essential nutrients for plant growth compared to organic fertilizer like manure. These salts and compounds cannot be absorbed by plants and remain on the soil surface. With the passage of time, these salts and compounds build up in the soil ultimately changing the soil chemistry.
- Application of too much fertilizer inclines to burn the plant.
- In cases of overwatering or heavy rains, the nutrients, salts, and other compounds drain into proximate water bodies. They not only pollute but also render them unhealthy for human or livestock consumption.

- Frequent fertilizer application not only causes an interruption in the natural biotic environment but also affects the structure and texture of the soil. Consistent inorganic fertilizers' application is a serious threat to the survival of soil organisms.

2.5.3 Indirect Impacts

2.5.3.1 Less Cleaner Agriculture

Fertilizers have indirect effects on environment and nature though these effects may be not that much clear on agriculture as it alters the biology of water bodies, depletes soil quality, and also affects human health. The main reason for the excessive use of fertilizer is demanding to achieve the goal of global food security. Some key points defining the side effects of excessive use of fertilizers on nature and environment are mentioned below.

Depletes the Quality of the Soil It is ironic to know that excessive use of chemical fertilizer results in an increment of acid levels in the soil, thus altering the fertility of the soil. Due to this reason, it is suggested to test soil at least once in every 3 years so that you can keep track whether you are using the right amount of fertilizers.

Biology of Water Bodies The use of too much chemical fertilizers in soil ultimately leads to eutrophication. Nitrates and phosphates being constituents of fertilizers are flooded into streams, lakes, rivers, and oceans through sewage and rain. As these substances are proven to be toxic for the aquatic life, thus it results in excessive algal growth in the water bodies, forming blooms and decreasing the oxygen levels.

Human Health Groundwater and drinking water are badly affected by nitrogen and other chemical constituents present in fertilizer. Contamination in drinking water results in the development of blue baby syndrome in infants (oxygen starvation in skin tissues), due to which their skin appears to be purplish or blue in color. Additionally, the use of pesticides and lawn fertilizers can cause health risks like chronic diseases in humans including cancer, especially in children.

2.5.3.2 Global Warming

Due to fertilizer overload, soil microbes unexpectedly belch high nitrous oxide levels, resulting in the emission of greenhouse gas with 300 times more heat-trapping power compared to carbon dioxide (Shcherbak et al. 2014). Soil microbes are known to involve in the conversion of nitrogen-rich fertilizers, including synthetic fertilizers and manure, into nitrous oxide (Fig. 4). Based on approximately 1000 field experiments, in the mid-2000s, the climate scientists are able to calculate that the dirt dwellers spew almost 1 kg of the greenhouse gas for every 100 kg of fertilizer, or roughly 1% (Mole 2014).

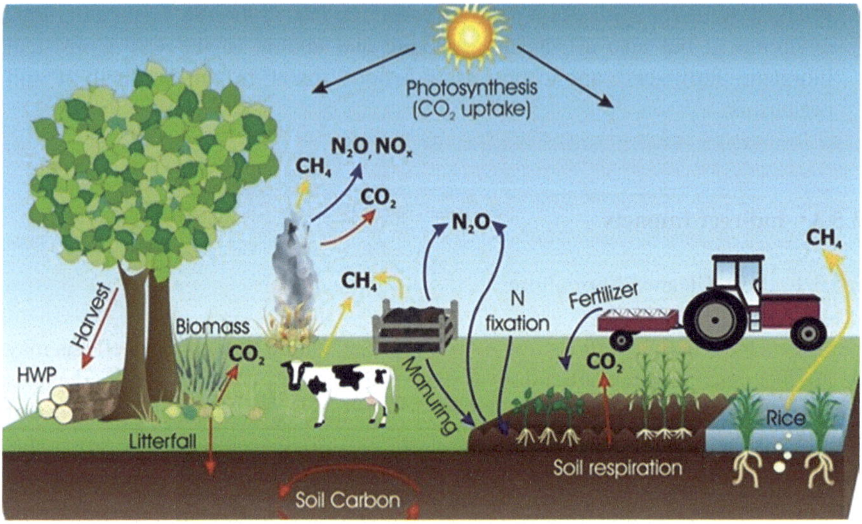

Fig. 4 Diagrammatical presentation of global warming due to fertilizers (Source: FAO 2014b)

2.5.3.3 Further Reduced Efficiency

Improving the fertilizer quality frequently used in Africa and South Asia is precarious from all perspectives such as social, environmental, and agronomic. The factors which affect this fertilizer in multiple dimensions cover a range of issues associated with diverse stages of fertilization. These can be classified into three groups: economic, technological and physical, and institutional factors (Raju 1989):

Economic factors:

- Fertilizer prices
- Other input prices
- Output prices

Technological and physical factors:

- Soil quality
- Other inputs' availability
- Climate
- Fertilizer use management
- Unstable use of various fertilizer nutrients
- Range of micronutrient deficiency

Institutional factors:

- Insufficient credit availability for not only farmers but also dealers
- Inadequate extension activities
- Insufficient infrastructures such as transportation and roads
- Scanty distribution facilities

- Ineffective domestic production
- Unavailability of quality fertilizers

All the abovementioned factors influence fertilizer use pattern in one or other way in different countries of Africa and South Asia through their comparative importance which varies crosswise the countries along with different regions, farmer groups, seasons, and site-specific characteristics of individual countries.

3 Nitrogen Management

3.1 Importance of Nitrogen for Crop Production

To meet the basic necessities, i.e., food and energy for increasing population, it is important to improve NUE in cereal crops at a low supply of fertilizer (White and Brown 2010). Cereal crops require a large amount of nitrogen to attain maximum production and yield and for which NUE is calculated, and its availability is >50% (Raun and Johnson 1999; Zhu 2000). Plants use nitrogen in the form of ammonium and nitrate, and as compared to other nutrients, N is taken up by the plants in the largest amount. To improve NUE in cereals, some basic strategies need to apply, for example, N management, plant breeding, and genetics (Table 4).

In preliminary studies, it is reported that before anthesis in wheat, 60–95% N comes from the remobilization of N stored in young roots and shoots (Palta and Fillery 1995). Post-flowering N uptake and N translocation to the grain are available in a very less quantity. Under N-deficient conditions, the size and the N content of the grain are significantly decreased (Dupont and Altenbach 2003). Until tillering, the soil supplies nitrogen to the plant. For nitrogen application, three applications are generally recommended: one at tillering, one at the beginning of stem elongation, and one at the second node stage (50–80 kg ha^{-1}, around 50 kg ha^{-1}, 40–50 kg h^{-1}, respectively). SPAD meter is mostly used for predicting grain N requirements in

Table 4 Typical inputs and cost for ammonia production

Inputs	Conventional reforming	Excess air reforming	Partial oxidation
Total energy of which inputs are used (Gj/TNH_3)[a]	32–35	32–35	39–45
Feedstock	24	26	32
Fuel	8–10	6–8	–
Water (t/t NH_3)[b]	0.7–1.5	0.7–1.5	1.2
Air (t/t NH_3)[c]	1.1	1.6	4

[a]High heat value (HHV). Total energy consumption is less
[b]Cooling water not included, variable from site to site
[c]Cooling air not included, variable from site to site
Source (UNEP et al. 1998)

Table 5 Typical inputs and cost for modern urea production

Inputs	CO_2 stripping	NH_3 stripping	ACES process	IDR process
NH_3 (t)	0.570	0.567	0.570	0.570
CO_2 (t)	0.750	0.735	0.740	0.740
Steam (t)	0.770 (ST), 0.800 (E)	0.760 (ST)	0.700 (ST), 0.570 (E)	0.760 (ST), 0.600 (E)
Cooling water (m^3)	60–70	80	51–60	60–75
Electricity (MJ)	54–396	76	108–436	72–432

Source (UNEP et al. 1998)
E electrometer, *ST* steam turbine

cereals (Lopez-Bellido et al. 2004). In other cereals, i.e., rice crop, the same N management strategies are used, but the plant mostly uses ammonium instead of nitrate, and vegetative organs use the remobilized N 70–90% of the total panicle N (Tabuchi et al. 2005; Vonk and Shackelford 2012; Datta et al. 2017b).

3.2 Typical Inputs and Cost of Production for Nitrogenous Fertilizers

The inputs and products of every fertilizer industry are different, and for the manufacturing of the product, there are different production procedures (Holzhaider et al. 2011). Fertilizer production industries include much input such as the major raw materials, i.e., energy, mineral phosphate, potassium salts, and sulfur, intermediate products such as ammonia and acids, and blending mixtures, and there are many small companies which purchase primary inputs to make mixtures, blends, and compounds to make fertilizers. In some minerals, i.e., phosphate, potassium salts, and ammonia, their production process is very difficult, so these minerals are used in the soil directly, so that one company input may be another company product (Table 5).

3.2.1 Typical Inputs and Cost for Nitric Acid Production

Apart from air and water, ammonia is the only external input, at a typical rate of 150–190 kg NH_3/t of fertilizer grade nitric acid. The commonest energy system involves initial start-up steam and power, energy recovery from tail gas in a gas turbine, electric pumps, and a condensing steam turbine to drive the compressors (UNEP et al. 1998).

3.2.2 Typical Inputs and Cost for Ammonium Nitrate (AN) and Calcium Ammonium Nitrate (CAN) Production

For the production, AN's mostly used inputs are 0.21 t NH_3, 0.78 t 100% HNO_3, and 25–40 kWh of power, and much larger amounts of power are needed to attain BAT standards for the production of AN. Steam requirements depend on the product concentration from the evaporator, so the net steam requirement may vary from 0 to 50 kg/t (UNEP et al. 1998).

3.3 A Major Player in Plant Nutrition: Soil

The soil is the key factor affecting nitrogen use efficiency which depends upon the capacity of the plant to extract inorganic nitrogen, assimilate nitrate and ammonium, and recycle organic nitrogen. Hodge and his coworkers (2000) reported that uptake of nitrogen by plant roots depends on soil conditions, microenvironment, and plant species, and estimated losses of nitrogen from soils are 50–70%. It is the need of the hour to develop agriculture on a sustainable basis to increase nutrient use efficiency by checking losses of nitrogen from the soil and therefore require minimum fertilizer. To achieve these goals, we need to make collaborative efforts by combining plant physiology, crop uptake ability, plant metabolism, and soil physical and chemical properties as a subject of interest.

Nitrogen mineralization and NO_3 levels in soil (being components of total nitrogen budget) are measured in the start of crop season (Koukoulakis and Papadopoulos 2001). Bolger and his team (2003) explained that organic content and soil texture determine the rate of nitrogen mineralization in soil. Mostly, in agricultural soils very less amount of plant-available nitrogen is present, so heavy doses of fertilizers are applied per annum to feed the plants for proper production. Most agricultural soils contain too little naturally occurring plant-available nitrogen to meet the needs of a crop throughout the growing season. Supplementary nitrogen applications are normally made each year to meet crop demand.

3.3.1 Important Soil Properties

Soil properties (geophysical, chemical, biological, etc.) define the productivity of soils. These including pH, organic contents, texture, structure, depth, and water holding capacity all boost soil fertility.

Soil Texture Precise application of fertilizer and lime depends on the understanding of the soil traits in each field. Consideration of topsoil texture is crucial to find out lime requirement.

Soil Structure Roots are the site of the first contact of any kind of nutrient and water for plant uptake. Better root growth is necessary for economical yield. It is, therefore, essential to sustain good soil structure, so that root development is not adversely affected by poor physical soil conditions, such as compaction.

Soil Organic Matter Plants use organic matter in soil as a major nitrogen source. It is composed of stable product humus, collected over a long period of time. Simply decayed portions of organic material vanish quickly, leaving at the back residues more resistant to decay. It has key agricultural, environmental, and ecological advantages, e.g., reduction in fertilizer requirement and improvement of soil conditions and biological activities.

Nitrogen Loss from the Soil System Leaching and runoff are two major processes for nitrogen loss in the environment (Follett and Delgado 2002; Meisinger and Delgado 2002), although ammonia volatilization is a source of a significant amount of NH3-N loss if susceptible sources of N volatilization are used without following suggested BMPs (Meisinger and Randall 1991). In anaerobic conditions, denitrification is also a major source of nitrogen loss (Mosier et al. 2002; Meisinger and Randall 1991; Peoples et al. 1995). Ample amount of research has been done to prove that leaching, nitrification, and denitrification are key sources of nitrogen losses (Delgado and Follett 2010; Follett et al. 1991; Meisinger and Delgado 2002; Mosier et al. 2002).

Denitrification is activated in the existence of reactive carbon, including residue carbon, and a small oxygen concentration in the soil. Ammonia losses occurred in alkaline conditions where human and livestock feces are applied.

Soil Quality and NUE Quality of agricultural soil depends on efficient cycling of nutrients, either direct uptake of nutrients to assimilate them or retain them ready to uptake in available form or in some way by prevailing the productivity and harvestability of crops (Brussaard et al. 2007; Giller et al. 1997; Harris et al. 2011; Keesstra et al. 2016). Schulte et al. (2014) explained soil quality is the ability of soil to prolong primary production, to regulate water availability, to decrease and adjust the emission of greenhouse gases, and to maintain biodiversity on a sustainable basis by efficient nutrient cycling. Coyle et al. (2016) endorsed their concept of soil quality. Schulte et al. (2014) identified five key functions as "ecosystem services" (Haines-Young and Potschin-Young 2018) interact with the other functions, e.g., quality, health, or fertility of the soil.

Consequently, approved site-specific management practices and evaluation methods of soil properties need considerations because the plasticity of functions changes with respect to time and place.

- Assess the effects of management scenarios on N dynamics and losses (Delgado et al. 2006; Shaffer and Delgado 2002).
- Develop and conduct national workshops to train technical personnel in nitrogen management techniques and procedures that protect soil quality.

- Develop and implement the concept of a national nitrogen trading tool (Delgado et al. 2008).
- Apply the concept of a tiered approach, as described by Shaffer and Delgado (2002).

3.4 *The Growing Challenges in Nitrogen Management (Nutrient Management Challenges)*

- Sustainable production of food to feed the ever-increasing population of the world is a big challenge for the agriculture sector by keeping the environment safe.
- Resources (soil, water, and air) for agricultural production are diminishing and face huge challenges of drought and flood due to a rapidly changing climate.
- Best nutrient management (especially N) using a durable approach is part of a sustainable agricultural system that is tough to meteorological change.
- Limited inorganic nitrogen sources are available and need ample amount of energy to produce fertilizers, so alternate sources and making organic matter available for plant production are big challenges in agricultural soils that make it the efficient beneficiary of resources on an economic basis.
- A nutrient gap occurs when the crop is removed from the field at harvesting time; that gap can be minimized by vigilant application of organic and inorganic fertilizer. Optimizations of nutrient uptake capacity of plants minimize the losses of nitrogen in the air, water, and soil.
- Interest in nonfood crops like biomass and biofuel crops is increasing for nonfarming purposes.

BMPs (*best management practices*) are particular cultural practices that endeavor maximum crop productivity by minimizing nutrient demand and losses. For nitrogen improving efficiency, the following points should be considered: (1) time of application; (2) timing and possibility of N loss; (3) crop stage-specific application of nitrogen; (4) application of N fertilizer in fall season; (5) application dose and rate; (6) sources of fertilizer; (7) site-specific placement method; (8) consistency in application with topography; (9) fortification of fertilizer to reduce losses; (10) organic nitrogen managements, e.g., manure; and (11) particular considerations to boost uptake of nutrients by plants.

3.4.1 Site-Specific Agricultural Soil Management

- Overapplication of nutrient is checked based on recommendations made on the specific soil conditions of each field parcel.
- Public health safety can be ensured because products are produced meeting the requirements for nitrate levels in plant tissue.

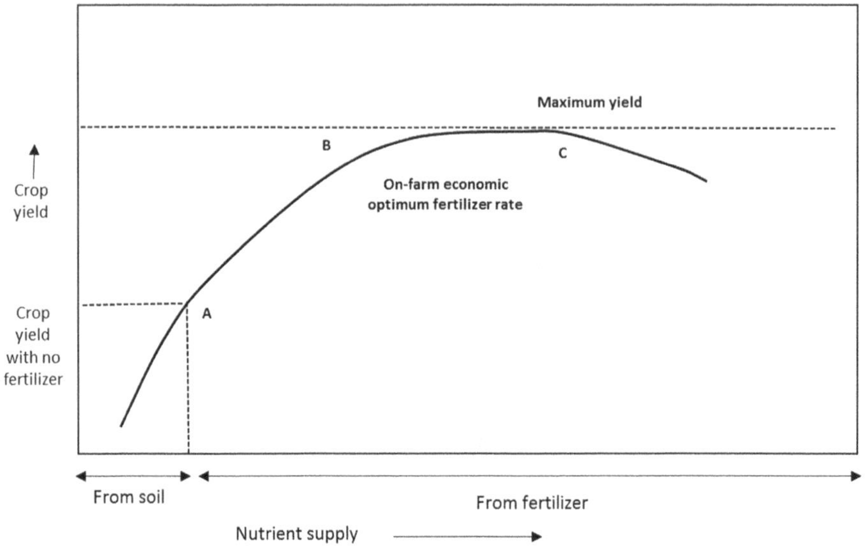

Fig. 5 Typical nitrogen response curve

- The environment can be protected because growers are advised to follow the least. Reduced polluting fertilization practice can be used to protect the environment from nitrate pollution of underground water, minimize eutrophication of surface waters, and augment the competitiveness of the agricultural foodstuffs.

3.4.2 Basis of the Recommendations

Provided there are sufficient supplies of water and another nutrient, N generally has a large impact on crop growth, yield, and quality. Figure 5 explained a typical nitrogen response curve. Applying nitrogen amplifies yield, but overdose application can reduce yield by creating problems such as lodging, foliar infections, and poor silage fermentation. When excessive nitrogen is applied, a larger proportion is unused by the crop. This causes an increase in financial cost and can also increase the risk of nitrogen losses.

Recommendations:

- Encourage efficient and environmentally safe management of crop nutrients.
- The unbiased and proficient use of plant nutrients from both organic N and inorganic N sources, at the farm and community levels, should be emphasized; use of local organic nitrogen sources should be encouraged.
- Inventive approaches to support and promote integrated plant nutrient management should be promoted.

- A comprehensive database desires to be developed for all inorganic and organic bases of nutrients, counting their amount, composition, processing techniques, their economic value, and their accessibility.
- The effects of micro- and macroeconomic policies on plant nutrient management should be evaluated.
- Maintain essential measures to lower expenses, reuse urban waste, protect land tenure, and augment production capacity.
- All measures should be required to lower the cost of fertilizers at farm-gate and to shrink the farmers' perception of the risk in the use of fertilizers by:

  1. Investing in distribution infrastructure
  2. Researching modern ways to share risks and to provide investment
  3. Cheering subregional cooperation for country-level fertilizer production facilities and/or procurement
  4. Improving dialogue between different sectors and agencies to arrive at a common approach to improve plant nutrition

Nitrogen management for intensification of cereals is a widely researched topic across the globe. Excellent papers and monographs are published regularly in international peer-reviewed journals. Besides those resources, readers may explore the following institutions for further information on nutrient management research trends or for collaboration opportunities:

- International Maize and Wheat Improvement Center www.cimmyt.org
- International Plant Nutrition Institute www.ipni.net
- International Institute of Tropical Agriculture www.iita.org
- Food and Agriculture Organization of the United Nations www.fao.org.
- United Nations Environment Programme Industry and the Environment www.unep.org
- United Nations Industrial Development Organization www.unido.org

4 Phosphorus Management: Focusing on Small Farmers

Phosphorus (P) exists in the Earth's crust in the form of phosphate rock containing around 5–10 Mt. of P, and it is only through the processes of weathering and leaching that P is mobilized into terrestrial systems (Suh and Yee 2011). The most abundant of the phosphate minerals is apatite, which is the most common naturally occurring P-containing mineral in the Earth's crust (over 95% of P) (Smil 2000). Phosphorus use exploded in the 1940s and 1950s, supported by the development of the modern phosphate industry. It has only been over the last 50 years that the human mobilization of P increased dramatically in order to cope with the unprecedented increase in the demand for food by rapidly growing global population (Tilman 2001; Datta et al. 2017c).

4.1 What Is P Management?

Phosphorus management is the process used to handle the application of P in the form of manure or fertilizers to crops, including their quantity, form, placement, and timing. On most farms, the nutrients predominantly come from manure and commercial fertilizer, but other sources, such as compost, can also be important. The purpose of nutrient management is to supply sufficient plant nutrients for optimum forage and crop yields and thus prevent excess applications that can contaminate water quality. A nutrient management plan (NMP) documents the source, rate, method, and timing of nutrient applications.

In 2009 the ENSET (Efficient Nutrient Supply in East Africa) project was implemented in Ethiopia and Kenya. The present report refers to the achievements made in Ethiopia. It evaluates nutrient management strategies for smallholder farms in Holetta (high-potential cereal-livestock mixed system) and Melkassa (semiarid low-potential area) in the central highlands of Ethiopia. In general, nutrient flow rates were higher in Holetta compared to Melkassa, but this was not reflected in the overall nutrient balances at the farm level, which was about −35 kg N/ha/season for both sites.

The 4R Nutrient Stewardship guidelines were developed by the fertilizer industry as a process to guide fertilizer best management practices (BMPs) in all regions of the world. Given that farmers purchase fertilizers at world prices in most regions, and these prices have been steadily increasing over time, most users are very cautious about the rates of nutrients they apply. To avoid unnecessary policy intervention by governments, the fertilizer industry needs to be unified in their promotion of BMPs that support improved nutrient use efficiency and environmental sustainability while supporting the farmer's profitability. Finally, some fertilizer products are preferred to others based on soil properties, like pH. Right rate means matching the fertilizer applied to the crop need – simple as that. Right place means making every effort to keep nutrients where crops can use them. This is an issue which poses the greatest challenge in smallholder agricultural systems, where most fertilizer is broadcast-applied, and in many cases without incorporation. Research indicates that fertilizer placement can not only improve crop response but also improve fertilizer use efficiency significantly by lowering nutrient application rates (Johnston and Bruulsema 2014; Meena et al. 2017).

The principles of nutrient management on small farms are similar to those of large farms. However, small farms may have a greater challenge when it comes to managing manure nutrients because of (1) the different characteristics of manure from the multiple animal species they keep, (2) the limited land area available to apply manure, and/or (3) a lack of equipment for manure management and spreading. The purpose of this publication is to outline how to quantify and determine the use of manure nutrients in small farms as a crop fertilizer in order to prevent environmental damage that may be caused by nutrient losses.

The sustainability of small-scale farming systems on marginal lands is still a topic of debate in scientific and institutional communities (Fig. 6). Global food security is a priority for the future development agenda of the United Nations. Given the high dependence of the modern global food production system on the continuous supply

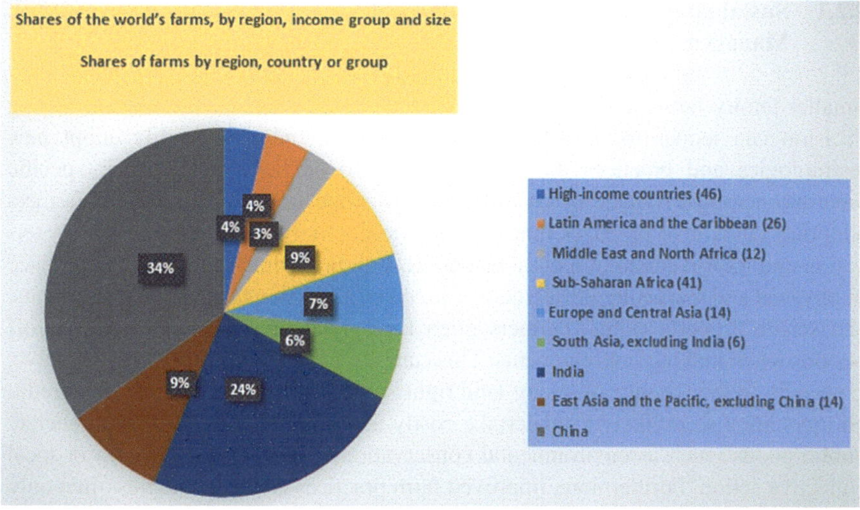

Fig. 6 Share of world's farms by region country or group. (Source: FAO 2013)

of commercial phosphorus (P) fertilizers, the goal of achieving global food security could be hampered by any form of the paucity of the global P resource. P is a finite, non-substitutable, nonrenewable, and geographically restricted resource. The anthropogenic influences on this critical resource are likely to pose a number of challenges to its sustainability.

Small farm systems are dependent on following the traditional P management practices. Nonetheless, due to the increase in population and poverty, these systems seem to be not viable and sustainable and eventually lead to soil and environment degradation. Among the main biophysical constraints to food production in traditional small farms is P deficiency caused by various factors, e.g., low level of inherent available P and low input of P fertilizer in highly weathered and acidic soils. It is estimated that 95% of acid soils in Asia, tropical Africa, America, Australia, and the Pacific Islands are deficient of P (Fairhust 1999).

4.2 *Why P Management Is Important?*

Demand for food is growing, while land and water resources are becoming ever more scarce and degraded. Climate change will make these challenges yet more difficult. Over the coming decades, farmers need to produce significantly larger amounts of food, mostly on land already in production. The large gaps between actual and potential yields for major crops show that there is significant scope for increased production through productivity growth on family farms. This can be achieved by developing new technologies and practices or through overcoming barriers and constraints to the adaptation and adoption of existing technologies and practices.

4.2.1 Sustainable Productivity on Small Farming Through P Management

Smaller family farms tend to rely on tried and trusted methods because one wrong decision can jeopardize an entire growing season, but they readily adopt new technologies and practices that they perceive to be beneficial in their specific circumstances. Nevertheless, several obstacles often stand in the way of farmers adopting innovative practices that combine productivity increases with the preservation and improvement of natural resources. Key impediments include the absence of physical and marketing infrastructure, financial and risk management instruments, and secure property rights. Farmers often face high initial costs and long payoff periods when making improvements. This can prove to be a prohibitive disincentive, especially in the absence of secure land rights and of access to financing and credit. Farmers are also unlikely to undertake costly activities and practices that generate public goods (such as environmental conservation) without compensation or local collective action. Furthermore, improved farm practices and technologies often only work well in the agroecological and social contexts for which they were designed, and if solutions are not adapted to local conditions, this can be a serious impediment to adoption. Local institutions, such as producers' organizations, cooperatives, and other community-based organizations, have a key role to play in overcoming some of these barriers (FAO 2014a).

4.3 Major Players Involved

- Agricultural research and advisory services generate public goods.
- Productivity, improved sustainability, lower food prices, poverty reduction, etc.
- Calling for strong government involvement.
- Local institutions, such as producers' organizations, cooperatives, and other community-based organizations, have a key role to play in overcoming some of these barriers.
- The effective functioning of local institutions and their coordination with the public and private sectors and with farmers themselves, both men and women, can determine whether or not small family farms can introduce innovative, sustainable improvements suited to their needs and local conditions.

4.4 Advances in P Management

P-Recycling Phosphorus recycling is addressed in detail as a key route to improving phosphorus sustainability, covering both agriculture reuse of sewage biosolids and manure through more technical P-recovery process routes. Success factors for P-recovery pathways are identified:

- The product should be familiar with the market and offer reliable quality and performance.
- Operational and investment costs should be competitive.
- Maximize P-recovery rates and ensure environmentally compatible operation, in order to ensure social acceptance.
- Preferably, the recovery process should be compatible with existing infrastructure to facilitate rollout.

5R Strategy Recently, several reviews (Crowther et al. 2016; Withers et al. 2015) have examined the P management issue from a holistic perspective. For instance, the 5R strategy (realign P inputs, reduce P losses, recycle P in bioresources, recover P in wastes, and redefine P in food systems) proposed by Withers et al. (2015) embraces aspects from the technical management of P resources on fields including management of the P soil legacy to the redesign of P use in society and agro-food systems, while Crowther et al. (2016) outlined how the biophysical management and behavioral factors should be taken into account in order to better utilize the soil legacy P as part of a more sustainable management. The analysis and assessment of P management are more often addressed from a system perspective at the global or national and regional.

Use of Cropping Sequences Cropping sequences are a primary factor impacting phosphorus management. Judicious implementation of crop rotations can improve soil P resources, efficiency of crop P use, and crop yield and yield stability (Łukowiak et al. 2016).

5 Challenges and Opportunities

Intensified agriculture scenario is requiring a need for holistic approach, looking at the full range of drivers and issues, as indicated above. This should include cost-effectiveness comparisons between phosphorus recycling or efficiency measures and resource import dependency. Perhaps the most critical and challenging way of initiating real and lasting changes in agricultural production and nutrient imbalances is to focus on consumer-driven programs and education rather than on farm production. Farmers are at the bottom of the food chain, and their decisions often are influenced by regional and even global economic pressures over which they have little or no control. Therefore, we have to look at new ways of using incentives to help farmers implement innovative measures to minimize on-farm nutrient surpluses. The challenge is to recognize how social policy and economic factors influence the nutrient management agenda. Equally important is that everyone is affected by and can contribute to a resolution of nutrient-related concerns. Extension and advisory services should focus on closing yield gaps and raising the labor productivity of small- and medium-sized farmers. Partnering with producers' organizations can help ensure that R&D and extension services are inclusive and responsive to farmers' needs.

References

Abele S, Frohberg K (2003) Subsistence agriculture in Central and Eastern Europe: how to break the vicious circle? Studies on the agricultural and food sector in Central and Eastern Europe, vol 22. IOMA, Halle

Adamopoulos T, Restuccia D (2014) The size distribution of farms and international productivity differences. Am Econ Rev 104:1667–1697

Adesemoye AO, Torbert HA, Kloepper JW (2009) Plant growth-promoting rhizobacteria allow reduced application rates of chemical fertilizers. Microb Ecol 58:921–929

Alexander M (1977) Introduction to soil microbiology-2, vol 2. Krieger Publishing Company, Melbourne

Atılgan A, Coşkan A, Saltuk B, Erkan M (2007) Antalya yöresindeki seralarda kimyasal ve organik gübre kullanım düzeyleri ve olası çevre etkileri. Ekoloji 15:37–47

Bandaogo A, Bidjokazo F, Youl S, Safo E, Abaidoo R, Andrews O (2014) Effect of fertilizer deep placement with urea supergranule on nitrogen use efficiency of irrigated rice in Sourou Valley (Burkina Faso). Nutr Cycl Agroecosyst 102:79–89

Barrett CB, Bellemare MF, Hou JY (2010) Reconsidering conventional explanations of the inverse productivity–size relationship. World Dev 38:88–97

Bolger TP, Angus JF, Peoples MB (2003) Comparison of nitrogen mineralisation patterns from root residues of Trifolium subterraneum and *Medicago sativa*. Biol Fertil Soils 38:296–300

Bosc P-M, Berdegué JA, Goïta M, JDvd P, Sekine K, Zhang L (2013) Investing in smallholder agriculture for food security: a report by the high level panel of experts on food security and nutrition. FAO, Rome

Brussaard L, de Ruiter PC, Brown GG (2007) Soil biodiversity for agricultural sustainability Agriculture. Ecotoxicol Environ 121:233–244

Carr P (2017) Guest editorial: conservation tillage for organic farming. Agriculture 7:19

Chaparro JM, Sheflin AM, Manter DK, Vivanco JM (2012) Manipulating the soil microbiome to increase soil health and plant fertility. Biol Fertil Soils 48:489–499

Cornia GA (1985) Farm size, land yields and the agricultural production function: an analysis for fifteen developing countries. World Dev 13:513–534

Coyle C, Creamer RE, Schulte RPO, O'Sullivan L, Jordan P (2016) A functional land management conceptual framework under soil drainage and land use scenarios. Environ Sci Pol 56:39–48

Crowder DW, Northfield TD, Strand MR, Snyder WE (2010) Organic agriculture promotes evenness and natural pest control. Nature 466:109–112

Crowther TW et al (2016) Quantifying global soil carbon losses in response to warming. Nature 540:104–108

Datta R, Anand S, Moulick A, Baraniya D, Pathan SI, Rejsek K, Vranova V, Sharma M, Sharma D, Kelkar A (2017a) How enzymes are adsorbed on soil solid phase and factors limiting its activity: a review. Int Agrophys 31(2):287–302

Datta R, Kelkar A, Baraniya D, Molaei A, Moulick A, Meena R, Formanek P (2017b) Enzymatic degradation of lignin in soil: a review. Sustainability 9(7):1163

Datta R, Baraniya D, Wang Y-F, Kelkar A, Meena RS, Yadav GS, Teresa Ceccherini M, Formanek P (2017c) Amino acid: its dual role as nutrient and scavenger of free radicals in soil. Sustainability 9(8):1402

Deininger K, Byerlee D (2011) Rising global interest in farmland. The World Bank, Washington, DC. https://doi.org/10.1596/978-0-8213-8591-3

De-la-Peña C et al (2010) Root secretion of defense-related proteins is development-dependent and correlated with flowering time. J Biol Chem 285:30654–30665. https://doi.org/10.1074/jbc.m110.119040
Delgado J, Follett R (2010) Advances in nitrogen management for water quality. Soil and Water Conservation Society, Ankeny
Delgado J, Alva A, Fares A, Paramasivam S, Mattos D Jr, Sajwan K (2006) Numerical modeling to study the fate of nitrogen in cropping systems and best management case studies. J Crop Improv 15:421–470
Delgado J, Shaffer M, Lal H, McKinney S, Gross C, Cover H (2008) Assessment of nitrogen losses to the environment with a Nitrogen Trading Tool (NTT). Comput Electron Agric 63:193–206
Dupont FM, Altenbach SB (2003) Molecular and biochemical impacts of environmental factors on wheat grain development and protein synthesis. J Cereal Sci 38:133–146
Eastwood R, Lipton M, Newell A (2010) Chapter 65: Farm size. In: Handbook of agricultural economics. Elsevier, Amsterdam. https://doi.org/10.1016/s1574-0072(09)04065-1
Ellis F (2005) Small farms, livelihood diversification, and rural-urban transitions: strategic issues in Sub-Saharan Africa. In: The future of small farms. IFPRI, Overseas Dev. Ins. and Imperial College, Kent, p 135
Fairhust T (1999) The importance, distribution and causes of phosphorus deficiency as a constraint to crop production in the tropics. In: Agroforestry forum, pp 2–8
FAO (2012) The state of food and agriculture 2012: investing in agriculture for a better future. Food and Agriculture Organization of the United Nations, Rome
FAO (2013) 2000 world census of agriculture methodological review (1996–2005). Food and Agriculture organization of the United Nations, Rome
FAO (2014a) Putting family farmers first to eradicate hunger. Food and Agriculture organization of the United Nations, Rome
FAO (2014b) The State of Food and Agriculture 2014: innovation in family farming. Food and Agriculture organization of the United Nations, Rome
Fließbach A, Winkler M, Lutz MP, Oberholzer H-R, Mäder P (2009) Soil amendment with *Pseudomonas fluorescens* CHA0: lasting effects on soil biological properties in soils low in microbial biomass and activity. Microb Ecol 57:611–623
Flores H (1999) 'Radicle' biochemistry: the biology of root-specific metabolism. Trends Plant Sci 4:220–226
Follett R, Delgado J (2002) Nitrogen fate and transport in agricultural systems. J Soil Water Conserv 57:402–408
Follett RF, Keeney DR, Cruse RM, Meisinger JJ, Randall GW (1991) Estimating nitrogen budgets for soil-crop systems. Soil Science Society of America, Madison
Freitas IF, Novais RF, Villani EMA, Novais SV (2013) Phosphorus extracted by ion exchange resins and mehlich-1 from oxisols (latosols) treated with different phosphorus rates and sources for varied soil-source contact periods. Rev Bras Ciênc Solo 37:667–677
Giller KE, Beare MH, Lavelle P, Izac AMN, Swift MJ (1997) Agricultural intensification, soil biodiversity and agroecosystem function. Appl Soil Ecol 6:3–16
Gojon A (2017) Nitrogen nutrition in plants: rapid progress and new challenges. J Exp Bot 68:2457–2462
Guiñazú LB, Andrés JA, Del Papa MF, Pistorio M, Rosas SB (2009) Response of alfalfa (*Medicago sativa* L.) to single and mixed inoculation with phosphate-solubilizing bacteria and Sinorhizobium meliloti. Biol Fertil Soils 46:185–190
Haines-Young R, Potschin-Young M (2018) Revision of the common international classification for ecosystem services (CICES V5.1): a policy brief. One Ecosystem 3:e27108
Harris JA et al (2011) Does soil biology hold the key to optimized slurry management? A manifesto for research. Soil Use Manag 27:464–469
Hazell P, Poulton C, Wiggins S, Dorward A (2010) The future of small farms: trajectories and policy priorities. World Dev 38:1349–1361

Heidari M, Karami V (2014) Effects of different mycorrhiza species on grain yield, nutrient uptake and oil content of sunflower under water stress. J Saudi Soc Agric Sci 13:9–13
Heltberg R (1998) Rural market imperfections and the farm size— productivity relationship: evidence from Pakistan. World Dev 26:1807–1826
Hirel B, Le Gouis J, Ney B, Gallais A (2007) The challenge of improving nitrogen use efficiency in crop plants: towards a more central role for genetic variability and quantitative genetics within integrated approaches. J Exp Bot 58:2369–2387
Hochmuth G, Mylavarapu R, Hanlon E (2014) The four Rs of fertilizer management University of Florida, Electronic Data Information Source
Hodge A, Robinson D, Fitter A (2000) Are microorganisms more effective than plants at competing for nitrogen? Trends Plant Sci 5:304–308
Holzhaider JC, Sibley MD, Taylor AH, Singh PJ, Gray RD, Hunt GR (2011) The social structure of New Caledonian crows. Anim Behav 81:83–92. https://doi.org/10.1016/j.anbehav.2010.09.015
IFAD U (2013) Smallholders, food security and the environment. International Fund for Agricultural Development, Rome
International F (2013) Powering up smallholder farmers to make food fair: a five point agenda
Jayne TS, Chamberlin J, Headey DD (2014) Land pressures, the evolution of farming systems, and development strategies in Africa: a synthesis. Food Policy 48:1–17
Johnston AM, Bruulsema TW (2014) 4R nutrient stewardship for improved nutrient use efficiency. Procedia Eng 83:365–370
Jones DL, Dennis PG, Owen AG, van Hees PAW (2003) Organic acid behavior in soils misconceptions and knowledge gaps. Plant Soil 248:31–41
Keesstra SD et al (2016) The significance of soils and soil science towards realization of the United Nations Sustainable Development Goals. Soil 2:111–128
Khabarov N, Obersteiner M (2017) Global phosphorus fertilizer market and national policies: a case study revisiting the 2008 price peak. Front Nutr 4:22. https://doi.org/10.3389/fnut.2017.00022
Khalil S, Loynachan TE, Tabatabai MA (1994) Mycorrhizal dependency and nutrient uptake by improved and unimproved corn and soybean cultivars. Agron J 86:949
Kirankumar R, Jagadeesh K, Krishnaraj P, Patil M (2010) Enhanced growth promotion of tomato and nutrient uptake by plant growth promoting rhizobacterial isolates in presence of tobacco mosaic virus pathogen. Karnataka J Agric Sci 21:309–311
Kogbe JOS, Adediran JA (2003) Influence of nitrogen, phosphorus and potassium application on the yield of maize in the savanna zone of Nigeria. Afr J Biotechnol 2:345–349
Koukoulakis P, Papadopoulos A (2001) Soil analysis interpretation. A. Stamoulis Publications, Athens. (in Greek)
Larson DF, Otsuka K, Matsumoto T, Kilic T (2013) Should African rural development strategies depend on smallholder farms? An exploration of the inverse-productivity hypothesis. Agric Econ 45:355–367
Liu B, Tu C, Hu S, Gumpertz M, Ristaino JB (2007) Effect of organic, sustainable, and conventional management strategies in grower fields on soil physical, chemical, and biological factors and the incidence of Southern blight. Appl Soil Ecol 37:202–214
Liu TQ, Fan DJ, Zhang XX, Chen J, Li CF, Cao CG (2015) Deep placement of nitrogen fertilizers reduces ammonia volatilization and increases nitrogen utilization efficiency in no-tillage paddy fields in central China. Field Crop Res 184:80–90
Lopez-Bellido R, Shepherd C, Barraclough P (2004) Predicting post-anthesis N requirements of bread wheat with a Minolta SPAD meter. Eur J Agron 20:313–320
Lowder SK, Skoet J, Raney T (2016) The number, size, and distribution of farms, smallholder farms, and family farms worldwide. World Dev 87:16–29
Lu C, Tian H (2017) Global nitrogen and phosphorus fertilizer use for agriculture production in the past half century: shifted hot spots and nutrient imbalance. Earth Syst Sci Data 9:181–192
Lucas JA (2010) Advances in plant disease and pest management. J Agric Sci 149:91–114
Łukowiak R, Grzebisz W, Sassenrath GF (2016) New insights into phosphorus management in agriculture—a crop rotation approach. Sci Total Environ 542:1062–1077

Marles RJ (2017) Mineral nutrient composition of vegetables, fruits and grains: the context of reports of apparent historical declines. J Food Compos Anal 56:93–103
Masso C, Baijukya F, Ebanyat P, Bouaziz S, Wendt J, Bekunda M, Vanlauwe B (2017) Dilemma of nitrogen management for future food security in sub-Saharan Africa – a review. Soil Res 55:425
Masters WA, Djurfeldt AA, De Haan C, Hazell P, Jayne T, Jirström M, Reardon T (2013) Urbanization and farm size in Asia and Africa: implications for food security and agricultural research. Glob Food Sec 2:156–165
Matheyarasu R, Sheshadri B, Bolan NS, Naidu R (2017) Nutrient budgeting as an approach to assess and manage the impacts of long-term irrigation using abattoir wastewater water. Air Soil Pollut 228:1–14
Maxwell S (2005) Six characters (and a few more) in search of an author: how to rescue rural development before it's too late. Agric Econ 32:61–73
Meena RS, Yadav RS (2014) Phonological performance of groundnut varieties under sowing environments in hyper arid zone of Rajasthan, India. J App and Nat Sci 6(2):344–348
Meena RS, Kumar S, Pandey A (2017) Response of sulfur and lime levels on productivity, nutrient content and uptake of sesame under guava (*Psidium guajava* L.) based agri-horti system in an acidic soil of eastern Uttar Pradesh, India. J Crop and Weed 13(2):222–227
Meena RS, Kumar V, Yadav GS, Mitran T (2018) Response and interaction of *Bradyrhizobium japonicum* and *Arbuscular mycorrhizal* fungi in the soybean rhizosphere: a review. Plant Growth Regul 84:207–223
Meisinger JJ, Randall GW (1991) Estimating nitrogen budgets for soil–crop systems. In: Follett, et al. (ed) Managing nitrogen for groundwater quality and farm profitability. SSSA, Madison, pp 85–124
Meisinger J, Delgado J (2002) Principles for managing nitrogen leaching. J Soil Water Conserv 57:485–498
Mendes R et al (2011) Deciphering the rhizosphere microbiome for disease-suppressive bacteria. Science 332:1097–1100
Mole B (2014) Earth & environment: nutrients may drop as CO2rises: crops' iron, zinc and protein may fall 5 to 10 percent by 2050. Sci News 185:12–12
Mosier A, Doran J, Freney J (2002) Managing soil denitrification. J Soil Water Conserv 57:505–512
Nagayets O (2005) Small farms: current status and key trends. In: The future of small farms, vol 355. International Food Policy Research Institute, Washington, DC
Nihorimbere V, Ongena M, Smargiassi M, Thonart P (2011) Beneficial effect of the rhizosphere microbial community for plant growth and health. Biotechnol Agron Soc Environ 15:327–337
Njoroge S, Zingore S (2015) 4R practices for efficient phosphorus management in Western Kenya, vol 99. IPNI, Norcross
Noor MA (2017) Nitrogen management and regulation for optimum NUE in maize–a mini review. Cogent Food Agric 3(1):1348214
Palta JA, Fillery IRP (1995) N application increases pre-anthesis contribution of dry matter to grain yield in wheat grown on a duplex soil. Aust J Agric Res 46:507
Pathak H, Bhatia A, Jain N, Aggarwal P (2010) Greenhouse gas emission and mitigation in Indian agriculture–a review. In: ING bulletins on regional assessment of reactive nitrogen, bulletin, vol 19. ING, SCON, New Delhi, pp 1–34
Pathak H, Jain N, Bhatia A, Kumar A, Chatterjee D (2016) Improved nitrogen management: a key to climate change adaptation and mitigation. Indian J Fertil 12:151–162
Paul EA (2014) Soil microbiology, ecology, and biochemistry. In: Perspective soil microbiology, ecology and biochemistry. Elsevier Science, Saint Louis, pp 3–24
Peoples MB, Mosier AR, Freney JR (1995) Minimizing gaseous losses of nitrogen. In: Bacon PE (ed) Nitrogen fertilization in the environment. Dekker, New York, pp 565–602
Pidwirny M (2002) Fundamentals of physical geography. Introduction to biogeography and ecology. The nitrogen cycle British Columbia Canada

Raju S (1989) Fertiliser use in Andhra Pradesh: an analysis of factors affecting consumption. Artha Vijnana: J Gokhale Inst Polit Econ 31:313

Raun WR, Johnson GV (1999) Improving nitrogen use efficiency for cereal production. Agron J 91:357

Reeve JR, Schadt CW, Carpenter-Boggs L, Kang S, Zhou J, Reganold JP (2010) Effects of soil type and farm management on soil ecological functional genes and microbial activities. ISME J 4:1099–1107

Rodelas B, González-López J, Martínez-Toledo MV, Pozo C, Salmerón V (1999) Influence of Rhizobium/Azotobacter and Rhizobium/Azospirillum combined inoculation on mineral composition of faba bean (*Vicia faba* L.). Biol Fertil Soils 29:165–169

Roe M, Pinchen H, Church S, Finglas P (2015) McCance and Widdowson's the composition of foods seventh summary edition and updated composition of foods integrated dataset. Nutr Bull 40:36–39

Savci S (2012) An agricultural pollutant: chemical fertilizer. Int J Environ Sci Dev 3:73–80

Schadt CW, Classen AT (2007) Soil microbiology, ecology, and biochemistry. Soil Sci Soc Am J 71:1420

Schulte RPO, Creamer RE, Donnellan T, Farrelly N, Fealy R, O'Donoghue C, O'hUallachain D (2014) Functional land management: a framework for managing soil-based ecosystem services for the sustainable intensification of agriculture. Environ Sci Pol 38:45–58

Shaffer MJ, Delgado JA (2002) Essentials of a national nitrate leaching index assessment tool. J Soil Water Conserv 57:327–335

Shaharoona B, Naveed M, Arshad M, Zahir ZA (2008) Fertilizer-dependent efficiency of Pseudomonads for improving growth, yield, and nutrient use efficiency of wheat (*Triticum aestivum* L.). Appl Microbiol Biotechnol 79:147–155

Shcherbak I, Millar N, Robertson GP (2014) Global metaanalysis of the nonlinear response of soil nitrous oxide (N2O) emissions to fertilizer nitrogen. Proc Natl Acad Sci 111:9199–9204

Smil V (2000) Phosphorus in the environment: natural flows and human interferences. Annu Rev Energy Environ 25:53–88

Smith T, Rana RS, Missiaen P, Rose KD, Sahni A, Singh H, Singh L (2007) High bat (*Chiroptera*) diversity in the Early Eocene of India. Naturwissenschaften 94:1003–1009

Solanki RL, Naagar KC, Agarwal SK (2017) Effect of fertilizer based on soil testing for better production of maize (*Zea mays* L.) in South Rajasthan, India. Int J Curr Microbiol App Sci 6:725–728. https://doi.org/10.20546/ijcmas.2017.608.092

Souza RM, Sobral LF, Viégas PRA, Oliveira A Jr, Carvalho MCS (2014) Eficiência agronômica de fosfatos de rocha em solo com elevado teor de cálcio trocável. Rev Bras Ciênc Solo 38:1816–1825

Stotzky G, Norman AG (1961) Factors limiting microbial activities in soil. Arch Mikrobiol 40:341–369

Suh S, Yee S (2011) Phosphorus use-efficiency of agriculture and food system in the US. Chemosphere 84:806–813

Sun X et al (2017) Subsoiling practices change root distribution and increase post-anthesis dry matter accumulation and yield in summer maize. PLoS One 12:e0174952

Sylvia DM, Fuhrmann JJ, Hartel PG, Zuberer DA (2005) Principles and applications of soil microbiology, vol QR111 S674 2005. Pearson Prentice Hall, Upper Saddle River

Tabuchi M, Sugiyama K, Ishiyama K, Inoue E, Sato T, Takahashi H, Yamaya T (2005) Severe reduction in growth rate and grain filling of rice mutants lacking OsGS1;1, a cytosolic glutamine synthetase 1;1. Plant J 42:641

Tamene L, Amede T, Kihara J, Tibebe D, Schulz S (2017) A review of soil fertility management and crop response to fertilizer application in Ethiopia: towards development of site-and context-specific fertilizer recommendation

Teeling EC, Springer MS, Madsen O, Bates P, O'Brien SJ, Murphy WJ (2005) A molecular phylogeny for bats illuminates biogeography and the fossil record. Science 307:580–584

Tilman D (2001) Forecasting agriculturally driven global environmental change. Science 292:281–284. https://doi.org/10.1126/science.1057544

Tisdale S, Nelson W, Beaton J, Havlin J (1993) Soil fertility and fertilizers. Macmillan Publishing Company, New York 5th

UNEP, UNIDO, IFA (1998) Mineral fertilizer production and the environment part 1. United Nations Enviornment Programme, Industry and the Environment, Paris

Vanlauwe B et al (2015) Integrated soil fertility management in sub-Saharan Africa: unravelling local adaptation. Soil 1:491–508

Vonk J, Shackelford TK (eds) (2012) The Oxford handbook of comparative evolutionary psychology. Oxford Library of Psychology. Oxford University Press, New York

White PJ, Brown PH (2010) Plant nutrition for sustainable development and global health. Ann Bot 105:1073–1080

Wiggins S, Kirsten J, Llambí L (2010) The future of small farms. World Dev 38:1341–1348

Withers PJA et al (2015) Stewardship to tackle global phosphorus inefficiency: the case of Europe. Ambio 44:193–206

World Bank (2007) World development report 2008 (overview). The World Bank. https://doi.org/10.1596/978-0-8213-7297-5

Wu M, Li G, Li W, Liu J, Liu M, Jiang C, Li Z (2017) Nitrogen fertilizer deep placement for increased grain yield and nitrogen recovery efficiency in rice grown in Subtropical China. Front Plant Sci 8:1227

Zahoor W, Khanzada H, Bashir U, Aziz K, Zahir S, Faheem A (2014) Role of nitrogen fertilizer in crop productivity and environmental pollution. Int J Agric For 4:201–206

Zaidi A, Khan MS, Amil M (2003) Interactive effect of rhizotrophic microorganisms on yield and nutrient uptake of chickpea (*Cicer arietinum* L.). Eur J Agron 19:15–21

Zhu Z (2000) Loss of fertilizer N from plants-soil system and the strategies and techniques for its reduction. Soil Environ Sci 9:1–6

Metal Toxicity and Nitrogen Metabolism in Plants: An Overview

Saddam Hussain, Abdul Khaliq, Mehmood Ali Noor, Mohsin Tanveer, Hafiz Athar Hussain, Sadam Hussain, Tariq Shah, and Tariq Mehmood

Abstract Heavy metal pollution has emerged as a severe threat to the environment as well as global food security. Exposure of plants to the heavy metals could cause perturbations in various physiological, biochemical, and metabolic processes including nitrogen (N) uptake and assimilation. Here, we discussed the effects of metal toxicity on N uptake, N forms, mechanism of metal toxicity, and nitrogen assimilation in plants. We provided a detailed description on the behavior of various enzymes including nitrate reductase (NR), nitrite reductase (NiR), glutamine synthetase (GS), glutamate synthase (GOGAT), and glutamate dehydrogenase (GDH) under metal toxicity. We highlighted the response of various nitrogenous compounds and their special role under metal toxicity. In addition, we discussed the effects of excess metals on N fixation in plants and provided the guidelines for further studies.

Keywords Amino acids · Ammonium · Heavy metals · Nitrogen fixation · Nitrogen metabolism

---

S. Hussain (✉) · A. Khaliq · S. Hussain
Department of Agronomy, University of Agriculture, Faisalabad, Punjab, Pakistan

M. A. Noor
Institute of Crop Sciences, Chinese Academy of Agricultural Sciences, Key Laboratory of Crop Physiology and Ecology, Ministry of Agriculture, Beijing, China
e-mail: mnoor@caas.cn

M. Tanveer
Tasmania Institute of Agriculture, University of Tasmania, Hobart, Australia
e-mail: mohsin.tanveer@utas.edu.au

H. A. Hussain
Institute of Environment and Sustainable Development in Agriculture, Chinese Academy of Agricultural Sciences, Beijing, China

T. Shah
Oil Crops Research Institute, Chinese Academy of Agriculture Science, Wuhan, China
e-mail: tariqshah@aup.edu.pk

T. Mehmood
Institute of Cotton Research, Chinese Academy of Agricultural Sciences, Beijing, China

R. Datta et al. (eds.), *Carbon and Nitrogen Cycling in Soil*,
https://doi.org/10.1007/978-981-13-7264-3_7

Abbreviations

As	Arsenic
Cd	Cadmium
Cu	Copper
Fe	Iron
GB	Glycine betaine
GDH	Glutamate dehydrogenase
Gln	Glutamine
Glu	Glutamate
GOGAT	Glutamate synthase
GS	Glutamine synthetase
Hg	Mercury
MDA	Malondialdehyde
Mn	Manganese
Mo	Molybdenum
N	Nitrogen
NH_4^+	Ammonium
Ni	Nickel
NiR	Nitrite reductase
NO_3^-	Nitrate
NR	Nitrate reductase
Pb	Lead
ROS	Reactive oxygen species
SH	Sulfhydryl
V	Vanadium
Zn	Zinc

1 Introduction

Heavy metal toxicity has become a serious environmental threat because of rapid industrialization, technological advancement, and disturbance of natural ecosystem due to a rapid increase in human population (Sarwar et al. 2010; Anjum et al. 2017; Ayangbenro and Babalola 2017). Heavy metals enter the agroecosystem by natural (such as erosion, volcanic eruptions, weathering of minerals, comets, etc.) and anthropogenic (such as coating, biosolids, alloy production, atmospheric depositions, pesticides, etc.) processes (Ayangbenro and Babalola 2017; Sarwar et al. 2017; Meena et al. 2017). Some types of soils inherit these heavy metals from parent material they are being originated having a high concentration of some of these metals which are hazardous to plants. Anthropogenic sources include sewage sludge, phosphate fertilizers, urban traffic, anthropogenic emissions from power stations, battery production, explosive manufacturing, cement industries, smelting, improper stacking of industrial solid waste, etc. (Wu et al. 2004; Sarwar et al. 2017; Ayangbenro and Babalola 2017; Marfo et al. 2015; Lojkova et al. 2015). These sources result in the buildup of higher concentrations of heavy metals in the agricultural soils.

Keeping in view their functions, heavy metals are categorized into two groups, essential and nonessential. Essential metals include zinc (Zn), manganese (Mn), copper (Cu), iron (Fe), and nickel (Ni) which have important roles in physiochemical and metabolic processes of living organism, such as in electron-transferring proteins and functioning as cofactors for some enzymes (Fageria et al. 2009; Chaffai and Koyama 2011; Fashola et al. 2016), while nonessential metals (like cadmium (Cd), mercury (Hg), and lead (Pb)) are not needed by plants for any biological functions (Dabonne et al. 2010). Plants exposed to high concentrations of heavy metals respond by altering the cellular metabolic process and gene expression (Hussain et al. 2004; Chaffai and Koyama 2011; Choppala et al. 2014). Heavy metals can cause the growth reduction in plants by decreasing photosynthetic activity and hindering the chlorophyll synthesis (Fig. 1; Gumpu et al. 2015). Heavy metals may cause hazardous effects in plants by disturbing essential groups of enzymes, blocking of essential functional groups and damaging the stability of important biomolecules, and disrupting the antioxidative defense mechanism as a result of higher generations of reactive oxygen species (ROS) (Chaffai and Koyama 2011; Choppala et al. 2014; Sarwar et al. 2015; Anjum et al. 2016; Ashraf et al. 2017).

Nitrogen (N) is among the most abundant essential macronutrients on earth (Cesco et al. 2010), and it is the critical limiting element for plant growth due to its unavailability (Graham and Vance 2000; Hussain et al. 2016). Plants can acquire N from different molecules (like nitrate, ammonium, urea, and amino acids) and utilized for different metabolic purposes, including the production of proteins, nucleic acids, as well as storage and signaling molecules (McAllister et al. 2012). Nitrate (NO_3^-) and ammonium (NH_4^+) are the major N sources for plants, and optimum conditions for activities of enzymes involved in the conversion of inorganic N into organic N are pivotal to plant biomass accumulation, growth, and final productivity. In plants, NO_3^- being readily mobile can be stored in the vacuole and is the main source of N under well-aerated aerobic soil conditions. However, it must be reduced to NH_4^+ for the synthesis of proteins and other organic compounds in plants (Garnett et al. 2009). Nitrate reductase (NR) converts NO_3^- into nitrite in the non-organelle portions of the cytoplasm. All the living plant cells have the ability to reduce NO_3^- to nitrite, using the energy and reductant (NADH, NADPH) of photosynthesis and/or respiration in green tissues and of respiration in nongreen tissues and roots. Ammonium is the only reduced form of N available to plant for assimilation into N-carrying amino acids such as glutamate (Glu), glutamine (Gln), aspartate, and asparagine (Ireland and Lea 1999; Antunes et al. 2008; Dadhich and Meena 2014). For the biosynthesis of these N-containing amino acids, various enzymes such as Glu synthase, Gln synthetase (GS), glutamate dehydrogenase (GDH), alanine amino transferase, aspartate amino transferase, and asparagine synthetase are important (Garnett et al. 2009).

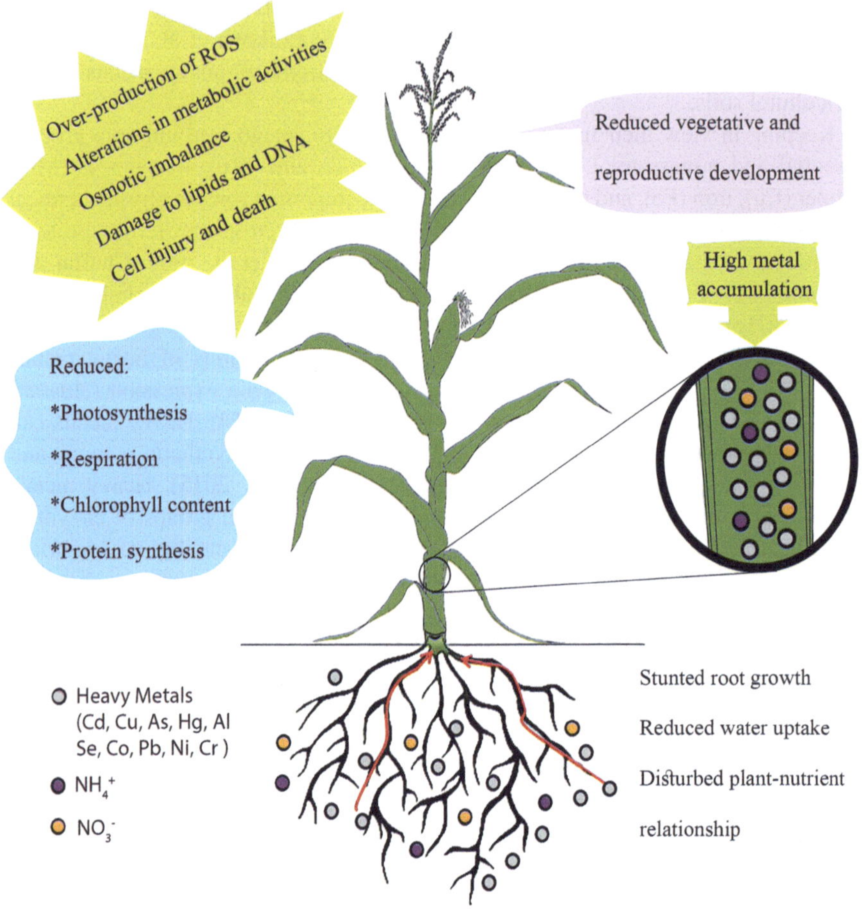

Fig. 1 Morphophysiological responses of plants to metal toxicity in soil

Heavy metals severely hamper the N metabolism by reducing the NO_3^- uptake and altering the enzyme activity. In the past, the effect of Cd, Pb, Hg, Cu (copper), Ni (nickel), Mo (molybdenum), and V (vanadium) on N metabolism in different plant species has been reported; however, the effects of heavy metals on N assimilatory enzymes vary with the sensitivity level of enzymes and their localization in the cells/organs, mobility of metals, metal concentration in soil, and time of plant exposure to metal toxicity. Heavy metals can alter the activity of various N assimilatory enzymes by binding to the vital sulfhydryl (SH) groups (Prasad and Strzalka 2013). The metal-induced decrease in N assimilation process can also reveal the disruption of the general homeostasis of metabolic activities in plants. Heavy metals that induce the alterations in plant water status (Belimov et al. 2015); uptake, transfer, and supply of essential nutrients (Zia-ur-Rehman et al. 2015; Sarwar et al. 2015); photosynthesis (Emamverdian et al. 2018); and generation of ROS (Malar et al. 2016; Riaz et al. 2018b) ultimately affect the N metabolism in plant.

In the present article, we synthesized the effects of heavy metal on N uptake, N forms, mechanism of metal toxicity, and nitrogen assimilation. We provided a detailed description on the behavior of various enzymes including NR, nitrite reductase, Glu synthase, GS, and GDH under heavy metal stresses. We also highlighted the response of various nitrogenous compounds and their special role under metal toxicity. In addition, we discussed the effects of excess metals on N fixation in plants and provided future guidelines for further studies.

2 Heavy Metals in the Environment: Challenge for Plants

Heavy metals are pollutants in the environment, and their higher concentration creates the problems for ecological, evolutionary, nutritional, and environmental reasons. Although some metals (Zn, Cu, Fe, and Ni) are essential micronutrients and are required for normal growth and development of plants (Verkleij et al. 2009; Fageria et al. 2009; Imran et al. 2015), nevertheless, these metals are harmful to plants at higher levels and may lead to soil quality deterioration, plant growth inhibition, significant yield reduction, and poor quality of food (Ayangbenro and Babalola 2017), while some metals and metalloids (intermediate between metals and nonmetals) such as Cd, Cr, Pb, Se, As, and Hg have no known biological functions in living organisms (Seth 2012; Ayangbenro and Babalola 2017; Datta et al. 2017b).

Metal uptake from soil is carried out by cortical root tissues because of their resemblance with some essential micronutrients and transfer to the aerial plant parts via xylem vessels (Ali et al. 2013; Sarwar et al. 2017). Heavy metals disrupt several physiological, biochemical, and molecular processes in plant and could inhibit plant growth, disrupt the cellular functions, and ultimately cause cell death (Fig. 1; Xu et al. 2009; Fashola et al. 2016; Ashraf et al. 2017; Riaz et al. 2018a, b, c). Metal-induced growth inhibition might be explained on the basis of damage cell membrane, decreased synthesis of metabolites, photosynthetic rate, and chlorophyll content (Chibuike and Obiora 2014; Anjum et al. 2016; Ashraf et al. 2017). In some plants, metal toxicity may also cause drought stress because of significant decreases in stomatal conductivity, transpiration, and leaf relative water contents owing to less number of xylem vessels and poor cell enlargement (Saifullah et al. 2014). Metal-induced toxicity can cause cell membrane damage and destruction of biomolecules (like proteins and lipids) and cellular organelles in plants due to increase in the production of reactive oxygen species (ROS) under stress (Ekmekci et al. 2008; Ashraf et al. 2017). Moreover, nonessential heavy metals inhibit various biological mechanisms by altering the structure of biomolecules and important stress regulatory proteins (Sarwar et al. 2010).

3 Nitrogen Nutrition in Plants: No Alternative Lies

Even though N is among the most abundant elements on earth (about 80% of the earth's atmosphere), it is the critical limiting element for the growth of plants (Greenwood and Earnshaw 2012; Wang et al. 2017; Li et al. 2018). In plants, N is required in huge amount, as it constitutes 1.5–2.0% total dry matter of plants (Frink et al. 1999; Xiong et al. 2006a). In most agricultural soils, NO_3^- is the major form of the inorganic N available to the plants (Xiong et al. 2006a). Uptake of N followed by N reduction and N assimilation in plants is the only way through which plants may convert inorganic N into organic N form. The available N forms vary with organic (urea, amino acids, etc.) and inorganic (NH_4^+, NO_3^-, dinitrogen) compounds and the habitat of the plant. Nitrogen is a key constituent of proteins, amino acids, enzymes, alkaloids, vitamins, and some growth hormones. Increased NO_3^- concentration in plants has a beneficial impact as it regulates the synthesis of proteins and amino acids, while the excess NH_4^+ content can be toxic as it promotes the formation of amides (Britto and Kronzucker 2002; Ram and Meena 2014). Geiger et al. (1999) studied the effects of different concentrations of ammonium nitrate and potassium nitrate on tobacco and reported that alanine, aspartate, Gln, arginine, and serine were increased considerably at higher N inputs. Terce´-Laforgue et al. (2004) also examined the influence of N fertilization (NO_3^-, NH_4^+, and low nitrogen) on the amino acid profile in tobacco leaves and concluded that Glu and total amino acids were increased from the old to young leaves under low N supply.

4 Interplay of Nitrogen Uptake and Heavy Metal Stress

Nitrogen (N) is mainly taken up by the plants in the organic form (NO_3^- and NH_4^+) from the soil solution. The plant's preference to uptake either N form depends on various factors, including the plant species, the concentrations of NO_3^- and NH_4^+ in soil, and the soil pH (Perkins et al. 2011; Miller 2016; Aziz et al. 2018). The soil N pool is driven mainly through mineralization, nitrification, and ammonification, which are believed to be influenced by soil microbiota. However, among the various abiotic stresses affecting the overall N cycle in the soil, heavy metal stress is considered to influence the nitrification and ammonification processes through modulating the activities of microbiota (Kapoor et al. 2015). Soil moisture, aeration, temperature, and the pH are the key drivers controlling these nitrifying agents and the respective processes, and therefore the applied N source in the soil fluctuates widely the pH of soil which in turn regulates the net proton extrusion rates. Generally, the NO_3^- fertilizer source is preferred over the NH_4^+ due to its readily available form for direct adsorption by plant roots (no need to undergo any further conservation) and due to its less volatile nature. Therefore, the soil pH is crucial in determining the availability of heavy metals for plant uptake (Bravo et al. 2017; Datta et al. 2017a). NO_3^- source decreases the soil acidification, and it prevents the accumulation and uptake of heavy metals by plants, while this case is opposite for

NH_4^+ nutrition. Here we discussed the synergistic effects of both the N source and the heavy metal stress influencing the plant uptake mechanism.

4.1 Nitrate Uptake

Soil type (structure and texture) is the key determinant of the N fertilizer source to be applied, which is readily available for plant uptake in the rhizosphere (Li et al. 2018; Aziz et al. 2018). Nitrate is believed to be the best source in well-aerated agricultural soils as compared to NH_4^+. Nitrification is the process in which ammonia is converted to NO_3^- through biological oxidation, and these conversion steps are mediated by aerobic nitrifying bacteria requiring sufficient soil aeration and oxygen for the oxidation process. Therefore, the oxidation of NH_4^+ is restricted in water-logged soils (Ma et al. 2016). Established reports have indicated that the NO_3^- absorption process by the plant is the active process, as the NO_3^- influx to plant roots is proton-coupled depending upon the H^+ pumping (Miller and Smith 1996; Zhou et al. 2016). The presence of the excessive heavy metals in soil affects the NO_3^- uptake by root cells (Fig. 1). Among these heavy metals, Cd is considered the most effective NO_3^- inhibitor which is reported to repress the NO_3^- absorption process very sharply even at very low Cd level (Boussama et al. 1999a; Gouia et al. 2000b; Huang and Xiong 2009). It was observed that excluding Cd from the uptake solution lasted its impact of NO_3^- inhibition till 96 h (Gouia et al. 2000b). However, this inhibitory effect was observed for several plant species. One vital reason for the reduced NO_3^- uptake is the damaged plasma membrane in roots due to excessive membrane leakage and the MDA contents (Huang and Xiong 2009). Such reduced NO_3^- contents were observed more for roots versus shoot. Wang et al. (2008) observed better shoot growth under high Cd stress, but the NO_3^- concentrations were much decreased for *Solanum nigrum.* Similarly, besides the Cd stress, nitrate uptake is also limited by Pb excess in plant uptake solution (Singh et al. 2017). Aluminum (Al) is the third most abundant metal known to cause the inhibitory effects upon NO_3^- uptake by plant roots (Schmitt et al. 2016; Riaz et al. 2018a, b). But the extent of Al toxicity depends upon the ionic form, its concentration, and the plant exposure to stress. Acceleratory effect of short-term Al exposure to plants (barley, cucumber) had been observed with increased nitrate uptake (Nichol et al. 1993; Jerzykiewicz 2001). However, higher external Al inputs had decreased dramatically the net NO_3^- uptake in *Vigna unguiculata* upon prolonged plant exposure to imposed stress (Cruz et al. 2014).

4.2 Ammonium Uptake

Ammonium ion is an intermediary during the N metabolism in plants, which is produced several times during NO_3^- assimilation, deamination of amino acids, and photorespiration (Andrews et al. 2013). NH_4^+ may accumulate in the soil, when this conversion process of N is limited or completely stopped due to any of the soil conditions, viz., low soil pH, lack of soil oxygen due to waterlogging, lack of organic matter for carbon stick, dry soils, and low soil temperature, affecting microbiota activity (Mengel and Kirkby 1987; Cheng et al. 2013; Ashoka et al. 2017). Therefore, all these features may affect the soil microclimate to enhance or depress the soil microorganism activities determining the NH_4^+ concentration in soil varying from micromolar to hundreds of millimolar. The optimum level nitrification process is attained at 20–40 °C, while this is 50–70 °C for optimum ammonification (Myers 1975; Yao et al. 2013). Hence, in tropical region soils, NH_4^+ may accumulate in the soil even under neutral pH conditions (Mengel and Kirkby 1987). As the NH_4^+ transport in plants is considered to be passive as well as active uptake depending upon the NH_4^+ availability in the soil solution (Wang et al. 1993; Cheng et al. 2018), studies on the restricted NH_4^+ uptake due to heavy metals stress suggested significantly decreased uptake of NH_4^+ by plant roots in various crops, mainly due to the perturbations in active uptake process of ATPase (Weber et al. 1991). Relatively decreased or suppressive NH_4^+ uptake was seen in *Silene vulgaris* at the presence of Cu (Weber et al. 1991). Toxicity of other heavy metals (Cd^{2+}, Pb^{2+}, and Fe^{2+}) was also observed with the reduced or diminished uptake of NH_4^+ in cucumber seedlings (Burzynski and Buczek 1998). Several other stress agents including Pb, Cd, Ni, Zn, and Mn are also reported to influence the uptake of NH_4^+ in different plant species. However, Cu-dependent alterations are known to affect the NH_4^+ uptake to the greatest extent as compared to other metal ions (Kubik-Dobosz et al. 2001).

4.3 Mechanisms of Metal Action on the N Uptake

Heavy metal stress is considered to induce several alterations, either direct or indirect, during the active as well as passive uptake of both inorganic forms of N. Physiological studies suggest that metal toxicity affects the NO_3^- and NH_4^+ uptake through disrupting constitutive and the inductive components of NO_3^- transportation system (Burzynski and Buczek 1994; Shruti and Dubey 2006). This effect was observed for metal toxicity with Cd, Pb, Cu, and Ni in cucumber seedlings which inhibited the NO_3^- transporters with higher affinities (Burzynski and Buczek 1994; Ashraf et al. 2015). The similar effect was also observed for inhibition of high-affinity NH_4^+ transporters (Burzynski and Buczek 1998). The explanations for restricted N absorption could be the direct interaction of metal ions and the NH_4^+ and NO_3^- transporter proteins of both low and high affinities (Shruti and Dubey 2006). Besides this direct interaction of metals and proteins, heavy metal alterations in these transporter regions are due to the downregulation of *NRT* and *AMTl* genes (Quesada et al. 1997, Krapp et al. 1998; Singh et al. 2008). This diminishing effect is attributed due to the lowered intercellular nitrates and the enhanced amino acids. Plants treated with external Cd inputs showed fluctuations in the NO_3^- as well as NH_4^+ and amino acid tissue content (Hernandez et al. 1996, 1997; Gouia et al. 2000b). Other effects of metal stress include the Cu and Fe bonding with phosphate residues of nucleic acids thus damaging DNA (Lloyd and Phillips 1999). Furthermore, heavy metals including Cu, Cd, and Ni are reported to accelerate the free radical generation, thus oxidizing several vital organic molecules (Shaw et al. 2004; Michalak 2006).

Among the indirect effects of heavy metal stress on ion uptake in plants, some key metal ions, viz., Cu, Pb, Hg, Cd, Ni, and Zn, are believed to interact with membrane components thus affecting the net ion transport system in plant body (Devi and Prasad 1999; Michalak 2006; Riaz et al. 2018c). This membrane interaction includes altered membrane lipids, total lipid amount, and their composition and saturation; in this process, the main damage is associated with the lipid peroxidation (Demidchik et al. 1997; Hernandez et al. 1997; Singh et al. 2008; Meena et al. 2018). Besides this, heavy metals especially Cu, Hg, Cd, Zn, and Al also induce the potassium leakage leading to increased cell membrane permeability (Demidchik et al. 1997; Hall 2002; Riaz et al. 2018c). Therefore, it is evident that the indirect influence of heavy metal stress during the NO_3^- and NH_4^+ uptake is the alteration in membrane permeability. In addition, metal ions also interact with the plasma membrane proton pump by affecting the activity of H^+ ATPase activity (Rengel et al. 2016).

5 Mineral Nitrogen Assimilation and Heavy Metals

Nitrogen assimilation is an important plant metabolic process, which not only controls plant growth and development but also plays an important role in plant survival under stress conditions. For plants, NH_4^+ and NO_3^- are major nitrogen sources and are required during different metabolic processes (Burger and Jackson 2003). Nitrate is converted into NH_4^+ via two-step process; during the first step, NO_3^- is converted into nitrite with the action of nitrate reductase, and, in the second step, nitrite is converted into NH_4^+ with the action of nitrite reductase. Nitrate reductase is located in the cytoplasm, while nitrite reductase is located in chloroplast and uses energy and some reductants such as NADH or NADPH from photosynthesis or respiration process to carry out such NO_3^- to NH_4^+ conversion reaction (Pérez-Tienda et al. 2014). Moreover, the first step of this reaction occurs in the cytosol, while the second step takes place in the plastid. Following NH_4^+ production, it has to be incorporated in carbon skeleton, and this process takes place primarily via GOGAT cycle. Moreover, there are two isoforms of GOGAT and GS, and their localization has been found in a tissue-specific manner, e.g., in roots, NADH-GOGAT and GS1 are primarily involved in nitrogen assimilation, while in leaves, reduced ferredoxin GOGAT and GS2 are predominantly involved in nitrogen assimilation (Mevel and Prieur 2000). Heavy metal toxicity significantly reduces nitrogen assimilation process (Fig. 2). However, the level of reduction depends on the localization and sensitivity of enzymes to heavy metal toxicity. Moreover, duration, concentration, and mobility of heavy metal ions in growth medium further aggravate alterations in nitrogen assimilation process. In this section, heavy metal-induced alterations in mineral nitrogen assimilation and associated enzymes have been discussed.

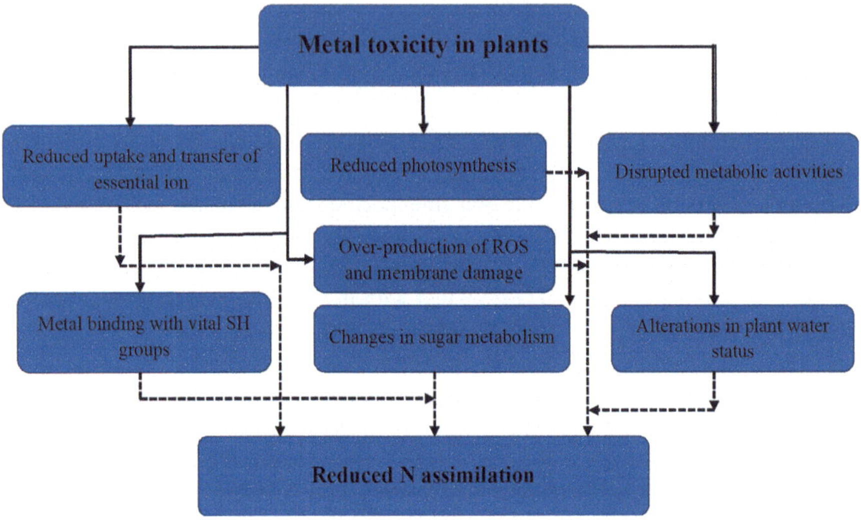

Fig. 2 Possible mechanisms of heavy metal-induced reductions in N assimilation

5.1 Nitrate Reductase Activity

Nitrate reductase (NR) enzyme is a very important enzyme and controls the initiation of NO_3^- assimilation process in plants (Silveira et al. 2001). NR enzyme activity is severely affected by numerous environmental factors including hyperaccumulation of heavy metal ions in soil or in plant tissues (Silveira et al. 2001; Sharma and Dubey 2005). It has been shown that heavy metal stress reduces NR activity by altering numerous associated physiological processes. Sharma and Dubey (2005) showed that inhibition of NR activity under Al toxicity was due to the direct interaction of Al with functional SH groups present in the active sites of NR. Contrarily, Huang and Xiong (2009) noted that the reduction in NR activity under Cd stress was due to alteration in sugar metabolism and declined photosynthetic activity. According to Barroso et al. (2006), reduction in photosynthesis can induce indirect effects of Cd toxicity on NR activity. However, Vajpayee et al. (2000) reported that under Cr stress, NR activity was decreased because of reduced chlorophyll biosynthesis, which led to lower photosynthesis and poor supply of photosynthates. Similarly, in pea plants, Cd toxicity not only inhibited NO_3^- uptake and transportation but also detrimentally affected the activity of NR and increased endogenous NH_4^+ level through deamination of free amino acids and other N forms (Hernandez et al. 1997; Datta et al. 2017c). An in vitro examination of Cd toxicity on NR activity showed that decline in NR activity was associated with the reduction of 80% in the production of NR protein and alteration in the molybdenum cofactor-binding domain of NR (Gouia et al. 2000a). In tomato, Chaffei et al. (2004) also reported similar results that Cd toxicity significantly reduced NR activity. Wang et al. (2008)

found that NR activity could also be influenced by Cd-induced reduction in NO_3^- concentration.

Similarly, the detrimental effects of other heavy metals on NR activity have also been reported elsewhere. Xiong et al. (2006b) showed that NR activity was reduced by 106% due to Pb stress, and this was mainly due to Pb-induced reduction in the shoot NO_3^- concentration (71–80%) and free amino acid (81–82%). Similarly, Cu stress-induced decline in NR activity was suggested as a direct effect of Cu on NR by attaching Cu^{2+} to SH groups, with subsequent inactivation of NR enzyme (Xiong et al. 2006a). Similarly, Kevrešan et al. (2001) studied the effects of Ni, Cd, Pb, and Mo (applied at various concentrations) on N metabolism in pea plants. They found that all the concentrations of Ni, Cd, and Pb caused a significant decrease in the activity of NR, and the highest decrease in NR activity was observed in the presence of Cd followed by Pb and Ni. However, the presence of Mo results in increased activity of the NR. The NR activity was less dependent on NO_3^- contents present in the cell and to a greater extent on the intensity of uptake and transport of NO_3^- to the sites of their reduction (Carillo et al. 2005; Varma et al. 2017). Moreover, a high concentration of heavy metals in the growth medium causes dehydration in plant tissues thus reduces NR synthesis process (Chaffei et al. 2004).

Heavy metal-induced oxidative stress (Khan et al. 2018; Shahzad et al. 2018) and enhanced lipid peroxidation may also influence the NR activity in plants. In a study, Jha and Dubey (2004) showed that As decreased NR activity by decreasing the enzyme affinity toward their substrate, thus resulting in considerable reduction in NR activity and N assimilation rate in rice. Likewise, Gajewska and Skłodowska (2009) substantiated that under Ni stress, NO_3^- contents were decreased with the subsequent reduction of 40% in NR activity in wheat seedlings. Rai et al. (2004) investigated the toxic effects of Cr on *Ocimum tenuiflorum* and found that excess Cr in the growing media severely reduced the NR activity through impaired substrate utilization. Hg stress has been reported as the most toxic stress in reducing NR activity as Hg can inactivate NR enzyme by binding to a thiol group (Sharma and Subhadra 2010; Meena and Meena 2017). Similar results have been found in maize under Hg stress that Hg not only competes with thiol group but also disturbs the production of cysteine (Pandey and Srivastava 1993). Thus it can be inferred from above results that heavy metal stress reduces NR activity by altering cysteine production, by increasing ROS production, by disturbing the balance between the substrate for NR reaction, and by affecting sugar metabolism and photosynthesis.

5.2 *Nitrite Reductase Activity*

Nitrite enters into plastids or chloroplast and reduced to NH_4^+ by the action of NiR—a nuclear-encoded enzyme (Shah and Dubey 2003). In general, NiR is more resistant to stress conditions than NR. As this enzyme is localized in plastids, therefore, metals have less access to NiR, compared with cytoplasmic NR. The NiR has not been reported to limit the N assimilation in plants. Therefore, the studies regarding the effect of heavy metals on NiR activity are rare. However, studies showed that NiR decreases under heavy metals stress. Ghosh et al. (2013) showed that exposure of arsenate to wheat seedling significantly reduced NiR activity, and this was associated with a reduction in NR activity and total nitrite contents, which is the first primary reason (Llorens et al. 2000). Heavy metals can also influence the NO_3^- uptake and NO_3^- homeostasis in the plant cell. Therefore, NO_3^- also induces the expression of NiR genes along with NR genes (Stitt 1999). Therefore, the regulation of NiR genes' expression by NO_3^- could be the main reason of NiR inhibition by heavy metals. In a study, Dinakar et al. (2009) showed that Cd stress reduced NiR activity up to 75%, and this was due to a significant reduction in NR activity.

5.3 Glutamine Synthetase and Glutamate Synthase Activities

In higher plants, NH_4^+ is assimilated into nontoxic glutamine and glutamate, and this reaction is carried out by glutamine synthetase (GS) and glutamate synthase (GOGAT) (Lam et al. 1996). The effects of metal toxicity on GS and GOGAT are often regarded as a mean of the effect of metal on all isoenzymes of GS or GOGAT localized either in the root or shoot. The decrease of one form of the enzyme may trigger the other. Therefore, the variations recorded in the activities of enzymes after metal exposure are different and sometimes difficult to infer. In the past, reductions in activities of GS and GOGAT were found after the exposure of plants to Cu (Yu et al. 2005), Cd (Chaffei et al. 2004), Al and Se (Ruiz et al. 2007), and As (Jha and Dubey 2004) toxicities by different researchers.

Heavy metals are known to disturb the NO_3^- absorption by the plants. Intracellular NO_3^- level affects the gene expression of GS and GOGAT along with NR and NiR (Oaks 1994; Stitt 1999). Possibly, this dependence of GS and GOGAT genes on NO_3^- is the cause of metal-induced inhibitions in the activities of these enzymes along with the direct metal action on the SH groups of enzyme proteins. Huang and Xiong (2009) reported that NH_4^+ content in shoot and root of rice seedlings was significantly increased, while GS activity was decreased after Cd stress. Chien and Kao (2000) also reported that the NH_4^+ accumulation in shoot and root was ascribed to the decline in GS and GOGAT activities in rice. Likewise, Cd reduced the activities of GS and GOGAT and consequently increased the level of NH_4^+ in maize seedlings (Boussama et al. 1999a). Chugh et al. (1992) studied the effect of Cd on GS and GOGAT activities in pea seedlings and reported that GS activity in leaves was practically unchanged; however, it was markedly reduced in the roots. Both GOGAT and GS were suppressed by the Cd toxicity in leaves of barley (Boussama et al. 1999b) and bean seedlings (Gouia et al. 2000a). In cucumber, Cd and Pb treatment for 24 or 48 h decreased the GS activity in roots but did not alter GS activity in the cotyledons (Burzynski 1990). He further reported considerable increases in the activity of GOGAT-NADH activity in the cotyledons under Pb as well as Cd toxicity (Burzynski 1990). Application of 100 μM Cu/Cd/Fe and 500 μM Pb to cucumber seedlings for 1 h decreased the activity of GS in the root (Burzynski and Buczek 1998). In the in vitro experiments, metals only reduced the enzyme activity at high concentrations, Cu at 1000 μM, Cd at 500 μM, and Fe at 1000 μM, while Pb was even detrimental at 10 μM (Burzynski and Buczek 1997), and the minor effect of Cu, Cd, and Fe on GS activity indicated the indirect action on enzyme activity. Correspondingly, high concentrations (500–1000 μM) of Co, Ni, Cd, Zn, and Cu repressed both GS forms in triticale seedlings (Bielawski 1994). A study on *Chlamydomonas reinhardtii* by Domínguez et al. (2003) demonstrated that the 24 h exposure to 150 μM Cd did not inhibit the NO_3^- uptake but reduced the GS activity by 45%, which enhanced the intracellular accumulation of NH_4^+.

Kevrešan et al. (2001) stated that the exposure of Cd, Ni, Mo, and Pb to young pea plants significantly decreased the GS activity at all the concentrations used, especially at the highest level, and the metal-induced reductions were in the order of Cd > Pb > Ni > Mo. The presence of Cd in the medium strongly inhibited (about 78%) the GS activity in young pea plants than the other metals (Kevrešan et al. 2001). Likewise, Popović et al. (1996) reported a complete inhibition of GS enzyme in the young sugar beet plants under Cd toxicity. Gajewska and Skłodowska (2009) found that Ni application did not significantly change the GS activity in wheat shoots. However, a significant decline in GS activity was reported in the leaves of Ni-stressed sugar beet (Kevrešan et al. 1998). Miflin et al. (1980) stated that the presence of Cu, Hg, Cd, Zn, Ni, and Fe resulted in a pronounced inhibition of GS in rice. However, Hg was more toxic for GS even at 10 μM. Orzechowski and Bielawski (1997) studied the effects of Cd, Zn, and Pb toxicity on NH_4^+ assimilation in Triticale crop and reported that the metal sensitivity of GS varied in root and shoot. Cd and Zn toxicity reduced the GS activity in shoots but increased in root compared with control. Gajewska and Skłodowska (2009) concluded that a decrease in the activities of both GOGAT and GS in metal-stressed plants was because of alterations in the oxidative metabolism of plants (Balestrasse et al. 2006). As metal toxicity leads to the overproduction of ROS in a plant cell, thus decrease in both GS and GOGAT activities might be related, at least partly, to oxidative destruction of these enzymes.

5.4 Glutamate Dehydrogenase Activity

In higher plants, GS/GOGAT is the main pathway of NH_4^+ assimilation under normal conditions; nevertheless, with the increase in endogenous NH_4^+ concentration, an alternate pathway, controlled by glutamate dehydrogenase (GDH), contributes in decreasing this internal NH_4^+ concentration. The exact role of alternate GDH pathway is still poorly understood in plants. The up- or downregulation of enzyme often varies depending on the species, cultivars, tested tissue, and intensity and duration of metal exposure in plants (Miflin and Habash 2002). Masclaux-Daubresse et al. (2002) stated that Glu is synthesized by the combined action of GOGAT and GS in both young and old leaves of tobacco, while GDH is responsible for deamination of Glu. GDH is mainly located in mitochondria, occasionally in the cytoplasm, within the phloem companion cells of the shoot (Terce´-Laforgue et al. 2004; Fontaine et al. 2006). Till date, controversy exists as to the role of the enzyme GDH in higher plants. It has been suggested that GDH has a role in NH_4^+ reassimilation under stress conditions (Gouia et al. 2000a, 2003). Skopelitis et al. (2006) reported that the formation of ROS under stress induced the synthesis of the a-subunit of GDH, when GS was inhibited. For instance, it has been reported that the GDH was induced under Cd stress (Miflin and Habash 2002; Astolfi et al. 2004; Yadav et al. 2017). Chaffei et al. (2003) also stated that GDH has a major role in cellular NH_4^+ detoxification. They noted that deaminating activity of GDH was inhibited by Cd (due to a direct action of Cd on the enzymatic protein or to a retroinhibitory effect of ammonia), while the reducing aminating activity of GDH was enhanced. Lee et al. (1976) found the decreased activity of GDH and low NO_3^- contents in soybean plants under Pb toxicity, whereas, malate dehydrogenase activity remained unaffected by Pb. Mittal and Sawhney (1990) reported that Pb retarded the utilization of N reserves from cotyledons and decreased the activity of GDH in germinating pea seeds compared with control, which disturbed the respiratory activity because of restricted generation of organic acids from amino acids. Gouia et al. (2000a) reported that under Cd stress, GDH fulfills the Glu pool required for synthesis of Cd-binding peptides and detoxify the recycling of the high NH_4^+ originating from a decline in the GS/GOGAT activities. Higher GDH-NADH activity in root and shoot of triticale under Cd, Zn, and Pb toxicity was also recorded by Orzechowski et al. (1997). A short-term (1 h) exposure of cucumber plants to Cd, Cu, Pb, and Fe inhibited the NADH-GDH activity in root because NH_4^+ in tissues was in excess (Burzynski and Buczek 1997, 1998). However, Cd or Pb toxicity significantly increased the root NADH-GDH activity after 24 and 48 h. Gouia et al. (2003) reported that Cd stress triggered the activity of GDH in bean seedling concomitant with the dramatic buildup of an NH_4^+ pool. The Cd-induced enhancement in GDH activity has also been reported in maize (Boussama et al. 1999a) and pea (Chugh et al. 1992), which were attributed to de novo synthesis and/or activation of specific isoenzymes that eliminate excess NH_4^+ (Syntichaki et al. 1996; Loulakakis and Loulakakis-Roubelakis 1996). Boussama et al. (1999a) have demonstrated that in Cd-stressed maize plants, Cd treatment triggered the NADH-GDH activity. Even though the role of GDH in higher plants is controversial, GDH isoenzymes are known to remove, in part, the excess of NH_4^+ under Cd toxicity.

Exposure of Ni or Cd enhanced the activity of GDH in the shoot of *Silene italica* (Mattioni et al. 1997). Van Assche et al. (1988) reported that toxicity of Zn or Cd beyond threshold level increased the GDH activity in *Phaseolus vulgaris* leaves, while Domínguez et al. (2003) reported that aminating GDH activity was increased by 75% in *Chlamydomonas reinhardtii* after 24 h exposure to 150 μM Cd and suggested that aminating GDH activity could support the biosynthesis rate of the Glu in Cd-treated plants.

While studying the effects of Cu on N metabolism in grapevines, Llorens et al. (2000) observed dramatic changes especially in the root system and found that NADH-GDH was active to keep the root function. Gajewska and Skłodowskahe (2009) found that activity of NADH- and NAD-GDH in wheat shoots increased after

7 days of Ni exposure to wheat seedlings. They further stated that under unfavorable environmental conditions like Ni toxicity, GDH could play an important role in detoxification of NH_4^+ released in response to stress as well as in the replenishment of Glu pool (Jha and Dubey 2004). Jha and Dubey (2004) reported that the activities of aminating and deaminating GDH increased at moderately toxic level (25 μM) of As_2O_3, whereas a higher As level of 50 μM was inhibitory to the enzymes. They suggested that under As toxicity, GDHs play an important role by helping in NH_4^+ assimilation and sustaining general C metabolism.

6 Accumulation of Nitrogenous Compounds in Metal-Stressed Plants

As an adaptive mechanism, plant accumulates numerous nitrogenous compounds such as proline or polyamines to protect the plant from detrimental effects of heavy metals (Alcázar et al. 2006; Anjum et al. 2017). These compounds exhibit a specific role in numerous metabolic processes and also vary in response to plant tissue, heavy metal concentration, and plant species. There are different soluble nitrogenous compounds which can significantly accumulate in the plant under heavy metal stress to alleviate heavy metal stress-induced adversities in plants. These soluble compounds include amino acids, polyamines, proline, or glycine betaine. Generally, these compounds regulate osmotic potential in plant cells and thus improve heavy metal stress tolerance (Wang et al. 2008). The following sections provide a brief discussion on the behavior of some important nitrogenous compounds in response to heavy metals.

6.1 Proline and Glycine Betaine Accumulation Under Heavy Metal Stress

Proline is a very important amino acid with multiple roles in different metabolic processes. Generally, under heavy metal stress, proline accumulation increases, thus improving stress tolerance in plants (Szabados and Savoure 2010). Higher accumulation of proline and its role in stress alleviation can be partially explained by its unique chemical properties as compared with other amino acids: proline being a water-soluble amino acid exists as in a zwitterionic state having both positive charge and weak negative charge in N groups and carboxylic acid, respectively (Lehmann et al. 2010). Proline also plays a key role as an osmolyte, electron sink, radical scavenger, component of the cell wall, and stabilizer of macromolecules (Matysik et al. 2002; Yancey 2005; Kumar et al. 2017). Significant accumulation of proline was noted under Cd stress in brassica (Singh and Tewari 2003). Similarly, Dinakar et al. (2008) showed that under Cd stress, proline accumulation increased by 160%, and this was accompanied with high glutathione reductase and NR activity. This can be supported with the findings of Sharma et al. (1998) who reported that proline protects NR and glucose-6-phosphate dehydrogenase against the Zn and Cd toxicity, and such protection occurs due to a reduction of the free metal ion activity owing to the formation of proline-metal complexes. Under Pb stress, Yang et al. (2011) showed that Pb tolerance in two wheat cultivars was associated with higher proline contents. Likewise, detrimental effects of Ni toxicity were mediated by higher proline accumulation in wheat (Parlak 2016). Under Cd toxicity, upregulation of proline biosynthesis resulted in higher Cd tolerance associated with higher GSH/GSSG ratio and higher glutathione *S*-transferase accumulation.

Glycine betaine (GB, the trimethylglycine) is a common osmolyte and N-based compound that accrues in plants under drought and salinity stresses (Sharma and Dietz 2006). It not only protects membranes and proteins but also reduces stress-induced alteration in enzymes (Ashraf and Foolad 2007). Moreover, GB helps in ROS scavenging, thus regulating redox regulation under stress conditions (Chen et al. 2006). Betaine has been regarded as the most promising osmolyte in protecting plant metabolism from Cd toxicity (Islam et al. 2010). Bharwana et al. (2014) showed that exogenous GB application protected cotton seedling from Pb-induced oxidative damage by increasing antioxidant defense system.

6.2 Polyamine Accumulation and Heavy Metal Tolerance

Besides the protective role of proline and glycine betaine, some other nitrogenous compounds such as amino acids and polyamines also play an important role in heavy metal stress alleviation and tolerance in plants. Polyamines such as spermine or spermidine or putrescine reduce heavy metal-induced adversities in the plant (especially N metabolism) by playing a multifaceted role in different physiological and biochemical processes (Kakkar and Sawhney 2002). In a study, it was found that spermine application increased Cd stress tolerance in mung bean by increasing glutathione *S*-transferase, glutathione reductase, and glutathione contents. Generally, levels of polyamines increase in different plant tissues under heavy metals stress; however, it has also been seen that different polyamines showed different responses under different heavy meal stresses in different plant species. For instance, Groppa et al. (2007) also showed that spermidine content was not affected by Cd and Cu, while spermine was significantly reduced. They also noted that putrescene was increased under Cd and Cu stress in wheat. Similarly, it has been seen that spermidine increased Al toxicity by increasing photosystem II D1 protein gene and glutathione reductase activity, thus improving photosynthesis and N metabolism (Sen et al. 2014). In another study, Mandal et al. (2013) documented that putrescene can reduce Al toxicity by downregulating H^+-ATPase activity and by increasing flavonoid production.

Under metal stress, the specific role of polyamines in plants is poorly known. However, these may effectively protect and stabilize the membranes against the toxicity of metal ions especially the redox active metals. Shi and Chan (2014) reported that polyamines enhance the plant stress tolerance by improving antioxidant production and osmolyte production (especially amino acids). In a study, it was found that spermidine improves Al stress tolerance by increasing GSH pool and Gly II activity. Besides activating defense system in plants, polyamines also increase NR activity. Rosales et al. (2012) showed that putrescence could increase NR activity by 63%, while spermine or spermidine can increase NR activity by 114%. Thus these results are clearly highlighting the protective role of polyamines in improving N metabolism under heavy metal stress conditions.

7 Nitrogen Fixation and Heavy Metal Toxicity

Nitrogen fixation is a very important process in plants (especially legumes), which helps in fixing molecular nitrogen into organic nitrogen via symbiosis process. Biological nitrogen fixation accounts for about 70% in fixing elemental N and is very crucial for plant growth (Matiru and Dakora 2004; Meena and Yadav 2015). Heavy metals considerably reduce the nitrogen fixation process. However some metalloids are beneficial in low concentration and promote growth and activity of different *Rhizobium* and *Bradyrhizobium* strains. The presence of high concentrations of heavy metal ions in soil restricts nitrogen fixation by reducing infection process, growth of *Rhizobium*, nodule development, or host plant growth (Broos et al. 2005); however it was found that the extent of reduction depends on heavy metal and plant species. For instance, Athar and Ahmad (2002b) found that Cd is more toxic to *Bradyrhizobium* and *Rhizobium* as compared with Ni, Zn, or Co. Similarly, Kalyanaraman and Sivagurunathan (1993) also noted that Cd is more toxic than Zn and reduces nitrogen fixation process significantly. In another study, it was found that a number of free-living *Azotobacter chroococcum* cells were severely influenced by Cd as compared with Zn, Ni, Cr, and Pb (Athar and Ahmad 2002a). Moreover, Paudyal et al. (2007) tested the growth and N_2 fixation process by two strains of *Rhizobium* in response to Al, Fe, and Mo toxicity and found that Al was the most lethal heavy metal for them, while Fe promoted the growth of both strains. Some other studies also reported toxic effects on different heavy metals such as Th, Ni, As, Cd, and Cu on the growth and morphology of numerous symbiotic N_2-fixing organisms including *R. leguminosarum*, *Sinorhizobium*, and *Mesorhizobium ciceri* (Santamarı'a et al. 2003; Wani 2008; Arora et al. 2010; Stan et al. 2011; Bianucci et al. 2011). All these studies suggested that differential toxicity levels of different heavy metals on nitrogen-fixing organisms were due to experimental conditions, heavy metal ion concentration, plant species, and solubility of heavy metals ions in the growth medium.

Heavy metal stress significantly alters nitrogen fixation in both symbiotic and nonsymbiotic system (Giller et al. 1998; Selosse et al. 2004; Molaei et al. 2017a, b). It has been reported in Wani et al. (2007) that Cd stress decreased symbiosis process and root N content by 34%, compared with the control. They also reported that such decline in nitrogen fixation and nitrogen contents in root was associated with a decline in a number of nodules per plant and dry mass of nodules. Moreover, Balestrasse et al. (2001) reported that Cd toxicity causes oxidative damage to nodule by increasing ROS production in nodules, thus reducing nitrogen fixation in nodules. Broos et al. (2005) showed that sludge treatment in white clover reduced nitrogen fixation up to 50%, and this was associated with dysfunctional nodules. In another study, Balestrasse et al. (2004) showed that Cd stress considerably reduces leghemoglobin levels, nitrogenase activity, and protease activity in nodules, and most importantly Cd also reduces the number of bacteroides per symbiosome and total effective area for N_2 fixation in soybean. In white lupin, Cd stress decreases N-amino compounds, malate, succinate, and soluble protein in nodule (Carpena et al. 2003). Arsenic (As) is another detrimental heavy metal, which reduces nitrogen fixation in nodules and activity of different nitrogen-fixing microorganism significantly. Pajuelo et al. (2008) studied the effects of As toxicity on the symbiotic interaction between *Sinorhizobium* and *Medicago sativa* and reported that As toxicity reduced the total number of nodules by 75% and a number of rhizobial infections by 90%. These reductions were attributed to damage of root hair and shorter infective root zone. Exposure of As toxicity in black gram reduced the nitrogenase activity in nodule by twofolds and restricted the rate of nitrogen fixation (Mandal et al. 2011; Verma et al. 2015). In soybean, though *B. japonicum* E109 stain showed tolerance to arsenic, however, there was still reduction in nodulation, and this was associated with reduced motility (swarming and swimming) of the microorganism in the presence of arsenic (Talano et al. 2013). In pea, chromium application decreased the ability of pea plants to fix atmospheric nitrogen symbiotically, and this was associated with a decline in total nitrogenase activity and nodule formation (Bishnoi et al. 1993).

8 Conclusions and Perspectives

Heavy metal toxicity has become a serious environmental threat worldwide. Heavy metals severely hamper the N metabolism by reducing the NO_3^- uptake and altering the enzyme activity; however, the effects of heavy metals on N assimilatory enzymes vary with the sensitivity level of enzymes and their localization in the cells/organs, mobility of metals, metal concentration in soil, and time of plant exposure to metal toxicity. Heavy metals can alter the activity of various N assimilatory enzymes (e.g., NR, GS, GOGAT) by binding to the vital SH groups. Heavy metals that induce the alterations in plant water status; sugar metabolism; reductions in uptake and supply of essential nutrients; decrease in photosynthesis; generation of ROS; and membrane damage ultimately affect the N metabolism in the plant (Fig. 2).

In higher plants, GS/GOGAT is the main pathway of NH_4^+ assimilation under normal conditions, nevertheless, with the increase in endogenous NH_4^+ concentration, an alternate pathway, controlled by GDH contributes in decreasing this internal NH_4^+ concentration. Therefore, the activity of GDH is generally linked with the internal NH_4^+ concentration in plants. Heavy metals, particularly at higher concentrations, hamper the rate of nitrogen fixation in plants, by reducing infection process, growth of *Rhizobium*, nodule development, or host plant growth; however, such effects depend on heavy metal and plant species. Different soluble nitrogenous compounds such as amino acids, polyamines, proline, or glycine betaine can significantly accumulate in the plant under heavy metal stress to alleviate metal-induced adversities in plants. Generally, these compounds regulate osmotic potential in plant cells and thus improve heavy metal stress tolerance.

In the past, the majority of the research work was concerned only with the effect of metal toxicity on the uptake of N or activities of enzymes involved in N assimilation. Future studies should focus on the possible interferences between N transport and its acquisition under metal toxicity. Moreover, use of recent genomic techniques at transcriptomic or proteomic level for expression of specific genes encoding the N assimilatory enzyme proteins and N transporters will allow exploring the molecular intricacies of metal action. This will also lead to the development of strategies for improving plant resistance against metal toxicity. Efforts should also be made to unravel the mechanisms of possible antagonisms/synergisms among metal ions and determine the interaction of one metal ion in the transport and/or homeostasis of the other(s) and their effects on the N uptake and assimilation. The responses of N metabolism to multi-metal stress should also be examined in future studies because single metal toxicity is virtually nonexistent in the environment.

References

Alcázar R, Marco F, Cuevas JC et al (2006) Involvement of polyamines in plant response to abiotic stress. Biotechnol Lett 28(23):1867–1876

Ali H, Khan E, Sajad MA (2013) Phytoremediation of heavy metals, concepts and applications. Chemosphere 91(7):869–881

Andrews M, Raven JA, Lea PJ (2013) Do plants need nitrate? The mechanisms by which nitrogen form affects plants. Ann Appl Biol 163(2):174–199

Anjum SA, Tanveer M, Hussain S et al (2016) Osmoregulation and antioxidant production in maize under combined cadmium and arsenic stress. Environ Sci Pollut Res 23(12):11864–11875

Anjum SA, Tanveer M, Hussain S et al (2017) Alteration in growth, leaf gas exchange, and photosynthetic pigments of maize plants under combined cadmium and arsenic stress. Water Air Soil Pollut 228(1):13

Antunes F, Aguilar M, Pineda M et al (2008) Nitrogen stress and the expression of asparagine synthetase in roots and nodules of soybean (*Glycine max*). Physiol Plant 133:736–743

Arora NK, Khare E, Singh S et al (2010) Effect of Al and heavy metals on enzymes of nitrogen metabolism of fast and slow growing rhizobia under explanta conditions. World J Microbiol Biotechnol 26:811–816

Ashoka P, Meena RS, Kumar S, Yadav GS, Layek J (2017) Green nanotechnology is a key for eco-friendly agriculture. J Clean Prod 142:4440–4441

Ashraf M, Foolad MR (2007) Roles of glycinebetaine and proline in improving plant abiotic resistance. Environ Exp Bot 59:206–216

Ashraf U, Kanu AS, Mo Z et al (2015) Lead toxicity in rice: effects, mechanisms, and mitigation strategies-a mini review. Environ Sci Pollut Res 22(23):18318–18332

Ashraf U, Hussain S, Anjum SA et al (2017) Alterations in growth, oxidative damage, and metal uptake of five aromatic rice cultivars under lead toxicity. Plant Physiol Biochem 115:461–471
Astolfi S, Zuchi S, Passera C (2004) Role of sulphur availability on cadmium induced changes of nitrogen and sulphur metabolism in maize (*Zea mays* L.) leaves. J Plant Physiol 161:795–802
Athar R, Ahmad M (2002a) Heavy metal toxicity: effect on plant growth and metal uptake by wheat, and on free living Azotobacter. Water Air Soil Pollut 138(1–4):165–180
Athar R, Ahmad M (2002b) Heavy metal toxicity in legume-microsymbiont system. J Plant Nutr 25 (2):369–386
Ayangbenro AS, Babalola OO (2017) A new strategy for heavy metal polluted environments: a review of microbial biosorbents. Int J Environ Res Public Health 14:1–94
Aziz O, Hussain S, Rizwan M et al (2018) Increasing water productivity, nitrogen economy, and grain yield of rice by water saving irrigation and fertilizer-N management. Environ Sci Pollut Res 25:1–15. https://doi.org/10.1007/s11356-018-1855-z
Balestrasse KB, Gardey L, Gallego SM et al (2001) Response of antioxidant defence system in soybean nodules and roots subjected to cadmium stress. Funct Plant Biol 28(6):497–504
Balestrasse KB, Gallego SM, Tomaro ML (2004) Cadmium-induced senescence in nodules of soybean (*Glycine max* L.) plants. Plant Soil 262(1–2):373–381
Balestrasse KB, Gallego SM, Tomaro ML (2006) Oxidation of the enzymes involved in nitrogen assimilation plays an important role in the cadmium-induced toxicity in soybean plants. Plant Soil 284(1–2):187–194
Barroso JB, Francisco JC, Alfonso C et al (2006) Localization of S-nitrosoglutathione and expression of S-nitrosoglutathione reductase in pea plants under cadmium stress. J Exp Bot 57 (8):1785–1793
Belimov AA, Dodd IC, Safronova VI et al (2015) The cadmium-tolerant pea (*Pisum sativum* L.) mutant SGECdt is more sensitive to mercury: assessing plant water relations. J Exp Bot 66 (8):2359–2369
Bharwana SA, Ali S, Farooq MA et al (2014) Glycine betaine-induced lead toxicity tolerance related to elevated photosynthesis, antioxidant enzymes suppressed lead uptake and oxidative stress in cotton. Turk J Bot 38(2):281–292
Bianucci E, Fabra A, Castro S (2011) Cadmium accumulation and tolerance in *Bradyrhizobium* spp. (Peanut Microsymbionts). Curr Microbiol 62:96–100
Bielawski W (1994) Effect of some compounds on glutamine synthetase isoform activity from Triticale seedling leaves. Acta Physiol Plant 16:303–308
Bishnoi NR, Chugh LK, Sawhney SK (1993) Effect of chromium on photosynthesis, respiration and nitrogen fixation in pea (*Pisum sativum* L.) seedlings. J Plant Physiol 142(1):25–30
Boussama N, Ouariti O, Suzuki A (1999a) Cd-stress on nitrogen assimilation. J Plant Physiol 159:310–317
Boussama N, Ouariti O, Ghorbal MH (1999b) Changes in growth and nitrogen assimilation in barley seedlings under cadmium stress. J Plant Nutr 22:731–752
Bravo S, Amorós JA, Pérez-de-los-Reyes C et al (2017) Influence of the soil pH in the uptake and bioaccumulation of heavy metals (Fe, Zn, Cu, Pb and Mn) and other elements (Ca, K, Al, Sr and Ba) in vine leaves, Castilla-La Mancha (Spain). J Geochem Explor 174:79–83
Britto DT, Kronzucker HJ (2002) NH4+ toxicity in higher plants: a critical review. J Plant Physiol 159(6):567–584
Broos K, Beyens H, Smolders E (2005) Survival of rhizobia in soil is sensitive to elevated zinc in the absence of the host plant. Soil Biol Biochem 37:573–579
Burger M, Jackson LE (2003) Microbial immobilization of ammonium and nitrate in relation to ammonification and nitrification rates in organic and conventional cropping systems. Soil Biol Biochem 35(1):29–36
Burzynski M (1990) Activity of some enzymes involved in N03’ assimilation in cucumber seedlings treated with lead or cadmium. Acta Physiol Plant 12:105–110

Burzynski M, Buczek J (1994) The influence of Cd, Pb, Cu and Ni on NO-uptake by cucumber seedlings. II. In vivo and in vitro effects of Cd, Pb, Cu and Ni on the plasmalemma ATPase and oxidoreductase from cucumber seedling roots. Acta Physiol Plant 16:297–302

Burzynski M, Buczek J (1997) The effect of Cu2+ on uptake and assimilation of ammonium by cucumber seedlings. Acta Physiol Plant 19:3–8

Burzynski M, Buczek J (1998) Uptake and assimilation of ammonium ions by cucumber seedlings from solutions with different pH and addition of heavy metals. Acta Soc Bot Polon 67:197–200

Carillo P, Mastrolonardo G, Nacca F et al (2005) Nitrate reductase in durum wheat seedlings as affected by nitrate nutrition and salinity. Funct Plant Biol 32(3):209–219

Carpena RO, Vázquez S, Esteban E et al (2003) Cadmium-stress in white lupin: effects on nodule structure and functioning. Plant Physiol Biochem 41(10):911–919

Cesco S, Neumann G, Tomasi N et al (2010) Release of plant-borne flavonoids into the rhizosphere and their role in plant nutrition. Plant Soil 329(1–2):1–25

Chaffai R, Koyama H (2011) Heavy metal tolerance in *Arabidopsis thaliana*. In: Advances in botanical research, vol 60. Academic, Dordrecht, pp 1–49

Chaffei C, Gouia H, Ghorbel MH (2003) Nitrogen metabolism in tomato plants under cadmium stress. J Plant Nutr 26:1617–1634

Chaffei C, Pageau K, Suzuki A et al (2004) Cadmium toxicity induced changes in nitrogen management in *Lycopersicon esculentum* leading to a metabolic safeguard through an amino acid storage strategy. Plant Cell Physiol 45:1681–1693

Chen C, Wabduragala S, Becker DF et al (2006) Tomato QM-like protein protects *Saccromyces cerevisiae* cells against oxidative stress by regulation intracellular proline levels. Appl Environ Microbiol 72:4001–4006

Cheng Y, Wang J, Mary B et al (2013) Soil pH has contrasting effects on gross and net nitrogen mineralizations in adjacent forest and grassland soils in central Alberta, Canada. Soil Biol Biochem 57:848–857

Cheng Y, Wang C, Chai S et al (2018) Ammonium N influences the uptakes, translocations, subcellular distributions and chemical forms of Cd and Zn to mediate the Cd/Zn interactions in dwarf polish wheat (*Triticum polonicum* L.) seedlings. Chemosphere 193:1164–1171

Chibuike G, Obiora S (2014) Heavy metal polluted soils: effect on plants and bioremediation methods. Appl Environ Soil Sci 2014:1–12

Chien HF, Kao CH (2000) Accumulation of ammonium in rice leaves in response to excess cadmium. Plant Sci 156:111–115

Choppala G, Saifullah BN et al (2014) Cellular mechanisms in higher plants governing tolerance to cadmium toxicity. Crit Rev Plant Sci 33:374–391

Chugh LK, Gupta VK, Sawhney SK (1992) Effect of cadmium on enzymes of nitrogen metabolism in pea seedlings. Phytochemistry 31:395–400

Cruz FJR, de Almeida HJ, dos Santos et al (2014) Growth, nutritional status and nitrogen metabolism in '*Vigna unguiculata*' (L.) Walp is affected by aluminum. Aust J Crop Sci 8 (7):1132

Dabonne S, Koffi B, Kouadio E et al (2010) Traditional utensils: potential sources of poisoning by heavy metals. Br J Pharmacol Toxicol 151(2):90–92

Dadhich RK, Meena RS (2014) Performance of Indian mustard (*Brassica juncea* L.) in response to foliar spray of thiourea and thioglycolic acid under different irrigation levels. Indian J Ecol 41 (2):376–378

Datta R, Anand S, Moulick A, Baraniya D, Pathan SI, Rejsek K, Vranova V, Sharma M, Sharma D, Kelkar A (2017a) How enzymes are adsorbed on soil solid phase and factors limiting its activity: a review. Int Agrophys 31(2):287–302

Datta R, Kelkar A, Baraniya D, Molaei A, Moulick A, Meena RS, Formanek P (2017b) Enzymatic degradation of lignin in soil: a review. Sustain MDPI 1163(9):1–18

Datta R, Baraniya D, Wang Y-F, Kelkar A, Meena RS, Yadav GS, Teresa Ceccherini M, Formanek P (2017c) Amino acid: its dual role as nutrient and scavenger of free radicals in soil. Sustainability 9(8):1402

Demidchik V, Sokolik A, Yurin V (1997) The effect of Cu2+ on ion transport systems of the plant cell plasmalemma. Plant Physiol 114:1313–1325

Devi SR, Prasad MNV (1999) Membrane lipid alterations in heavy metal exposed plants. In: Prasad MNV, Hagemeyer J (eds) Heavy metal stress in plants, from molecules to ecosystems. Springer, New York, pp 99–116

Dinakar N, Nagajyothi PC, Suresh S et al (2008) Phytotoxicity of cadmium on protein, proline and antioxidant enzyme activities in growing *Arachis hypogaea* L. seedlings. J Environ Sci 20 (2):199–206

Dinakar N, Nagajyothi PC, Suresh S et al (2009) Cadmium induced changes on proline, antioxidant enzymes, nitrate and nitrite reductases in *Arachis hypogaea* L. J Environ Biol 30(2):289–294

Domínguez MJ, Gutiérrez F, León R (2003) Cadmium increases the activity levels of glutamate dehydrogenase and cysteine synthase in *Chlamydomonas reinhardtii*. Plant Physiol Biochem 41:828–832

Ekmekci Y, Tanyolac D, Ayhan B (2008) Effects of cadmium on antioxidant enzyme and photosynthetic activities in leaves of two maize cultivars. J Plant Physiol 165:600–611

Emamverdian A, Ding Y, Mokhberdoran F et al (2018) Growth responses and photosynthetic indices of bamboo plant (Indocalamus latifolius) under heavy metal stress. Sci World J 2018:1

Fageria NK, Filho MPB, Moreira A et al (2009) Foliar fertilization of crop plants. J Plant Nutr 32:1044–1064

Fashola M, Ngole-Jeme V, Babalola O (2016) Heavy metal pollution from gold mines: environmental effects and bacterial strategies for resistance. Int J Environ Res Public Health 13:1047

Fontaine JX, Saladino F, Agrimonti C et al (2006) Control of the synthesis and subcellular targeting of the two GDH gene products in leaves and stems of Nicotiana plumbaginifolia and *Arabidopsis thaliana*. Plant Cell Physiol 47:410–418

Frink CR, Waggoner PE, Ausubel JH (1999) Nitrogen fertilizer: retrospect and prospect. Proc Natl Acad Sci 96(4):1175–1180

Gajewska E, Skłodowska M (2009) Nickel-induced changes in nitrogen metabolism in wheat shoots. J Plant Physiol 166:1034–1044

Garnett T, Conn V, Kaiser BN (2009) Root based approaches to improving nitrogen use efficiency in plants. Plant Cell Environ 32(9):1272–1283

Geiger M, Haake V, Ludewig F (1999) The nitrate and ammonium nitrate supply have a major influence on the response of photosynthesis, carbon metabolism, nitrogen metabolism and growth to elevated carbon dioxide in tobacco. Plant Cell Environ 22(10):1177–1199

Ghosh S, Saha J, Biswas AK (2013) Interactive influence of arsenate and selenate on growth and nitrogen metabolism in wheat (*Triticum aestivum* L.) seedlings. Acta Physiol Plant 35 (6):1873–1885

Giller KE, Witter E, McGrath SP (1998) Toxicity of heavy metals to microorganisms and microbial process in agricultural soils. A review. Soil Biol Biochem 30:1389–1414

Gouia H, Ghorbal MH, Meyer C (2000a) Effects of cadmium on activity of nitrate reductase and on other enzymes of the nitrate assimilation pathway in bean. Plant Physiol Biochem 38:629–638

Gouia H, Suzuki A, Brulfert J et al (2000b) Effects of cadmium on the co-ordination of nitrogen and carbon metabolism in bean seedlings. J Plant Physiol 160:367–376

Gouia H, Suzuki A, Brulfert J et al (2003) Effects of cadmium on the co-ordination of nitrogen and carbon metabolism in bean seedlings. J Plant Physiol 160(4):367–376

Graham PH, Vance CP (2000) Nitrogen fixation in perspective: an overview of research and extension needs. Field Crops Res 65(2–3):93–106

Greenwood NN, Earnshaw A (2012) Chemistry of the elements. Elsevier, Amsterdam/London/New York

Groppa MD, Tomaro ML, Benavides MP (2007) Polyamines and heavy metal stress: the antioxidant behavior of spermine in cadmium-and copper-treated wheat leaves. Biometals 20 (2):185–195

Gumpu MB, Sethuraman S, Krishnan UM et al (2015) A review on detection of heavy metal ions in water? An electrochemical approach. Sens Actuators B Chem 213:515–533

Hall JL (2002) Cellular mechanisms for heavy metal detoxification and tolerance. J Exp Bot 53 (366):1–11

Hernandez LE, Giirate A, Carpenta-Ruiz R (1996) Influence of cadmium on the assimilation of two cultivars of *Zea mays*. In: Cleemput OV, Hofman G, Vermoesen A (eds) Progress in nitrogen cycling studies. Kluwer Academic Publishers, Dordrecht, pp 115–132

Hernandez LE, Garate A, Carpena-Ruiz R (1997) Effects of cadmium on the uptake, distribution and assimilation of nitrate in *Pisum sativum*. Plant Soil 189:97–106

Huang H, Xiong ZT (2009) Toxic effects of cadmium, acetochlor and bensulfuron-methyl on nitrogen metabolism and plant growth in rice seedlings. Pest Biochem Physiol 94:64–67

Hussain D, Haydon MJ, Wang Y et al (2004) P-type ATPase heavy metal transporters with roles in essential zinc homeostasis in Arabidopsis. Plant Cell 16:1327–1339

Hussain S, Khan F, Cao W et al (2016) Seed priming alters the production and detoxification of reactive oxygen intermediates in rice seedlings grown under sub-optimal temperature and nutrient supply. Front Plant Sci 7. https://doi.org/10.3389/fpls.2016.00439

Imran M, Kanwal S, Hussain S et al (2015) Efficacy of zinc application methods for concentration and estimated bioavailability of zinc in grains of rice grown on a calcareous soil. Pak J Agric Sci 52(1):169–175

Ireland RJ, Lea PJ (1999) The enzymes ofglu-nine, glutamate, asparagine and aspartate metabolism. In: Singh BK (ed) Plant amino acids: biochemistry and biotechnology, vol 109. Marcel Dekker, New York

Islam MM, Hoque MA, Okuma E et al (2010) Exogenous proline and glycinebetaine increase antioxidant enzyme activities and confer tolerance to cadmium stress in cultured tobacco cells. J Plant Physiol 166:1587–1159

Jerzykiewicz J (2001) Aluminium effect on the nitrate assimilation in cucumber (*Cucumis sativus* L.) roots. Acta Physiol Plant 23:213–219

Jha AB, Dubey RS (2004) Arsenic exposure alters activity behavior of key nitrogen assimilatory enzymes in growing rice plants. Plant Growth Regul 43:259–268

Kakkar RK, Sawhney VK (2002) Polyamine research in plants–a changing perspective. Physiol Plant 116(3):281–292

Kalyanaraman SB, Sivagurunathan P (1993) Effect of cadmium, copper and zinc on the growth of blackgram. J Plant Nutr 16:2029–2042

Kapoor V, Li X, Elk M et al (2015) Impact of heavy metals on transcriptional and physiological activity of nitrifying bacteria. Environ Sci Technol 49(22):13454–13462

Kevrešan S, Petrović N, Popović M et al (1998) Effect of heavy metals on nitrate and protein metabolism in sugar beet. Biol Plant 41(2):235–240

Kevrešan S, Petrović N, Popović M et al (2001) Nitrogen and protein metabolism in young pea plants as affected by different concentrations of nickel, cadmium, lead, and molybdenum. J Plant Nutr 24(10):1633–1644

Khan F, Hussain S, Tanveer M et al (2018) Coordinated effects of lead toxicity and nutrient deprivation on growth, oxidative status, and elemental composition of primed and non-primed rice seedlings. Environ Sci Pollut Res 19:1–10

Krapp A, Fraisier V, Schleible WR et al (1998) Expression studies ofNrt2: INp, a putative high affinity nitrate transporter: evidence for its role in nitrate uptake. Plant J 6:723–732

Kubik-Dobosz G, Halajko T, Gbrska A (2001) Expression of amtl gene in the presence of some toxic ions. Acta Physiol Plant 23:187–192

Kumar S, Meena RS, Yadav GS, Pandey A (2017) Response of sesame (*Sesamum indicum* L.) to sulphur and lime application under soil acidity. Int J Plant Soil Sci 14(4):1–9

Lam HM, Coschigano KT, Melo-Oliveira O et al (1996) The molecular genetics of nitrogen assimilation into amino acids in higher plants. Annu Rev Plant Physiol Plant Mol Biol 47:569–593

Lee KC, Cunningham BA, Paulsen GM et al (1976) Effects of cadmium on respiration rate and activities of several enzymes in soybean seedlings. Physiol Plant 36(1):4–6

Lehmann S, Funck D, Szabados L et al (2010) Proline metabolism and transport in plant development. Amino Acids 39(4):949–962

Li P, Lu J, Wang Y et al (2018) Nitrogen losses, use efficiency, and productivity of early rice under controlled-release urea. Agric Ecosyst Environ 251:78–87

Llorens N, Arola L, Blade C et al (2000) Effects of copper exposure upon nitrogen metabolism in tissue cultured *Vitis vinifera*. Plant Sci 160:159–163

Lloyd DR, Phillips DH (1999) Oxidative DNA damage mediated by copper (II), iron (II) and nickel (II) Fenton reactions: evidence for site-specific mechanisms in the formation of double-strand breaks, 8-hydroxydeoxyguanosine and putative intrastrand cross-links. Mutat Res/Fundam Mol Mech Mutagen 424(1):23–36

Lojkova L, Datta R, Sajna M, Marfo TD, Janous D, Pavelka M, Formanek P (2015) Limitation of proteolysis in soils of forests and other types of ecosystems by diffusion of substrate. In: Amino acids, vol 8. Springer, Wien, pp 1690–1691

Loulakakis KA, Loulakakis-Roubelakis KA (1996) The seven NAD(H)-glutamate dehydrogenase isoenzymes exhibit similar anabolic and catabolic activities. Physiol Plant 96:29–35

Ma B, Wang S, Cao S et al (2016) Biological nitrogen removal from sewage via anammox: recent advances. Bioresour Technol 200:981–990

Malar S, Vikram SS, Favas PJ et al (2016) Lead heavy metal toxicity induced changes on growth and antioxidative enzymes level in water hyacinths [*Eichhornia crassipes* (Mart.)]. Bot Stud 55 (1):54

Mandal SM, Gouri SS, De D et al (2011) Effect of arsenic on nodulation and nitrogen fixation of blackgram (*Vigna mungo*). Indian J Microbiol 51(1):44–47

Mandal C, Ghosh N, Dey N et al (2013) Physiological responses of *Salvinia natans* L to aluminium stress and its interaction with putrescine. J Stress Physiol Biochem 9(4):1

Marfo TD, Datta R, Lojkova L, Janous D, Pavelka M, Formanek P (2015) Limitation of activity of acid phosphomonoesterase in soils. In: Amino acids, vol 8. Springer, Wien, pp 1690–1691

Masclaux-Daubresse C, Valadier MH, Carrayol E et al (2002) Diurnal changes in the expression of glutamate dehydrogenase and nitrate reductase are involved in the C/N balance of tobacco source leaves. Plant Cell Environ 25:1451–1462

Matiru VN, Dakora FD (2004) Potential use of rhizobial bacteria as promoters of plant growth for increased yield in landraces of African cereal crops. Afr J Biotechnol 3(1):1–7

Mattioni C, Gabbrielli R, Vangronsveld J et al (1997) Nickel and cadmium toxicity and enzymatic activity in intolerant and non-tolerant populations of *Silene italica* Pers. J Plant Physiol 150 (1–2):173–177

Matysik J, Alia BB et al (2002) Molecular mechanisms of quenching of reactive oxygen species by proline under stress in plants. Curr Sci 10:525–532

McAllister CH, Beatty PH, Good AG (2012) Engineering nitrogen use efficient crop plants: the current status. Plant Biotechnol J 10(9):1011–1025

Meena H, Meena RS (2017) Assessment of sowing environments and bio-regulators as adaptation choice for clusterbean productivity in response to current climatic scenario. Bangladesh J Bot 46 (1):241–244

Meena RS, Yadav RS (2015) Yield and profitability of groundnut (*Arachis hypogaea* L) as influenced by sowing dates and nutrient levels with different varieties. Leg Res 38(6):791–797

Meena RS, Kumar S, Pandey A (2017) Response of sulfur and lime levels on productivity, nutrient content and uptake of sesame under guava (*Psidium guajava* L.) based Agri-horti system in an acidic soil of eastern Uttar Pradesh, India. J Crop Weed 13(2):222–227

Meena RS, Kumar V, Yadav GS, Mitran T (2018) Response and interaction of *Bradyrhizobium japonicum* and *Arbuscular mycorrhizal* fungi in the soybean rhizosphere: a review. Plant Growth Regul 84:207–223

Mengel, Kirkby (1987) Further elements of importance. Principle of plant nutrition, 4th edn. IPI Bern, Switzerland, pp 577–582

Mevel G, Prieur D (2000) Heterotrophic nitrification by a thermophilic Bacillus species as influenced by different culture conditions. Can J Microbiol 46(5):465–473

Michalak A (2006) Phenolic compounds and their antioxidant activity in plants growing under heavy metal stress. Pol J Environ Stud 15(4):523–530

Miflin BJ, Habash DZ (2002) The role of glutamine synthetase and glutamate dehydrogenase in nitrogen assimilation and possibilities for improvement in the nitrogen utilization of crops. J Exp Bot 53:979–987

Miflin BJ, Lea PJ, Wallsgrove RM (1980) The role of glutamine in ammonia assimilation and reassimilation in plants. In: Glutamine: metabolism, enzymology, and regulation. Academic Press, New York, pp 213–234

Miller JO (2016) Soil PH and nutrient availability. http://hdl.handle.net/1903/18519

Miller AJ, Smith SJ (1996) Nitrate transport and compartmentation in cereal root cells. J Exp Bot 47:1455–1463

Mittal S, Sawhney SK (1990) Influence of lead on enzymes of nitrogen metabolism in germinating pea seeds. Plant Physiol Biochem 17:73

Molaei A, Lakzian A, Datta R, Haghnia G, Astaraei A, Rasouli-Sadaghiani M, Ceccherini MT (2017a) Impact of chlortetracycline and sulfapyridine antibiotics on soil enzyme activities. Int Agrophys 31(4):499–505

Molaei A, Lakzian A, Haghnia G, Astaraei A, Rasouli-Sadaghiani M, Ceccherini MT, Datta R (2017b) Assessment of some cultural experimental methods to study the effects of antibiotics on microbial activities in a soil: an incubation study. PLoS One 12(7):e0180663

Myers RJK (1975) Temperature effects on ammonification and nitrification in a tropical soil. Soil Biol Biochem 7:83–86

Nichol BE, Oliveira LA, Glass ADM (1993) The effects of aluminum on the influx of calcium, potassium, ammonium, nitrate and phosphate in an aluminum-sensitive cultivar of barley (*Hordeum vulgare* L.). Plant Physiol 101:1263–1266

Oaks A (1994) Primary nitrogen assimilation in higher plants and its regulation. Can J Bot 72 (6):739–750

Orzechowski S, Bielawski W (1997) Heavy metals and ammonium assimilation in triticale. J Appl Genet 38:265–270

Orzechowski S, Kwinta J, Gworek B et al (1997) Biochemical indicators of environmental contamination with heavy metals. Pol J Environ Stud 6:47–50

Pajuelo E, Rodríguez-Llorente ID, Dary M et al (2008) Toxic effects of arsenic on Sinorhizobium–Medicago sativa symbiotic interaction. Environ Pollut 154(2):203–211

Pandey M, Srivastava HS (1993) Inhibition of nitrate reductase activity and nitrate accumulation by mercury in maize leaf segments. Indian J Environ Health 39:473–481

Parlak KU (2016) Effect of nickel on growth and biochemical characteristics of wheat (*Triticum aestivum* L.) seedlings. NJAS-Wageningen J Life Sci 76:1–5

Paudyal SP, Aryal RR, Chauhan SVS et al (2007) Effect of heavy metals on growth of rhizobium strains and symbiotic efficiency of two species of tropical legumes. Sci World 5:27–32

Pérez-Tienda J, Corrêa A, Azcón-Aguilar C et al (2014) Transcriptional regulation of host NH4+ transporters and GS/GOGAT pathway in arbuscular mycorrhizal rice roots. Plant Physiol Biochem 75:1–8

Perkins LB, Johnson DW, Nowak RS (2011) Plant-induced changes in soil nutrient dynamics by native and invasive grass species. Plant Soil 345(1–2):365–374

Popović M, Kevrešan S, Kandrač J et al (1996) The role of sulphur in detoxification of cadmium in young sugar beet plants. Biol Plant 38(2):281–287

Prasad MN, Strzalka K (eds) (2013) Physiology and biochemistry of metal toxicity and tolerance in plants. Springer, Dordrecht

Quesada A, Hidalgo J, Femandez E (1997) Three nrt genes are differentially regulated in Chlamydomonas. Mol Genet 258:373–377

Rai V, Vajpayee P, Singh SN et al (2004) Effect of chromium accumulation on photosynthetic pigments, oxidative stress defense system, nitrate reduction, proline level and eugenol content of *Ocimum tenuiflorum* L. Plant Sci 167(5):1159–1169

Ram K, Meena RS (2014) Evaluation of pearl millet and mungbean intercropping systems in arid region of Rajasthan (India). Bangladesh J Bot 43(3):367–370

Rengel Z, Bose J, Chen Q et al (2016) Magnesium alleviates plant toxicity of aluminium and heavy metals. Crop Pasture Sci 66(12):1298–1307

Riaz M, Yan L, Wu X et al (2018a) Boron alleviates the aluminum toxicity in trifoliate orange by regulating antioxidant defense system and reducing root cell injury. J Environ Manag 208:149–158

Riaz M, Yan L, Wu X et al (2018b) Boron reduces aluminum-induced growth inhibition, oxidative damage and alterations in the cell wall components in the roots of trifoliate orange. Ecotoxicol Environ Saf 153:107–115

Riaz M, Yan L, Wu X et al (2018c) Mechanisms of organic acids and boron induced tolerance of aluminum toxicity: a review. Ecotoxicol Environ Saf 165:25–35

Rosales EP, Iannone MF, Groppa MD et al (2012) Polyamines modulate nitrate reductase activity in wheat leaves: involvement of nitric oxide. Amino Acids 42(2–3):857–865

Ruiz JM, Rivero RM, Romero L (2007) Comparative effect of Al, Se, and Mo toxicity on NO_3^- assimilation in sunflower (*Helianthus annuus* L.) plants. J Environ Manag 83(2):207–212

Saifullah N, Sarwar S, Bibi M et al (2014) Effectiveness of zinc application to minimize cadmium toxicity and accumulation in wheat (*Triticum aestivum* L). Environ Earth Sci 71:1663–1672

Santamarı'a MM, Marrero ARD, Herna'ndez J et al (2003) Effect of thorium on the growth and capsule morphology of Bradyrhizobium. Environ Microbiol 5:916–924

Sarwar N, Malhi SS, Zia MH et al (2010) Role of mineral nutrition in minimizing cadmium accumulation by plants. J Sci Food Agric 90(6):925–937

Sarwar N, Ishaq W, Farid G et al (2015) Zinc-cadmium interactions: impact on wheat physiology and mineral acquisition. Ecotoxicol Environ Saf 122:528–536

Sarwar N, Imran M, Shaheen MR et al (2017) Phytoremediation strategies for soils contaminated with heavy metals: modifications and future perspectives. Chemosphere 171:710–721

Schmitt M, Watanabe T, Jansen S (2016) The effects of aluminium on plant growth in a temperate and deciduous aluminium accumulating species. AoB Plants 8:plw065

Selosse MA, Baudoin E, Vandenkoornhuyse P (2004) Symbiotic microorganisms, a key for ecological success and protection of plants. C R Biol 327(7):639–648

Sen G, Eryilmaz IE, Ozakca D (2014) The effect of aluminium-stress and exogenous spermidine on chlorophyll degradation, glutathione reductase activity and the photosystem II D1 protein gene (psbA) transcript level in lichen *Xanthoria parietina*. Phytochemistry 98:54–59

Seth CS (2012) A review on mechanisms of plant tolerance and role of transgenic plants in environmental clean-up. Bot Rev 78(1):32–62

Shah K, Dubey RS (2003) Environmental stresses and their impact on nitrogen assimilation in higher plants. In: Hemantranjan A (ed) Advances in plant physiology, vol 5. Scientific Publishers, Jodhpur, pp 397–431

Shahzad B, Tanveer M, Rehman A (2018) Nickel; whether toxic or essential for plants and environment: a review. Plant Physiol Biochem 132:641. https://doi.org/10.1016/j.plaphy.2018.10.014

Sharma SS, Dietz KJ (2006) The significance of amino acids and amino acid-derived molecules in plant responses and adaptation to heavy metal stress. J Exp Bot 57(4):711–726

Sharma P, Dubey RS (2005) Modulation of nitrate reductase activity in rice seedlings under aluminium toxicity and water stress: role of osmolytes as enzyme protectant. J Plant Physiol 162(8):854–864

Sharma JOT, Subhadra AV (2010) The effect of mercury on nitrate reductase activity in bean leaf segments (*Phaseolus vulgaris*) and its chelation by phytochelatin synthesis. Life Sci Med Res 2010:1–8

Sharma SS, Schat H, Vooijs R (1998) In vitro alleviation of heavy metal-induced enzyme inhibition by proline. Phytochemistry 49(6):1531–1535

Shaw BP, Sahu SK, Mishra RK (2004) Heavy metal induced oxidative damage in terrestrial plants. In: Presad MNV (ed) Heavy metal stress in plants: from biomolecules to ecosystems, 2nd edn. Springer, Berlin, pp 84–126

Shi H, Chan Z (2014) Improvement of plant abiotic stress tolerance through modulation of the polyamine pathway. J Integr Plant Biol 56(2):114–121

Shruti M, Dubey RS (2006) Heavy metal uptake and detoxification mechanisms in plants. Int J Agric Res 1:122–141

Silveira JAG, Melo ARB, Viégas RA et al (2001) Salinity-induced effects on nitrogen assimilation related to growth in cowpea plants. Environ Exp Bot 46(2):171–179

Singh PK, Tewari RK (2003) Cadmium toxicity induced changes in plant water relations and oxidative metabolism of *Brassica juncea* L. plants. J Environ Biol 24(1):107–112

Singh RP, Singh DP, Jaiwal PK (2008) Nitrate and ammonium transporters in plants. Plant membrane and vacuolar transporters. CAB International, Wallingford, pp 83–103

Singh H, Singh A, Hussain I et al (2017) Oxidative stress induced by lead in *Vigna radiata* L. seedling attenuated by exogenous nitric oxide. Trop Plant Res 4(2):225–234

Skopelitis DS, Paranychianakis NV, Paschalidis KA et al (2006) Abiotic stress generates ROS that signal expression of anionic glutamate dehydrogenases to form glutamate for proline synthesis in tobacco and grapevine. Plant Cell 18:2767–2781

Stan V, Gament E, Corena CP et al (2011) Effects of heavy metal from polluted soils on the rhizobium diversity. Not Bot Hort Agrobot Cluj 39:88–95

Stitt M (1999) Nitrate regulation of metabolism and growth. Curr Opin Plant Biol 2:178–186

Syntichaki KM, Loulakakis KA, Loulakakis-Roubelakis KA (1996) The amino-acid sequence similarity of plant glutamate dehydrogenase to the extremophilic archaeal enzyme conforms to its stress-related function. Gene 68:87–92

Szabados L, Savoure A (2010) Proline: a multifunctional amino acid. Trends Plant Sci 15:89–97

Talano MA, Cejas RB, González PS et al (2013) Arsenic effect on the model crop symbiosis Bradyrhizobium–soybean. Plant Physiol Biochem 63:8–14

Terce-Laforgue´ T, Dubois F, Ferrario-Me'ry S et al (2004) Glutamate dehydrogenase of tobacco is mainly induced in the cytosol of phloem companion cells when ammonia is provided either externally or released during photorespiration. Plant Physiol 136:4308–4317

Vajpayee P, Tripathi RD, Rai UN et al (2000) Chromium (VI) accumulation reduces chlorophyll biosynthesis, nitrate reductase activity and protein content in *Nymphaea alba* L. Chemosphere 41:1075–1082

Van Assche F, Cardinaels C, Clijsters H (1988) Induction of enzyme capacity in plants as a result of heavy metal toxicity: dose-response relations in *Phaseolus vulgaris* L., treated with zinc and cadmium. Environ Pollut 52(2):103–115

Varma D, Meena RS, Kumar S (2017) Response of mungbean to fertility and lime levels under soil acidity in an alley cropping system in Vindhyan region, India. Int J Chem Stud 5(2):384–389

Verkleij JA, Golan-Goldhirsh A, Antosiewisz DM et al (2009) Dualities in plant tolerance to pollutants and their uptake and translocation to the upper plant parts. Environ Exp Bot 67 (1):10–22

Verma SK, Singh SB, Prasad SK, Meena RN, Meena RS (2015) Influence of irrigation regimes and weed management practices on water use and nutrient uptake in wheat (*Triticum aestivum* L. Emend. Fiori and Paol.). Bangladesh J Bot 44(3):437–442

Wang MY, Siddiqi MY, Ruth TJ et al (1993) Ammonium uptake by rice roots. Plant Physiol 103:1259–1267

Wang L, Zhou Q, Ding L et al (2008) Effect of cadmium toxicity on nitrogen metabolism in leaves of *Solanum nigrum* L. as a newly found cadmium hyperaccumulator. J Hazard Mater 154:818–825

Wang Y, Lu J, Ren T et al (2017) Effects of nitrogen and tiller type on grain yield and physiological responses in rice. AoB Plants 9. https://doi.org/10.1093/aobpla/plx012

Wani PA (2008) Heavy metal toxicity to plant growth promoting rhizobacteria (PGPR) and certain legume crops. Ph.D. Thesis, Aligarh Muslim University, Aligarh, India

Wani PA, Khan MS, Zaidi A (2007) Impact of heavy metal toxicity on plant growth, symbiosis, seed yield and nitrogen and metal uptake in chickpea. Aust J Exp Agric 47(6):712–720

Weber MB, Schat H, Ten Bookum-Van Der Maarel WM (1991) The effect of copper toxicity on the contents of nitrogen compounds in Silvena vulgaris (Moench) Garcke. Plant Soil 133:101–109

Wu LH, Luo YM, Xing XR et al (2004) EDTA-enhanced phytoremediation of heavy metal contaminated soil with Indian mustard and associated potential leaching risk. Agric Ecosyst Environ 102(3):307–318

Xiong ZT, Liu C, Geng B (2006a) Phytotoxic effects of copper on nitrogen metabolism and plant growth in *Brassica pekinensis* Rupr. Ecotoxicol Environ Saf 64(3):273–280

Xiong ZT, Zhao F, Li MJ (2006b) Lead toxicity in *Brassica pekinensis* Rupr.: effect on nitrate assimilation and growth. Environ Toxicol Int J 21(2):147–153

Xu J, Yin H, Li X (2009) Protective effects of proline against cadmium toxicity in micropropagated hyperaccumulator, *Solanum nigrum* L. Plant Cell Rep 28(2):325–333

Yadav GS, Babu S, Meena RS, Debnath C, Saha P, Debbaram C, Datta M (2017) Effects of godawariphosgold and single supper phosphate on groundnut (*Arachis hypogaea*) productivity, phosphorus uptake, phosphorus use efficiency and economics. Indian J Agric Sci 87 (9):1165–1169

Yancey PH (2005) Organic osmolytes as compatible, metabolic and counteracting cytoprotectants in high osmolarity and other stresses. J Exp Biol 208:2819–2830

Yang Y, Zhang Y, Wei X et al (2011) Comparative antioxidative responses and proline metabolism in two wheat cultivars under short term lead stress. Ecotoxicol Environ Saf 74(4):733–740

Yao S, Ni JR, Ma T et al (2013) Heterotrophic nitrification and aerobic denitrification at low temperature by a newly isolated bacterium, Acinetobacter sp. HA2. Bioresour Technol 139:80–86

Yu CC, Hung KT, Kao CH (2005) Nitric oxide reduces Cu toxicity and Cu-induced NH4+ accumulation in rice leaves. J Plant Physiol 162(12):1319–1330

Zhou X, Gu Z, Xu H et al (2016) The effects of exogenous ascorbic acid on the mechanism of physiological and biochemical responses to nitrate uptake in two rice cultivars (*Oryza sativa* L.) under aluminum stress. J Plant Growth Regul 35(4):1013–1024

Zia-ur-Rehman M, Sabir M, Nadeem M (2015) Remediating cadmium-contaminated soils by growing grain crops using inorganic amendments. In: Soil remediation and plants: prospects and challenges. Elsevier Inc/Academic, London, pp 367–396

Soil Microbial Ecology and Its Role in Soil Carbon Sequestration in Sustainable Agroecosystems Under Climate Change

Julio Alves Cardoso Filho and Gildemberg Amorim Leal Junior

Abstract To help the sustainable intensification of food production systems, and minimize the levels of the greenhouse gases anthropogenic emissions, the current agriculture needs to create and use methodologies (e.g., soil carbon sequestration) that minimize loss of terrestrial biodiversity on agroecosystems. The development of the soil microbial ecology in the last 30 years are related to the role of soil microorganisms in the maintenance of soil health. Therefore in this chapter, we have provided the current information on soil microbial ecology management of agro-ecosystems for carbon sequestration under global climate change.

Keywords Soil microbial ecology·carbon sequestration · Agro-ecosystems · Climate change

Abbreviations

AMF	Arbuscular Mycorrhizal Fungi
a_w	Water activity
CCX	Chicago Climate Exchange
CEC	Cation Exchange Capacity
CMNs	Common Mycorrhizal Network
DOM	Derived Organic Matter
ESD	Education for Sustainable Development
FA	Fulvic Acid
FAO	Food and Agricultural Organization
FCPF	Forest Carbon Partnership Facility
GHGs	Greenhouse Gases
GPP	Gross Primary Production
GSOCS	Global Soil Organic Carbon Stocks

J. A. C. Filho (✉) · G. A. L. Junior
Center of Agricultural Sciences, Sector of Plant Pathology, Federal University of Alagoas, Rio Largo, Alagoas, Brazil

R. Datta et al. (eds.), *Carbon and Nitrogen Cycling in Soil*,
https://doi.org/10.1007/978-981-13-7264-3_8

HA	Humic Acid
HANPP	Human Appropriation of Net Primary Production
$HANPP_{harv}$	NPP consumed + unused NPP by humans
$HANPP_{luc}$	loss of potential NPP due to land use change
IPCC	Intergovernmental Panel on Climate Change
MAMPs	Microbe-Associated Molecular Patterns
NEP	Net Ecosystem Production
NPP	Net Primary Production
NPP_{act}	the actual NPP in anthropogenic-altered system
NPP_{pot}	the potential NPP in undisturbed system
OM	Organic Matter
Pa	Pascals
Pg	1 Petagram (Pg) = 10^{15} g
p_o	pure water
PRRs	Pattern Recognition Receptors
Q_{10}	Temperature Sensitivity
R_A	Autotrophic Respiration
REDD	Deforestation and Forest Degradation
R_H	Heterotrophic Respiration
R_m	Microbial Respiration
RMUs	Removal Units
ROM	Recalcitrant Organic Matter
R_S	Soil Respiration
SOC	Soil Organic Carbon
SOM	Soil Organic Matter
UNEP	United Nations Environment Programme
UNFCCC	United Nations Framework Convention on Climate Change
VOCs	Volatile Organic Compounds
WMO	World Meteorological Organization
Ψ	Water potential
Ψ_g	Water gravitational potential
Ψ_m	Water matric potential
Ψ_o	Water osmotic potential

1 Introduction

Soil microbiology is defined as the study of soil microorganisms (Alexander 1961; Clark 1977). In 1885, Martinus Beijerinck (1851–1931), and Sergei Nikolaevitch Winogradsky (1856–1953) established the basis of the discipline of soil microbial ecology (Allison 1961; Moreira and Siqueira 2006; Cheeke et al. 2013; Balestrini et al. 2015; Cardoso and Andreote 2016). Soil microbial ecology studies the composition and structure of microbial communities aiming to uncover the origin of their diversity (Yadav 2017; Tecon and Or 2017). The most relevant aspects of soil microbial ecology, increasingly researched is the quantity of environmental services or soil services (e.g., microbes roles in energy flow and in nutrient cycling) that are provided to a wide variety of soil microorganisms, aiding the agricultural and natural ecosystems (Lucas et al. 2017; Tecon and Or 2017). The challenge for soil microbiology is to connect the microbial communities structure with their ecosystem function (Stein and Nicol 2011; Larsen et al. 2012; Saleem 2015), and to identify the biotic and abiotic environmental factors that stimulate, or limit the microbial communities survival in the habitat where they live (Evans and Wallenstein 2014; Jørgensen and Marshall 2016).The organic matter (OM) is an important soil pool of carbon, and it is affected by anthropogenic activities (Rousk and Bengtson 2014). The rate of OM degradation in soil is controlled by the native microbial community and the edaphic conditions (e.g., temperature and pH), which controls the activities of the soil microbiota (Yadav 2017).

The carbon (C) -induced global warming has a crucial role in human history evolution (in the last 650,000 years) on the Earth (Steven 2010; Diffenbaugh and Field 2013; Bauska et al. 2015). The global C pools (terrestrial ~ 6.2% abiotic + biotic combined), ocean pool ~ 77.4% (marine sediments + carbonate rocks combined), and atmospheric C pool contain ~1.5% of a total ~50,400 billion metric tons Pg C (Gt = 1 billion tonnes = 1 Petagram (Pg) = 10^{15} g) (Lorenz and Lal 2018; Lal 2010). The atmospheric C pool is easily overwhelmed by human-induced perturbation (e.g., burning of fossil fuels (deforestation), soil degradation and desertification) of the terrestrial pools, since ~1850.Currently, the Climate change presents a real challenge ever to have confronted all human social, political, and economic systems (IPCC 2014a, b; Hawkins and Sutton 2016). The recognizing of the problem of global climate change, by the WMO (World Meteorological Organization) and the UNEP (United Nations Environment Programme) was established in 1988 in the IPCC (the Intergovernmental Panel on Climate Change) to assess climate change based on the latest science (Davenport 2008; Mazzoleni et al. 2012; IPCC 2017). Carbon soil sequestration refers to the uptake of C-containing substances from the atmosphere (e.g., carbon dioxide (CO_2) in terrestrial or marine reservoirs) and its storage into the soil carbon stocks (Zomer et al. 2017). Carbon enters soil organic matter (SOM) pool via C-containing compounds founded in animals, microorganisms, plants, soil, and water (Cheeke et al. 2013). Terrestrial plants contain ~650 Pg C (Canadell and Schulze 2014), soil microbiota contain ~110 Pg C (Lal 2011a, b), soil are estimated to contain ~3300 Pg C (C organic + C inorganic combined) stored to 1-m depth (Lal 2010; Meena and Meena 2017). The soils can behave simultaneously, as sources and sinks of C, depending upon land-use, biomass input levels, edaphic and climatic change conditions (Zomer et al. 2017). It has been suggested that, over the next century, improved soils and their management strategies could sequester up to 150 Pg CO_2 (Lal 2011a, b; Gougoulias et al. 2014), reduce the atmospheric concentrations of greenhouse gases (GHGs), and the global warming (reviewed by Havlík et al. 2014). Nowadays, the microbial ecology studies in these Earth system models (e.g., network ecological indicators) are used to estimate the soil organic carbon stocks, because it has become a key issue over recent years, for studies of climate change, and food security (Batjes 2011; Gougoulias et al. 2014; Frank et al. 2015; Tian et al. 2015).

2 Soil as a Habitat for Microbiota Life

The purpose of this topic is to provide a holistic view to soil microbial ecology and includes the nature of soil microenvironments, and the physiological ecology of microbiota in these soil microhabitats, without a distinction between bulk soil (non-rhizosphere soil) and soil under the influence of the roots of plants (rhizosphere soil) (Standing and Killham 2007). The terms soil microbiota refers to all microorganisms, including predators, saprophytes, endophytes, epiphytes, commensals, non-symbionts, symbionts, rhizospheric and non-rhizospheric, non-opportunistic and opportunistic pathogens, that inhabit the soil, animals, and plants (for further details see the reviews by Berg and Smalla 2009; Redford et al. 2010; Larsen et al. 2012; Vorholt 2012; Mwajita et al. 2013; Udvardi and Poole 2013; de Vries et al. 2013; Berg and McClaugherty 2014; Osono 2014; Hardoim et al. 2015; Poudel et al. 2016; Ofaim et al. 2017). On the basis of the ecological diversity, soil microhabitats can be separated roughly into phyllosphere (the aerial parts), rhizoplane (the root surface), rhizosphere (the zone around the root), which are linked forming a fractal of the Bulk Soil Food Web and Detritus-Based Food Chain (reviewed by Lemanceau et al. 2017; Glavatska et al. 2017). This microbiota is summoned from the environment (soil, atmosphere) and from the plant via the seed (Maude 1996). All these microhabitats provide crucial biotic and abiotic environmental conditions for Earth soil microbial life (Hoehler and Jørgensen 2013; Nazir et al. 2017).

2.1 General Soil Concepts

Soils (also called lithosphere) are composed of three phases of clastic particles (mineral matter), organic materials in various stages of decay, living organisms, water (or ice), and gases within pores of various sizes (Martin 1950; Alexander 1961; Stotzky 1972, 1997; Nannipieri et al. 2003; Bastida et al. 2009; Marschner and Marschner 2012; Myrold et al. 2014; Cardoso and Andreote 2016; Kumar et al. 2016; Datta et al. 2017b). The solid phase forms the soil matrix; the liquid phase is the water in the soil, which always contains dissolved substances, so it should be called the soil solution, and the gaseous phase is the soil atmosphere (Cardoso and Andreote 2016). The soil is a specific weathered product that results from physical (Mechanical), chemical and biologic activity, with its first parental rock, climate variations, innumerable topography combinations, time, the history of land use and its management regimes (Arnold 1983; Brady 1984; Ranger et al. 1992; Wilding 1994; Stiles et al. 2003).The soil is the principal reserve of microbial life that influences the plant-microbe interactions, making the host plant resistant to abiotic and biotic stress (Filho et al. 2017; Weil and Brady 2016).

2.1.1 Soil Physicochemical Properties

The physicochemical soil properties have effects on the activity, ecology, and population dynamics of microbiota in the soil. Any conceptual inference to study soil microbial ecology should address the organisms, biological processes, and the spatial and dynamic of soil microhabitats and microhabitats. A brief consideration about the nature of the physicochemical of soil properties, microenvironments and the spatial distribution of soil microbiota will be addressed in the next section.

2.1.1.1 Soil Structure and Texture

All soils have a specific morphology (identification and description of soil profiles and pedons), defined as its structure or form (Ball 2013). Soil structure is dynamic and continuously altered by edaphic soil conditions (Azcón-Aguilar and Barea 2015). Soil structure is correlated to the spatial arrangement of soil particles (Churchman 2013; Schlüter and Vogel 2016) into complex of aggregates (Jackson 2014) and of pores (Miedema 1997; Cooper et al. 2016). Usually, there are three recognized categories of soil structure, single grained, massive, and aggregated (Ball 2013). The aggregated is the soil structure that is the ideal condition for plant growth, mainly in the early stages of the seed germination and seedling adaptation to soil environmental conditions (Maffei 2014). Soil aggregates are originated by the physicochemical interaction between derived organic matter (DOM) of the plant, animals (Foster 1994; Weil and Brady 2016; Meena et al. 2017) and microbial filamentous growths (Peng et al. 2013; Voroney and Heck 2015), with mineral (clay) primary and secondary particles (Churchman and Lowe 2012; Churchman 2013) via bivalent ions and water activity (Bagheri et al. 2012).

The soil contains clay, silt, and sand particles (Weil and Brady 2016: called the soil texture). Sand and silt are primary minerals (weathering-resistant) originated from the rock, and Clays (particles finer than 2 micrometers (μm) in diameter) are secondary minerals (e.g., Kaolinite, Montmorillonite, Illite, Limonite, Goethite, Vermiculite) formed by decomposition of the primary minerals (Tisdale et al. 1985). In the temperate region soils, the commonly found clay fraction is silicate clays whereas, in tropical regions, hydrated oxides of iron and aluminum (called sesquioxides) are most commonly found (Fassbender 1987). Only four clay minerals occur in all parts of the world (e.g., Kaolinite, Montmorillonite Illite, and Vermiculite). It is safe to say that kaolinitic soils occur usually in tropical conditions than elsewhere (Weil and Brady 2016). The clay fraction is very important in imparting specific physical properties to soils, soil microbiota, and to plant growth and activity via nutrient availability (Marques et al. 2015). Colloid (the particle with 0.1–0.001 μm in diameter) is a state of matter that approach but never reach molecular sizes (Tan 2011). SOM materials (non-humified and humified) and plants solids (e.g., carbohydrates, amino acids, proteins, lipids, lignins) also occur in the colloidal state (Tan 1982; Datta et al. 2017c).

The soil structure and texture are fundamental qualitative measures of the biota habitat (Havlicek and Mitchell 2014), and it drives the distribution of the spatial patterns of soil biota occupying that habitat, and to provide many different micro-habitats for soil microbial life (Lehmann 2007; Maron et al. 2011; Bevivino et al. 2014; Constancias et al. 2015).

2.1.1.2 Soil Organic Matter, Humus Formation, Microbial-Organo-Mineral Complex Interactions

Soil organic matter is best known and least understood component of soils (Tan 2011). In fact, we still do not know how to define what is soil organic matter (reviewed by Horwath 2015; Paul 2016). The fraction of SOM for the global carbon cycle has been estimated to ~1550 Pg of C organic to a depth of 1 m and ~2450 Pg of C to a depth of 2 or 3 m (Lal 2004a, b; Wang et al. 2016; Lorenz and Lal 2018). The current consensus is that microbial constituents (e.g., polysaccharides) are essential to stabilizing the SOM, and this new conceptual and several quantitative SOM models are added to this point of view (Feng et al. 2014; Kallenbach et al. 2016). The composition and global distribution of SOM is a storehouse of the database on vegetation, parent materials, climate, and environmental disturbance (Paul 2016). After 50 years, it is still difficult to define humus (Yarrow 2015). The characteristics and functions of SOM can be summarized as biological properties (e.g., soil reservoir of metabolic energy, source of nutrients and ecosystem resilience), physical properties (e.g., stabilization of soil structure and water retention), chemical properties (e.g., Cation Exchange Capacity – CEC), Buffering Capacity and pH effects, chelation of metals and interactions with xenobiotics (Baldock 2007; Baldock and Broos 2012). Humus is a colloidal complex organic substance (reviewed by Berg and McClaugherty 2014), that occurs naturally in soil and, increases the soil Cation Exchange Capacity (Tan 2011: CEC = Σ mEq exchangeable cations per 100 g of soil, presently as cmol/(pt) kg^{-1}), hence it has the ability to store nutrients by chelation (Baldock and Broos 2012). While these nutrients (cations and anions) are accessible to plants, they are held in the soil safe from being leached by rain or irrigation (Hinsinger et al. 2012). Humus can be fractioned (Schnitzer 1990) into non-humified substances, such as fulvic acid (FA), hymatomelanic acid, humic acid (HA) and its hydroxybenzoic acid derivatives (Tan 2011). FA and HA may form metal complexes by chelation and contributes to soil organo-mineral complex formation (Berg and McClaugherty 2014).

The legacy of the ecology for SOM decomposition and the humus formation is important for two reasons. Firstly, because significant amounts of the GHGs, including CO_2, methane (CH_4) and nitrous oxide (N_2O) are released as products of the SOM decomposition into the atmosphere. Secondly, the soils represent a sink for carbon, due to the improvement in the SOM decomposition rate by soil biota, and to enhance organic carbon sequestration in agricultural soils (Lal 2011a, b; Piccolo 2012; Ram and Meena 2014). The soil organo-mineral complex is a blend of primary particles (e.g., clay, silt, and sand) and organic compounds (e.g., cationic, anionic, and polar nonionic in nature) stabilized by organic and inorganic chemical bonds (reaction of a metal ion: electron-pair acceptor and ligands: electron-pair donor), electrostatic forces, and biological agents (plants, micro, meso and macrobiota of soil) and is the main basic fraction of soil aggregation (Li et al. 2016).

2.1.1.3 Water Activity (a_w) and Soil Water Availability

The availability of water to soil microbiota is more crucial than the total soil water content (Uhlirová et al. 2005; Azcón-Aguilar and Barea 2015). Water availability can be express in terms of its free energy ("potential energy") per unit mass, and soil-water content is a measure of the quantity of water (volume or mass) contained in a unit volume or mass of soil (Richards 1965; Voroney and Heck 2015). The water activity can be express as the amount of water available for microbial use (Reid 1980), and depends on the number of moles of water, the number of moles of solute, and the activity coefficients for water and the particular solute (Atlas and Bartha 1993).

Water activity (a_w) or the relative humidity (soil atmosphere) is defined as the ratio of the vapor pressure of any solution (p) of saturated air, to that of pure water (p_o) at a specified temperature (Fassbender 1987):

$$a_w = \frac{P}{P_o}$$

The soil water potential (Ψ), can be expressed (for equilibrium conditions) in pascals (Pa: energy units per mass). Considering the relation between a_w and water potential, where −1 bar = −100 KPa = −0,1 MPa (Papendick and Campbell 1981).

$$\Psi = \left(\frac{RT}{V_w}\right) \ln a_w$$

where

R = universal gas constant (8.21×10^{-5} m^3 bar mol^{-1} K^{-1}).
T = absolute temperature (K).
V_w = partial molal volume of water (1.8×10^{-5} m^3 mol^{-1} at 4 °C).

Total soil water potential (Ψ) components includes the sum of the *matric potential* (Ψ_m), *osmotic potential* (Ψ_o), *pressure potential* (Ψ_p) and *gravitational potential* (Ψ_g) (Metting 1993). Water usually moves from sites of high potential to sites of lower potential (Harris 1981). The availability of water influences the moisture available to soil microbiota, soil aeration status (Soil Porosity) the type and amount of soluble materials, the osmotic pressure, and the pH of the soil solution (Voroney and Heck 2015). At extreme soil low water conditions and nutrient availability limiting, several soil microbial processes (e.g., complete organic matter decomposition), are restricted or do not occur (Papendick and Campbell 1981; Prosser 2007;

Gougoulias et al. 2014). Plants and the soil microbiota are able to obtain water from the soil as long as, and maintain their internal water potential below that of the soil (Kaisermann et al. 2015). If the soil water availability is too low for any microbial activity to occur. Consequently, the soil microbial community will be unable to conductany ecosystem processes (Schiraldi et al. 2012; Verma et al. 2015).

2.1.1.4 Soil Air Composition and Soil Porosity

The soil air is composed of carbon dioxide (CO_2), methane (CH_4), ethylene (C_2H_4), oxygen O_2, dinitrogen (N_2), nitrous oxide (N_2O), nitric oxide (NO), nitrogen dioxide (NO_2), ammonia (NH_3), aliphatic amines, hydrogen sulfide (H_2S), dimethyl sulfide (CH_3SH), dimethyl disulfide (CH_3SSCH_3), methyl metacaptan (CH_3SH), carbon disulfide (CS_2), carbonyl sulfide (COS) and many others gases that change in function of both the sources and sinks of the gas concentration (Rolston and Møldrup 2002). Gas transport in soils occurs due to the diffusion and advection process (Roland et al. 2015). Diffusion is the main mechanism in the trade of gases between the soil and the atmosphere. However, advection may be responsible for displacing higher volume of gas but usually over short time periods (Lee et al. 2010).

Fick's First law is used to describe the gas diffusion in soils (steady-state condition), due to its simplicity (Currie 1960):

$$J_i = \frac{-\boldsymbol{D}_i}{\boldsymbol{RT}} \frac{C_i \partial(\mu_i)}{\partial x} \quad (\boldsymbol{i} = \mathbf{1}, \mathbf{2}, \ldots, \boldsymbol{n})$$

where J_i is the flux of gas species i, $-\boldsymbol{D}_i$ is bulk diffusity, $\boldsymbol{R}$ is gas constant, $\boldsymbol{T}$ is temperature, $\boldsymbol{x}$ is diffusion path (one-dimensional), $\boldsymbol{C}_i$ is molar fraction of gas species $\boldsymbol{i}$, and μ_i is the chemical potential of species $\boldsymbol{i}$ at a given state (Jaynes and Rogowski 1983), μ_i is a function of the concentration/density of mass species and is expressed by,

$$\mu_i = \mu_o + \boldsymbol{RT} \ln \boldsymbol{C}_i$$

where μ_o is the standard chemical potential of gas species $\boldsymbol{i}$ under 1 atm and mol^{-L} (Jaynes and Rogowski 1983).

Fick's second law describes the unsteady diffuse flux to concentration gradient (Moldrup et al. 2004), is given as

$$\frac{\partial C_i}{\partial t} = D_i \frac{\partial^2 C_i}{\partial x^2}$$

and predicts the concentration change with time (t) on gases diffusion mechanism (Kirkham and Powers 1972).

The gaseous fluxes by diffusion is due to the gradients concentration established within the soil by respiration of soil biota and plant roots (Kai et al. 2011), by accumulation of gases due to the biological reactions (e.g., fermentation, nitrification, and denitrification) and by input of chemicals (e.g., fumigants and pesticides), and volatile organic compounds (VOCs) from toxic waste habitats (Roland et al. 2015).

The water flows in the soil is a relevant hydrological process that occurs at the interface of the soil-plant-atmosphere system (reviewed by Assouline 2013; Assouline and Selker 2017), because it is related to water availability to the plants, and the soil biota (Zonta et al. 2012).The pore space is the total space of soil not filled by soil particles (Anon 1991), whereas density is the mass per unit volume including por space (Culley 1993). Soil pores are classified by size (Luxmoore 1981) such as macropores (diameter > 0.1 mm); are responsible for aeration and gravity flow; mesopores (diameter 30–100 μm) conduct water by capillary flow; micropores (diameter < 30 μm) are responsible for water retention and slow capillary (Bouma 1981).

Total porosity (sum of all pore size volumes) (Badorreck et al. 2013) is measured as

$$\%\mathbf{pore\ space} = \mathbf{100} - \left[\left(\frac{\mathbf{D_b}}{\mathbf{D_p}}\right) x\ \mathbf{100}\right]$$

where:

$$\mathbf{D}_b = \mathbf{soil\ bulk\ density,\ (Mg\ m^{-3})} \therefore = \frac{\mathbf{Soil\ mass\ (Mg)}}{\mathbf{soil\ bulk\ volume}(\mathrm{m}^3)}$$

$$\mathbf{D}_p = \mathbf{soil\ particle\ density,\ (Mg\ m^{-3})} \therefore = \frac{\mathbf{soil\ mass\ (Mg)}}{\mathbf{soil\ particle\ volume}\ (\mathrm{m}^3)}$$

Currently applied in microcomputerized X-ray tomography (CT scanning) allow the study of soil physicochemical properties, such as the soil porosity on soil's intact 3D-dimensional structure (Chen et al. 2015; Hapca et al. 2015; Rogers et al. 2016). These methods are used to analyze the micro computerized X-ray tomography image of the soil porosity under numerous soil management conditions (Vaz et al. 2011).

The soil compaction plays a critical role in microbial activity since it increases the abundance of anaerobic microsites and the reduction in the aerobic microbial activity, due to the lower O_2 and CO_2 diffusion rates and (Silva et al. 2011), and to induce N losses by denitrification (Voroney and Heck 2015).The O_2 takes control of the rate of aerobic decomposition in its role as an electron acceptor in the mineralization of SOM (Keiluweit et al. 2016).The soil respiration is a good indicator of soil SOM decomposition (Bond-Lamberty and Thomson 2010), by both aerobic and anaerobic microbes, that is an advantage over techniques based on the O_2 uptake. Unfortunately, our limited knowledge of the CO_2 transport through the soil restricts our understanding of the role of the various abiotic and biotic controls on soil respiration (Rolston and Møldrup 2002). The knowledge of the water and gaseous fluxes is important to management practices that affect the physical quality of the soil (Assouline 2013; Cardoso et al. 2013; Assouline and Selker 2017; Dhakal et al. 2015).

2.1.1.5 Temperature

In Temperate and tropical climate conditions, the temperature and water are the soil-forming factors of most importance to microbial ecology (Atlas and Bartha 1993). All life forms on the Earth have a critical growth temperature, at which they exhibit their highest growth and reproduction rates (Voroney and Heck 2015), they also have minimal growth temperatures below that any metabolic activity is observed (Bradford et al. 2008). The discovery of extremophilic and extremotrophic micro-organisms in permafrost soil represents the most recent knowledge about the ecology and the physiology of cold-adapted microorganisms, plants and animals (reviewed by Weber et al. 2007; Battistuzzi and Hedges 2009; Dopson et al. 2016) and may be contributed to learn how microbial activity may be part of a positive feedback loop that also links shrub growth and warmer soil temperatures (Classen et al. 2015) and explains the mechanisms of cold adaptation on the enzymatic and molecular level, through enzymes of cold-adapted organisms (Vandenkoornhuyse et al. 2015; Nikrad et al. 2016). The climatic change (e.g., global warming) have direct and indirect impacts on the soil microbial function (reviewed by Blankinship et al. 2011; Henry 2012; Manzoni et al. 2012; A'Bear et al. 2014, Chen et al. 2014; Joshi and Shekhawat 2014; Kumar et al. 2016). The global warming alters microbial species distributions and thus interactions among microbial communities (DeAngelis et al. 2010; van der Putten 2012; DeAngelis et al. 2015).

Recently, the temperature sensitivity (Q_{10}) of SOM decomposition has been used to estimate the potential negative and positive contributions of soil microbes to the soil carbon (C) sequestration (C is fixed from the atmosphere via plants or organic residues and stored in the soil) and global warming in terrestrial ecosystems scenarios (von Lützow and Kögel-Knabner 2009; Zhou et al. 2009; Balser et al. 2010; Nianpeng et al. 2013; Karhu et al. 2014; Classen et al. 2015). The Q_{10} is the rate of increase in an activity or process over a 10 °C increase in temperature (Atlas and Bartha 1993) which:

$$\frac{\mathbf{Q_{10}{=}Activity\ at\ tempT{+}10\,_{0}C}}{\mathbf{Activity\ at\ tempT}}$$

The knowledge and understanding of SOM decomposition rate and Q_{10} under long-term grazing exclusion (GE) conditions (He et al. 2008; Wang et al. 2016), would be useful for efficiently and accurately evaluating long-term soil C sequestration (Sierra 2012; Nianpeng et al. 2013; Datta et al. 2017a).

2.1.1.6 Hydrogen Ion Concentration (pH)

In terms of acid-base equilibria in aqueous solution (e.g., as soil solution), three constants must be considered: the acid dissociation (K_a), the base dissociation (K_b), and the self-ionization of water (K_w). Usually, for a weak acid (HA), K_a is $[H_3O^+] \times [A^-]/[HA]$, and for a weak base (B), K_b is $[BH^+] \times [OH^-]/[B]$, considering aqueous solutions. Then, $pK_a = -\log K_a$ and $pK_b = -\log K_b$ or $K_a = 10^{-pK_a}$, $K_b = 10^{-pK_b}$. On the other hand, we can represent the self-ionization of water as $K_w = [H_3O^+] \times [OH^-] = 1.00 \times 10^{-14}$ (at 298 K), where $pK_w = -\log K_w = 14.00$ or $K_w = K_a \times K_b$. In aqueous solution, it is correct to write $[H_3O^+]$ (aq) than H^+ (aq), because the protons are solvated, and the pH is given by $pH = -\log[H_3O^+]$ (Novozamsky et al. 1976). Based on the Brønsted concept, the water can act as both a proton donor (a Brønsted acid) and a proton acceptor (a Brønsted base) (Hadzi et al. 1968; Reed 2013). In this case, the role of water depends on the relative strengths of the numerous species found in the aqueous solution (Hadzi et al. 1968). For example, when a strong acid (e.g., the hydrogen chloride (HCl)) is diluted into water, the water behaves as a Brønsted base (accepts a proton) to form $[H_3O^+]$. In the opposite direction, $[H_3O^+]$ acts as a weak acid and Cl^- acts as a weak base. In other words, the ions $[H_3O^+]$ and Cl^- are, respectively, the conjugate acid and conjugate base of H_2O and HCl (Hadzi et al. 1968). Considering the above, the "soil reaction" (pH) is used to evince the acid-base reactions that occur in the soils (Tan 1982; Marschner and Marschner 2012). Based on the relative degree of acidity, the soils are grouped into many acidity or alkalinity gradations scales (Tan 1982). In this context, the most well-studied chemical property of the soil and its effects on the microbiota activity is pH (reviewed by Stotzky 1972; Magdoff and Bartlett 1984; Hinsinger et al. 2003; Hinsinger et al. 2009; Babauta et al. 2012; Hinsinger et al. 2012; Husson 2013; Muthukumar et al. 2014). It is very well established that several soil chemical (e.g., formation of clays minerals) and biochemical (e.g., rate of decomposition of organic matter) reactions are driven by the soil pH (Cardoso and Andreote 2016). The soil pH-buffering capacity (pHBC) is related to the soil to resist pH change (Voroney and Heck 2015). The soil pH-buffering capacity is provided by organo-mineral complex (Busari et al. 2015). Large part of the soil microbiota acts within a narrow optimal range of pH and their activity is inhibited in extreme acidic or alkaline enviromental conditions (Stotzky 1997). The distribution of soil fungi and protists into the soil microenvironment are less impacted to soil pH and more targeted by climate conditions and latitudinal gradients (Bates et al. 2013; Tedersoo et al. 2014; Peay et al. 2016). The soil pH is the best predictor of bacterial community composition at the continental scale (Andam et al. 2016). Soil organic matter (SOM) contributes to the neutral to slightly acid soil pH (Brady 1984; Brady and Weil 2002; Brady and

Weil 2010) by buffering soil pH (Magdoff and Bartlett 1984), forming complexes with Al (low pH: Skyllberg et al. 2001), and soluble complexes with calcium (Ca) and magnesium (Mg) (high pH: Brady and Weil 2010). Modern agricultural practices are intensifying soil acidification (e.g., repeated applications of nitrogen fertilizers) in amounts that exceed the soil pH-buffering capacity (Tisdale and Nelson 1985). In this regard, these modern agricultural systems (unsustainable agricultural systems) should be avoided and replaced by methods that are environmentally friendly.

2.1.1.7 Redox Potential (E_h)

The oxidation-reduction potential or Redox Potential (E_h) has received little or no attention in agricultural sciences, unlike pH, is regarded as a primer parameter (reviewed by Husson 2013; Tokarz and Urban 2015). E_h is the tendency for a substance to lose or gain electrons under specific conditions, and in soil collective reactions of that is quantifiable as the electrical potential expressed in volts or millivolts, is defined by the Nernst Equation (Metting 1993), where:

$E_h = E_o + [RT/nF]$ ln [oxidized species/reduced species] , in which
E_o = standard potential of the system in reference to a standard H electrode
R = universal gas constant (8.317 J/mol K)
T = temperature (K)
F = Faraday constant (96,000 coulombs).
n = number of electrons participating in the reaction.

A high positive $E_h$value indicates an environment that favors oxidations reactions (e.g., well-drained soils), and a low negative E_h indicates a strongly reducing environment (e.g., flooded and water logging or compaction soils) (Atlas and Bartha 1993). Microorganisms are able to change the E_h and the pH of their surrounding medium, to supply their nutritional requirements (Husson 2013). For instance, fungi usually grow under moderately reducing conditions ($E_h > +250$ mV), while bacteria grow under highly reducing conditions ($E_h < 0$ mV) (Seo and DeLaune 2010).

In a simplified way, the $E_h$can be calculated as which (Fassbender 1987):

$$pe = \frac{E_h}{\frac{2.3\ RT}{F}}$$

where pe is called negative logarithm decimal of electronegativity, at normal environmental pressure conditions and 25 °C, so

$$pe = \frac{E_h}{0.59}$$

The dynamics and mobility of sulfur (S), phosphorus (P), iron (Fe) and Nitrogen (N) transformations (e.g., nitrification, mineralization, and immobilization) in flooded, waterlogging or compacted soils is influenced by changes in E_h and pH, which is related to the reducing activity of soil microbiota (Huang et al. 2016). The sms able to use a particular electron acceptor are expected to outcompete organisms using an alternative metabolic pathway, less favorable electron acceptors (Sidhu et al. 2017; Ucar et al. 2017; Molaei et al. 2017a, b). A change in E_h status of a soil indicates changing the availability of electron acceptors, requiring adaptations (within a wider or a narrower E_h-range) in microbial metabolic lifestyles (Jewell et al. 2017), which can lead to losses of soil biodiversity (Lamers et al. 2012).

3 Spatial Distribution of Soil Microbiota

The spatial distribution of soil microbiota vary vertically through the soil vadose zone (the zone between the groundwater level and the land surface) and horizontally, across the uncountable soil topography combinations (Voroney and Heck 2015; Orgiazzi et al. 2016; Römbke et al. 2016; Lukac 2017). Soil spatial heterogeneity is an essential factor for coexistence and species interactions in soil microbiota populations (Griffiths and Philippot 2013) stimulating biodiversity by the recalcitrance of specific and individualized microbial communities (Wisz et al. 2013; Voroney and Heck 2015; Larkin and Martiny 2017). Previous studies confirmed that soil biota is frequently not randomly distributed (Hughes et al. 2001; Horner-Devine et al. 2007; Bradford et al. 2008; Bates et al. 2010; Barberán et al. 2012; Bradford and Fierer 2012; Vos et al. 2013; Barnes et al. 2016; Meena and Yadav 2014).

Recent evidence suggests that the use of soil's Network Models and plant microbiomes can supply new goals for spatial distribution, and co-occurrence patterns in soil microbial ecology studies (Chaffron et al. 2010; Barberán et al. 2012; Faust and Raes 2012; Lupatini et al. 2014; Faust et al. 2015; Poudel et al. 2016; Kuang et al. 2017), and have been used to a diversity of research inquiries regarding biological interactions between soil organisms (Williams et al. 2014). Network inference methods (Faust and Raes 2012; Ding et al. 2015; Kurtz et al. 2015) can be used to reveal the coexistence of patterns encompassing from pairs of microbial taxa in a manifoldness of ecosystems (Eiler et al. 2012; Kittelmann et al. 2013; Moll et al. 2016) and functional groups (Duran-Pinedo et al. 2011; Embree et al. 2015; Cordero and Datta 2016; Li et al. 2017) to plant-microbe interactions (Shaw and Pautasso 2014; Poudel et al. 2016), and identify the keystone species within an specific ecosystem (Steele et al. 2011; Eiler et al. 2012; Kovács 2014; Berry and Widder 2014).

3.1 Ecological Roles of the Soil Microbiota Under Climate Change

The soil microbiota is a complex and dynamic fraction of the planet biodiversity and play an essential role in ecosystem processes and services, that includes the soil global nutrient cycling process (Lupatini et al. 2014), organic matter decomposition and climate regulation (Delgado-Baquerizo et al. 2017), by reducing the GHGs emissions (Bhattacharya et al. 2016), and to provide the functions to sustain all forms of life (Pajares et al. 2016; Pajares and Bohannan 2016). Soil microbial communities are engineers (Jones et al. 1994; Wright and Jones 2006; Cuddington et al. 2011; Singh 2015; Graham et al. 2016) and the soil architects (Rajendhran and Gunasekaran 2008). Soil microbial biodiversity is declining worldwide with anthropogenic and natural deleterious effects on ecosystems and agroecosystems (Delgado-Baquerizo et al. 2016a, b), and this alters ecosystem processes and changes the resilience of ecosystems to environmental change (Chapin et al. 2000). There exists a gap of quantitative knowledge of the shape of the correlation between soil microbiota biodiversity and their ecosystem function (Delgado-Baquerizo et al. 2017; Kim et al. 2016). This limits the understanding of how microbial diversity depletion affects the soil microbiota key functions for human well-being in sustainable agricultural systems (Barnes et al. 2016) and represents not only an irreversible loss to the Earth but also threatens humanity's life support system (Cardinale et al. 2012; Lladó et al. 2017).

3.1.1 Soil Respiration and Its Role in Carbon Sequestration

The soil respiration (R_S) refers to the CO_2 released from metabolic activity of soil organisms (humans, animals, plant roots and microbiota) that contributes naturally to GHGs emissions (Bond-Lamberty and Thomson 2010). It has commonly been assumed that soil CO_2 efflux is derivative of plant respiration (R_P), also called autotrophic respiration (R_A) and microbial respiration (R_m), also called heterotrophic respiration (R_H), but recent studies showed that pedochemical and geological activities (e.g., as geothermal and volcanic CO_2 degassing), are fundamental in some areas (reviewed by Liu et al. 2014; Rey 2015; Bhattacharya et al. 2016; Decina et al. 2016; Abbas et al. 2017; Mariotte et al. 2018). The microbial production of recalcitrant organic matter (ROM), and their implications for productivity and climate change strategies (e.g., C sequestration) was analyzed by Liang and Balser (2010). For the authors, there is a severe ignorance about the microbial C sequestration, and the understanding of the real microbial role in soil C stabilization will stimulate the current status quo of our knowledge about the global C-cycling models. The understanding of the role of Rs in related to the global soil organic carbon stocks (GSOCS) is a critical step in order to evaluate the unpredictability in global climate change future projections (Gougoulias et al. 2014; Thornton et al. 2014; Wang et al. 2016; Erb et al. 2018; Yadav et al. 2018).

3.1.2 Soil Carbon Stocks and Carbon Sequestration

Soil contains the main C reservoir (420–620 Pg) than the atmosphere (829 Pg) and vegetation (3500–4800 Pg) combined (Lal 2004a, b; Mazzoleni et al. 2012; Wang et al. 2016; FAO 2017). The soil organic carbon (SOC) is a critical natural resource, and it is reducing because to the anthropogenic land use such as the conversion of natural ecosystems to food or bioenergy production agricultural ecosystems (Nziguheba et al. 2015). Usually, the quantification of SOC stocks needs the estimation of the C content, bulk density, rock fragment content and depth of a respective soil layer (Poeplau et al. 2017). The soil C stocks are related to the equilibrium between C sequestration by plant photosynthesis and C release to atmosphere through soil microbiota respiration, it is a crucial contributing key factor in the soil pathway energy flows, recycling of SOM and biodiversity (Johnson et al. 2014; Banwart et al. 2014). Numerous studies showed that SOC stocks quantifications (e.g., Tucker 1997; Homann et al. 1998: Soil Profile-based; Alvaro-Fuentes et al. 2014: Model-Based; Zhi et al. 2014: Soil Profile Statistics; Zhang et al. 2008: GIS-based Soil Type methodologies) are frequently overestimated by mistakes and misuse of the parameters bulk density and rock fragment content (Wang and Dalal 2006; Lozano-García and Parras-Alcántara 2013; Poeplau and Don 2013; Friedlingstein et al. 2014; Beem-Miller et al. 2016; Henkner et al. 2016). The accuracy in SOC stock estimation has become a primer of current research on global climate change studies, and it is fundamental for evaluating soil quality, global C modeling cycle, and evaluation global climate change (Zhi et al. 2014). The GSOCS on croplands, particularly within the context of a warming climate (Scharlemann et al. 2014; Lal et al. 2015; Zomer et al. 2017), are central to important discussions within several international fora (Zomer et al. 2017), and are including within mitigation strategies (of climate change) and protocols (e.g., International Research Program, the '4 per mille Soils for Food Security and Climate' of the Lima-Paris Action Agenda) at the Paris climate conference (COP21) in December 2015, where 195 countries adopted the first-ever universal, legally binding global climate deal (Minasny et al. 2017). In this context, the anthropogenic intervention (directly or indirectly) to reduce the sources (e.g., reduction of particulate matter emissions) or enhance the sinks (e.g., fertilization with organic manures) of GHGs is called Mitigation (Brassard et al. 2016). Nowadays, some agricultural strategies are used to mitigation the GHGs emissions, include those that produce biomass (e.g., cover cropping, mulching, rotational grazing regime, conversion of degraded croplands and native vegetation to improved pastures, sowing of leguminous and grass pasture species, fire management, and water conservation), even under high-impact climate conditions (e.g., higher temperatures, lower precipitation), or under extreme events (e.g., tropical storms, seasonal hurricanes, drought and flooding) (Lal 2010; Powlson et al. 2011; Becker and Lawrence 2014).

3.2 Global Climate Change and World's Carbon Market

Indisputably, the global climate change (or "global warming") is not only a scientific question but also one of the main challenges facing modern humanity. The Kyoto Protocol established a specific time period (commencing in 2008 and ending in 2012) and sets binding targets for developed countries (known as "Annex I Parties") to limit or reduce GHGs emissions (Moss et al. 2010; IPCC 2017). The Kyoto Protocol also approved the "Emissions Trading "(The World's Carbon Market) thatallows developed countries that exceed their GHGs emissions to offset them by buying 'C credits' from others countries (e.g., developing countries) that stay below their GHGs emissions targets (Hodge and Clasen 2014).

The markets for C are negotiated in terms of carbon dioxide equivalent (CO_2e) where the GHGs more frequently listed in Annex B of the Kyoto Protocol (IPCC 2014a, b), are expressed in terms of their equivalence to CO_2 in the C market mechanisms for mitigating climate change (Nabuurs et al. 2007). Any type of GHGs removed from the atmosphere through mensurable sink activities generates C credits known as removal units (RMUs) (Hunt and Baum 2009).The European Union emissions trading scheme (EU ETS) is a good example of a regional C market system operating under the Kyoto Protocol umbrella (Falloon and Betts 2010). Nowadays, C prices range from less than US$1 to up to US$140/tCO_2e (World Bank 2017). The C Market is key to mobilizing the US$700 billion in financial services annually, and that could grow to U$ 1 trillion by 2030 (World Bank 2017). The C pricing revenues from emerging initiatives such as bilateral entities (e.g., United States Climate Action Partnership), sub-national tiers of government (e.g., China Certified Emission Reduction – CCER), grass-roots groups (e.g., World Bank's Forest Carbon Partnership Facility -FCPF, Reducing Emissions from Deforestation and Forest Degradation – REDD), private enterprises (e.g., Chicago Climate Exchange – CCX) and individuals are being used to accelerate the transition to a low-carbon economy (Krug 2018). The Kyoto Protocol is a good indicator of the fragmented nature of the international mitigation framework and has led to an intensive debate on how to continue the negotiation of future treaties for the C market (reviewed by Smith et al. 2013; Bustamante et al. 2014; Summers et al. 2015; Creutzburg et al. 2017; Pickering et al. 2017; Stevanović et al. 2017; World Bank 2017; Krug 2018).

4 Improving Activity Mycorrhizosphere for Carbon Sequestration Management in Sustainable Agroecosystems

Plant-microbe interactions have driven and modeled the origin, organization, and evolution of all organic life forms on Earth Planet (for recent reviews, see Gilbert et al. 2012; Delaux et al. 2013; Goh et al. 2013; Coats and Rumpho 2014; Rosenberg and Zilber-Rosenberg 2014; Remigi et al. 2016; Rosenberg and Zilber-Rosenberg

2016; Hamel and Plenchete 2017) symbiosis is the best known and studied mutualist association formed by Glomeromycotan fungi and plant roots. (reviewed by Smith and Read 2008; Solaiman 2014; Zuccaro et al. 2014; Pagano and Gupta 2016; Filho et al. 2017). Mycorrhiza symbiosis implements changes in the biological and edaphic soil properties in the mycorrhizosphere (also the so-called zone of mycorrhizal influence), the term coined by Oswall and Ferchau in 1968 (cited by Azcón-Aguilar and Barea 2015). Arbuscular mycorrhizal fungi (AMF) belong to Glomeromycota phylum (Schüßler et al. 2001) and contain four orders, 11 families, 25 genera (Redecker et al. 2013), with more than 240 species (Krüger et al. 2012). Before physical contact, diffusible signal molecules (strigolactones and flavonoids) are secreted by plant roots can be perceived by AM fungi (Akiyama et al. 2005). Posteriorly, the plant root system changes in response to pre-symbiotic signaling, and the host plants recognize the AM fungi through pattern recognition receptors (PRRs) that perceive microbe-associated molecular patterns (MAMPs) (Zhang and Zhou 2010).The development of AM symbiosis is established and separated into distinct steps that are characterized by the progression of fungal hyphae during root colonization (Gutjahr and Parniske 2013), the internal and external root colonization (Smith and Read 2008). Phytohormones regulate the functioning of the AM symbiosis (Gutjahr 2014).AM fungi are obligate symbionts and form a morpho-physiological interaction with the host plant. In this the symbiont receives the carbon from their host plant and provides to plant water and mineral nutrients, low mobility or diffusion ions such as P (phosphorus) as $H_2PO_4{}^{-2}$ or H_2PO^{4-}, Cu (copper) as Cu^{+2} or Cu-chelate, and Zn (zinc) as Zn^{+2} or Zn-chelate from the soil volumes, called depletion zones (reviewed by Smith et al. 2011; Fellbaum et al. 2012; Smith and Smith 2012; Lehmann et al. 2014; Lehmann and Rillig 2015; Berruti et al. 2016; Walder et al. 2016) and other benefits, such as alleviate the toxicity of heavy metals (Ferrol et al. 2016), acidity stress (Muthukumar et al. 2014), drought stress (Rapparini and Peñuelas 2014; Pagano 2014), salinity stress (Porcel et al. 2012; Hajiboland 2013; Abdel Latef and Miransari 2014; Chen et al. 2017) and plant disease protection (Filho et al. 2016). AM symbioses are able to host plant uptake inorganic phosphorus (Pi) by two pathways: directly by root epidermal cells and root hairs, and via AM fungi (Smith and Smith 2012; Meena et al. 2015; Meena and Meena 2017). The mechanism of the P come from to the endosymbiont to the plant, and the regulation of AM development by Pi availability has not yet been elucidated for any mycorrhizal type (Smith and Smith 2012). It is also still not clear whether the P uptake from soil solution (by external mycelium) to AM fungi, is made by passive or an active transport system (Pagano and Gupta 2016). Currently, the advances in knowledge got from new methods (e.g., biofertilizers with AM fungi, maintenance of AM fungi by using bioactive molecules) for soil microbiology research opened new perspectives to AM symbiosis management in crop nutrition and production for Sustainable Agriculture (Zhuang et al. 2013; Ellouze et al. 2014; Pagano and Dhar 2015). Evidence indicates that common mycorrhizal network (CMNs) can link multiple plant species and transfer all elochemicals (bioactive molecules) among plants (Barto et al. 2011; File et al. 2012; Walder et al. 2012, 2016; Engelmoer and Kiers 2015; Dadhich and Meena 2014). Those networks increase the bioactive zones of all elochemicals in natural ecosystems and have positive consequences for interspecies chemical interactions and facilitation in plants in sustainable agroecosystems (Barto et al. 2011).

Research proposes that AM fungi act as a component of the SOC pool, and their hyphae could contribute to soil C sequestration by different pathways such as rhizodeposition (Nguyen 2003) or by stabilizing soil aggregates with glomalin (Wright and Upadhyaya 1998; Rillig and Steinberg 2002) glycoprotein (produced by AM fungi) formation (Fellbaum et al. 2012; Solaiman and Mickan 2014; Averill et al. 2014; Wang et al. 2016). In this context, there are some open questions (e.g., with respect to genes and proteins that mediated the C fluxes between the host plant and their symbiont fungi), and future studies need to consider the AM fungal-mediated role on processes that lead to long-term soil C gains (in the decadal timescale), that contribute to soil C sequestration (recently reviewed by Jansa and Treseder 2017).

4.1 *Improving Biochar Application for Soil Carbon Sequestration Management in Sustainable Agroecosystems*

Biochar is charcoal made from the carbonization of biomass (e.g., wood, manure, or leaves), heated (300 °C and 1000 °C), in the absence of oxygen (O_2) or under low O_2 concentration, and used as a soil amendment (Downie et al. 2009; Lehmann 2009; Atkinson et al. 2010; Anderson et al. 2011; Bruun et al. 2012; Mukherjee and Lal 2017; Sánchez-García et al. 2014; Speratti et al. 2017). Biochar is known as "Indian Black Earth" ("Terra Preta de Índio") soils by Ecologists (Petersen et al. 2001; Lehmann et al. 2003; Glaser et al. 2015; Omondi et al. 2016). The use of biochar to improve agricultural soils phycochemical and biological properties is not novel since it has been practiced for centuries by the pre-Columbian and Amerindian inhabitants (e.g., Kuikuro Indians, Brazilian indigenous people from Xingu National Park) of the Amazonia region (Neves et al. 2003; Kim et al. 2007; Glaser and Birk 2012; Lehmann et al. 2011). Recent studies show the roles of biochar on soil C sequestration (Spokas et al. 2012; Lal 2016; Brassard et al. 2016), GHG emissions (Cayuela et al. 2014; Sánchez-García et al. 2014; Brassard et al. 2016), soil fertility (van Zwieten et al. 2010; Schulz et al. 2013; Ding et al. 2015), crop production (Biederman and Harpole 2013), soil biota (Lehmann et al. 2011; Ducey et al. 2015; Liao et al. 2016), soil chemical properties (Brassard et al. 2016; Mukherjee and Lal, 2017; Kumar et al. 2018), and remediation of contaminated soils (Park et al. 2011). Nowadays, advances in knowledge may lead to the application of biochar to improve the soil fertility in sustainable agroecosystems, however its current use is still limited (Atkinson et al. 2010) and a great number of research gaps as well as uncertainties (Ennis et al. 2012, Ameloot et al. 2013, Prayogo et al. 2014; Lanza et al. 2016) still exist as discussed. The application of the biochar provides a long-term sink for C, but the process is not readily available due to innumerable and bureaucratic of regulatory and economic reasons (Diacono and Montemurro 2010). These uncertainties and controversies reported in the literature need to be evaluated carefully in consideration and regarding the use of biochar as a soil amendment.

4.2 Net Primary Production, Net Ecosystem Production, Human Appropriation of Net Primary Production and Soil Carbon Fluxes

The conventional cultivated agricultural soils contain 25% to 75% less of SOC than their quantification in undisturbed or natural ecosystems (Lal 2004a, b). The conventional agricultural practices create a decline in soil quality because of C losses (e.g., by soil degradation and desertification) and the reduced rate of SOM mineralization that leads to a low input of biomass C and reduces the Net Primary Production (NPP) (Haberl et al. 2007). The NPP refers to the creation of new OM by Gross Primary Production (GPP = All CO_2 fixed by the plants and other autotrophs in the photosynthesis process (Leith 1975; Heimann and Reichstein 2008). It refers to the primary food energy source for all earth ecosystems and is quantified in units of elemental carbon (Krausmann et al. 2013). In terms of energy efficiency (or organic matter), the global NPP open systems (e.g., oceans, tropical rain forests, savannas, and tropical seasonal forests) are more productive (Ohtsuka et al. 2007). However, the open systems have low NPP per unit area (Krausmann et al. 2013). The estimate of the Net Ecosystem Production (NEP) formation can be calculated by the following expression (Erb et al. 2009) in which

$$\mathrm{NEP} = \mathrm{GPP} - (\mathrm{R_A} + \mathrm{R_H} + \mathrm{R_D})$$

R_A = Autotrophic Respiration;
R_H = Heterotrophs Respiration;
R_D = Decomposers Respiration.

According to Kindler et al. (2011), NEP is expressed as energy (expressed in calories m^{-2} year $^{-1}$) or organic matter (expressed in grams m^{-2} year $^{-1}$). NEP varies among ecosystems due to climate and nutrient availability (Yashiro et al. 2010). NEP is used to estimate, whether an ecosystem is net autotrophic (production exceeds respiration) or net heterotrophic (respiration exceeds production), it is usually used if the quantifications are based on ecosystem-carbon stock changes (Chapin et al. 2006).

The Human Appropriation of Net Primary Production (HANPP) is a good socioecological indicator that quantifies the anthropogenic effects (Crutzen 2002: also called "the Anthropocene Era") induced by changes in the land use (such as deforestation or intensive agriculture) and harvest on ecological biomass flows (into the terrestrial biosphere (Krausmann et al. 2013; Pritchard et al. 2018). According to Krausmann et al. (2013), the global HANPP doubled in the twentieth century. These anthropogenic effects can influence the land to produce biomass via human inputs and management (Ellis 2011), through ecological and social systems (Haberl et al. 2007).

The HANPP can be quantified by (Haberl et al. 2007, 2014):

$$HANPP = HANPP_{LUC} + HANPP_{HARV}$$

where $HANPP_{luc}$ = loss of potential NPP due to land use change;
$HANPP_{harv}$ = NPP consumed + unused NPP by humans.

$$HANPP_{LUC} = NPP_{pot} - NPP_{act}$$

where

NPP_{pot} = the potential NPP in undisturbed system (hypothetical),
and NPP_{act} = the actual NPP in anthropogenic-altered system.

HANPP is an effective new methodology and can be useful to improving understanding of land use intensification in Earth complex socioecological systems, for improving land use decisions and landscape management, objectifying to minimize the anthropogenic impact on the ecosystem and ensure the flow of agroecosystem services (Pritchard et al. 2018). HANPP can be expressed as flows of biomass, C or energy (Canadell et al. 2010; Canadell and Schulze 2014). HANPP can be linked with data on ecological C stocks (in biota and soils) to form a comprehensive view of soil C accounting system (Haberl et al. 2014).

5 The Challenge of Sustainable Agricultural Intensification Under Climate Change

"Sustainable Development seeks to meet the needs of the current generation without compromising the ability of future generations to meet their own needs" (Cerin 2006). Currently, undoubtedly there is a realistic choice to develop eco-agricultural systems that are less harmful to the environment, based on environmentally sustainable practices (e.g., ensuring a sustainable level of population and conserving and enhancing the resource base) and less aggregate on agricultural inputs (e.g., synthetic agrochemicals), aiming to reduce its deleterious effects on the environment, to conserve and improve soil health, and to guarantee food security (Moonen and Bàrberi 2008; Azcón-Aguilar and Barea 2015). To achieve these aims, it is necessary to improve the plant and soil health characteristics that are the result of the interactions among physicochemical and biological properties, and allow the small-holder farmers to maintain the soil healthy, and productive soil for crops without depauperate the agroecosystem and ecosystem (Trivedi et al. 2016).

The world's population will exceed the landing of 9 billion by 2050 (Alexandratos and Bruinsma 2012; Garnett and Godfray 2012) and 11 billion by

2100 (OECD/FAO 2017). Today, the world's population to grow more slowly than in past decades. This slower world population growth is due to the lower fertility rates, and there is a global higher life expectancy worldwide (Kim et al. 2016). The conventional agricultural practices are overloaded and needs to provide increased yields to feed the continued growth of the world population (Timmusk et al. 2017). The global agricultural production may need to be improved by ~ 60% to meet these increasing demands (OECD/FAO 2017).

Unequivocally, the warming in the climate system is happening, and since the 1950s, there are innumerable observed changes (e.g., warming of the atmosphere and the ocean, diminishing snow and ice, rising sea levels and increasing concentrations of greenhouse gases) unprecedented over decades to millennia, and making achievement of food security even more challenging, especially in the most vulnerable parts of the developing world (e.g., Sub-Saharan Africa, North Africa, and the Middle East) (Alexandratos and Bruinsma 2012; IPCC 2013). In these areas, in particular, adapting agricultural and others farmers systems (e.g., agriculture, fisheries, aquaculture, and forestry) to the effects of climate change will be obligatory for survival (Kim et al. 2016). In 2010, the United Nations Framework Convention on Climate Change (UNFCCC) indicated that adaptation should be focused at the same level of priority as mitigation (Henderson-Sellers and McGuffie 2012).

Finally, in this debate the academics agree that agricultural intensification needs to establish the culture of sustainability, and bringing the term "sustainable intensification of agriculture" (Andres and Bhullar 2016: also called "agriculture ecological intensification") on the agenda of agricultural research and policy to increase the full range of the fundamental life-sustaining benefits that soils provide.

6 Conclusion and Future Prospects

Increasing demand for food, fiber, and fuel is exercising an intense pressure on the global agroecosystems. To help the sustainable intensification of food production systems, and minimize the levels of greenhouse gases (GHGs) anthropogenic emissions, the current agriculture needs to create and use methodologies (e.g., improving biochar application) that minimize loss of terrestrial biodiversity on agroecosystems. Undoubtedly the soil carbon sequestration is a fundamental climate mitigation approach, but only a very small slice of the decrement can be achieved via this sink. This limitation has been succinctly summarized by W.H. Schlesinger as 'trying to sequester the geosphere in the biosphere' (reviewed by Schlesinger 1999). Despite these limitations (e.g., displacement and difficulties in verification), the soil C sequestration can be useful to meet short- to medium-term targets and provides innumerable co-benefits on soils (e.g., increase the soil stocks of organic N, P, and other nutrients). In terms of future prospects, the carbon markets are coordinating networks to enable humanity to integrate nature into neo-modernity infrastructure with a socio-political and economic organization to organize the use of the environment in space and time (Pickering et al. 2017; Meena et al. 2018), while contributing to sustainable development. Also, in terms of future prospects, the insertion of Education for Sustainable Development (ESD) in university classrooms, must be as one of the primary objective for action in higher education because of its direct reflex on the formation of soil science (soil biology, physical and chemistry) future professionals, with a broad and real holistic sense of sustainability (García-González et al. 2017). Finally, an excellent example of the ESD insertion is the EU Masters program at the Ecole des Mines, Nantes in France (Boron et al. 2017).

Acknowledgments We thank the editors and anonymous reviewers for their constructive comments, which helped us to improve the manuscript.

References

A'Bear AD, Jones TH, Boddy L (2014) Potential impacts of climate change on interactions among saprotrophic cord-forming fungal mycelia and grazing soil invertebrates. Fungal Ecol 10:34–43. https://doi.org/10.1016/j.funeco.2013.01.009

Abbas F, Hammad HM, Fahad S, Cerdà A, Rizwan M, Farhad W, Ehsan S, Bakhat HF (2017) Agroforestry: a sustainable environmental practice for carbon sequestration under the climate change scenarios-a review. Environ Sci Pollut Res Int 24:11177–11191. https://doi.org/10.1007/s11356-017-8687-0

Abdel Latef AAH, Miransari M (2014) The role of arbuscular mycorrhizal fungi in salt stress. In: Miransari M (ed) Use of microbes for the alleviation of soil stresses, vol 2. Springer, New York, pp 23–39. https://doi.org/10.1007/978-1-4939-0721-2_2

Akiyama K, Matsuzaki K, Hayashi H (2005) Plant sesquiterpenes induce hyphal branching in arbuscular mycorrhizal fungi. Nature 435:824–827

Alexander M (1961) Introduction to soil microbiology. Wiley, New York

Alexandratos N, Bruinsma J (2012) World agriculture towards 2030/2050: the 2012 revision. ESA working paper no. 12–03. FAO, Rome

Allison F (1961) Twenty-five years of soil microbiology and a look to the future. Soil Sci Soc Am Proc 26:432–439

Alvaro-Fuentes J, Plaza-Bonilla D, Arrue JL, Lampurlanes J, Cantero-Martinez C (2014) Soil organic carbon storage in a no-tillage chronosequence under Mediterranean conditions. Plant Soil 376:31–41. https://doi.org/10.1007/s11104-012-1167-x

Ameloot N, Graber ER, Verheijen FGA, De Neve S (2013) Interactions between biochar stability and soil organisms: review and research needs. Eur J Soil Sci 64:379–390. https://doi.org/10.1111/ejss.12064

Andam CP, Doroghazi JR, Campbell AN, Kelly PJ, Choudoir MJ, Buckley DH (2016) A latitudinal diversity gradient in terrestrial bacteria of the genus Streptomyces. MBio 7(2):e02200–e02215. https://doi.org/10.1128/mBio.02200-15

Anderson CR, Condron LM, Clough TJ, Fiers M, Stewart A, Hill RA, Sherlock RR (2011) Biochar induced soil microbial community change: implications for biogeochemical cycling of carbon, nitrogen and phosphorus. Pedobiologia 54:309–320. https://doi.org/10.1016/j.pedobi.2011.07.005

Andres C, Bhullar GS (2016) Sustainable intensification of tropical agro-ecosystems: need and potentials. Front Environ Sci 4:5. https://doi.org/10.3389/fenvs.2016.00005

Anon (1991) Soil physics. In: Landon JR (ed) Booker tropical soil manual. Longman, Harlow, pp 59–71

Arnold RW (1983) Concepts and interactions. In: Wilding LP, Smeck NE, Hall GF (eds) Pedogenesis and soil taxonomy. Elsevier Science, Amsterdam, pp 1–21

Assouline S (2013) Infiltration into soil: conceptual approaches and solutions. Water Resour Res 49:1755–1772. https://doi.org/10.1002/wrcr.20155

Assouline S, Selker J (2017) Introduction and evaluation of a Weibull hydraulic conductivity-pressure head relationship for unsaturated soils. Water Resour Res 53:4956–4964. https://doi.org/10.1002/2017WR020796

Atkinson CJ, Fitzgerald JD, Hipps NA (2010) Potential mechanisms for achieving agricultural benefits from biochar application to temperate soils: a review. Plant Soil 337:1–18. https://doi.org/10.1007/s11104-010-0464-5

Atlas RM, Bartha R (1993) Microbial ecology: fundamentals and applications. The Benjamim/Cummings Publishing Company, California, pp 213–245

Averill C, Turner BL, Finzi AC (2014) Mycorrhiza-mediated competition between plants and decomposers drives soil carbon storage. Nature 505:543–545. https://doi.org/10.1038/nature12901

Azcón-Aguilar C, Barea JM (2015) Nutrient cycling in the mycorrhizosphere. J Soil Sci Plant Nutr 25:372–396. https://doi.org/10.4067/S0718-95162015005000035

Babauta JT, Nguyen HD, Harrington TD, Renslow R, Beyenal H (2012) pH, redox potential and local biofilm potential microenvironments within *Geobacter sulfurreducens* biofilms and their roles in electron transfer. Biotechnol Bioeng 109:2651–2662. https://doi.org/10.1002/bit.24538

Badorreck A, Gerk HH, Hüttl RF (2013) Morphology and physical soil crusts and infiltration patterns in an artificial catchment. Soil Till Res 129:1–8. https://doi.org/10.1016/j.still.2013.01.001

Bagheri I, Kalhori SB, Akef M, Khormali F (2012) Effect of compaction on physical and micromorphological properties of forest soils. Am J Plant Sci 3:159–163. https://doi.org/10.4236/ajps.2012.31018

Baldock JA (2007) Composition and cycling of organic carbon in soil. In: Marschner P, Rengel Z (eds) Nutrient cycling in terrestrial ecosystems. Soil biology, vol 10. Springer, Berlin, Heidelberg, pp 1–35. https://doi.org/10.1007/978-3-540-68027-7_1

Baldock JA, Broos K (2012) Soil organic matter. In: Huang PM, Li Y, Sumner M (eds) Handbook of soil sciences: resource management and environmental impacts, 2nd edn. CRC Press, Boca Raton, pp 11-1–11-37

Balestrini R, Lumini E, Borriello R, Bianciotto V (2015) Plant-soil biota interactions. In: Paul EA (ed) Soil microbiology, ecology and biochemistry. Academic/Elsevier, London, pp 311–338

Ball BC (2013) Soil structure and greenhouse gas emissions: a synthesis of 20 years of experimentation. Eur J Soil Sci 64:357–373. https://doi.org/10.1111/ejss.12013

Balser TC, Liang C, Gutknecht MJL (2010) How will climate change impact soil microbial communities? In: Dixon GR, Tilston EL (eds) Soil microbiology and sustainable crop production. Springer, Dordrecht, pp 373–397. https://doi.org/10.1007/978-90-481-9479-7_10

Banwart S, Black H, Cai Z, Gicheru P, Joosten H, Victoria R, Milne E, Noellemeyer E, Pascual U, Nziguheba G, Vargas R, Bationo A, Buschiazzo D, Delphine de-Bogniez, Melillo J, Ritcher D, Termansen M, van Noordwijk M, Goverse T, Ballabio C, Bhattacharyya T, Goldhaber M, Nikolaidis N, Zhao Y, Funk R, Duffy C, Pan G, la Scala N, Gottschalk P, Batjes N, Six J, van Wesemael B, Stocking M, Bampa F, Bernoux M, Feller C, Lemanceau P, Montanarella L (2014) Benefits of soil carbon: report on the outcomes of an international scientific committee on problems of the environment rapid assessment workshop. Carbon Manag 5:185–192. https://doi.org/10.1080/17583004.2014.913380

Barberán A, Bates ST, Casamayor EO, Fierer N (2012) Using network analysis to explore co-occurrence patterns in soil microbial communities. ISME J 6:343–351. https://doi.org/10.1038/ismej.2011.119

Barnes AD, Weigelt P, Jochum M, Ott D, Hodapp D, Haneda NF, Brose U (2016) Species richness, and biomass explain spatial turnover in ecosystem functioning across tropical and temperate ecosystems. Philos Trans R Soc B 371:20150279. https://doi.org/10.1098/rstb.2015.0279

Barto EK, Hilker M, Müller F, Mohney BK, Weidenhamer JD, Rillig MC (2011) The fungal fast lane: common mycorrhizal networks extend bioactive zones of allelochemicals in soils. PLoS One. https://doi.org/10.1371/journal.pone.0027195

Bastida F, Moreno JL, Nicolas C, Hernandez T, Garcia C (2009) Soil metaproteomics: a review of an emerging environmental science. Significance, methodology, and perspectives. Eur J Soil Sci 60:845–859. https://doi.org/10.1111/j.1365-2389.2009.01184.x

Bates ST, Berg-Lyons D, Caporaso JG, Walters WA, Knight R, Fierer N (2010) Examining the global distribution of dominant archaeal populations in soil. ISME J 5:908–917. https://doi.org/10.1038/ismej.2010.171

Bates ST, Clemente JC, Flores GE, Walters WA, Parfrey LW, Knight R, Fierer N (2013) Global biogeography of highly diverse protistan communities in soil. ISME J 7:652–659. https://doi.org/10.1038/ismej.2012.147

Batjes NH (2011) Soil organic carbon stocks under native vegetation: revised estimates for use with the simple assessment option of the carbon benefits project system. Agric Ecosyst Environ 142:365–373. https://doi.org/10.1016/j.agee.2011.06.007

Battistuzzi FU, Hedges SB (2009) A major clade of prokaryotes with ancient adaptations to life on land. Mol Biol Evol 26:335–343. https://doi.org/10.1093/molbev/msn247

Bauska TK, Joos F, Mix AC, Roth R, Ahn J, Brook EJ (2015) Links between atmospheric carbon dioxide, the land carbon reservoir and climate over the past millennium. Nat Geosci 8:383–387. https://doi.org/10.1038/NGEO2422

Becker K, Lawrence P (2014) Carbon farming: the best and safest way forward? Carbon Manag 5:31–33. https://doi.org/10.4155/cmt.13.79

Beem-Miller JP, Kong AYY, Ogle S, Wolfe D (2016) Sampling for soil carbon stock assessment in rocky agricultural soils. Soil Sci Soc Am J 80:1411–1423. https://doi.org/10.2136/sssaj2015.11.0405

Berg B, McClaugherty C (2014) Introduction In: Plant litter. Decomposition, humus formation, carbon sequestration, 3rd edn. Springer, Berlin/Heidelberg, pp 1–10. https://doi.org/10.1007/978-3-642-38821-7_1

Berg G, Smalla K (2009) Plant species and soil type cooperatively shape the structure and function of microbial communities in the rhizosphere. FEMS Microbiol Ecol 68:1–13. https://doi.org/10.1111/j.1574-6941.2009.00654.x

Berruti A, Lumini E, Balestrini R, Bianciotto V (2016) Arbuscular mycorrhizal fungi as natural biofertilizers: let's benefit from past successes. Front Microbiol 6:1559. https://doi.org/10.3389/fmicb.2015.01559

Berry D, Widder S (2014) Deciphering microbial interactions and detecting keystone species with co-occurrence networks. Front Microbiol 5:219. https://doi.org/10.3389/fmicb.2014.00219

Bevivino A, Paganin P, Bacci G, Florio A, Pellicer MS, Papaleo MC, Mengoni A, Ledda L, Fani R, Benedetti A, Dalmastri C (2014) Soil bacterial community response to differences in agricultural management along with seasonal changes in a Mediterranean region. PLoS One. https://doi.org/10.1371/journal.pone.0105515

Bhattacharya SS, Kim KH, Das S, Uchimiya M, Jeon BH, Kwon E, Szulejko JE (2016) A review on the role of organic inputs in maintaining the soil carbon pool of the terrestrial ecosystem. J Environ Manag 167:214–827. https://doi.org/10.1016/j.jenvman.2015.09.042

Biederman LA, Harpole WS (2013) Biochar and its effects on plant productivity and nutrient cycling: a meta-analysis. GCB Bioenergy 5:202–214. https://doi.org/10.1111/gcbb.12037

Blankinship JC, Niklaus PA, Hungate BA (2011) A meta-analysis of responses of soil biota to global change. Oecologia 165:553–565. https://doi.org/10.1007/s00442-011-1909-0

Bohn HL, McNeal BL, O'Conner GA (1979) Soil chemistry. Wiley-Interscience, New York

Bond-Lamberty B, Thomson A (2010) Temperature-associated increases in the global soil respiration record. Nature 464:579–582. https://doi.org/10.1038/nature08930

Boron S, Murray KR, Thomson GB (2017) Sustainability education: towards total sustainability management teaching. In: Leal Filho W, Brandli L, Castro P, Newman J (eds) Handbook of theory and practice of sustainable development in higher education, World sustainability series. Springer, Cham, pp 37–51. https://doi.org/10.1007/978-3-319-47868-5_3

Bouma J (1981) Comment on micro, meso, and macroporosity of soil. Soil Sci AM J 45:1244–1245

Bradford MA, Fierer N (2012) The biogeography of microbial communities and ecosystem processes: implications for soil and ecosystem models. In: Wall DH (ed) Soil ecology and ecosystem services. Oxford University Press, Oxford, pp 189–198. https://doi.org/10.1093/acprof:oso/9780199575923.003.0017

Bradford MA, Davies CA, Frey SD, Maddox TR, Melillo JM, Mohan JE, Reynolds JF, Treseder KK, Wallenstein MD (2008) Thermal adaptation of soil microbial respiration to elevated temperature. Ecol Lett 11:1316–1327. https://doi.org/10.1111/j.1461-0248.2008.01251.x

Brady NC (1984) The nature and properties of soil, 10th edn. Macmillan, New York

Brady NC, Weil RR (2002) The nature and properties of soils, 13th edn. Prentice Hall, Upper Saddle River

Brady NC, Weil RR (2010) Elements of the nature and properties of soils. Pearson Education International, Hoboken

Brassard P, Stephane G, Vijaya R (2016) Soil biochar amendment as a climate change mitigation tool: key parameters and mechanisms involved. J Environ Manag 181:484–497. https://doi.org/10.1016/j.jenvman.2016.06.063

Bruun EW, Ambus P, Egsgaard H, Hauggaard-Nielsen H (2012) Effects of slow and fast pyrolysis biochar on soil C and N turnover dynamics. Soil Biol Biochem 46:73–79. https://doi.org/10.1016/j.soilbio.2011.11.019

Busari AB, Kukal SS, Kaur A, Bhatt R, Dulazi AA (2015) Conservation tillage impacts on soil, crop and the environment. Int Soil Water Conserv Res 3:119–129. https://doi.org/10.1016/j.iswcr.2015.05.002

Bustamante M, Robledo-Abad C, Harper R, Mbow C, Ravindranat NH, Sperling F, Haberl H, Pinto Ade S, Smith P (2014) Co-benefits, trade-offs, barriers and policies for greenhouse gas mitigation in the agriculture, forestry and other land use (AFOLU) sector. Glob Chang Biol 20:3270–3290. https://doi.org/10.1111/gcb.12591

Canadell JG, Schulze ED (2014) Global potential of biospheric carbon management for climate mitigation. Nat Commun 5:5282. https://doi.org/10.1038/ncomms6282

Canadell JG, Ciais P, Dhakal S, Dolman H, Friedlingstein P, Gurney KR, Held A, Jackson RB, Le Quéré C, Malone EL, Ojima DS, Patwardhan A, Peters GP, Raupach MR (2010) Interactions of the carbon cycle, human activity, and the climate system: a research portfolio. Curr Opin Environ Sustain 2:301–311. https://doi.org/10.1016/j.cosust.2010.08.003

Cardinale BJ, Duffy JE, Gonzalez A, Hooper DU, Perrings C, Venail P, Narwani A, Mace GM, Tilman D, Wardle DA, Kinzig AP, Daily GC, Loreau M, Grace JB (2012) Biodiversity loss and its impact on humanity. Nature 486:59–67. https://doi.org/10.1038/nature11148

Cardoso EJBN, Andreote FF (2016) Microbiologia do solo, 2nd edn. ESALQ, Piracicaba

Cardoso EJBN, Vasconcellos RLF, Bini D, Miyauchi MYH, Santos CA, Alves PRL, Paula AM, Nakatani AS, Pereira JM, Nogueira MA (2013) Soil health: looking for suitable indicators. What should be considered to assess the effects of use and management on soil health? Sci Agric 70:274–289. https://doi.org/10.1590/S0103-90162013000400009

Cayuela ML, van Zwieten L, Singh BP, Jeffery S, Roig A, Sánchez-Monedero MA (2014) Biochar's role in mitigating soil nitrous oxide emissions: a review and meta-analysis. Agric Ecosyst Environ 191:5–16. https://doi.org/10.1016/j.agee.2013.10.009

Cerin P (2006) Bringing economic opportunity into line with environmental influence: a discussion on the Coase theorem and the porter and van der Linde hypothesis. Ecol Econ 56:209–225. https://doi.org/10.1016/j.ecolecon.2005.01.016

Chaffron S, Rehrauer H, Pernthaler J, von Mering C (2010) A global network of coexisting microbes from environmental and whole-genome sequence data. Genome Res 20:947–959. https://doi.org/10.1101/gr.104521.109

Chapin FS III, Zavaleta ES, Eviner VT, Naylor RL, Vitousek PM, Reynolds HL, Hooper DU, Lavorel S, Sala OE, Hobbie SE, Mack MC, Díaz S (2000) Consequences of changing biodiversity. Nature 405:234–242

Chapin III FS, Woodwell GM, Randerson JT, Rastetter EB, Lovett GM, Baldocchi DD, Clark DA, M. E. Harmon ME, Schimel DS, Valentini R, Wirth C, Aber JD, Cole JJ, Goulden ML, Ryan

MG, Harden JW, Heimann M, Howarth RW, Matson PA, McGuire AD, Melillo JM, Mooney HA, Neff JC, Houghton RA, Pace ML, Running SW, Sala OE, Schlesinger WH, Schulze E-D (2006) Reconciling carbon-cycle concepts, terminology, and methods. Ecosystems 9:1041–1050. https://doi.org/10.1007/s10021-005-0105-7

Cheeke TC, Coleman DC, Wall DH (2013) Microbial ecology in sustainable agroecosystems. CRC Press, Boca Raton

Chen S, Zou J, Hu Z, Chen H, Lu Y (2014) Global annual soil respiration in relation to climate, soil properties and vegetation characteristics: summary of available data. Agric For Meteorol 198–199:335–346. https://doi.org/10.1016/j.agrformet.2014.08.020

Chen C, Hu K, Li H, Yun A, Li B (2015) Three-dimensional mapping of soil organic carbon by combining Kriging method with profile depth function. PLoS One. https://doi.org/10.1371/journal.pone.0129038

Chen G, Liu C, Gao Z, Zhang Y, Jiang H, Zhu L, Ren D, Yu L, Xu G, Qian Q (2017) OsHAK1, a high-affinity potassium transporter, positively regulates responses to drought stress in rice. Front Plant Sci 8:1885. https://doi.org/10.3389/fpls.2017.01885

Churchman GJ (2013) The key role of micromorphology in studies of the genesis of clay minerals and their associations in soils and its relevance to advances in the philosophy of soil science. Turkish J Earth Sci 22:376–390. https://doi.org/10.3906/yer-1201-11

Churchman GJ, Lowe DJ (2012) Alteration, formation, and occurrence of minerals in soils. In: Huang PM, Li Y, Sumner ME (eds) Handbook of soil sciences, properties and processes, vol 1, 2nd edn. CRC Press/Taylor and Francis, Boca Raton, pp 20.1–20.72

Clark FE (1977) Soil microbiology – it's a small world. Soil Sci Am J 41:238–241

Classen AT, Sundqvist MK, Henning JA, Newman GS, Moore JAM, Cregger MA, Moorhead LC, Patterson CM (2015) Direct and indirect effects of climate change on soil microbial and soil microbial-plant interactions: what lies ahead? Ecosphere 6:130. https://doi.org/10.1890/ES15-00217.1

Coats VC, Rumpho ME (2014) The rhizosphere microbiota of plant invaders: an overview of recent advances in the microbiomics of invasive plants. Front Microbiol 5:368. https://doi.org/10.3389/fmicb.2014.00368

Constancias F, Terrat S, Saby NP, Horrigue W, Villerd J, Guillemin JP, Biju-Duval L, Nowak V, Dequiedt S, Ranjard L, Chemidlin Prévost-Bouré N (2015) Mapping and determinism of soil microbial community distribution across an agricultural landscape. Microbiology 4:505–517. https://doi.org/10.1002/mbo3.255

Cooper M, Boschi RS, Silva VBS, Silva LFS (2016) Software for micromorphometric characterization of soil pores obtained from 2D image analysis. Sci Agric 73:388–393. https://doi.org/10.1590/0103-9016-2015-0053

Cordero OX, Datta MS (2016) Microbial interactions and community assembly at microscales. Curr Opin Microbiol 31:227–234. https://doi.org/10.1016/j.mib.2016.03.015

Creutzburg MK, Scheller RM, Lucash MS, LeDuc SD, Johnson MG (2017) Forest management scenarios in a changing climate: trade-offs between carbon, timber, and old forest. Ecol Appl 27:503–518. https://doi.org/10.1002/eap.1460

Crutzen P (2002) Geology of mankind. Nature 415:23. https://doi.org/10.1038/415023a

Cuddington K, Byres JE, Wilson WG, Hastings A (2011) Ecosystem engineers: plants to protists. Academic Press, London

Culley JLB (1993) Density and compressibility. In: Carter MR (ed) Soil sampling and methods of analysis. Lewis Publishers, Boca Raton, pp 529–539

Currie JA (1960) Gaseous diffusion in porous media. Part 1. A non-steady state method. Br J Appl Phys 11:314–317. https://doi.org/10.1088/0508-3443/11/8/302

Dadhich RK, Meena RS (2014) Performance of Indian mustard (*Brassica juncea* L.) in response to foliar spray of thiourea and thioglycollic acid under different irrigation levels. Indian J Ecol 41:376–378

Datta R, Anand S, Moulick A, Baraniya D, Pathan SI, Rejsek K, Vranova V, Sharma M, Sharma D, Kelkar A (2017a) How enzymes are adsorbed on soil solid phase and factors limiting its activity: a review. Int Agrophys 31:287–302. https://doi.org/10.1515/intag-2016-0049

Datta R, Kelkar A, Baraniya D, Molaei A, Moulick A, Meena R, Formanek P (2017b) Enzymatic degradation of lignin in soil: a review. Sustainability 9:1163. https://doi.org/10.3390/su9071163

Datta R, Baraniya D, Wang Y-F, Kelkar A, Meena RS, Yadav GS, Teresa Ceccherini M, Formanek P (2017c) Amino acid: its dual role as nutrient and scavenger of free radicals in soil. Sustainability 9:1402. https://doi.org/10.3390/su9081402

Davenport D (2008) The international dimension of climate policy. In: Compston H, Bailey I (eds) Turning down the heat: the politics of climate policy in affluent democracies. Palgrave Macmillan, London, pp 48–62

de Vries FT, Thébault E, Liiri M, Birkhofer K, Tsiafouli MA, Bjørnlund L, Jørgensen HB, Brady MV, Christensen S, de Ruiter PC, d'Hertefeldt T, Frouz J, Hedlund K, Hemerik L, Hol WHG, Hotes S, Mortimer SR, Setälä H, Sgardelis SP, Uteseny K, van der Putten WH, Wolters V, Bardgett RD (2013) Soil food web properties explain ecosystem services across european land use systems. PNAS USA 110:14296–14301. https://doi.org/10.1073/pnas.1305198110

DeAngelis KM, Silver WL, Thompson AW, Firestone MK (2010) Microbial communities acclimate to recurring changes in soil redox potential status. Environ Microbiol 12:3137–3149. https://doi.org/10.1111/j.1462-2920.2010.02286.x

DeAngelis KM, Pold G, Topçuoğlu BD, van Diepen LTA, Varney RM, Blanchard JL, Melillo J, Frey SD (2015) Long-term forest soil warming alters microbial communities in temperate forest soils. Front Microbiol 6:104. https://doi.org/10.3389/fmicb.2015.00104

Decina SM, Hutyra LR, Gately CK, Getson JM, Reinmann AB, Short Gianotti AG, Templer PH (2016) Soil respiration contributes substantially to urban carbon fluxes in the greater Boston area. Environ Pollut 212:433–439. https://doi.org/10.1016/j.envpol.2016.01.012

Delaux PM, Séjalon-Delmas N, Bécard G, Ané JM (2013) Evolution of the plant-microbe symbiotic 'toolkit'. Trends Plant Sci 18:298–304. https://doi.org/10.1016/j.tplants.2013.01.008

Delgado-Baquerizo M, Giaramida L, Reich PB, Khachane AN, Hamonts K, Edwards C, Lawton LA, Singh BK (2016a) Lack of functional redundancy in the relationship between microbial diversity and ecosystem functioning. J Ecol 104:936–946. https://doi.org/10.1111/1365-2745.12585

Delgado-Baquerizo M, Grinyer J, Reich PB, Singh BK (2016b) Relative importance of soil properties and microbial community for soil functionality: insights from a microbial swap experiment. Funct Ecol 30:1862–1873. https://doi.org/10.1111/1365-2435.12674

Delgado-Baquerizo M, Trivedi P, Trivedi C, Aldridge DJ, Jeffries TC, Reich P, Singh BK (2017) Microbial richness and composition independently drive soil multifunctionality. Funct Ecol 31:2330–2343. https://doi.org/10.1111/1365-2435.12924

Dhakal Y, Meena RS, De N, Verma SK, Singh A (2015) Growth, yield and nutrient content of mungbean (*Vigna radiata* L.) in response to INM in eastern Uttar Pradesh, India. Bangladesh J Bot 44:479–482

Diacono M, Montemurro F (2010) Long-term effects of organic amendments on soil fertility. A review. Agron Sustain Dev 30:401–422. https://doi.org/10.1051/agro/2009040

Diffenbaugh NS, Field CB (2013) Changes in ecologically critical terrestrial climate conditions. Science 341:486–492. https://doi.org/10.1126/science.1237123

Ding J, Zhang Y, Deng Y, Cong J, Lu H, Sun X, Yang C, Yuan T, Van Nostrand JD, Li D, Zhou J, Yang Y (2015) Integrated metagenomics and network analysis of soil microbial community of the forest timberline. Sci Rep 5:7994. https://doi.org/10.1038/srep07994

Dopson M, Ni G, Sleutels TH (2016) Possibilities for extremophilic microorganisms in microbial electrochemical systems. FEMS Microbiol Rev 40:164–181. https://doi.org/10.1093/femsre/fuv044

Downie A, Crosky A, Munroe P (2009) Physical properties of biochar. In: Lehmann J, Joseph S (eds) Biochar for environmental management: science and technology. Earthscan, London, pp 13–32

Ducey TF, Novak JM, Johnson MG (2015) Effects of biochar blends on microbial community composition in two coastal plain soils. Agriculture 5:1060–1075. https://doi.org/10.3390/agriculture5041060

Duran-Pinedo AE, Paster B, Teles R, Frias-Lopez J (2011) Correlation network analysis applied to complex biofilm communities. PLoS One. https://doi.org/10.1371/journal.pone.0028438

Eiler A, Heinrich F, Bertilsson S (2012) Coherent dynamics and association networks among lake bacterioplankton taxa. ISME J 6:330–342. https://doi.org/10.1038/ismej.2011.113

Ellis E (2011) Anthropogenic transformation of the terrestrial biosphere. Philos Trans R Soc A 369:1010–1035. https://doi.org/10.1098/rsta.2010.0331

Ellouze W, Esmaeili Taheri A, Bainard LD, Yang C, Bazghaleh N, Navarro-Borrell A, Hanson K, Hamel C (2014) Soil fungal resources in annual cropping systems and their potential for management. Biomed Res Int 531824. https://doi.org/10.1155/2014/531824

Embree M, Liu JK, Al-Bassam MM, Zengler K (2015) Networks of energetic and metabolic interactions define dynamics in microbial communities. Proc Natl Acad Sci 112:15450–15455. https://doi.org/10.1073/pnas.1506034112

Engelmoer DJ, Kiers ET (2015) Host diversity affects the abundance of the extraradical arbuscular mycorrhizal network. New Phytol 205:1485–1491. https://doi.org/10.1111/nph.13086

Ennis CJ, Evans AG, Islam M, Ralebitso-Senior K, Senior E (2012) Biochar: carbon sequestration, land remediation, and impacts on soil microbiology. Crit Rev Environ Sci Technol 42:2311–2364. https://doi.org/10.1080/10643389.2011.574115

Erb KH, Krausmann F, Gaube V, Gingrich S, Bondeau A, Fischer-Kowalski M, Haberl H (2009) Analyzing the global human appropriation of net primary production – processes, trajectories, implications. An introduction. Ecol Econ 69:250–259. https://doi.org/10.1016/j.ecolecon.2009.07.001

Erb KH, Kastner T, Plutzar C, Bais ALS, Carvalhais N, Fetzel T, Gingrich S, Haberl H, Lauk C, Niedertscheider M, Pongratz J, Thurner M, Luyssaert S (2018) Unexpectedly large impact of forest management and grazing on global vegetation biomass. Nature 553:73–76. https://doi.org/10.1038/nature25138

Evans SE, Wallenstein MD (2014) Climate change alters ecological strategies of soil bacteria. Ecol Lett 17:155–164. https://doi.org/10.1111/ele.12206

Falloon P, Betts R (2010) Climate impacts on European agriculture and water management in the context of adaptation and mitigation-the importance of an integrated approach. Sci Total Environ 408:5667–5687. https://doi.org/10.1016/j.scitotenv.2009.05.002

FAO (2017) Soil organic carbon: the hidden potential. Food and Agriculture Organization of the United Nations (FAO), Rome

Fassbender HW (1987) Química de suelos com énfasis em suelos de America Latina, 2nd edn. Rev. Insituto Interamericano de Cooperación para La agricultura – IICA, San José

Faust K, Raes J (2012) Microbial interactions: from networks to models. Nat Rev Microbiol 10:538–550. https://doi.org/10.1038/nrmicro2832

Faust K, Lima-Mendez G, Lerat J-S, Sathirapongsasuti JF, Knight R, Huttenhower C, Lenaerts T, Raes J (2015) Cross-biome comparison of microbial association networks. Front Microbiol 6:1200. https://doi.org/10.3389/fmicb.2015.01200

Fellbaum CR, Mensah JA, Pfeffer PE, Kiers ET, Bücking H (2012) The role of carbon in fungal nutrient uptake and transport: implications for resource exchange in the arbuscular mycorrhizal symbiosis. Plant Signal Behav 7:1509–1512. https://doi.org/10.4161/psb.22015

Feng W, Plante AF, Six J (2014) Soil organic matter stability in organo-mineral complexes as a function of increasing C loading. Soil Biol Biochem 69:398–405. https://doi.org/10.1016/j.soilbio.2013.11.024

Ferrol N, Tamayo E, Vargas P (2016) The heavy metal paradox in arbuscular mycorrhizas: from mechanisms to biotechnological applications. J Exp Bot 67:6253–6265. https://doi.org/10.1093/jxb/erw403

File AL, Klironomos J, Maherali H, Dudley SA (2012) Plant kin recognition enhances abundance of symbiotic microbial partner. PLoS One. https://doi.org/10.1371/journal.pone.0045648

Filho JAC, Pascholati SF, Sabrinho RR (2016) Mycorrhizal association and their role in plant disease protection. In: Hakeem K, Akhtar M (eds) Plant, soil and microbes. Springer, Cham, pp 95–143. https://doi.org/10.1007/978-3-319-29573-2_6

Filho JAC, Sobrinho RR, Pascholati SF (2017) Arbuscular mycorrhizal symbiosis and its role in plant nutrition in sustainable agriculture. In: Meena V, Mishra P, Bisht J, Pattanayak A (eds) Agriculturally important microbes for sustainable agriculture. Springer, Singapore, pp 129–164. https://doi.org/10.1007/978-981-10-5343-6_5

Foster RC (1994) Microorganisms and soil aggregates. In: Pankhurst CE, Doube BM, Gupta VVSR, Grace PR (eds) Soil biota. CSIRO Publications, Melbourne, pp 144–155

Frank S, Schmid E, Havlık P, Schneider UA, Bottcher H, Balkovic Ĵ, Obersteiner M (2015) The dynamic soil organic carbon mitigation potential of European cropland. Glob Environ Change 35:269–278. https://doi.org/10.1016/j.gloenvcha.2015.08.004

Friedlingstein P, Meinshausen M, Arora VK, Jones CD, Anav A, Liddicoat SK, Knutti R (2014) Uncertainties in CMIP5 climate projections due to carbon cycle feedbacks. J Clim 27:511–526. https://doi.org/10.1175/JCLI-D-12-00579.1

García-González E, Jiménez-Fontana R, Azcárate Goded P, Cardeñoso JM (2017) Inclusion of sustainability in university classrooms through methodology. In: Leal Filho W, Brandli L, Castro P, Newman J (eds) Handbook of theory and practice of sustainable development in higher education, World sustainability series. Springer, Cham, pp 3–19. https://doi.org/10.1007/978-3-319-47868-5_1

Garnett T, Godfray C (2012) Sustainable intensification in agriculture. Navigating a course through competing food system priorities, food climate research network and the Oxford Martin programme on the future of food. University of Oxford, Oxford

Gilbert SF, Sapp J, Tauber AI (2012) A symbiotic view of life: we have never been individuals. Q Rev Biol 87:325–341

Glaser B, Birk JJ (2012) State of the scientific knowledge on properties and genesis of anthropogenic dark earths in Central Amazonia (Terra Preta de Índio). Geochim Cosmochim Acta 82:39–51. https://doi.org/10.1016/j.gca.2010.11.029

Glaser B, Wiedner K, Seelig S, Schmidt HP, Gerber H (2015) Biochar organic fertilizers from natural resources as substitute for mineral fertilizers. Agron Sustain Dev 35:667–678. https://doi.org/10.1007/s13593-014-0251-4

Glavatska O, Müller K, Butenschoen O, Schmalwasser A, Kandeler E, Scheu S, Totsche KU, Ruess L (2017) Disentangling the root- and detritus-based food chain in the micro-food web of an arable soil by plant removal. PLoS One. https://doi.org/10.1371/journal.pone.0180264

Goh C-H, Veliz Vallejos DF, Nicotra AB, Mathesius U (2013) The impact of beneficial plant-associated microbes on plant phenotypic plasticity. J Chem Ecol 39:826–839. https://doi.org/10.1007/s10886-013-0326-8

Gougoulias C, Clark JM, Shaw LJ (2014) The role of soil microbes in the global carbon cycle: tracking the below-ground microbial processing of plant-derived carbon for manipulating carbon dynamics in agricultural systems. J Sci Food Agric 94:2362–2371. https://doi.org/10.1002/jsfa.6577

Graham EB, Knelman JE, Schindlbacher A, Siciliano S, Breulmann M, Yannarell A, Beman JM, Abell G, Philippot L, Prosser J, Foulquier A, Yuste JC, Glanville HC, Jones DL, Angel R, Salminen J, Newton RJ, Bürgmann H, Ingram LJ, Hamer U, Siljanen HMP, Peltoniemi K, Potthast K, Bañeras L, Hartmann M, Banerjee S, Yu R-Q, Nogaro G, Richter A, Koranda M, Castle SC, Goberna M, Song B, Chatterjee A, Nunes OC, Lopes AR, Cao Y, Kaisermann A, Hallin S, Strickland MS, Garcia-Pausas J, Barba J, Kang H, Isobe K, Papaspyrou S, Pastorelli R, Lagomarsino A, Lindström ES, Basiliko N, Nemergut DR (2016) Microbes as engines of ecosystem function: when does community structure enhance predictions of ecosystem processes? Front Microbiol 7:214. https://doi.org/10.3389/fmicb.2016.00214

Griffiths BS, Philippot L (2013) Insights into the resistance and resilience of the soil microbial community. FEMS Microbiol Rev 37:112–129. https://doi.org/10.1111/j.1574-6976.2012.00343.x

Gutjahr C (2014) Phytohormone signaling in arbuscular mycorrhiza development. Curr Opin Plant Biol 20:26–34. https://doi.org/10.1016/j.pbi.2014.04.003

Gutjahr C, Parniske M (2013) Cell and developmental biology of arbuscular mycorrhiza symbiosis. Annu Rev Cell Dev Biol 29:593–617. https://doi.org/10.1146/annurev-cellbio-101512-122413

Haberl H, Erb K-H, Krausmann F (2014) Human appropriation of net primary production: planets, trends, and planetary boundaries. Annu Rev Environ Resour 39:363–691. https://doi.org/10.1146/annurev-environ-121912-094620

Haberl H, Erb KH, Krausmann F, Gaube V, Bondeau A, Piutzar C, Gingrich S, Lucht W, Fischer-Kowalski M (2007) Quantifying and mapping the human appropriation of net primary production in earth's terrestrial ecosystems. Proc Natl Acad Sci U S A 104:12942–12947. https://doi.org/10.1073/pnas.0704243104

Hadzi D, Klofutar C, Oblak S (1968) Hydrogen bonding in some adducts of oxygen bases with acids. Part IV. basicity in hydrogen bonding and ionization. J Chem Soc (A):905–918

Hajiboland R (2013) Role of arbuscular mycorrhiza in amelioration of salinity. In: Ahmad P, Azooz MM, Prasad MNV (eds) Salt stress in plants: signalling, omics and adaptations. Springer, New York, pp 301–354. https://doi.org/10.1007/978-1-4614-6108-1_13

Hamel C, Plenchete C (2017) Implications of past, current, and future agricultural practices for mycorrhiza-mediated nutriente flux. In: Johnson NC, Gehring C, Jansa J (eds) Mycorrhizal mediation of soil fertility, structure and carbono storage. Elsevier, Amsterdam, pp 175–186. https://doi.org/10.1016/B978-0-12-804312-7.00010-3

Hapca S, Baveye PC, Wilson C, Lark RM, Otten W (2015) Three-dimensional mapping of soil chemical characteristics at micrometric scale by combining 2D SEM-EDX data and 3D X-ray CT images. PLoS One. https://doi.org/10.1371/journal.pone.0137205

Hardoim PR, van Overbeek LS, Berg G, Pirttilä AM, Compant S, Campisano A, Döring M, Sessitsch A (2015) The hidden world within plants: ecological and evolutionary considerations for defining functioning of microbial endophytes. MMBR 79:293–320. https://doi.org/10.1128/MMBR.00050-14

Harris RF (1981) Effect of water potential on microbial growth and acitivity. In: Parr JF, Gardner WR, Elliot LF (eds) Water potential relations in soil microbiology. Soil Sci Soc Am, Madison, pp 23–95

Havlicek E, Mitchell EAD (2014) Soils supporting biodiversity. In: Dighton J, Krumins JA (eds) Interactions in soil: promoting plant growth, biodiversity, community and ecosystems 1. Springer, Dordrecht, pp 28–58. https://doi.org/10.1007/978-94-017-8890-8_2

Havlík P, Valin H, Herrero M, Obersteiner M, Schmid E, Rufino MC, Mosnier A, Thornton PK, Böttcher H, Conant RT, Frank S, Fritz S, Fuss S, Kraxner F, Notenbaert A (2014) Climate change mitigation through livestock system transitions. Proc Natl Acad Sci U S A 111:3709-3714. https://doi.org/10.1073/pnas.1308044111

Hawkins E, Sutton R (2016) Connecting climate model projections of global temperature change with the real world. Bull Am Meteorol Soc 97:963–980. https://doi.org/10.1175/BAMS-D-14-00154.1

He NP, Yu Q, Wu L, Wang YS, Han XG (2008) Carbon and nitrogen store and storage potential as affected by land-use in a *Leymus chinensis* grassland of northern China. Soil Biol Biochem 40:2952–2959. https://doi.org/10.1016/j.soilbio.2008.08.018

Heimann M, Reichstein M (2008) Terrestrial ecosystem carbon dynamics and climate feedbacks. Nature 451:289–292. https://doi.org/10.1038/nature06591

Henderson-Sellers A, McGuffie K (2012) The future of the world's climate, 2nd edn. Elsevier Science, Amsterdam

Henkner J, Scholten T, Kühn P (2016) Soil organic carbon stocks in permafrost-affected soils in West Greenland. Geoderma 282:147–159. https://doi.org/10.1016/j.geoderma.2016.06.021

Henry HAL (2012) Soil extracellular enzyme dynamics in a changing climate. Soil Biol Biochem 47:53–59. https://doi.org/10.1016/j.soilbio.2011.12.026

Hinsinger P, Plassard C, Tang C, Jaillard B (2003) Origins of root-mediated pH changes in the rhizosphere and their responses to environmental constraints: a review. Plant Soil 248:43–59. https://doi.org/10.1023/A:1022371130939

Hinsinger P, Bengough AG, Vetterlein D, Young IM (2009) Rhizosphere: biophysics, biogeochemistry and ecological relevance. Plant Soil 321:117–152. https://doi.org/10.1007/s11104-008-9885-9

Hinsinger P, Jones DL, Marschner P (2012) Biogeochemical, biophysical and biological processes in the rhizosphere. In: Huang PM, Li Y, Sumner M (eds) Handbook of soil sciences: resource management and environmental impacts, 2nd edn. CRC Press, Boca Raton, pp 6-1–6-30

Hodge JM, Clasen TF (2014) Carbon financing of household water treatment: background, operation and recommendations to improve potential for health gains. Environ Sci Technol 48:12509–12515. https://doi.org/10.1021/es503155m

Hoehler TM, Jørgensen BB (2013) Microbial life under extreme energy limitation. Nat Rev Microbiol 11:83–94. https://doi.org/10.1038/nrmicro2939

Homann PS, Sollins P, Fiorella M, Thorson T, Kern JS (1998) Regional soil organic carbon storage estimates for Western Oregon by multiple approaches. Soil Sci Soc Am J 62:789–796. https://doi.org/10.2136/sssaj1998.03615995006200030036x

Horner-Devine MC, Silver JM, Leibold MA, Bohannan BJ, Colwell RK, Fuhrman JA, Green JL, Kuske CR, Martiny JB, Muyzer G, Ovreås L, Reysenbach AL, Smith VH (2007) A comparison of taxon co-occurrence patterns for macro- and microorganisms. Ecology 88:1345–1353. https://doi.org/10.1890/06-0286

Horwath WR (2015) Carbon cycling: the dynamics and formation of organic matter. In: Paul EA (ed) Soil microbiology, ecology and biochemistry, 4th edn. Academic, San Diego, pp 339–359

Huang X, Zhu-Barker X, Horwath WR, Faeflen SJ, Luo H, Xin X, Jiang X (2016) Effect of iron oxide on nitrification in two agricultural soils with different pH. Biogeosciences 13:5609–5617. https://doi.org/10.5194/bg-13-5609-2016

Hughes JB, Hellmann JJ, Ricketts TH, Bohannan BJ (2001) Counting the uncountable: statistical approaches to estimating microbial diversity. Appl Environ Microbiol 67:4399–4406. https://doi.org/10.1128/AEM.67.10.4399-4406.2001

Hunt C, Baum S (2009) The "hidden" costs of forestry off sets. Mitig Adapt Strat Glob Change 14:107–120. https://doi.org/10.1007/s11027-008-9153-6

Husson O (2013) Redox potential (E_h) and pH as drivers of soil/plant/microorganism systems: a transdisciplinar overview pointing to integrative opportunities for agronomy. Plant Soil 362:389–417. https://doi.org/10.1007/s11104-012-1429-7

IPCC (2013) Summary for policymakers. In: Stocker TF, Qin D, Plattner G-K, Tignor M, Allen SK, Boschung J, Nauels A, Xia Y, Bex V, Midgley PM (eds) Climate change 2013: the physical science basis. Contribution of working group I to the fifth assessment report of the intergovernmental panel on climate change. Cambridge University Press, Cambridge/New York

IPCC (2014a) In: Field CB, Barros VR, Dokken DJ, Mach KJ, Mastrandrea MD, Bilir TE, Chatterjee M, Ebi KL, Estrada YO, Genova RC, Girma B, Kissel ES, Levy AN, MacCracken S, Mastrandrea PR, White LL (eds) Climate change 2014: impacts, adaptation, and vulnerability. Part a: global and sectoral aspects. Contribution of working group II to the fifth assessment report of the intergovernmental panel on climate change. Cambridge University Press, Cambridge/New York

IPCC (2014b) In: Barros VR, Field CB, Dokken DJ, Mastrandrea MD, Mach KJ, Bilir TE, Chatterjee M, Ebi KL, Estrada YO, Genova RC, Girma B, Kissel ES, Levy AN, MacCracken S, Mastrandrea PR, White LL (eds) Climate change 2014: impacts, adaptation, and vulnerability. Part B: regional aspects. Contribution of working group II to the fifth assessment report of the intergovernmental panel on climate change. Cambridge University Press, Cambridge/New York

IPCC (2017) In: Shukla PR, Skea J, Diemen R, Huntley E, Pathak M, Portugal-Pereira J, Scull J, Slade R (eds) Meeting report of the intergovernmental panel on climate change expert meeting on mitigation, sustainability and climate stabilization scenarios. IPCC Working Group III Technical Support Unit, Imperial College London, London

Jackson RS (2014) Site selection and climate. In: Jackson RS (ed) Wine science: principles and applications; a volume in food science and technology, 4th edn. Elsevier Science, Amsterdam, pp 307–346

Jansa J, Treseder KK (2017) Introduction: mycorrhizas and the carbon cycle. In: Johnson NC, Catherine Gehring C, Jansa J (eds) Mycorrhizal mediation of soil: fertility, structure, and carbon storage. Elsevier, Amsterdam, pp 343–355

Jaynes DB, Rogowski AS (1983) Applicability of Fick's law to gas diffusion. Soil Sci Am J 47:425–430

Jewell TNM, Karaoz U, Bill M, Chakraborty R, Brodie EL, Williams KH, Beller HR (2017) Metatranscriptomic analysis reveals unexpectedly diverse microbial metabolism in a biogeochemical hot spot in an alluvial aquifer. Front Microbiol 8:40. https://doi.org/10.3389/fmicb.2017.00040

Johnson JA, Runge CF, Senauer B, Foley J, Polasky S (2014) Global agriculture and carbon tradeoffs. Proc Natl Acad Sci U S A 111:12342–12347. https://doi.org/10.1073/pnas.1412835111

Jones CG, Lawton JH, Shachak M (1994) Organisms as ecosystem engineers. Oikos 69:373–386. https://doi.org/10.2307/3545850

Jørgensen BB, Marshall IP (2016) Slow microbial life in the seabed. Annu Rev Mar Sci 8:311–332. https://doi.org/10.1146/annurev-marine-010814-015535

Joshi PA, Shekhawat DB (2014) Microbial contributions to global climate changes in soil en-vironments: impact on carbon cycle: a short review. Ann Appl Biol-Sci 1:R7–R9

Kai FM, Tyler SC, Randerson JT, Blake DR (2011) Reduced methane growth rate explained by decreased northern hemisphere microbial sources. Nature 476:194–197. https://doi.org/10.1038/nature10259

Kaisermann A, Maron PA, Beaumelle L, Lata JC (2015) Fungal communities are more sensitive indicators to non-extreme soil moisture variations than bacterial communities. Appl Soil Ecol 86:158–164. https://doi.org/10.1016/j.apsoil.2014.10.009

Kallenbach CM, Frey SD, Grandy AS (2016) Direct evidence for microbial-derived soil organic matter formation and its ecophysiological controls. Nat Commun 7:13630. https://doi.org/10.1038/ncomms13630

Karhu K, Auffret MD, Dungait JA, Hopkins DW, Prosser JI, Singh BK, Subke JA, Wookey PA, Agren GI, Sebastià MT, Gouriveau F, Bergkvist G, Meir P, Nottingham AT, Salinas N, Hartley IP (2014) Temperature sensitivity of soil respiration rates enhanced by microbial community response. Nature 513:81–84. https://doi.org/10.1038/nature13604

Keiluweit M, Nico PS, Kleber M, Fendorf S (2016) Are oxygen limitations under recognized regulators of organic carbon turnover in upland soils? Biogeochemistry 127:157–171. https://doi.org/10.1007/s10533-015-0180-6

Kim JS, Sparovek G, Longo RM, De Melo WJ, Crowley D (2007) Bacterial diversity of terra preta and pristine forest soil from the Western Amazon. Soil Biol Biochem 39:684–690. https://doi.org/10.1016/j.soilbio.2006.08.010

Kim S, Lee S, McCormick M, Kim JG, Kang H (2016) Microbial community and greenhouse gas fluxes from abandoned rice paddies with different vegetation. Microb Ecol 72:692–703. https://doi.org/10.1007/s00248-016-0801-1

Kindler R, Siemens J, Kaiser K, Walmsley DC, Bernhofer C, Buchmann N, Cellier P, Eugster W, Gleixner G, Grünwald T, Heim A, Ibrom A, Joness SK, Jones M, Klumpp K, Kutsch W, Larsen KS, Lehuger S, Loubet B, McKenzie R, Moors E, Osborne B, Pilegaard K, Rebmann C, Saunders M, Schmidt MWI, Schrumpf M, Seyfferth J, Skiba U, Soussana J-F, Sutton MQA, Tefs C, Vowinckel B, Zeeman MJ, Kaupenjohann M (2011) Dissolved carbon leaching from soil is a crucial component of the net ecosystem carbon balance. Glob Chang Biol 17:1167–1185. https://doi.org/10.1111/j.1365-2486.2010.02282.x

Kirkham D, Powers WL (1972) Advanced soil physics. Wiley-Interscience, New York

Kittelmann S, Seedorf H, Walters WA, Clemente JC, Knight R, Gordon JI, Janssen PH (2013) Simultaneous amplicon sequencing to explore co-occurrence patterns of bacterial, archaeal and eukaryotic microorganisms in rumen microbial communities. PLoS One. https://doi.org/10.1371/journal.pone.0047879

Kovács ÁT (2014) Impact of spatial distribution on the development of mutualism in microbes. Front Microbiol 5:649. https://doi.org/10.3389/fmicb.2014.00649

Krausmann F, Erb KH, Gingrich S, Haberl H, Bondeau A, Gaube V, Lauk C, Plutzar C, Searchinger T (2013) Global human appropriation of net primary production doubled in the 20th century. Proc Natl Acad Sci U S A 110:10324–10329. https://doi.org/10.1073/pnas.1211349110

Krug JHA (2018) Accounting of GHG emissions and removals from forest management: a long road from Kyoto to Paris. Carbon Balance Manag 13:1. https://doi.org/10.1186/s13021-017-0089-6

Krüger M, Krüger C, Walker C, Stockinger H, Schüßler A (2012) Phylogenetic reference data for systematics and phylotaxonomy of arbuscular mycorrhizal fungi from phylum to species level. New Phytol 193:970–984. https://doi.org/10.1111/j.1469-8137.2011.03962.x

Kuang J, Cadotte MW, Chen Y, Shu H, Liu J, Chen L, Hua Z, Shu W, Zhou J, Huang L (2017) Conservation of species-and trait-based modeling network interactions in extremely acidic microbial community assembly. Front Microbiol 8:1486. https://doi.org/10.3389/fmicb.2017.01486

Kumar M, Singh DP, Prabha R, Rai AK, Sharma L (2016) Role of microbial inoculants in nutrient use efficiency. In: Singh D, Singh H, Prabha R (eds) Microbial inoculants in sustainable agricultural productivity. Springer, New Delhi, pp 133–142. https://doi.org/10.1007/978-81-322-2644-4_9

Kumar S, Meena RS, Bohra JS (2018) Interactive effect of sowing dates and nutrient sources on dry matter accumulation of Indian mustard (*Brassica juncea* L.). J Oilseed Brassica 9:72–76

Kurtz ZD, Müller CL, Miraldi ER, Littman DR, Blaser MJ, Bonneau RA (2015) Sparse and compositionally robust inference of microbial ecological networks. PLoS Comput Biol. https://doi.org/10.1371/journal.pcbi.1004226

Lal R (2004a) Soil carbon sequestration to mitigate climate change. Geoderma 123:1–22. https://doi.org/10.1016/j.geoderma.2004.01.032

Lal R (2004b) Agricultural activities and the global carbon cycle. Nutr Cycl Agroecosyst 70:103–116. https://doi.org/10.1023/B:FRES.0000048480.24274.0f

Lal R (2010) Managing soils and ecosystems for mitigating anthropogenic carbon emissions and advancing global food security. Bioscience 60:708–721. https://doi.org/10.1525/bio.2010.60.9.8

Lal R (2011a) Soil health and climate change: an overview. In: Singh B, Cowie A, Chan K (eds) Soil health and climate change. Soil biology, vol 29. Springer, Berlin/Heidelberg, pp 3–24. https://doi.org/10.1007/978-3-642-20256-8_1

Lal R (2011b) Sequestering carbon in soils of agro-ecosystems. Food Policy 36:S33–S39. https://doi.org/10.1016/j.foodpol.2010.12.001

Lal R (2016) Biochar and soil carbon sequestration. In: Guo M, He Z, Uchimiya M (eds) Agricultural and environmental applications of biochar: advances and barriers. Soil Sci Soc Am, SSSA, Madison, pp 175–198

Lal R, Nagassa W, Lorenz K (2015) Carbon sequestration in soils. Curr Opin Environ Sustain 15:79–86. https://doi.org/10.1016/j.cosust.2015.09.002

Lamers LPM, van Diggelen JMH, Op den Camp HJM, Visser EJW, Lucassen ECHET, Vile MA, Jetten MSM, Smolders AJP, Roelofs JGM (2012) Microbial transformations of nitrogen, sulfur, and iron dictate vegetation composition in wetlands: a review. Front Microbiol 3:156. https://doi.org/10.3389/fmicb.2012.00156

Lanza G, Rebensburg P, Kern J, Lentzsch P, Wirt S (2016) Impact of biochars and readily available carbon on soil microbial respiration and microbial community composition in a dynamic incubation experiment. Soil Till Res 164:18–24. https://doi.org/10.1016/j.still.2016.01.005

Larkin AA, Martiny AC (2017) Microdiversity shapes the traits, niche space, and biogeography of microbial taxa. Environ Microbiol Rep 9:55–70. https://doi.org/10.1111/1758-2229.12523

Larsen PE, Gibbons SM, Gilbert JA (2012) Modeling microbial community structure and functional diversity across time and space. FEMS Microbiol Lett 332:91–98. https://doi.org/10.1111/j.1574-6968.2012.02588.x

Lee JY, Santamarina JC, Ruppel C (2010) Volume change associated with formation and dissociation of hydrate in sediment. Geochem Geophys Geosyst. https://doi.org/10.1029/2009GC002667

Lehmann RM (2007) Microbial distributions and their potential controlling factors in terrestrial subsurface environments. In: Franklin RB, Mills AL (eds) The spatial distribution of microbes in the environment. Springer, Dordrecht, pp 135–178. https://doi.org/10.1007/978-1-4020-6216-2_6

Lehmann J (2009) Terra preta Nova – where to from here? In: Woods WI, Teixeira WG, Lehmann J, Steiner C, Winkler Prins A (eds) Amazonian Dark Earths: Wim Sombroek's Vision. Springer, Dordrecht, pp 473–486. https://doi.org/10.1007/978-1-4020-9031-8_28

Lehmann A, Rillig MC (2015) Arbuscular mycorrhizal contribution to copper, manganese and iron nutrient concentrations in crops-a meta-analysis. Soil Biol Biochem 81:147–158. https://doi.org/10.1016/j.soilbio.2014.11.013

Lehmann J, da Silva JP Jr, Steiner C, Nehls T, Zech W, Glaser B (2003) Nutrient availability and leaching in an archaeological Anthrosol and a Ferralsol of the Central Amazon basin: fertilizer, manure and charcoal amendments. Plant Soil 249:343–357. https://doi.org/10.1023/A:1022833116184

Lehmann J, Rillig MC, Thies J, Masiello CA, Hockaday WC, Crowley D (2011) Biochar effects on soil biota: a review. Soil Biol Biochem 43:1812–1836. https://doi.org/10.1016/j.soilbio.2011.04.022

Lehmann A, Veresoglou SD, Leifheit EF, Rillig MC (2014) Arbuscular mycorrhizal influence on zinc nutrition in crop plants—a meta-analysis. Soil Biol Biochem 69:123–131. https://doi.org/10.1016/j.soilbio.2013.11.001

Leith H (1975) Modeling the primary productivity of the world. In: Leith H, Whittaker R (eds) Primary production of the biosphere. Springer, New York, pp 237–263

Lemanceau P, Barret M, Mazurier S, Mondy S, Pivato B, Fort T, Vacher C (2017) Plant communication with associated microbiota in the spermosphere, rhizosphere and phyllosphere. Adv Bot Res 82:101–133. https://doi.org/10.1016/bs.abr.2016.10.007

Li L, Vogel J, He Z, Zou X, Ruan H, Huang W, Huang W, Wang J, Bianchi TS (2016) Association of soil aggregation with the distribution and quality of organic carbon in soil along an elevation gradient on Wuyi Mountain in China. PLoS One. https://doi.org/10.1371/journal.pone.0150898

Li X, Kappler U, Jiang G, Bond PL (2017) The ecology of acidophilic microorganisms in the corroding concrete sewer environment. Front Microbiol 8:683. https://doi.org/10.3389/fmicb.2017.00683

Liang C, Balser TC (2010) Microbial production of recalcitrant organic matter in global soils: implications for productivity and climate policy. Nat Rev Microbiol 8:593–599. https://doi.org/10.1038/nrmicro2386-c1

Liao N, Li Q, Zhang W, Zhou G, Ma L, Min W, Ye J, Hou Z (2016) Effects of biochar on soil microbial community composition and activity in drip-irrigated desert soil. Eur J Soil Biol 72:27–34. https://doi.org/10.1016/j.ejsobi.2015.12.008

Liu C, Lu M, Cui J, Li B, Fang C (2014) Effects of straw carbon input on carbon dynamics in agricultural soils: a meta-analysis. Glob Chang Biol 20:1366–1381. https://doi.org/10.1111/gcb.12517

Lladó S, López-Mondéjar R, Baldrian P (2017) Forest soil bacteria: diversity, involvement in ecosystem processes, and response to global change. Microbiol Mol Biol Rev. 81:e00063-16. https://doi.org/10.1128/MMBR.00063-16

Lorenz K, Lal R (2018) Agricultural land use and the global carbon cycle. In: Carbon sequestration in agricultural ecosystems. Springer, Cham, pp 1–37. https://doi.org/10.1007/978-3-319-92318-5_1

Lozano-García B, Parras-Alcántara L (2013) Land use and management effects on carbon and nitrogen in Mediterranean Cambisols. Agric Ecosyst Environ 179:208–214. https://doi.org/10.1016/j.agee.2013.07.009

Lucas J, Bill B, Stevenson B, Kaspari M (2017) The microbiome of the ant-built home: the microbial communities of a tropical arboreal ant and its nest. Ecosphere. https://doi.org/10.1002/ecs2.1639

Lukac M (2017) Soil biodiversity and environmental change in European forests. Cent Eur For J 63:59–65. https://doi.org/10.1515/forj-2017-0010
Lupatini M, Suleiman AKA, Jacques RJS, Antoniolli ZI, de Siqueira Ferreira A, Kuramae EE, Roesch LFW (2014) Network topology reveals high connectance levels and few key microbial genera within soils. Front Environ Sci 2:10. https://doi.org/10.3389/fenvs.2014.00010
Luxmoore RJ (1981) Micro, meso, and macroporosity of soil. Letter to editor. Soil Sci Soc Am J 45:671–672
Maffei ME (2014) Magnetic field effects on plant growth, development, and evolution. Front Plant Sci 5:445. https://doi.org/10.3389/fpls.2014.00445
Magdoff FR, Bartlett RJ (1984) Soil pH buffering revisited. Soil Sci Soc Am J 49:145–148
Manzoni S, Schimel JP, Porporato A (2012) Responses of soil microbial communities to water stress: results from a meta-analysis. Ecology 93:930–938. https://doi.org/10.1890/11-0026.1
Mariotte P, Mehrabi Z, Bezemer TM, De Deyn GB, Kulmatiski A, Drigo B, Veen GFC, van der Heijden MGA, Kardol P (2018) Plant-soil feedback: bridging natural and agricultural sciences. Trends Ecol Evol 33:129–142. https://doi.org/10.1016/j.tree.2017.11.005
Maron PA, Mougel C, Ranjard L (2011) Soil microbial diversity: methodological strategy, spatial overview and functional interest. C R Biol 334:403–411. https://doi.org/10.1016/j.crvi.2010.12.003
Marques JDO, Luizão FJ, Teixeira WG, Sarrazin M, Ferreira SJF, Beldini TP, Marques EMA (2015) Distribution of organic carbon in different soil fractions in ecosystems of Central Amazonia. R Bras Ci Solo 39:232–242. https://doi.org/10.1590/01000683rbcs20150142
Marschner H, Marschner P (2012) Marschner's mineral nutrition of higher plants, 3rd edn. Academic Press, New York
Martin JP (1950) Use of acid, rose bengal and streptomycin in the plate method for estimating soil fungi. Soil Sci 69:215–232
Maude RB (1996) Seedborne diseases and their control: principles and practice. CAB International, Wallinford
Mazzoleni S, Bonanomi G, Giannino F, Incerti G, Piermatteo D, Spaccini R, Piccolo A (2012) New modeling approach to describe and predict carbon sequestration dynamics in agricultural soils. In: Piccolo A (ed) Carbon sequestration in agricultural soils. Springer, Berlin/Heidelberg, pp 291–307. https://doi.org/10.1007/978-3-642-23385-2_11
Meena RS, Yadav RS (2014) Phonological performance of groundnut varieties under sowing environments in hyper arid zone of Rajasthan, India. J Appl Nat Sci 6:344–348. https://doi.org/10.31018/jans.v6i2.425
Meena VS, Verma JP, Meena SK (2015) Towards the current scenario of nutrient use efficiency in crop species. J Clean Prod 102:556–557. https://doi.org/10.1016/j.jclepro.2015.04.030
Meena SK, Meena VS (2017) Importance of soil microbes in nutrient use efficiency and sustainable food production. In: Meena V, Mishra P, Bisht J, Pattanayak A (eds) Agriculturally important microbes for sustainable agriculture. Springer, Singapore, pp 3–23. https://doi.org/10.1007/978-981-10-5343-6_1
Meena RS, Kumar S, Pandey A (2017) Response of sulfur and lime levels on productivity, nutrient content and uptake of sesame under guava (*Psidium guajava* L.) based Agri-horti system in an acidic soil of eastern Uttar Pradesh, India. J Crop Weed 13:222–227
Meena H, Meena RS, Lal R, Singh GS, Mitran T, Layek J, Patil SB, Kumar S, Verma T (2018) Response of sowing dates and bioregulators on yield of clusterbean under current climate in alley cropping system in eastern U.P. Indian Leg Res 41:563–571
Metting FB Jr (1993) Structure and physiological ecology of soil microbial communities. In: Metting FB (ed) Soil microbial ecology: applications in agricultural and environmental management. Marcel Dekker, New York, pp 3–25
Miedema R (1997) Application of micromorphology of relevance to agronomy. Adv Agron 59:119–169
Minasny B, Malone BP, McBratney AB, Angers DA, Arrouays D, Chambers A, Chaplot V, Chen Z-S, Cheng K, Das BS, Field DJ, Gimona A, Hedley CB, Hong SY, Mandal B, Marchant BP, Martin M, McConkey BG, Mulder VL, O'Rourke S, Richer-de-Forges AC, Odeh I, Padarian J, Paustian K, Pan G, Poggio L, Savin I, Stolbovoy V, Stockmann U, Sulaeman Y,

Tsui C-C, Vågen T-G, van Wesemael B, Winowiecki L (2017) Soil carbon 4 per mille. Geoderma 292:59–86. https://doi.org/10.1016/j.geoderma.2017.01.002
Molaei A, Lakzian A, Datta R, Haghnia G, Astaraei A, Rasouli-Sadaghiani M, Ceccherini MT (2017a) Impact of chlortetracycline and sulfapyridine antibiotics on soil enzyme activities. Int Agrophys 31:499–505
Molaei A, Lakzian A, Haghnia G, Astaraei A, Rasouli-Sadaghiani M, Ceccherini MT, Datta R (2017b) Assessment of some cultural experimental methods to study the effects of antibiotics on microbial activities in a soil: an incubation study. PLoS One. https://doi.org/10.1371/journal.pone.0180663
Moldrup P, Olesen T, Yoshikawa S, Komatsu T, Rolston D (2004) Three-porosity model for predicting the gas diffusion coefficient in undisturbed soil. Soil Sci Soc Am J 68:750–759. https://doi.org/10.2136/sssaj2004.7500
Moll J, Hoppe B, König S, Wubet T, Buscot F, Krüger D (2016) Spatial distribution of fungal communities in an arable soil. PLoS One. https://doi.org/10.1371/journal.pone.0148130
Moonen AC, Bàrberi P (2008) Functional biodiversity: an agroecosystem approach. Agric Ecosys Environ 127:7–21. https://doi.org/10.1016/j.agee.2008.02.013
Moreira FMS, Siqueira JO (2006) Microbiologia e bioquímica do solo. Editora UFLA, Lavras
Moss R, Edmonds JA, Hibbard KA, Manning MR, Rose SK, van Vuuren DP, Carter TR, Emori S, Kainuma M, Kram T, Meehl GA, Mitchell JFB, Nakicenovic N, Riahi K, Smith SJ, Stouffer RJ, Thomson AM, Weyant JP, Wilbanks TJ (2010) The next generation of scenarios for climate change research and assessment. Nature 463:747–756. https://doi.org/10.1038/nature08823
Mukherjee A, Lal R (2017) Biochar and soil characteristics. In: Lal R (ed) Encyclopedia of soil science, 3rd edn. Taylor and Francis, Abingdon, pp 183–188
Muthukumar T, Priyadharsini P, Uma E, Jaison S, Pandey RR (2014) Role of arbuscular mycorrhizal fungi in alleviation of acidity stress on plant growth. In: Miransari M (ed) Use of microbes for the alleviation of soil stresses, vol 1. Springer, New York, pp 43–71. https://doi.org/10.1007/978-1-4614-9466-9_3
Mwajita MR, Murage H, Tani A, Kahangi EM (2013) Evaluation of rhizosphere, rhizoplane and phyllosphere bacteria and fungi isolated from rice in Kenya for plant growth promoters. SpringerPlus 2:606. https://doi.org/10.1186/2193-1801-2-606
Myrold DD, Zeglin LH, Jansson JK (2014) The potential of metagenomic approaches for understanding soil microbial processes. Soil Sci Soc Am J 78:3–10. https://doi.org/10.2136/sssaj2013.07.0287dgs
Nabuurs GJ, Masera O, Andrasko K, Benitez-Ponce P, Boer R, Dutschke M, Elsiddig E, FordRobertson J, Frumhoff P, Karjalainen T, Krankina O, Kurz WA, Matsumoto M, Oyhantcabal W, Ravindranath NH, Sanz Sanchez MJ, Zhang X (2007) Forestry. In: Metz B, Davidson OR, Bosch PR, Dave R, Meyer LA (eds) Climate change 2007: mitigation. Contribution of working group III to the fourth assessment report of the intergovernmental panel on climate change. Cambridge University Press, Cambridge, pp 541–584. https://doi.org/10.1017/CBO9780511546013.013
Nannipieri P, Ascher J, Ceccherini MT, Landi L, Pietramellara G, Renella G (2003) Microbial diversity and soil functions. Eur J Soil Sci 54:655–670. https://doi.org/10.1046/j.1351-0754.2003.0556.x
Nazir R, Mazurier S, Yang P, Lemanceau P, van Elsas JD (2017) The ecological role of type three secretion systems in the interaction of bacteria with fungi in soil and related habitats is diverse and context-dependent. Front Microbiol 8:38. https://doi.org/10.3389/fmicb.2017.00038
Neves EG, Petersen JB, Bartone RN, Silva CAD (2003) Historical and socio- cultural origins of Amazonian dark earths. In: Lehmann J, Kern DC, Glaser B, Woods WI (eds) Amazonian dark earths: origin, properties, management. Kluwer Academic Publishers, Dordrecht, pp 29–50
Nguyen C (2003) Rhizodeposition of organic C by plants: mechanisms and controls. Agronomie 23:375–396. https://doi.org/10.1051/agro:2003011

Nianpeng H, Ruomeng W, Yang G, Jingzhong D, Xuefa W, Guirui Y (2013) Changes in the temperature sensitivity of SOM decomposition with grassland succession: implications for soil C sequestration. Ecol Evol 3:5045–5054. https://doi.org/10.1002/ece3.881

Nikrad MP, Kerkhof LJ, Häggblom MM (2016) The subzero microbiome: microbial activity in frozen and thawing soils. FEMS Microbiol Ecol. https://doi.org/10.1093/femsec/fiw081

Novozamsky I, Beek J, Bolt GH (1976) Chemical equilibria. In: Bolt GH, Bruggenwert MGM (eds) soil chemistry: part a. basic elements, vol 5. Elsevier Scientific, Amsterdam, pp 13–42

Nziguheba G, Vargas R, Bationo A, Black H, Buschiazzo DE, de Brogniez D, Joosten H, Melillo J, Richter D, Termansen M (2015) Soil carbon: a critical natural resource – wide-scale goals, urgent actions. In: Banwart SA, Noellemeyer E, Milne E (eds) Soil carbon: science, management, and policy for multiple. SCOPE, volume 7. Cab International, Wallingford, pp 10–25

OECD/FAO (2017) OECD-FAO agricultural outlook 2017–2026. OECD Publishing and FAO. https://doi.org/10.1787/agr_outlook-2017-en

Ofaim S, Ofek-Lalzar M, Sela N, Jinag J, Kashi Y, Minz D, Freilich S (2017) Analysis of microbial functions in the rhizosphere using a metabolic-network based framework for metagenomics interpretation. Front Microbiol 8:1606. https://doi.org/10.3389/fmicb.2017.01606

Ohtsuka T, Mo W, Satomura T, Inatomi M, Koizumi H (2007) Biometric basedcarbon flux measurements and net ecosystem production (NEP) in a temperate deciduous broad-leaved forest beneath a flux tower. Ecosystems 10:324–334. https://doi.org/10.1007/s10021-007-9017-z

Omondi MO, Xia X, Nahayo A, Liu X, Korai PK, Pan G (2016) Quantification of biochar effects on soil hydrological properties using meta-analysis of literature data. Geoderma 274:28–34. https://doi.org/10.1016/j.geoderma.2016.03.029

Orgiazzi A, Panagos P, Yigini Y, Dunbar MB, Gardi C, Montanarella L, Ballabio C (2016) A knowledge-based approach to estimating the magnitude and spatial patterns of potential threats to soil biodiversity. Sci Total Environ 545–546:11–20. https://doi.org/10.1016/j.scitotenv.2015.12.092

Osono T (2014) Diversity and ecology of endophytic and epiphytic fungi of tree leaves in Japan: a review. In: Verma VC, Gange AC (eds) Advances in endophytic research. Springer, New Delhi, pp 3–26. https://doi.org/10.1007/978-81-322-1575-2_1

Pagano MC (2014) Drought stress and mycorrhizal plants. In: Miransai M (ed) Use of microbes for the alleviation of soil stress, volume 1. Springer, New York, pp 97–110. https://doi.org/10.1007/978-1-4614-9466-9_5

Pagano MC, Dhar PP (2015) Arbuscular mycorrhizal fungi: association and production of bioactive compounds in plants. In: Gupta VK, Tuohy MG, Lohani M, O'Donovan A (eds) Biotechnology of bioactive compounds: sources and applications. Wiley, Chichester, pp 225–243

Pagano MC, Gupta VK (2016) Overview of the recent advances in mycorrhizal fungi. In: Pagano M (ed) Recent advances on mycorrhizal fungi. Fungal biology. Springer, Cham, pp 1–13. https://doi.org/10.1007/978-3-319-24355-9_1

Pajares S, Bohannan BJM (2016) Ecology of nitrogen fixing, nitrifying, and denitrifying microorganisms in tropical forest soils. Front Microbiol 7:1045. https://doi.org/10.3389/fmicb.2016.01045

Pajares S, Bohannan BJM, Souza V (2016) Editorial: the role of microbial communities in tropical ecosystems. Front Microbiol 7:1805. https://doi.org/10.3389/fmicb.2016.01805

Papendick RI, Campbell GS (1981) Theory and measurement of water potential. In: Parr JF, Gardner WR, Elliot LF (eds) Water potential relations in soil microbiology. Soil Science Society of America, Madison, pp 1–22

Park JH, Choppala GK, Bolan NS, Chung JW, Chuasavathi T (2011) Biochar reduces the bioavailability and phytotoxicity of heavy metals. Plant Soil 348:439–451. https://doi.org/10.1007/s11104-011-0948-y

Paul ED (2016) The nature and dynamics of soil organic matter: plant inputs, microbial transformations, and organic matter stabilization. Soil Biol Biochem 98:109–126. https://doi.org/10.1016/j.soilbio.2016.04.001

Peay KG, Kennedy PG, Talbot JM (2016) Dimensions of biodiversity in the earth mycobiome. Nat Rev Microbiol 14:434–447. https://doi.org/10.1038/nrmicro.2016.59

Peng S, Guo T, Liu G (2013) The effects of arbuscular mycorrhizal hyphal networks on soil aggregations of purple soil in Southwest China. Soil Biol Biochem 57:411–417. https://doi.org/10.1016/j.soilbio.2012.10.026

Petersen JB, Neves E, Heckenberger MJ (2001) Gift from the past: Terra Preta and prehistoric Amerindian occupation in Amazonia. In: McEwan C, Barreto C, Neves E (eds) Unknown Amazonia. British Museum Press, London, pp 86–105

Piccolo A (2012) The nature of soil organic matter and innovative soil managements to fight global changes and maintain agricultural productivity. In: Piccolo A (ed) Carbon sequestration in agricultural soils. Springer, Berlin/Heidelberg, pp 1–19. https://doi.org/10.1007/978-3-642-23385-2_1

Pickering AJ, Arnold BF, Dentz HN, Colford JM, Null C (2017) Climate and health co-benefits in low-income countries: a case study of carbon financed water filters in Kenya and a call for independent monitoring. Environ Health Perspect 125:278–283. https://doi.org/10.1289/EHP342

Poeplau C, Don A (2013) Sensitivity of soil organic carbon stocks and fractions to different land-use changes across Europe. Geoderma 192:189–201. https://doi.org/10.1016/j.geoderma.2012.08.003

Poeplau C, Vos C, Don A (2017) Soil organic carbon stocks are systematically overestimated by misuse of the parameters bulk density and rock fragment content. Soil 3:61–66. https://doi.org/10.5194/soil-3-61-201

Porcel R, Aroca R, Ruiz-Lozano JM (2012) Salinity stress alleviation using arbuscular mycorrhizal fungi. A review. Agron Sustain Dev 32:181–200. https://doi.org/10.1007/s13593-011-0029-x

Poudel R, Jumpponen A, Schlatter DC, Paulitz TC, McSpadden Gardener BB, Kinkel LL, Garret KA (2016) Microbiome networks: a systems framework for identiying candidate microbial assemblages for disease management. Phytopathology 106:1083–1096. https://doi.org/10.1094/PHYTO-02-16-0058-FI

Powlson DS, Whitmore AP, Goulding KWT (2011) Soil carbon sequestration to mitigate climate change: a critical re-examination to identify the true and the false. Eur J Soil Sci 62:42–55. https://doi.org/10.1111/j.1365-2389.2010.01342.x

Prayogo C, Jones JE, Baeyens J, Bending GD (2014) Impact of biochar on mineralisation of C and N from soil and willow litter and its relationship with microbial community biomass and structure. Biol Fertil Soils 50:695–702. https://doi.org/10.1007/s00374-013-0884-5

Pritchard R, Ryan CM, Grundy I, van der Horst D (2018) Human appropriation of net primary productivity and rural livelihoods: findings from six villages in Zimbabwe. Ecol Econ 146:115–124. https://doi.org/10.1016/j.ecolecon.2017.10.003

Prosser JI (2007) Microorganisms cycling soil nutrients and their diversity. In: Van Elsas JD, Jansson JK, Trevors JT (eds) Modern soil microbiology. CRC Press, New York, pp 237–261

Rajendhran J, Gunasekaran P (2008) Strategies for accessing soil metagenome for desired applications. Biotechnol Adv 26:576–590. https://doi.org/10.1016/j.biotechadv.2008.08.002

Ram K, Meena RS (2014) Evaluation of pearl millet and mungbean intercropping systems in arid region of Rajasthan (India). Bangladesh J Bot 43:367–370

Ranger J, Nys C, Robert M (1992) Intérêt de l'implantation de minéraux–test dans les sols pour caractériser le fonctionnement actuel des sols. Science du Sol 30:193–214

Rapparini F, Peñuelas J (2014) Mycorrhizal fungi to alleviate drought stress on plant growth. In: Miransari M (ed) Use of microbes for the alleviation of soil stresses, volume 1. Springer, New York, pp 21–42. https://doi.org/10.1007/978-1-4614-9466-9_2

Redecker D, Schüßler A, Stockinger H, Stürmer SL, Morton JB, Walker C (2013) An evidence-based consensus for the classification of arbuscular mycorrhizal fungi (Glomeromycota). Mycorrhiza 23:515–531. https://doi.org/10.1007/s00572-013-0486-y

Reed CA (2013) Myths about the proton. The nature of H^+ in condensed media. Acc Chem Res 46:2567–2575

Redford AJ, Bowers RM, Knight R, Linhart Y, Fierer N (2010) The ecology of the phyllosphere: geographic and phylogenetic variability in the distribution of bacteria on tree leaves. Environ Microbiol 12:2885–2893

Reid DS (1980) Water activity as a the criterion of water availability. In: Ellwood DC, Latham MJ, Hedger JN, Lynch JM (eds) Contemporary microbial ecology. Academic, London, pp 15–28

Remigi P, Zhu J, Young JPW, Masson-Boivin C (2016) Symbiosis within symbiosis: evolving nitrogen-fixing legume symbionts. Trends Microbiol 24:63–75. https://doi.org/10.1016/j.tim.2015.10.007

Rey A (2015) Mind the gap: non-biological processes contributing to soil CO_2 efflux. Glob Chang Biol 21:1752–1761. https://doi.org/10.1111/gcb.12821

Richards LA (1965) Physical condition of water in soil. In: Black CA (ed) Methods of soil analysis, part 1, physical and mineralogical properties, Agronomy monograph series, vol 9. ASA, Madison, pp 128–152

Rillig MC, Steinberg PD (2002) Glomalin production by an arbuscular mycorrhizal fungus: a mechanism of habitat modification? Soil Biol Biochem 34:1371–1374. https://doi.org/10.1016/S0038-0717(02)00060-3

Rogers ED, Monaenkova D, Mijar M, Nori A, Goldman DI, Benfey PN (2016) X-ray computed tomography reveals the response of root system architecture to soil texture. Plant Physiol 171:2028–2040. https://doi.org/10.1104/pp.16.00397

Roland M, Vicca S, Bahn M, Ladreiter-Knauss T, Schmitt M, Janssens IA (2015) Importance of nondiffusive transport for soil CO_2 efflux in a temperate mountain grassland. J Geophys Res Biogeosci 120:502–512. https://doi.org/10.1002/2014JG002788

Rolston DE, Moldrup P (2002) Gas diffusivity. In: Dane JH, Topp GC (eds) Methods of soil analysis. Part 4, SSSA book series, vol 5. ASA and SSSA, Madison, pp 1113–1139

Römbke J, Gardi C, Creamer R, Miko L (2016) Soil biodiversity data: actual and potential use in European and national legislation. Appl Soil Ecol 97:125–133. https://doi.org/10.1016/j.apsoil.2015.07.003

Rosenberg E, Zilber-Rosenberg I (2014) The hologenome concept: human, animal and plant microbiota. Springer, Heidelberg

Rosenberg E, Zilber-Rosenberg I (2016) Microbes drive evolution of animals and plants: the hologenome concept. MBio 7:e01395l5. https://doi.org/10.1128/mBio.01395-15

Rousk J, Bengtson P (2014) Microbial regulation of global biogeochemical cycles. Front Microbiol 5:103. https://doi.org/10.3389/fmicb.2014.00103

Saleem M (2015) Microbiome ecosystem ecology: unseen majority in an anthropogenic ecosystem. In: Microbiome community ecology. SpringerBriefs in Ecology. Springer, Cham, pp 1–11. https://doi.org/10.1007/978-3-319-11665-5_1

Sánchez-García M, Roig A, Sánchez-Monedero MA, Cayuela ML (2014) Biochar increases soil N2O emissions produced by nitrification-mediated pathways. Front Environ Sci 225. https://doi.org/10.3389/fenvs.2014.00025

Scharlemann JPW, VJ Tanner EVJ, Hiederer R, Kapos V (2014) Global soil carbon: understanding and managing the largest terrestrial carbon pool. Carbon Manage 5:81–91. https://doi.org/10.4155/cmt.13.77

Schiraldi A, Fessas D, Signorelli M (2012) Water activity in biological systems – a review. Polish J Food Nut Sci 62:5–13. https://doi.org/10.2478/v10222-011-0033-5

Schlesinger WH (1999) Carbon sequestration in soils. Science 284:2095. https://doi.org/10.1126/science.284.5423.2095

Schlüter S, Vogel H-J (2016) Analysis of soil structure turnover with garnet particles and X-ray microtomography. PLoS One. https://doi.org/10.1371/journal.pone.0159948

Schnitzer M (1990) Selected methods for the characterization of soil humic susbstances. In: MacCarthy P, Clapp CE, Malcolm RL, Bloom PR (eds) Soil and crops sciences-selected readings. Soc Agron Soil Sci Am, Madison, pp 65–89

Schulz H, Dunst G, Glaser B (2013) Positive effects of composted biochar on plant growth and soil fertility. Agron Sustain Dev 33:817–827. https://doi.org/10.1007/s13593-013-0150-0

Schüßler A, Schwarzott D, Walker C (2001) A new fungal phylum, the Glomeromycota: phylogeny and evolution. Mycol Res 105:1413–1421. https://doi.org/10.1017/S0953756201005196

Seo DC, DeLaune RD (2010) Effect of redox conditions on bacterial and fungal biomass and carbon dioxide production in Louisiana coastal swamp forest sediment. Sci Total Environ 408:3623–3631. https://doi.org/10.1016/j.scitotenv.2010.04.043

Shaw MM, Pautasso M (2014) Networks and plant disease management: concepts and applications. Annu Rev Phytopathol 52:477–493. https://doi.org/10.1146/annurev-phyto-102313-050229

Sidhu C, Vikram S, Pinnaka AK (2017) Unraveling the microbial interactions and metabolic potentials in pre- and post-treated sludge from a wastewater treatment plant using metagenomic studies. Front Microbiol 8:1382. https://doi.org/10.3389/fmicb.2017.01382

Sierra CA (2012) Temperature sensitivity of organic matter decomposition in the Arrhenius equation: some theoretical considerations. Biogeochemistry 108:1–15. https://doi.org/10.1007/s10533-011-9596-9

Silva SR, Silva IR, Barros NF, de Sá Mendonça E (2011) Effect of compaction on microbial activity and carbon and nitrogen transformations in two oxisols with different mineralogy. R Bras Ci Solo 35:1141–1149. https://doi.org/10.1590/S0100-06832011000400007

Singh SJ (2015) Microbes: the chief ecological engineers in reinstating equilibrium in degraded ecosystems. Agric Ecosyst Environ 203:80–82. https://doi.org/10.1016/j.agee.2015.01.026

Skyllberg U, Raulund-Rasmussen K, Borggaard OK (2001) pH buffering in acidic soils developed under *Picea abies* and *Quercus robur*—effects of soil organic matter, adsorbed cations and soil solution ionic strength. Biogeochemistry 56:51–74. https://doi.org/10.1023/A:1011988613449

Smith SE, Read DJ (2008) Mycorrhizal symbiosis, 3rd edn. Academic, London

Smith SE, Smith FA (2012) Fresh perspectives on the roles of arbuscular mycorrhizal fungi in plant nutrition and growth. Mycologia 104:1–13. https://doi.org/10.3852/11-229

Smith SE, Jakobsen I, Grønlund M, Smith FA (2011) Roles of arbuscular mycorrhizas in plant phosphorus nutrition: interactions between pathways of phosphorus uptake in arbuscular mycorrhizal roots have important implications for understanding and manipulating plant phosphorus acquisition. Plant Physiol 156:1050–1057. https://doi.org/10.1104/pp.111.174581

Smith P, Haberl H, Popp A, Erb KH, Lauk C, Harper R, Tubiello FN, de Siqueira Pinto A, Jafari M, Sohi S, Masera O, Böttcher H, Berndes G, Bustamante M, Ahammad H, Clark H, Dong H, Elsiddig EA, Mbow C, Ravindranath NH, Rice CW, Robledo Abad C, Romanovskaya A, Sperling F, Herrero M, House JI, Rose S (2013) How much land-based greenhouse gas mitigation can be achieved without compromising food security and environmental goals? Glob Chang Biol 19:2285–2302. https://doi.org/10.1111/gcb.12160

Solaiman ZM (2014) Contribution of arbuscular mycorrhizal fungi to soil carbon sequestration. In: Solaiman Z, Abbott L, Varma A (eds) Mycorrhizal fungi: use in sustainable agriculture and land restoration. Soil Biology, vol 41. Springer, Berlin/Heidelberg, pp 287–296. https://doi.org/10.1007/978-3-662-45370-4_18

Solaiman ZM, Mickan B (2014) Use of mycorrhiza in sustainable agriculture and land restoration. In: Solaiman Z, Abbott L, Varma A (eds) Mycorrhizal fungi: use in sustainable agriculture and land restoration, Soil biology, vol 41. Springer, Berlin/Heidelberg, pp 1–15. https://doi.org/10.1007/978-3-662-45370-4_1

Speratti AB, Johnson MS, Sousa HM, Torres GN, Couto EG (2017) Impact of different agricultural waste biochars on maize biomass and soil water content in a Brazilian Cerrado Arenosol. Agronomy 7:49–68. https://doi.org/10.3390/agronomy7030049

Spokas KA, Cantrell KB, Novak JM, Archer DW, Ippolito JA, Collins HP, Boateng AA, Lima IM, Lamb MC, McAloon AJ, Lentz RD, Nichols KZ (2012) Biochar: a synthesis of its agronomic impact beyond carbon sequestration. J Environ Qual 41:973–989. https://doi.org/10.2134/jeq2011.0069

Standing D, Killham K (2007) The soil environment. In: van Elsas JD, Trevors JT, Jansson JK (eds) Modern soil microbiology, 2nd edn. CRC Press, Boca Raton, pp 1–22

Steele JA, Countway PD, Xia L, Vigil PD, Beman JM, Kim DY, Chow CE, Sachdeva R, Jones AC, Schwalbach MS, Rose JM, Hewson I, Patel A, Sun F, Caron DA, Fuhrman JA (2011) Marine bacterial, archaeal and protistan association networks reveal ecological linkages. ISME J 5:1414–1425. https://doi.org/10.1038/ismej.2011.24

Stein LY, Nicol GW (2011) Grand challenges in terrestrial microbiology. Front Microbiol 2:6. https://doi.org/10.3389/fmicb.2011.00006
Stevanović M, Popp A, Bodirsky BL, Humpenöder F, Müller C, Weindl I, Dietrich JP, Lotze-Campen H, Kreidenweis U, Rolinski S, Biewald A, Wang X (2017) Mitigation strategies for greenhouse gas emissions from agriculture and land-use change: consequences for food prices. Environ Sci Technol 51:365–374. https://doi.org/10.1021/acs.est.6b04291
Steven F (2010) Unlocking the global warming toolbox: key choices for carbon restriction and sequestration. PennWell Corp, Oklahoma
Stiles CA, Mora CI, Driese SG (2003) Pedogenic processes and domain boundaries in a vertisol climosequence: evidence from titanium and zirconium distribution and morphology. Geoderma 116:279–299. https://doi.org/10.1016/S0016-7061(03)00105-8
Stotzky G (1972) Activity, ecology, and population dynamics of microorganisms in the soil. CRC Crit Ver Microbiol 2:59–137
Stotzky G (1997) Soil as an environment for microbial life. In: Van Elsas J, Trevors J, Wellington E (eds) Modern soil microbiology. Marcel Dekker, New York, pp 1–20
Summers SK, Rainey R, Kaur M, Graham JP (2015) CO_2 and H_2O: understanding different stakeholder perspectives on the use of carbon credits to finance household water treatment projects. PLoS One. https://doi.org/10.1371/journal.pone.0122894
Tan KH (1982) Principles of soil chemistry. Marcel Dekker, New York, pp 179–194
Tan KH (2011) Principles of soil chemistry, 4th edn. CRC Press/Taylor and Francis, Boca Raton/Broken Sound Parkway, pp 246–266
Tecon R, Or D (2017) Biophysical processes supporting the diversity of microbial life in soil. FEMS Microbiol Rev 41:599–623. https://doi.org/10.1093/femsre/fux039
Tedersoo L, Bahram M, Põlme S, Kõljalg U, Yorou NS, Wijesundera R, Villarreal Ruiz L, Vasco-Palacios AM, Thu PQ, Suija A, Smith ME, Sharp C, Saluveer E, Saitta A, Rosas M, Riit T, Ratkowsky D, Pritsch K, Põldmaa K, Piepenbring M, Phosri C, Peterson M, Parts K, Pärtel K, Otsing E, Nouhra E, Njouonkou AL, Nilsson RH, Morgado LN, Mayor J, May TW, Majuakim L, Lodge DJ, Lee SS, Larsson KH, Kohout P, Hosaka K, Hiiesalu I, Henkel TW, Harend H, Guo LD, Greslebin A, Grelet G, Geml J, Gates G, Dunstan W, Dunk C, Drenkhan R, Dearnaley J, De Kesel A, Dang T, Chen X, Buegger F, Brearley FQ, Bonito G, Anslan S, Abell S, Abarenkov K (2014) Global diversity and geography of soil fungi. Science 346:1256688. https://doi.org/10.1126/science.1256688
Thornton PK, Ericksen PJ, Herrero M, Challinor AJ (2014) Climate variability and vulnerability to climate change: a review. Glob Chang Biol 20:3313–3328. https://doi.org/10.1111/gcb.12581
Tian H, Lu C, Yang J, Banger K, Huntzinger DN, Schwalm CR, Michalak AM, Cook R, Ciais P, Hayes D, Huang M, Ito A, Jain AK, Lei H, Mao J, Pan S, Wilfred MP, Peng S, Poulter B, Schaefer K, Shi X, Tao B, Wang W, Wei Y, Yang Q, Zhang B, Zeng N (2015) Global patterns and controls of soil organic carbon dynamics as simulated by multiple terrestrial biosphere models: current status and future directions. Global Biogeochem 29:775–792. https://doi.org/10.1002/2014GB005021
Timmusk S, Behers L, Muthoni J, Muraya A, Aronsson A-C (2017) Perspectives and challenges of microbial application for crop improvement. Front Plant Sci 8:49. https://doi.org/10.3389/fpls.2017.00049
Tisdale SL, Nelson WL, Beaton JD (1985) Soil fertility and fertilizers, 4th edn. Macmillian Publishers, New York
Tokarz E, Urban D (2015) Soil redox potential and its impact on microorganisms and plants of wetlands. J Ecol Eng 16:20–30. https://doi.org/10.12911/22998993/2801
Trivedi P, Delgado-Baquerizo M, Anderson IC, Singh BK (2016) Response of soil properties and microbial communities to agriculture: implications for primary productivity and soil health indicators. Front Plant Sci 7:990. https://doi.org/10.3389/fpls.2016.00990
Tucker M (1997) Climate change and the insurance industry: the cost of increased risk and the impetus for action. Ecol Econ 22:85–96. https://doi.org/10.1016/S0921-8009(96)00556-3

Ucar D, Zhang Y, Angelidaki I (2017) An overview of electron acceptors in microbial fuel cells. Front Microbiol 8:643. https://doi.org/10.3389/fmicb.2017.00643

Udvardi M, Poole PS (2013) Transport and metabolism in legume-rhizobia symbioses. Annu Rev Plant Biol 64:781–805. https://doi.org/10.1146/annurev-arplant-050312-120235

Uhlirová E, Elhottova D, Triska J, Santruckova H (2005) Physiology and microbial community structure in soil at extreme water content. Folia Microbiol 50:161–166. https://doi.org/10.1007/BF02931466

van der Putten WH (2012) Climate change, aboveground- belowground interactions and species range shifts. Annu Rev Ecol Evol Syst 43:365–383. https://doi.org/10.1146/annurev-ecolsys-110411-160423

van Zwieten L, Kimber S, Morris S, Chan KY, Downie A, Rust J, Joseph S, Cowie A (2010) Effects of biochar from slow pyrolysis of papermill waste on agronomic performance and soil fertility. Plant Soil 27:235–246. https://doi.org/10.1007/s11104-009-0050-x

Vandenkoornhuyse P, Quaiser A, Duhamel M, Le Van A, Dufresne A (2015) The importance of the microbiome of the plant holobiont. New Phytol 206:1996–1206. https://doi.org/10.1111/nph.13312

Vaz CMP, De Maria IC, Lasso PO, Tuller M (2011) Evaluation of an advanced benchtop micro-computed tomography system for quantifying porosities and pore-size distributions of two brazilian oxisols. Soil Sci Soc Am J 75:832–841. https://doi.org/10.2136/sssaj2010.0245

Verma SK, Singh SB, Prasad SK, Meena RN, Meena RS (2015) Influence of irrigation regimes and weed management practices on water use and nutrient uptake in wheat (*Triticum aestivum* L. emend. Fiori and Paol.). Bangladesh. J Bot 44:437–442. https://doi.org/10.3329/bjb.v44i3.38551

von Lützow M, Kögel-Knabner I (2009) Temperature sensitivity of soil organic matter decomposition: what do we know? Biol Fertil Soils 46:1–15. https://doi.org/10.1007/s00374-009-0413-8

Vorholt JA (2012) Microbial life in the phyllosphere. Nat Rev Microbiol 10:828–840. https://doi.org/10.1038/nrmicro2910

Voroney RP, Heck RJ (2015) The soil habitat. In: Paul EA (ed) Soil microbiology, ecology and biochemistry, 4th edn. Academic, San Diego, pp 15–39

Vos M, Wolf AB, Jennings SJ, George A, Kowalchuk GA (2013) Micro-scale determinants of bacterial diversity in soil. FEMS Microbiol Rev 37:936–954. https://doi.org/10.1111/1574-6976.12023

Walder F, Niemann H, Natarajan M, Lehmann MF, Boller T, Wiemken A (2012) Mycorrhizal networks: common goods of plants shared under unequal terms of trade. Plant Physiol 159:789–797. https://doi.org/10.1104/pp.112.195727

Walder F, Boller T, Wiemken A, Courty PE (2016) Regulation of plants' phosphate uptake in common mycorrhizal networks: role of intraradical fungal phosphate transporters. Plant Signal Behav. https://doi.org/10.1080/15592324.2015.1131372

Wang W, Dalal R (2006) Carbon inventory for a cereal cropping system under contrasting tillage, nitrogen fertilisation and stubble management practices. Soil Till Res 91:68–74. https://doi.org/10.1016/j.still.2005.11.005

Wang Q, He T, Liu J (2016) Litter input decreased the response of soil organic matter decomposition to warming in two subtropical forest soils. Sci Rep 6:33814. https://doi.org/10.1038/srep33814

Weber APM, Horst RJ, Barbier GG, Oesterhelt C (2007) Metabolism and metabolomics of eukaryotes living under extreme conditions. Int Rev Cytol 256:1–34

Weil RR, Brady NC (2016) The nature and properties of soils, 15th edn. Pearson Education, London

Wilding LP (1994) Factors of soil formation: contributions to pedology. In: Amundson R, Harden J, Singer M (eds) Factors of soil formation: a fiftieth anniversary retrospective, vol 33. Soil Sci Soc Am Special Publ, Madison, pp 15–30

Williams RJ, Howe A, Hofmocke lKS (2014) Demonstrating micro- bial co-occurrence pattern analyses within and between ecosystems. Front Microbiol 5:358. https://doi.org/10.3389/fmicb.2014.00358

Wisz MS, Pottier J, Kissling WD, Pellissier L, Lenoir J, Damgaard CF, Dormann CF, Forchhammer MC, Grytnes JA, Guisan A, Heikkinen RK, Høye TT, Kühn I, Luoto M, Maiorano L, Nilsson MC, Normand S, Öckinger E, Schmidt NM, Termansen M, Timmermann A, Wardle DA, Aastrup P, Svenning JC (2013) The role of biotic interactions in shaping distributions and realised assemblages of species: implications for species distribution modelling. Biol Rev Camb Philos Soc 88:15–30. https://doi.org/10.1111/j.1469-185X.2012.00235.x

World Bank (2017) Results-based climate finance in practice : delivering climate finance for low-carbon development (English). World Bank Group, Washington, DC. Accessed in http://documents.worldbank.org/curated/en/410371494873772578/Results-based-climate-finance-in-practice-delivering-climate-finance-for-low-carbon-development

Wright JP, Jones CG (2006) The concept of organisms as ecosystem engineers ten years on progress, limitations, and challenges. Bioscience 56:203–209

Wright SF, Upadhyaya A (1998) A survey of soils for aggregate stability and glomalin, a glycoprotein produced by hyphae of arbuscular mycorrhizal fungi. Plant Soil 198:97–107. https://doi.org/10.1023/A:1004347701584

Yadav NA (2017) Agriculturally important micro biomes: biodiversity and multifarious PGP attributes for amelioration of diverse abiotic stresses in crops for sustainable agriculture. Biomed J Sci Tech Res 4:1–4. https://doi.org/10.26717/BJSTR.2017.01.000321

Yadav GS, Das A, Lal R, Babu S, Meena RS, Patil SB, Saha P, Datta M (2018) Conservation tillage and mulching effects on the adaptive capacity of direct-seeded upland rice (*Oryza sativa* L.) to alleviate weed and moisture stresses in the North Eastern Himalayan region of India. Arch Agron Soil Sci. 64:1254–1267. https://doi.org/10.1080/03650340.2018.1423555

Yarrow D (2015) Geology into biology: carbon, minerals, and microbes-tools to remineralize soil, sequester carbon, and restore the earth. In: Goreau TJ, Larson RW, Campe J (eds) Geotherapy: innovative methods of soil fertility restoration, carbon sequestration, and reversing CO_2 increase. CRC Press/Taylor and Francis Group, Boca Raton, pp 195–223

Yashiro Y, Lee NY, Ohtsuka T, Shizu Y, Saitoh TM, Koizumi H (2010) Biometric-based estimation of net ecosystem production in a mature Japanese cedar (*Cryptomeria japonica*) plantation beneath a flux tower. J Plant Res 123:463–472. https://doi.org/10.1007/s10265-010-0323-8

Zhang J, Zhou JM (2010) Plant immunity triggered by microbial molecular signatures. Mol Plant 3:783–793. https://doi.org/10.1093/mp/ssq035

Zhang Y, Zhao YC, Shi XZ, Lu XX, Yu DS, Wang HJ, Sun WX, Dailek JL (2008) Variation of soil organic carbon estimates in mountain regions: a case study from Southwest China. Geoderma 146:449–456. https://doi.org/10.1016/j.geoderma.2008.06.015

Zhi J, Jing C, Lin S, Zhang C, Liu Q, DeGloria SD, Wu J (2014) Estimating soil organic carbon stocks and spatial patterns with statistical and gis-based methods. PLoS One. https://doi.org/10.1371/journal.pone.0097757

Zhou T, Shi P, Hui D, Luo Y (2009) Global pattern of temperature sensitivity of soil heterotrophic respiration (Q10) and its implications for carbon-climate feedback. J Geophys Res. 114. https://doi.org/10.1029/2008JG000850

Zhuang X, Gao J, Ma A, Fu S, Zhuang G (2013) Bioactive molecules in soil ecosystems: masters of the underground. Int J Mol Sci 14:8841–8868. https://doi.org/10.3390/ijms14058841

Zomer RJ, Bossio DA, Sommer R, Verchot LV (2017) Global sequestration potential of increased organic carbon in cropland soils. Sci Rep 7:15554–15558. https://doi.org/10.1038/s41598-017-15794-8DO

Zonta JH, Martinez MA, Pruski FF, Silva DD, Santos MR (2012) Effect of successive rainfall with different patterns on soil water infiltration rate. Rev Bras Ci Solo 36:377–388. https://doi.org/10.1590/S0100-06832012000200007

Zuccaro A, Lahrmann U, Langen G (2014) Broad compatibility in fungal root symbioses. Curr Opin Plant Biol 20:135–145. https://doi.org/10.1016/j.pbi.2014.05.013

Sustainable C and N Management Under Metal-Contaminated Soils

Vasudev Meena, Mohan Lal Dotaniya, Jayanta Kumar Saha, Bharat Prakash Meena, Hiranmoy Das, Beena, and Ashok Kumar Patra

Abstract Across the world, a major challenge is deteriorating environmental health by increasing growth of industries with the unscientific management of industrial waste. Soil contamination with organic and inorganic pollutant is a major task during the production of healthy food. In the last three decades, the concentration of heavy metals in soil has increased drastically, posing a risk to the whole environment, human, as well as animal health. Soil contamination is a threat to sustainable agricultural development and food security in developing countries. Nowadays protection and preservation of the environment from further deterioration have drawn increasing research attention. In the present context, use of modern and traditional technologies aims to maintain the health of natural resources from contamination at economic feasibility. Another major concern is remediation or minimization of toxic metal entry in the food chain contamination of different ecosystems without affecting their functionality. There is a need to make land resources free from metal contamination for healthy and safe agricultural production, to increase food security, and to maintain land use pattern. Advanced remediation techniques are more focusing on in situ environment-friendly practices. Several organic and inorganic remediation technologies to treat heavy metal-contaminated soils are discussed in this chapter.

Keywords Crop production · Carbon management · Contaminated soil · Heavy metals · Inorganic material · Nitrogen management · Organic material · Soil amendments

V. Meena (✉) · M. L. Dotaniya · J. K. Saha · B. P. Meena · H. Das · A. K. Patra
ICAR-Indian Institute of Soil Science, Bhopal, Madhya Pradesh, India
e-mail: Vasudev.Meena@icar.gov.in; Mohan.Dotaniya@icar.gov.in; Jayant.Saha@icar.gov.in; Bharat.Meena@icar.gov.in; Hiranmoy.Das@icar.gov.in; Ashok.Patra@icar.gov.in

Beena
Department of Education, Banasthali Vidhyapith, Niwai (Tonk), Rajasthan, India

R. Datta et al. (eds.), *Carbon and Nitrogen Cycling in Soil*,
https://doi.org/10.1007/978-981-13-7264-3_9

Abbreviations

BECCS	Bioenergy with carbon capture and storage
C	Carbon
DNDC	Denitrification decomposition
GHGs	Greenhouse gases
GSH	Glutathione
Gt	Gega tons
HMs	Heavy metals
MSW	Municipal solid waste
N	Nitrogen
PC	Phytochelatins
Pg	Picogram
ppb	Parts per billion
ppm	Parts per million
PSB	Phosphorus solubilizing bacteria
SOC	Soil organic carbon
SOM	Soil organic matter
USPEA	United States Environmental Protection Agency

1 Introduction

Agricultural crop production system is mainly affected by the soil fertility, climatic events, genetic potential of a cultivar, and management. Among all, soil parameters play a crucial role and influence the crop yield productivity in a region. The essential plant nutrients like carbon (C) and nitrogen (N) are governing a vital role in biomass production and ultimately in crop yield production. Over the research period, many research experiments were conducted at the different agroclimatic zone to compute the N and C dynamics in different soil orders. The balance amount of N in soil enhances the metabolic activities of plant roots and synergistic effect on the soil properties (Dotaniya et al. 2016c). The C element in plant acts as a structural nutrient and the primary platform for microbial reaction in the soils. The reaction kinetics and balance of both the elements in the soil are controlled by each other. Plants absorb carbon dioxide (CO_2) from the air through their leaves and take part in photosynthesis. The CO_2 diffuses through small holes in the underside of the leaf called stomata. Nitrogen is quite unreactive with other elements, and it can't be used directly as a nutrient by plants and animals in the way oxygen or C can. In a process known as N fixation, special bacteria can convert elemental N from the atmosphere into ammonia in the plant parts, which is an essential mechanism in the plants.

Heavy metals (HMs) are toxic and ubiquitous in nature and create a severe threat to the human beings on the planet Earth. Usually metals are produced by the geogenic source, but due to the fast industrial growth and unscientific management of anthropogenic waste have enhanced their area of reaction (or are enhanced by their area of reaction). These metals are highly toxic even in meager amount and resistant to decomposition (Hu and Cheng 2013; Odukudu et al. 2014; Li et al. 2015a, b). Heavy metals and metalloids reach at natural bodies by means of irrigation in agriculture or by dumping in the outskirt. It affects the plant growth production and ecological system productivity (Duan et al. 2015; Wu et al. 2016; Li et al. 2016; Datta et al. 2017a). In natural systems, a few HMs are more common and having toxicity in the living systems, i.e., arsenic-As, cadmium-Cd, chromium-Cr, nickel-Ni, selenium-Se, lead-Pb, zinc-Zn, copper-Cu, mercury-Hg, aluminum-Al, etc. These metals reach human bodies via food chain contamination and malfunction with the biological process (Laidlaw and Filippelli 2008; Li et al. 2011; Okorie et al. 2011). Short period intake of metals may cause different diseases like a mental disorder, nerve system failure, and poor growth, and long period intake can cause cancer (Gong et al. 2010; Li et al. 2015a, b; Liang et al. 2017). Toxicity of metals in the soil also affects the soil microbial activity and diversity and restricts the plant nutrient mineralization rate and their uptake toward plants. Some HMs have a similar structure as essential plant nutrient and reach to plant metabolic process and retard the growth of plants. The persistence of HMs in an ecosystem is also affected by the type, toxicity, nature, and interaction with soil components. The bioavailability of the metal ions in the soil-plant system is mainly affected by the amount of organic matter present in the soil and its pH value (Thornton et al. 2008). Most of the metal availability increase with decreasing the soil pH, and the toxicity is reported under acidic soil conditions. Changing soil pH from alkaline to acidic range enhances the bioavailability of HMs and more chances to leach down in lower profile. Some of the HMs (Cu, Mn, and Zn) are necessary for the proper functioning of the plant, animal, and human system, but higher concentration poses a threat to survival. The elevated concentration of HMs under acidic condition may adsorb on organic matter or soil particles and reduce the toxicity potential of a metal. It also forms different metal-humus complexes and affects the HM availability to crop plants. However, anthropogenic activities enhance the metal distribution in non-contaminated areas mostly through waste dumping, sewage irrigation, emission through fossil fuels, poor agricultural inputs, and industrial waste disposal (Facchinelli et al. 2001; Li et al. 2013). The toxicity potential, persistence in the ecosystem, and biodegradation categorize different metals in the various classes as per the international organization, the US Environmental Protection Agency (USEPA), and it is educating the peoples about the harmful effects of HMs across the globe (Rodrigues et al. 2013).

2 Why Sustainable C and N Management

Nitrogen-containing fertilizers play a key to enhance the crop yield and improve the soil health. Higher net crop production and quantity of residues which get returned to soil due to N fertilization contribute to SOC stock (Dotaniya 2018; Dotaniya et al. 2013a). Acting as a storehouse of nutrients, SOC enhances the soil fertility, and its depletion can be a major cause of yield stagnation under intensive cultivation (Dotaniya and Datta 2015; Meena et al. 2016). Soil organic matter is also the substrate for soil microbes and, hence, significantly affects the C turnover in the soil (Jat et al. 2018). Therefore, field management practices mainly influence the C fixation capacity of the soil. Thus, appropriate management of C and N is the key to good soil health and sustainable crop production.

2.1 Role of C and N in Crop Production

2.1.1 Role of Carbon

Plants' consumption of CO_2 in the form of carbon during photosynthesis is a chemical process in which organisms use sunlight to convert CO_2 into a chemical carbohydrate molecule. Plants use this carbohydrate to grow and transfer sink to source as crop yield. Aquatic plants also generate oxygen, but they use CO_2 from water. The higher concentration of atmospheric CO_2 increases crop yield by increasing photosynthesis. The role of C in plants is to foster a healthier and more productive growth of the plants. Carbon and plant growth are then intrinsically linked.

2.1.2 Role of Nitrogen

Plant metabolism requires minerals from the soil to form complex molecules. The activity and healthiness of the plant or plant parts depend on the availability of essential nutrients like nitrogen in the form of nitrate ions. Nitrogen enhances plant growth and development. It is known as the structural backbone of green plants. It is required in the large quantity and plays important functions among the essential nutrients. It is the main constituent of all the amino acids which are the building blocks of plant protein, chlorophyll, plant tissues, cell membrane, etc. Nitrogen is the main constituent of nucleic acid that forms DNA, a genetic material. It also holds the genetic code in the plant nucleus.

2.2 C and N Cycle in Ecosystem

2.2.1 Carbon Cycle

The founder of the carbon cycle was Joseph Priestley and Antoine Lavoisier. Later on, it was popularized by Humphry Davy. In this process C exchanges among the biosphere, pedosphere, geosphere, hydrosphere, and atmosphere of the Earth and is mediated by different biogeochemical reactions. Carbon is a main element of the C compounds in the Universe. It is the main constituent of biological compounds and significant constituent of limestone. Recycling and reuse of the C throughout the biosphere and its sequestration by long-term processes as C sinks is known as carbon cycling. The C exchanges among the different ecosystems governed by the various physicochemical and biological processes. In terms of C storage as a form of active pool, the area of sea near to terrestrial land contains the largest volume of C.

2.2.2 Nitrogen Cycle

Like the carbon cycle, it is also a biogeochemical process where nitrogen gets converted into various forms. Within the biosphere, it moves through the atmosphere, hydrosphere, and lithosphere. The N cycle involves, i.e., N fixation by chemically or biologically process, the release of N from different inputs, leaching, volatilization, nitrification, denitrification, and plant uptake. Application of N fertilizers in the soil during the crop production converts in different N fractions and is further uptake by the plant roots. It is found that a minor amount of this application reached in groundwater by leaching in the atmosphere as volatilization and denitrification process and immobilized as NH_4^+in silicate minerals (Dotaniya et al. 2014b; Buragohain et al. 2017). The legume plant has the potential to fix the atmospheric N into the soil through root nodules. The symbiotic relationship between plant roots and N-fixing bacteria enhances the soil fertility. Microalgae also perform such types of association in nature in the aquatic ecosystem. In the whole process, two bacteria perform mainly: (1) *Nitrosomonas* convert NH_4^+ to NO_2^-; and (2) *Nitrobacter* convert NO_2^- to NO_3^-. After this process fixed N by the biological process is reached to the soil by degradation of plant roots or microbes. Simultaneously denitrification process goes on, and a portion of N fixation reaches the atmosphere with the help of denitrification bacteria.

2.3 *Advanced Techniques Related to C and N Recycling*

2.3.1 Carbon Sequestration

Carbon sequestration refers to a phenomenon when atmospheric CO_2 gets captured and stored in long-lived C pools in the different ecosystems. During C sequestration different physicochemical and biochemical processes are involved. The research finding suggests that a few methods enhance the C sequestration potential of an ecosystem. These are the following:

- Removing and capturing C from the atmospheric gases and fixed into long-lived pools
- Maintaining and minimizing the weathering biochemical reaction between atmospheric CO_2 and the rocks

Phytoplankton It is another approach to enhance natural oceanic uptake of CO_2 by microscopic plants. The theory is that fertilizing the world's oceans with iron would increase biological productivity and enable phytoplankton near the sea surface to absorb more CO_2, thereby removing it from the atmosphere and storing it.

2.3.1.1 Biological Processes

Carbon sequestration through biological processes is also called biosequestration. These processes adversely affect the global carbon cycle. After biochemical reaction manipulation, it gets converted into various C-containing materials like clathrate, limestone, and silicate minerals.

2.3.1.2 Peat Production

Peat bogs are the sources which accumulate partially decomposed biomass and act as a crucial carbon store. There is a discrepancy on peatlands which perform both C source and sink and are much affected by the local climatic conditions. Carbon sequestration can be enhanced by the modification in the properties of peat or change in its management strategies.

2.3.1.3 Forestry

Forestry is a key pillar of the Earth's ecosystem and has a potential to reduce the atmospheric C and mitigate the global climate change effect. Plantation of green plants on waste or marginal land may give a good economic return and improve the environmental quality by capturing atmosphere C. Forest enhances the natural cycle of carbon and significantly changes the net carbon sink-source balances (Dotaniya et al. 2018a; Varma et al. 2017). A managed forest can function as a net sink of atmospheric CO_2. If 100 million ha of land is afforested, it can lock 25–50 Gt of CO_2 in the atmosphere. In India, about 175 million ha of wastelands are suitable and potential for growing trees without affecting agricultural productivity. Hence, afforestation in India can offer greater scope in carbon sequestration. Forest plants have higher biomass production capacity in a short time span. The C recycling capacity of a natural ecosystem shows that it's again released of C from dead plant part to the atmosphere.

2.3.1.4 Urban Forestry

Most of the megacities are the mountain of concrete. Using fallow land or increasing the green areas is a way of sustainability; otherwise these areas will convert into gas chambers. It can be practiced and maintained on smaller as well as larger scales.

The results of urban forestry vary with the type and nature of vegetation. Further, it can function as a sink as well as a source of emission.

2.3.1.5 Wetland Restoration

It has huge potential to C sequestration and comprises 14.5% of global soil C. Across the world it has only 6% wetland area.

2.3.1.6 Agriculture

Agriculture crop production system holds vast potential for C sequestration and mitigates the adverse climate effect on the soil process and crop productivity. Intensive cropping system declines the SOC in the soil as compared to natural farming. Across the global world cleaning of forest land for agricultural practices has decreased SOC by 30–40% (Post and Kwon 2008). Apart from this, crop removal, crop management practices have also caused organic C depletion from the agriculture field and reduced the soil fertility and crop yield in succeeding crop. Additions of crop residue in the soil get stabilized after microbial decomposition and cause a significant portion of C loss to the environment (Dotaniya et al. 2014e). After a few months, added C equilibrium maintains the SOC status. Soil microbial population and diversity are greatly influenced by the amount of organic matter in the soil and its physiochemical properties (Dotaniya and Datta 2015). Different strategies, such as the addition of crop residue, microbial inputs, crop rotation, mixed cropping, and minimum tillage, maintain the SOC in the soil and balance fertilization and agronomic management options (Kundu et al. 2013; Kushwah et al. 2014; Meena et al. 2015a). Long duration or perennial crops consisting higher C in below ground portion are approximately 1500 Gt of organic carbon across the cultivated land up to 1 m of soil depth. Kumar (2015) has described the different agricultural practices to enhance the SOC stock in agricultural crop field to counter the climate change effect.

2.3.1.7 Deep Soil

Deeper soil with higher content of SOC in stabilized form has fewer chances of mineralization or loss. About 90% stored C is in the form of mineral-organic associations. Soils have approximately four times higher C than present in the atmosphere. The repeated use of sewage, irrigation enhances the dissolved organic C in the lower layer of the soil profile.

2.3.1.8 Reducing Emissions

Gas emission can be reduced by improving the nutrient inputs, using them efficiently, and maintaining the yield with minimum consumption of energy. The improved and recommended production practices should be followed with more precise use of nutrients, less tillage operation, efficient water management, and promising crop cultivars for increased yield (Yadav et al. 2017a, b, c; Datta et al. 2017b). More energy-efficient farming operations should be replaced with the less energy-efficient farming operation. Minimum tillage or zero tilth reduces the fossil fuel consumption as well as reduces the chances of organic C loss to the environment. Crop residue incorporation has fewer greenhouse gas emissions than residue burning.

2.3.1.9 Enhancing Carbon Removal

During the crop growth period, in general, all the crops absorb carbon dioxide and release carbon after the harvest of the crop. The removed agricultural carbon should sequester within the soil through the carbon cycle. This depends on the method of farming by recycling the crop biomass to the soil and increasing the plant nutrient concentration to the plant. This includes:

- Using cover crops, grasses, and bushes in fallow land
- Using deep-rooted crop with trap root system to take up nutrient from the deeper layer for profuse growth of the crops
- Protecting the soil to reduce the SOC losses from soil
- Enhancing the vegetation on degraded and marginal lands

2.3.2 Physical Processes

2.3.2.1 Bioenergy with Carbon Capture and Storage (BECCS)

Power stations and boilers mainly use biomass for energy production. In this process, atmospheric C is captured by the green plant as plant biomass and used for the energy generation to mitigate the electricity crisis at the industrial level. This process is widely known as bioenergy with carbon storage (BECS). These technologies are giving the platform to biomass use in an efficient way in most of the fields of human life. In the agricultural sector, the use of biochar is a classic example of BECS.

2.3.2.2 Burial

The processes of burying biomass into the soil for improving soil fertility or C sequestration are the common practices from the ancient period. Another common physical way of sequestration is landfilled in the urban areas.

2.3.2.3 Biochar Incorporation

Biochar refers to the process in which biomass of the crops, plants, and other woody material burns at limited oxygen condition. This process is widely known as pyrolysis. Such types of activities are also used for capturing the atmospheric CO_2 for minimizing the GHG effect. Landfilling of biochar or its use in the agricultural production system is a common practice in most of the countries. In long-term perspective it is fruitful technology for C sequestration. With the help of the biochar process, 9.5 Pg C annually is stored in the soil.

2.3.2.4 Organic Waste Recycling

This practice includes organic inputs like crop biomass, animal excreta, farm waste, sewage sludge, and municipal solid waste, decomposable organic waste coming from agro-industries, kitchen waste, and other related organic waste. The raw waste can be compost before application to the field to minimize the immobilization process in the soil during the growth. Rice industries generate a huge volume of rice husk containing 0.3%, 0.2%, and 0.3% of N, P, and K content, respectively. Organic waste is also important to reduce the salinity or alkalinity in the degraded soil. It is observed that the use of rick industrial waste improved the physical condition of saline degraded soils. Incorporation of sugarcane press mud and bagasse in the soil improved the N and P status of the soil and also enhanced the crop yield (Dotaniya sugarcane). It contained average 1.09% nitrogen in composted bagasses and press mud material when mean C:N was 20.1. NPK in cereal straw ranges 0.5% N, 0.6% P, and 1.5% K, respectively.

2.3.2.5 Vermicompost

Vermicompost is prepared from earthworm and has higher nutrient content than FYM. In common 1.6%, 4.0%, and 0.8% are N, P_2O_5, and K_2O, respectively. It also has a lower C:N ratio, and its addition in the soil improves the soil microbial properties. The passage of earthworm through the soil secret sticky substances is promoting bacterial and actinomycete growth. Earthworms consume 2–5 times more feedstock than their body weight. It is an appropriate technique for disposal of nontoxic solid and liquid organic wastes by reducing the time frame of composting.

2.3.2.6 Municipal and Sewage Wastes

It is one of the crucial components of organic wastes and hence considered a low price fertilizer because of its high NPKS content along with other micronutrients which are essential for a better crop performance. The total municipal refuse generation levels in the world are expected to double by 2025. In India, municipal and sewage waste is about 12 million tons per annum which have average nutrient contents: 0.5%, 0.3%, and 0.3% of N, P, and K, respectively. Sewage farming is very much predominant in the peri-urban belt of any country due to the cheap availability of water potential, organic matter, and plant nutrients. As per an estimate, one generation of sewage sludge in India is more than 4 million tons per year.

2.3.2.7 Biocultures

Several types of biocultures available in the global market have enhanced the microbial population in the soil by addition application during the crop production and mediated the soil fertility dynamics, such as PSB: *Bacillus*, *Pseudomonas*, *Aspergillus*, *Penicillium*, mycorrhiza, etc.

2.3.2.8 Green Manuring

Soil incorporation of any green manure crops in their green stage or soon after flower in wet condition is called as green manuring. The green manure may be forage or leguminous crops which are grown for their leafy materials to restore the soil fertility (Kumar et al. 2018). These are fast-growing plants, and when dug into the ground while still green, they add organic matter, N, or other valuable nutrients to the soil and improve the soil structure. The green manure crops like sun hemp, dhaincha, *Gliricidia*, *Sesbania*, alfalfa, cowpea, clover, etc. are more suitable in this context (Meena et al. 2018a, b, c).

2.3.2.9 Organic Farming

In India, organic farming is gaining popularity with the pace of time. It is a unique production system that restricts the use of chemical inputs like pesticides, fertilizers, growth products, etc. It allows utilization of crop residues, crop rotation, organic waste, farm manures, and rock additives for the nutrient mobilization and plant protection. It helps to improve the soil health by improving soil physicochemical properties. Organic faming improves the soil's properties, as it enhances the soil microbial biomass; minimizes the SOC loss, plant macro- and micronutrients, easy tillage operations, and nutrient losses into the water bodies and environment; and promotes favorable chemical reactions.

2.4 Measurement of C and N Pools

The distribution of SOM in different ecosystems can be categorized into different pools based on chemical function. Their reactions in the soil are the following:

- Structural litter fraction: These groups contain a higher amount of lignin and wider C:N ratio (>150:1) (Datta et al. 2017b). In this class mostly, woody plant parts are included.
  - Metabolic pool fraction: This group contains easily decomposable plant materials like leaves, flowers, as well as animal excreta. It is observed that during decomposition fraction of N and C loss occurred. It contained less C:N ratio (10–25:1) than the structural group.
  - Active pool: It is the most easily decomposable part of the soil C. It comprises soil microbial population biomass and their metabolites. Lower C:N ratio enhances the soil fertility easily without the plant nutrient immobilization. In this group, water-soluble fractions of organic matter and lower structural carbohydrates remain present in the soil.
- Slow decomposable class: This class of C compounds is mostly equilibrium with a natural system of decomposition having C:N ratio approximately 20:1. It is easily decomposable and remains present in organic compounds as a humus form.
- Passive fraction: It is a recalcitrant part of organic matter having very low C:N ratio (7–9:1). It is highly resistant to decomposition-oxidation reactions and doesn't take part in equilibrium C dynamics in the soil. Such types of fractions in the soil are found as an inert material without equilibrium relationship with the other fractions of the organic matter.

Depending upon its mineralization in the soil, SOC has been categorized into different pools. The easily mineralizable fractions are considered most labile and have a very low residence time in the soil. There are other fractions too that exist in the soil for a relatively long period and thus are termed as slow or resistant fractions or non-labile fractions.

The categorization of an active and passive pool of organic matter is necessary for the identification of the soil fertility status and their interpretation to calculate C mineralization dynamics in the soil. A key parameter in the soil health is to measure CO_2 evolution from the soil microbial biomass as an active pool, by the laboratory incubations methods. There are two methods used for determination of SOC: (1) combustion method (TOC analyzer) and (2) Walkley-Black method.

2.4.1 Resistant Carbon Pool Determination

In this method, the soil sample is hydrolyzed and *shaking* followed by centrifuging. This step is repeated for a number of times, and then residues are dried at 60 °C, and C in the residual soil is measured by TOC analyzer.

2.4.2 Active and Slow Carbon Pool Determination

The remaining two pools are estimated by double decomposition method. Identification of a particular pool calculates the CO_2 evolution rate for a particular time frame. In this method two-component first-order model is used as follows:

$$Ct = C_{\mathrm{a}}e^{-Ka\,t} + C_{\mathrm{s}}e^{-Ks\,t}$$

Ct is the total OC (sum of active + slow fractions) at a particular time *t*, C_a and C_s are the size fraction of active and slow C pools, and *K*a and *K*s are decay constant of active and slow pools, respectively.

$$\text{Slow} + \text{active C content} + \text{Total organic C} - \text{Resistant C}$$

Carbon is measured with the help of titrimetric method as described by Parr and Smith (1969). In this method, 10 ml of 2 M NaOH in a small vial is kept in a conical flask hung by a thread. Flask should be sealed with paraffin. The vials should be taken out on regular intervals (initially at 1–2 days interval for 15 days and then at the weekly interval) and titrated with 0.5MHCl after adding 1 ml of $BaCl_2$ using phenolphthalein indicator and amount of CO_2 calculated as:

$$\begin{aligned}&CO_2 - \text{C evolved}\left(\text{mg}\,\text{kg}^{-1}\right)\\&\quad = (\text{Final consumption of HCl} - \text{Initial consumption of HCl}) \times N \times 6\end{aligned}$$

N is normality of HCl.

2.4.3 Nitrogen Pools and Their Measurements

The inorganic forms (NH_4^+-N and NO_3^--N) are very important form for crop nutrition point of view because plant root takes up N from the soil mostly as NO_3^- and to some extent as NH_4^+-N. The NO_2^- form is unstable and is usually present in the soil to a lesser extent. However, heavy application of N fertilizers, anaerobic conditions of the soil, extreme pH values, salinity, and low temperature favor its accumulation in the soil. The aerobic soil, NO_3^-N, and the anaerobic soils, NH_4^+-N, are the major fraction of available N. Available N content (both NH_4^+ and NO_3^-) decreases with the soil depth.

2.4.4 Measurements of Available Nitrogen

The available N can be precisely determined by the Kjeldahl method both in the plant and the soil. This method involves three major steps, namely, (1) digestion to convert N into HNO_3, (2) distillation to capture ammonia formed during digestion, and (3) volumetric analysis to estimate NH_3. The apparatus required for the determination are Kjeldahl Digestion Assembly and Ammonia Distillation Assembly.

2.4.4.1 Digestion

In this process, a known quantity of soil or plant material is kept in digestion tubes. After this added known amount of concentrated H_2SO_4, add a pinch of $CuSO_4.H_2O$ as a catalyst and also K_2SO_4 to maintain the required temperature for the reaction during the estimation. After digestion organic compounds undergo into different fractions: C into CO_2, oxygen into water, and N into ammonia.

A major portion of NO_3^- from organic matter gets lost during the digestion except a few amounts. This loss can't give the true picture of the mineralized N in most of the soils, and researchers will get the lower value of NO_3^--N as compared to organic bound N.

2.4.4.2 Distillation

In this process NH_3 liberated during the digestion process is distilled with sodium hydroxide instead of getting absorbed in the hydrochloric acid. The content of ammonia from the digestion is determined by distillation with NaOH, and absorption of the evolved NH_3 is in the standard HCl.

2.4.4.3 Volumetric Analysis

The unconsumed amount of hydrochloride acid is back titrated with sodium hydrochloride. In this process, methyl red is being used as an indicator during analysis. At the end point, colour of the analyzing sample changes from pink to yellow.

Whereas the colorimetric method is used for estimation of ammoniacal N, the apparatus required for this is colorimeter or spectrophotometer and mechanical shaker. The required reagents are 10% sodium chloride, Nessler's reagent, 10% sodium tartrate, and standard NH_4Cl solution.

3 Effect of Climate Change on C and N Dynamics

3.1 Why Climate Change?

Climate change refers to long-term weather conditions of a particular region which greatly affect the services of ecosystems. These changes may differ from temporal and spatial from tens to hundreds or millions of years. Rapid industrialization, urbanization, deforestation, faulty agricultural practices, and modification in the land use pattern are potential sources of greenhouse gas (GHG) emission. Climate change event includes a temporal and spatial variation of a region, atmospheric pressure activities, modification in air composition, rainfall pattern, intensity, etc. (Meena and Dotaniya 2017; Meena et al. 2017). Global pattern of climate change over a period enhances the temperature of the Earth's surface by 1.4 °C–5.8 °C till 2100 (IPCC 2007). Global temperature enhances and mediates the plant and soil forming process due to the increased concentration of CO_2 and other human interventions.

3.2 Composition of GHGs

The compositions of atmospheric gases are the deciding factors of climate change. A few gases pose more warming potential known as global warming potential. Major benefits of GHGs in climate change phenomena are water vapor (H_2O), carbon dioxide (CO_2), methane (CH_4), nitrous oxide (N_2O), and other trace gases. These gases maintain the Earth's temperature for comfort stay of human and animals. Without the GHGs Earth's temperature will be frozen (−18 ° C). On the other side, increasing GHG concentration due to fast industrialization, unplanned urbanization, and faulty agricultural practices enhances the global temperature. According to the Intergovernmental Panel on Climate Change (IPCC), increasing concentration of GHGs will raise the global temperature by 1.4 °C–5.8 °C till 2100 (IPCC 2007). This situation is very much pathetic and enhances the rate of natural calamities like flood, drought, insect pest attack, lower crop yield, etc. The calculation of increasing concentration of GHGs from the pre-industrialization era is 280 to 395 ppm (or $\times 10^3$ ppb), 715 to 1882 ppb, and 227 to 323 ppb for CO_2, CH_4, and N_2O from the year 1750 and 2012, respectively. The warming of Earth's surface is directly proportioned to the global warming potential of GHGs. Throughout the warming potential, CO_2 is the base gases having warmed potential one, and CH4 is 21, and N_2O has 310 times more heating capacity than CO_2.

3.3 Effect of Climate Change on Carbon Sequestration Potential

Carbon sequestration is a process to capture and store atmospheric CO_2 into long-lived pools. The purpose of the process is to enhance the C stocks in the soil for improving the soil fertility status and reducing the C concentration from the atmosphere (Dotaniya et al. 2017b; Meena et al. 2015b). In this mechanism, green plants are the primary key component which utilizes the atmospheric CO_2 and converts it in the form of carbohydrates. Increasing the C sequestration potential by crops requires best management practices and the optimum conditions of crop growths. During the photosynthesis, plants synthesize different types of organic compounds, and part of this is released into the soil systems. These labile pools of C consist the food source for the microorganisms and improve the soil biodiversity.

Soil temperature mediates the soil microbial process and the soil fertility. Increasing the amount of root exudates enhances the availability of phosphorus, nitrogen, zinc, and other nutrients. Increasing climate change phenomena enhance the respiration rate of the crops and reduce the chances of conversion rate from source to sink in the tropical areas. This situation gets a little bit different in temperate regions of the world by a small increment in crop yield. Reducing the rate of C sequestration in the soil is affected by the global rate of C stock and physicochemical properties of the soils. Increasing the soil surface temperature enhances the C-burning rate in the soil and reduces the soil fertility status of the soil. According to the computation of C sequestration, global soils range from 0.4 to 1.2 Gt C/year (Lal 2001, 2003). The C sequestration potential is also affected by the duration and climatic conditions of a region. Among the different ecosystems, terrestrial ecosystem sequestered more amount of C is emitted during the time frame of preindustrial to the industrial period (Ruddiman 2003). According to the estimates from 1850 to 1998, C emission from the terrestrial system got almost double than the preindustrial period (Houghton 1999).

3.4 *Effect of Climate Change on Soil Processes*

Climate change with its hydrological consequences may result in the significant alteration of the soil conditions. The soil structure is heavily influenced by the change in climate (size and durability of soil aggregates) and its intricate phenomenon. The straight effect of climate change includes aggregate destruction, surface runoff, and increasing hazard which depends on their frequency and intensity, whereas indirect influences are caused due to the changes in the crop and land utilization pattern.

Climate change greatly affects and modifies the soil forming process and its nutrient dynamics. The reasons might be a change in the microorganism diversity or mineralization rate of that respective soil. The change in climate affects the rate of chemical reactions, physical process, and biological transformation of the soil. Due to climate change activities, higher precipitation led to the leaching of plant nutrient with the run of water and deposited into water bodies caused eutrophication (Fig. 1). The rising CO_2 concentration, directly and indirectly, affects the soil processes like organic matter dynamics, nutrient dynamics especially N, and biological parameters, i.e., soil biodiversity and enzyme activities; modifications in chemical reactions, i.e., pH, salinization, and alkalization; and physical process, i.e., soil water dynamics, runoff, variability in rainfall pattern and intensity, soil erosion, soil structure, etc. (Fig. 2). All these processes are directly or indirectly involved in plant growth and development. Soil process is greatly affected by the mineralization dynamics of plant nutrients. It involves physical, chemical, and biological process and affects the soil

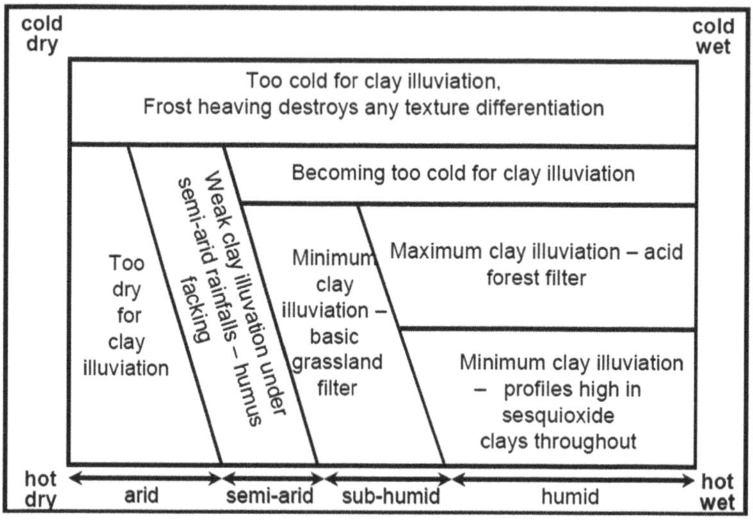

Fig. 1 The effect of climate scenarios on texture differentiation of soils. (Adopted from Karmakar et al. 2016)

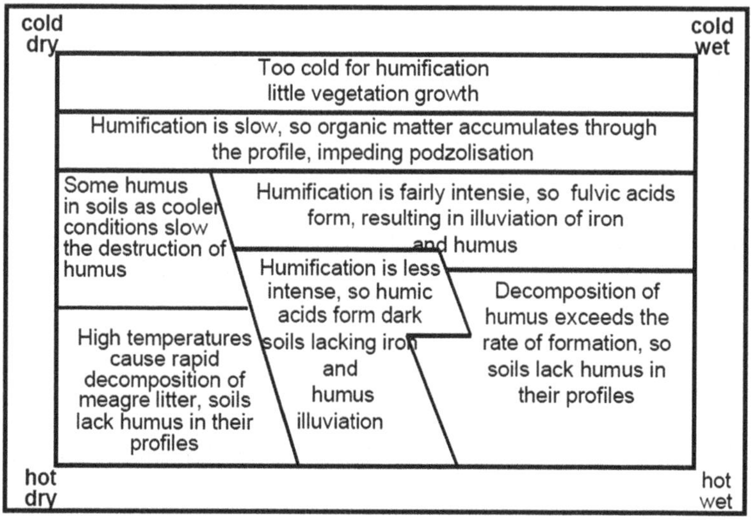

Fig. 2 The effect of climate change on soil physical properties. (Adopted from Karmakar et al. 2016)

health. Change in chemical and biological soil properties by modification in physical parameters declines soil health due to adverse climatic effect in developing countries. Marschner (1995) points out that changing the pattern of soil moisture level may increase or decrease the plant nutrient availability by modifying the soil microbial action on SOC.

3.4.1 Weathering

The concentration of CO_2 in the soil is 10–50 times higher than atmospheric. Change in the soil process and CO_2 respiration by plants enhances the soil weathering (Pendall et al. 2001; Karberg et al. 2005; Kumar et al. 2017). Soil contains more amount of CO_2 than the atmospheric concentration.

3.4.2 pH

Rising atmospheric CO_2 doesn't directly affect the pH of most of the soils. The indirect effect of global climate change affects the hydrological cycle. The increasing rate of precipitation leaches the cations from the soil and reduces the soil pH and fertility status of the soils. The continuous leaching of base cations from the soil causes lower soil pH mostly in the high rainfall areas.

3.4.3 Fertility

Increasing global temperature has mediated the plant and soil process by changing the mineralization dynamics of C and N in the ecosystems. Increasing CO_2 concentration under P stress condition enhances the root growth of the wheat by 30–41%. Increasing the microbial growth also enhances the mineralization kinetics of the nutrients. The root exudates enhance mineralization process and convert immobile P into plant available (Dotaniya et al. 2013b; Verma et al. 2015b). Staddon et al. (1999) mentioned that *Plantago lanceolata* and *Trifolium repens* are the potential strain to enhance the P use efficiency under the elevated condition of CO_2. The change in the climatic factors also alters the beneficial plant-microbe interactions. The survival and activity of microorganisms in the rhizosphere mainly depend on the flow of C in the root exudates (Dotaniya et al. 2014a, b, c, d, e, f; Dotaniya and Meena 2017; Shamina et al. 2018). Increasing CO_2 and temperature under the climate change scenario enhances the algal growth and indirectly improves the N fixation capacity of plants (Cheng et al. 2000). On the other side, increasing the temperature enhances the C and N losses from the terrestrial system.

3.4.4 Nutrient Loss

Climate changes adversely affect the hydrological cycle which may change the transformation and transportation characteristics of nutrients. A part of applied N and P fertilizers reaches water bodies during the heavy precipitation and causes water pollution. In this process overgrowth of algal bloom reduces the oxygen availability to the aquatic animals known as eutrophication. A few years back, the flood occurred in Karnataka state of India, and the 287 Mt fertile soil was washed out from the agricultural fields. On estimation, 8 hundred thousand tons of essential part nutrient and 39 hundred thousand tons of SOM were lost reducing the fertility status of the regions (Natarajan et al. 2010).

3.4.5 Soil Resilience

The ability of a soil to reach back to its original state in response to destabilizing influences is referred to as soil resilience. Burning of SOC under increasing soil surface temperature reduces the soil resilience capacity. In this process, SOM plays a crucial role in increasing resilience. More potential of soil to quick restoration of its original condition shows good soil health. It is also affected by soil and management factors.

3.4.6 Biodiversity

The initial effect of elevated CO_2 on the soil microbial population would take place in the rhizosphere. Elevated CO_2 concentration in the soil and atmosphere modifies the plant metabolic process. Rising CO_2 leads to accumulating a significant amount of C-containing compounds in the rhizosphere zone of the plant known as the highest microbial population zone of the soil. Root exudation gets enhanced by elevated atmospheric CO_2 concentration (Haase et al. 2007). Under high CO_2, fast-growing bacterial strains reduce the fungal infestation in crop plants by secreting root exudates (Tarnawski et al. 2006; Dotaniya et al. 2013b). The population and diversity of the soil microorganisms are highly varied in soil due to the availability of food materials. Research experiment showed that under higher root secretion, population of fungus gets more than bacteria. Elevated concentration of CO_2 promotes more competition between the species and within the species for water and mineral nutrition.

3.4.7 Enzyme Activities

The elevated CO_2 enhances the concentration of the enzymes, i.e., protease, xylanase, invertase, acid and alkaline phosphatase, and arylsulfatase in the surface soils. Studies reveal that elevated CO_2 can increase extracellular enzyme activities by increasing microbial demand for N and P (Dhillion et al. 1995; Dotaniya et al. 2018c; Marfo at el. 2015; Meena et al. 2017; Lojkova et al. 2015).

3.4.8 Soil Water Dynamics

Paradigm shifting in rainfall pattern and intensity governs a set of climatic events and its direct effect on the soil system across the world. Increase in global temperature increases evapotranspiration process which results in more dryness and decline in groundwater level (Dotaniya et al. 2016b). High rainfall increasing the rate of soil erosion as well as nutrient loss from agricultural fields reduces the fertility status of soil in long ways.

3.4.9 Salinization

Directly global climate change does not affect the salinity in agriculture fields. However, increasing demand of irrigation water enhances the salt content and soil pH in the long term. Increasing salt concentration reduces the crop biomass and crop yield under arid and semiarid climatic conditions.

3.5 Interpretation of C and N Dynamics Through Modeling

Nitrogen and C in the soil and their prediction help to compute the nutrient use efficiency as well as the soil health parameters. Both are important to crop yield and maintain soil diversity under different ecosystems. Prediction of C and N dynamics in contaminated soils is a little bit similar to healthy soil. HM concentration adversely affects the soil biodiversity. It affects the mineralization kinetics of metal in the soil solution. Most of the C and N model consist of temperature, rainfall, plant nutrient content, soil texture, as well as C and N concentration in the atmosphere and soil systems. The CENTURY model describes the dynamics of C and N in various soil and plant systems. It simulates the dynamics of agricultural crop system, agroforestry, and silvicultural systems. The input and output balance of N and C in the soil and their losses during different physiological condition are also considered during the computation of nutrient budgeting. The denitrification-decomposition (DNDC) model computes the different process involved in the biochemical processes. In these model soil properties, the effect of climate change of C and N process, crop management strategies, and deposition of C and N by the atmospheric events are considered during the estimation of C and N through DNDC. It consists of a long-time duration, i.e., days to centuries. RothC-26.3 model also calculates the SOC turnover under non-waterlogged soils. In this model, soil surface cover by plant residue, soil moisture, soil properties, and atmospheric temperature affect the value of C and N dynamics (Coleman and Jenkinson 1996). It calculates the SOC, C through microbial biomass, and related form of C in the soil. SOMM is a model dealing the SOM dynamics, i.e., humification, mineralization, and release of C and N in the system by various processes (Chertov and Komarov 1996). It has linear differential equations with variable coefficients for the computation of C mineralization (Chertov and Komarov 1995).

4 Waste Generation

4.1 Global Statistics of Waste Generation

According to a report by the World Bank, projected 70% more MSW would generate by 2025 at a global scale. The situation is more problematic with developing countries due to the low rate of awareness and poor disposal of generated waste. The generation rate would be increased by 1.2–2.2 times by the year 2025. It is a very pathetic situation for mankind on the Earth and needs huge monetary assistance to build the infrastructure for safe disposal and recycling of the waste. If we are talking about the per capita waste generation, it is going to increase 1.2 kg to 1.42 kg per day. The maintenance and running infrastructure cost would reach higher from $205 to $375 billion per year (WBWR 2015). The generation of waste is also increasing with the increase in lifestyle. Profuse growth of megacities is producing a higher rate of waste as compared to smaller towns. The OECD countries have higher waste generation potential than the low-income countries like Africa (Table 1).

4.2 Impact of Waste Generation on Natural Resources

4.2.1 Hazardous Gas Emissions

Among the GHGs, CH4 is more emissions from the dumping sites and shares 23% in 2006 (USEPA 2008). This gas has inflammable properties and sometimes causes blast in the landfill sites. It also adversely affects the ozone layer of the atmosphere.

Table 1 Waste generation projections for 2025

Region	Waste generation (kg capita^{-1} day^{-1})	Urban waste generation	Projected by 2025
Sub-Saharan Africa	0.65	169,119	441,840
East Asia and Pacific Region	0.95	738,958	1,865,379
Eastern and Central Asia	1.1	254,389	354.810
Latin America and Caribbean	1.1	437,545	728,392
Middle East and North Africa	1.1	173,545	369,320
OECD	2.2	1,566,286	1,742,417
South Asia	0.45	192,410	567,545
Total	–	3,532,252	6,069,703

4.2.2 Water Contamination

Dumping of municipal solid waste in virgin land contaminates the groundwater with organic and inorganic contaminants (Kelley et al. 2002). Long-term applications of sewage sludge in peri-urban area contaminate the heavy metal and microbial contamination to groundwater and cause different diseases during its consumption.

4.2.3 Energy Consumption

In the urban and semi-urban areas, landfills are a big problem. To avoid the risk of polluting environment, these landfills have to be shifted in the outskirts of the mega cities. It needs a huge amount to establish the transportation, waste segregation, use of decomposer, minimization of fossil fuel consumption, control of foul, and spread of contaminants in non-contaminated areas. It requires a lot of labor and energy.

4.2.4 Natural Habitat Degradation

Production of huge quantity of MSW requires more land to dump in the landfills which are not secure for many plants and wildlife health. Even this polluted and degraded land can't recover healthy soil for food crops growth.

4.2.5 Biodegradation

Landfills are the most common and economical method for waste disposal. This method of disposal of waste is under natural degradation and requires less manpower cost and skill. It requires turning for enhancing oxidation-reduction mechanism during the decomposition.

4.2.6 Soil Quality

Dumping of these MSW increases the acidity of soil near the garbage heaps, and it also pollutes the surrounding land area due to the flow of water during rainy season.

4.3 Types and Composition of Waste Generation

In simple term, waste refers to those materials which are the by-product of primary use or of less use. They can be categorized based on generation from the source based on contamination, biotoxicity, and reactability in nature. The waste generation can be classified as

Municipal waste which includes household waste, commercial waste, and sewage sludge:

- Industrial waste
- Biomedical waste
- Radioactive waste
- Electronic waste (e-waste)

4.3.1 Composition of Municipal Solid Waste

Municipal waste contains a huge amount of organic matter and a significant figure of plant nutrient useful for crop production, mostly in peri-urban areas. It is also the valuable source of water mostly in water scarcity regions. The compositions of MSW majorly contain total organic C (50–57%), recyclable fractions (16–19%), inert materials (28–30%), and moisture content (45–51%) (Kumar et al. 2016). The MSW comprises of bio-decomposable material, waste of household, vegetable residue, and kitchen gardening waste, which contains a larger portion as an organic part and can be utilized for compost preparation. Apart from these, it also includes inorganic C-containing items like paper, cloth waste, plastic containers, etc. In most of the low-income countries, poor facilities of waste segregation lead to a heavy metal concentration in the compost prepared from MSW. In common developing countries, MSW contains a greater amount of organic part as compared to developing countries. The average composition of MSW is shown in the Fig. 3.

4.3.2 Municipal Sewage Water

Sewage may be in the form of liquid or solid wastes generated from sewer channels. It comprises domestic waste, human and animal waste, municipal drainage and industrial effluent, etc. In developing countries, discharge of this untreated wastewater is a common practice due to lack of infrastructure, technical and institutional capacity, and financing. In and around urban and peri-urban areas, it is a major

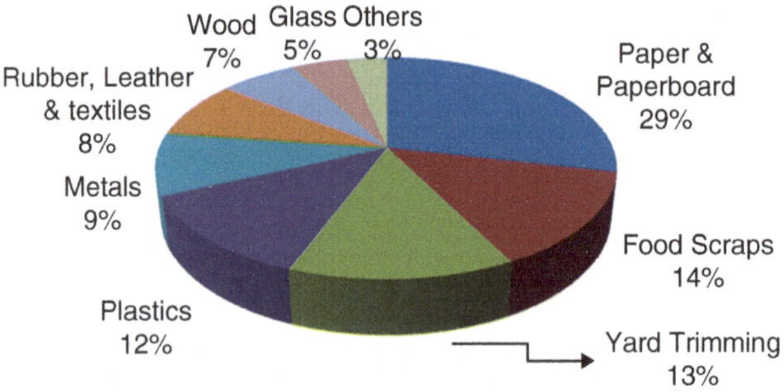

Fig. 3 Composition of municipal solid waste. (Adopted from Ozcan et al. 2016)

source of water pollution in India. The generated sewage is directly being utilized for the domestic purpose without any treatment. It contains heavy metals and biological loads, which are the major cause of health problem in the poor sanitary areas of developing countries (Dotaniya et al. 2014a, b, c, d, e, f). Heavy metals are carcinogenic in nature and reduce the life span of human beings. About 70% of wastewater is treated in high-income countries, whereas it is only 8% in low-income countries. The composition of sewage water is very much affected by the source of waste generation, intermixing of pollutant, duration, etc. Sewage water analyzed in the rainy season shows a lower amount of heavy metals as compared to samples collected in the summer season.

4.3.3 Chemical Composition

Sewage is a comprised mixture of water (99%) and organic and inorganic matter (1%) in suspended and soluble form. In common, it contains about 350 mg kg^{-1} organic matter, 52 mg kg^{-1} N, 45 mg kg^{-1} K, and 16 mg kg^{-1} P. It contains a significant amount of HMs.

4.3.4 Microbial Composition

The sewage contains a number of species of microorganisms which vary from lacs to millions per millimeter in population/number. The microorganisms like fungi, bacteria count, and protozoa mass are known as sewage fungus. The most frequently bacterial species found in sewage water are *coliforms*, *streptococci*, *clostridia*, *lactobacilli*, *Pseudomonas*, *Proteus*, and *micrococci*.

4.4 Heavy Metal Cycle in Agroecosystem

Heavy metal concentrations which exceed a certain level in the ecosystem cause imbalance in the nutrient cycling and adversely affect the crop yield. Accumulation of HM occurs through anthropogenic activities in the agroecosystem and makes it reach to various components of the soil and plant (Dotaniya et al. 2014b, 2016b; Rajendiran et al. 2015; Dadhich and Meena 2014). In soil, different HMs follow a path in the food web and make access from primary producer to consumer via food chain contamination (Pipalde and Dotaniya 2018). With the help of advanced tool and techniques, HM flows in agroecosystems to monitor the source and identification of particular metal toxicity on different soil process (Dotaniya et al. 2018a; Rajendiran et al. 2018). On the basis of analytical results, preventive measures and management options can be implemented during the growing of crops, such as using contaminated sites for remediation purpose and identifying the metal pathway and its accumulation efficiency by applying different agricultural amendments consisting organic and inorganic substances (Kabata-Pendias and Pendias 1992; Dotaniya et al. 2018b, c). Increasing the organic amendment reduces the HM uptake by crops (Dotaniya et al. 2016a). Some of the metals also consist the antagonistic effect on other metals and reduce the availability of metals in the soil for crops (Dotaniya and Pipalde 2018). Forest plants have higher tolerance capacity toward HM toxicity. Toxicity of metal in the soil is affected by soil properties like SOM, texture, presence of other ions, soil pH, as well as the genetic potential of the crop. Increasing concentration of Cr in soil reduces the soil enzymatic activities, i.e., DHA and acid and alkaline phosphatases, in the soil. Increasing concentration of HMs reduces the germination and root and shoot growth of wheat (Dotaniya et al. 2014a) and pigeon pea (Dotaniya et al. 2014b). Increasing the concentration of Cd in the soil adversely affects the uptake kinetics of Cr in spinach grown under Vertisol of Central India (Dotaniya et al. 2017a). Use of biochar application in Cu (Coumar et al. 2015a) and Cd (Coumar et al. 2015b)-contaminated soils improved the soil properties and spinach yield.

4.5 Heavy Metal Chemistry in Soil

Heavy metal dynamics in the soil are complex and have various paths from the origin to final disposal. Some metals have a higher potential risk to plant-human-animal continuum. Metal and metalloids are originated as geogenic sources and after the different anthropogenic activities; their intensity is upscaling and causing huge distribution in the environment. Use of poor natural resources with faulty management practices accumulates a significant amount of HM in soil and reaches the human body via food chain contamination (Dotaniya et al. 2017b). In an ideal condition, increasing level of metals in the soil increases metal concentration in the plant parts. The added concentration of metals in soil follows the equilibrium dynamics in soil between metal in solution and total metal in soil.

$$M_{\text{solution}} \leftrightarrow M_{\text{soil}} \tag{1}$$

There, M_{solution} and M_{soil} represent the metal concentration in soil solution and in the soil, respectively.

Further, elaboration of Eq. 1 with the help of partitioning coefficient of a particular metal in soil and in soil solution phases:

$$K_d = \frac{[M_{\text{soil}}]}{[M_{\text{solution}}]} \tag{2}$$

K_d is partitioning coefficient.

Further, the extensively elaborate soil constituents and their effect on the metal dynamics have a valuable place during computing bioavailability to plants. Soil is a complex system consisting of SOM, various oxides, i.e., Fe, Al, and Mn; and also, particulate organic matter and free ion concentration in solution phase along with its complexes with organic and inorganic substances (Impellitteri et al. 2000).

$$K_d = \frac{[M_{\text{soil}}]}{[M_{\text{solution}}]} = \frac{[M - POM] + [M - FeOx] + [M - MnOx] + \cdots}{[M^{2+}] + [MOH^{+}] + [MCO_3^0] + [M - DOM] + \cdots} \quad (3)$$

The accumulation of HM in the plant parts may be affected by the genetic characteristics of the plants. Some plants accumulate a higher amount of HM in different parts without affecting the growth and crop yield known as hyperaccumulators (Saha et al. 2017; Dotaniya et al. 2018a). In this food web, primary contamination of HM is green plant → primary consumer → secondary consumer → tertiary consumer → decomposer. The concentration of HMs in each trophic level decreases, and toxicity also depends on the genetic potential of organisms. This cycle is the same for all HMs. However, some metals have the volatilization characteristic along with having the different path of metal dynamics from soil to the final disposal. The Hg concentration in the living organism is affected by the fatty tissues and the environmental exposure concentration of Hg. The HM chemistry and mobility in the soil are affected by different soil and climatic factors, i.e., the presence of OM, other metal ions, salt concentration, soil pH, CEC, soil texture, soil moisture, etc. The higher amount of OM in the soil reduces the toxic form of Cr(VI) into nontoxic Cr(III) and reduces the availability of Cu in soil; indirectly these mechanisms are affected by the increasing population of the soil microorganism.

4.6 *Impact of Heavy Metal on Soil Biodiversity*

Soil biodiversity is necessary for maintaining the soil health as one of the key components. It includes soil macro- and micro-animals, plant roots, root exudates, and their interrelation products. Increasing the concentration of soil pollutant may be in the form of organic or HMs. Both adversely affect the soil biological properties in the soil (Roane and Kellogg 1996). An elevated concentration of HMs reduces the soil microbial population and diversity and adversely affects the crop yield. Soil biodiversity is affected by the absorption, migration, and transformation of plant nutrient in the soil from the plant system (Leyval et al. 1997). Xie et al. (2016) reported that application of Cd in the soil reduced essential metal interaction with plants, microbial biomass C, and type and diversity of microorganisms. Increasing the Cr concentration more than 20 mg kg^{-1} in the soil reduced the C mineralization rate and soil enzymatic activities in Vertisol of Central India (Dotaniya et al. 2017c). Biological N fixation is directly associated with the soil biodiversity, and conversion of environmental N into biological N in legume roots is much sensitive to HMs present in ecosystems (McGrath et al. 1988). Increasing concentration of HMs in the

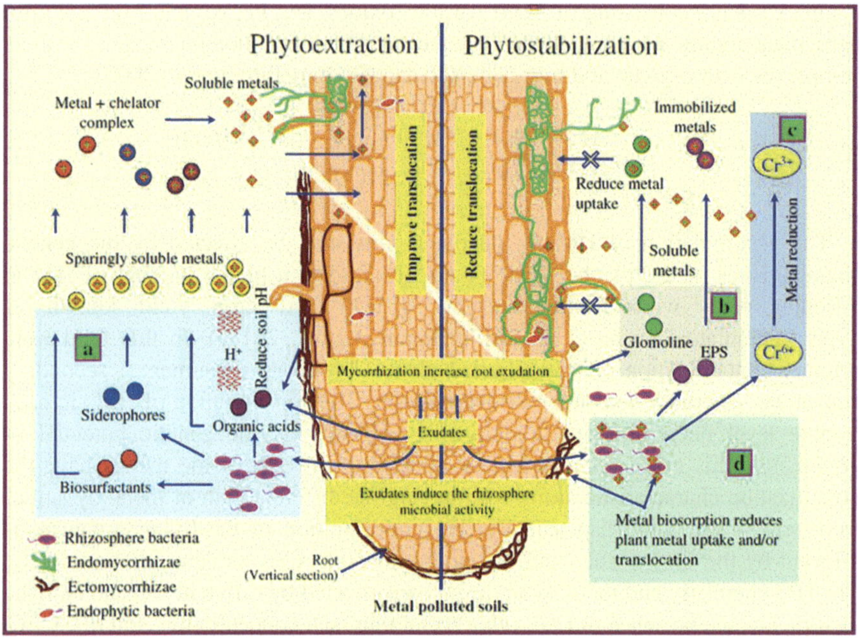

Fig. 4 Sketching diagram representing the role of microbes in metal dynamics in soil and plant system. (Adopted from Raj Kumar et al. 2012)

soil drastically reduces the nodulation in legume crops (Khan et al. 2008). Nowadays advanced tool and techniques are used for the isolation of microbial biomass to compute the effect of metal toxicity on the soil biodiversity (Rutgers et al. 2016). Identification of a microorganism for beneficial utilization in the sustainable crop production system is made with the help of genetic DNA sequencing in order to increase abiotic stress and climate change. Increasing the toxic metal concentration in the soil reduces the symbiosis relationship (Wani et al. 2008), reduces cell growth and elongation (Sresty and Madhava Rao 1999), and reduces photosynthesis rate of plant and crop yield (Sharma and Talukdar 1987). All abovementioned processes adversely affect the release of kinetic energy of root exudates and modify the soil biodiversity (Dotaniya et al. 2017c). Root exudates affect the HM dynamics in the soil by modifying the phytostabilization and phytoextraction mechanism of HM uptake by plants (Mandal et al. 2016; Ashoka et al. 2017). In these processes, soil microorganisms play a different type of role and immobilize the soluble metal from the soil solution (Fig. 4).

Khan et al. (2002) mentioned that increasing concentration of HMs adversely affects the other mineral metal uptake in the crop plant by affecting the plant nutrient dynamics through soil microorganisms. The soil fungi, bacteria, actinomycetes, and other macro- and microorganisms are much affected by the higher concentration of Cr, Hg, Cd, and Cu in the soil (Chu 2018). The declining growth of the microbial community in the soil adversely affects the ecological services, i.e., plant nutrient mineralization kinetics, secretion of plant growth-promoting substrates, root exudates, production of antibiotics, etc. Increasing the amount of pollutant through organic and inorganic reduces the synthesis of soil enzymes by the soil microorganisms and adversely affects the mineralization rate of P, N, and other micronutrients in the soil (Dotaniya et al. 2018a; Verma et al. 2015a; Molaei et al. 2017a, b).

5 Soil Degradation by Metal Contamination

The HMs are ubiquitous in nature having a higher potential of toxicity in the living organisms. The primary source of HM generation is geogenic, and the fast urbanization and extraction of metal lead natural resources contamination. Soil and water have a higher capacity to absorb these pollutants at a lower cost due to their resilient capacity. However, the rate or reclamation is more than the rate of contamination that causes pollution. Higher concentration of metal in the soil solution reduces the essential plant nutrient and reduces the nutrient-supplying capacity of the soil. Toxicity of Pb reduces the soil microbial population growth and diversity. Higher population pressure on limited natural resources causes HM pollution and reaches the stomach of the human being via food chain contamination (Xiong et al. 2016; Pierart et al. 2015). Different types of human and animal disease are reported in HM-contaminated areas. Majority of the disorders are mental retardation, hair fall, memory loss, cardiovascular diseases, cancer, fatigue, severe anemia, skin irritation, bone deformation, damage of kidneys, and nervous system failure or in extreme toxicity may lead death (Jarup 2003).

5.1 Management Through Organic Amendments

Heavy metals are toxic in nature, and their chemistry in the soil is much affected by the soil properties. Addition of organic substances in the soil forms a different type of bioabsorption materials and mediates the absorption kinetics in the soil (Shahid et al. 2014; Sabir et al. 2013). During the process of organic matter, decomposition in the soil microorganism breaks down the toxic compounds into less toxic form in the soil. An organic fraction in the soil makes the metal-humus complex and enhances the locking period of HMs in the soil. These processes are greatly affected by the oxidation-reduction kinetics, clay percent in the soil, OM, availability of other plant nutrients, CEC, and also climatic and management factors. The major organic substances which are being utilized for metal immobilization are crop residue, sewage sludge, animal waste, C-containing organic waste generated from industries, etc. Sewage and sludge application contain a significant amount of HMs in composition and affect the metal kinetics in the soil. Soil factors like pH highly affect the availability of HMs. Application of organic amendment in contaminated soil enhances the soil fertility status by increasing organic C in the soil (Bolan et al. 2014). A range of other plant materials like carbohydrates, starch, amino acid, and polysaccharides enhance the HM-binding capacity (Datta et al. 2017c). Application of organic substrates immobilizes the essential plant nutrients and reduces crop growth (Shahid et al. 2014). Humus-metal complex and its stability are influenced by the type of metal and organic compounds and the soil pH. Added organic substances enhance the microbial growth, and soil enzymatic activities in the soil reduce the metal uptake by crops. Gadd (2000) mentioned that increasing the S-containing organic substances in the soil reduces the HM availability in the soil. Different types of organic amendments which are used for immobilization of HMs are discussed below:

5.1.1 Crop Residues and Manures

Crop residue and manure application during crop production improve crop biomass by enhancing the plant nutrient concentration in the soil. Application of crop residue in the soil acts as a bio-absorber for HMs and reduces the metal availability to crop plants. It is playing a potential role in removing HMs from terrestrial ecosystem as a phytostabilization process. On the other hand, animal excreta play a significant impact on the phytoextraction of metal from the soil. Immobilization and phytoextraction techniques might be used to remediate the soil contaminated with metal. The results illustrated that the used organic residues increased the plant heights, leaf/plant number, and yield as compared to control treatment in both polluted and non-polluted soil. It also increases the NPK content and uptake by the plants. In a pot study, maize (*Zea mays*) is used as a test crop, sown in metal-contaminated soil (10 kg pot^{-1}), and soil amendments with different manure and dust were used each at 5 g $kg^{-1}$of soil. The findings pointed out that the uptake of heavy metals (Cr, Pb, Co, Ni, and Cd) by maize depended on the organic materials used. Addition of residues from water hyacinth, fern, mustard, and barnyard grass was likely to reduce metal content in the maize. The range of decline varied among them from 20.5% to 33.3% for fern dust, 17.3% to 22.0% for water hyacinth, 18.6% to 21.3% for mustard stover dust, and 17.33% to 20.5% for barnyard grass dust (Table 2). Manures and vermicompost dust were likely to increase metal content in the maize, and it varied from 6.80% to 18.7% for cow manure, 18.9% to 86.7% for poultry manure, and 17.4% to 16.0% for vermicompost

Table 2 Metal content of different applied organic materials

Materials	Concentration (μg g^{-1} dry wt)				
Animal excreta	Pb	Cd	Ni	Cr	Co
Cow manure dust	5.21	0.36	4.70	3.52	5.80
Poultry manure dust	9.3	0.60	6.20	6.78	8.60
Vermicompost dust	10.3	0.42	11.4	8.32	9.15
Fern dust	2.40	0.25	5.90	2.50	3.22
Water hyacinth dust	2.56	0.39	4.22	3.10	4.43
Mustard stover dust	1.14	0.36	6.10	1.98	3.30
Barnyard grass dust	1.16	0.22	3.50	2.00	2.71

Adopted from Naser et al. (2017)

(Naser et al. 2017; Yadav et al. 2017a). Application of organic residues to the soil positively reduced Cd, Pb, and Zn concentration in the plant. Decomposition of organic residue in soil produced a different type of organic acids and a chelating agent which adsorbed or immobilized the metal ions in the soil and reduced the toxicity of metal in the soil solution.

5.1.2 FYM/Compost/Vermicompost

Use of organic substances in the form of vermicompost or FYM reduces the HM availability and improves the microbial population in the soil. The FYM highly absorbs Fe, Cr, Ni, Mn, and Pb and declines the availability in the soil solution. A study was conducted to estimate the bioavailability fractions of heavy metals under different organic amendment practices with *Amaranthus viridis*, and it was found that FYM applied field reduced the effective concentration of HMS as compared to red mud, sewage, and vermicompost treatment (Nwoko et al. 2012). Addition of compost reduced 79–88% and 86–96% As concentration in carrot and in lettuce, respectively. Use of organic substances in potato tubers reduced Cd contamination (Angelova et al. 2010). Different types of compost have a wide range of metal remediation and detoxification potential due to variability in composition (Gadepalle et al. 2007). Use of olive husk, vegetable residue and compost showed that olive husk performed better over the other in relation to immobilization of Pb, Cd, and Zn.

5.1.3 Biochar

It partially burns material of organic residue in the presence of limited oxygen supply known as pyrolysis. In the last two decades, research on biochar and its positive effect on the soil's physicochemical properties were widely studied. Its application during crop production has higher crop yield and improves soil chemical properties like pH, CEC, organic C, and N. After the positive signs in the crop production, it is also used for the remediation purpose mostly in contaminated soils. It has higher C sequestration potential and captures the C for a longer duration. It had a higher surface area and worked on adsorption principle during the immobilization of HMs. Application of biochar enhances the soil microbial biomass C and reduces the bioavailability fractions of HMs in the soil (Puga et al. 2015; Sihag et al. 2015; Datta et al. 2017a). Incorporation of biochar enhances the soil pH and reduces the availability of toxic metals like Cd, Cu, and Pb. Park et al. (2011) mentioned that use of chicken manure biochar in the contaminated soil reduces the phyto-availability of Cd and Pb. Lehmann et al. (2011) reported that different types of biochar prepared from a different type of organic produced have wide variability in the adsorption of metal ions. Uchimiya et al. (2012) mentioned that biochar efficiency is much affected by the type of raw material, preparation time, metal type and concentration of metals, etc.

5.1.4 Bamboo Charcoal

It is a cheap and renewable bio-resources. It has larger biomass and a higher number of pores in micro size which enhance the many fold reactive surface area of produce charcoal. It also has a higher bio-adsorption rate toward HM contaminates and is potentially used for the purification of impurities from water. As an organic nature of bamboo charcoal, it can be potentially utilized for enhancing the fertilizer use efficiency and C sequestration in soils (Shazia et al. 2015).

5.1.5 Sludge

The use of sludge was reported to increase vegetation growth at contaminated sites. It is also utilized as an amendment in degraded soils. Higher amount of organic matter enhances the biological activities in the soil and provides a base for mineralization of plant nutrients. Use of sludge in combination with flash enhances the crop yield in the physically constrained soils by improving soil physicochemical properties. It acts as a source and sinks for HM contamination. Application of sludge in contaminated soils reduces the availability of HMs during the crop cultivation.

5.1.6 Genetically Modified Plants

Advancement in science depicts its significant role in the metal remediation. Use of advanced tools and technologies is creating a scientific base for developing the genetic engineered transgenic plants which are more capable of accumulating the metal concentration in different parts without affecting the contamination in economic part. Several hyperaccumulator genes have been identified and utilized for the hyperaccumulation of HMs from the contaminated sites. Different genes are isolated like *Arabidopsis thaliana* (Guo et al. 2012), *Pb* genes are identified from *Nicotiana glauca* (Gisbert et al. 2003), Cd genes are identified from *Nicotiana tabacum* (Khoudi et al. 2013), and Zn hyperaccumulator genes are identified from *Lactuca sativa* and Indian mustard. In another study, transgenic *Brassica* and *Nicotiana* plants have shown higher accumulation and tolerance toward Cd. These foreign genes enhance the level of GSH and PC in the plants and modify the expression pattern of cysteine synthase.

5.2 Management Through Inorganic Amendments

Like the organic source of metal decontamination, inorganic sources are also playing a crucial function in the soil for detoxification of toxic substances. Most of the cases, the addition of inorganic substances modifies the soil pH, which highly affects the

HM dynamics in the soil. These products which are waste or by-product of industries can be utilized for remediation of contaminated lands. These materials are low cost, and their availability is ample at industrial location (Puschenreiter et al. 2005). Here, a list of inorganic substances is described which is mostly being utilized for decontamination of the metal-contaminated soil.

5.2.1 Lime

It is a valuable source for reclamation of acid soils mainly in the high rainfall and coastal soils. But it is also used for immobilization of HM availability in the soil. Addition of lime in high metal availability soil reduces the toxicity. Fang and Wong (1999) incorporation of lime with sludge composted crop field showed poor availability of Mn, Cu, and Zn. Further research studies showed that application of sludge in every 2 years at 2–10 t ha^{-1} enhanced the immobilization percent of HMs (Puschenreiter et al. 2005). Oliver et al. (1996) conducted research studies and observed that addition of lime during the wheat crop reduced the Cd concentration by modifying the soil pH from 4 to 5.

5.2.2 Industrial Waste

Industrial waste disposal is a big problem across the world and needs safe disposal (Dotaniya and Datta 2014). The agriculture sector has vast potential to utilize these wastes for various purposes. These are cheaper and easily available. Use of termitaries, egg waste (Soares et al. 2015), red mud (Smiciklas et al. 2014), and sugarcane waste is utilized for detoxification of contaminated soils. Red mud is an industrial waste by-product, produced during the extraction of alumina. Incorporation of mud reduces the metal availability in the soil and improves the crop yield in long-term (Singh and Kalamdhad 2012; Meena et al. 2018a). Similar findings were by Qiao (1997) reported the metal availability reduction during the addition of red mud. The effectiveness of red mud on HM availability, toxicity, and dynamics is affected by the type of metal, species, and intensity of a particular metal in the contaminated soil.

5.2.3 Zeolite

It is a hydrated aluminosilicate having the capacity to adsorb the metal ions. It is used for enhancing the nutrient use efficiency in the crop plants. It has a huge number of micropores which enhance the surface area for cation adsorption. Adsorption dynamics of zeolite shows higher affirmation toward HMs and reduces the availability. Across the global research area, a range of synthetic zeolitic materials are being used for reducing the HM toxicity and improving the nutrient status of the soil during crop growth stages (Puschenreiter et al. 2005).

Application of zeolite in the rape crop reduces the Pb dynamics in the soil and lowers down the Pb concentration in the economic part of the crop (Li et al. 2009).

5.2.4 Phosphate Salt

On Earth, 95% of all P is present within apatite minerals [$Ca_5(PO_4)_3(F,OH,Cl)$] (Smil 2000), and the biogeochemical interactions of this mineral not only play a role in global P cycling but also influence the mobility of the soluble metals. It is observed that the application of various phosphate compounds (e.g., orthophosphate solutions, soluble polyphosphates, organophosphates, phosphatidic clays, apatite reactive barriers, and vivianite) to immobilize contaminants (e.g., Cd, Pb, U, and Zn) detected contaminant sequestration via a combination of microbial-mediated mechanisms and abiotic reactions. Soluble P, in the forms of phosphoric acid, phytic acid, and tripolyphosphate, has been examined for sequestration of Cd, Cu, Pb, U, and Zn from polluted sites (Wellman et al. 2006, 2007; Rijkenberg and Depree 2010). Compounds such as phytic acid and polyphosphate have been examined as chelating agents that minimize the bioavailability of cations within contaminated environments (Stefano et al. 2006, 2010). Laboratory studies examining the immobilization of various cations (Ba, Co, Mn, Ni, Pb, U, and Zn) within contaminated sediments have shown phytic acid to be effective in reducing soluble contaminant concentrations. Specifically, concentrations of U(VI) were reduced from 2242 to 76 $g\,kg^{-1}$ and of Ni from 58 to 9.6 $mg\,kg^{-1}$ in sediment batch incubations treated with calcium phytate (i.e., the salt of phytic acid) (Nash et al. 1998; Seaman et al. 2003; Meena et al. 2018a).

5.2.5 Fly Ash

It is the waste generated during the generation of electricity from coal-based thermal power plants. Consumption of electricity by the country is the sign of prosperity and the economic development. Most of the developing countries have a higher consumption of electricity, and generated fly ash is being used for brick and another construction purpose. Rational use of fly ash in an agricultural crop production system without affecting the food chain contamination and ecosystem services is the prime need of today's research. Fly ash contains a higher amount of fine clay and a meager amount of essential plant nutrients. It also contains HMs in trace amount and chances to contaminate the food crop. Fly ash is used for remediation of mining sites. The continuous application of fly ash can lead to the HM contamination in the soil water system. Use of zeolites with fly ash treatment is an effective immobilized agent to reduce metal contamination in groundwater bodies. Some studies suggested that use of fly ash can reduce the approximately more than 80%, 48–56%, 27–36%, and 6–24% of Pb, Cd, Zn, and Cu, respectively (Liu et al. 2015).

5.2.6 Bentonite

Peri-urban areas, used for vegetable crop cultivation, represent a sink for HMs because of the excessive accumulation of HMs by sewage water irrigation and industrial effluent due to the paucity of good quality water and availability of polluted water at no cost. Clay minerals modify the soil solution pH, decrease the chemical-extractable fractions and bioavailability of HMs in soils, and decrease HM concentration in the plant parts. It is observed that amendment of the soil with bentonite at 2.5% positively enhanced the plant biomass by 76.5% and 41.7% in the first and second harvest, respectively. The metal concentration of amaranth was reduced by 6.5% and 21.2% for Zn, 30.5% and 29.9% for Cu, and 34.4% and 40.2% for Ni at the first and second harvest, respectively, on the application of 2.5% bentonite (Varrault and Bermond 2011). Amending the soil with bentonite (2.5%) significantly decreased the bioconcentration factor (BCF) of metals by 74% and 28% for Zn, 38% and 36% for Cu, and 44% and 34% for Ni at the first and second harvest, respectively. The hazard quotient (HQ) for metal uptake of *Amaranthus viridis* was notably reduced through application of bentonite at 2.5% in polluted soil (Kumararaja et al. 2016).

5.2.7 Hydroxyapatite

Across the global research, studies showed that modification in the clay composition with a chelating agent by changing the chemical properties of soil with the help of different types of apatite reduces the metal availability in the soil solution. An alternative to introducing soluble orthophosphate into contaminated sediment and groundwater is the application of apatite [$Ca_5(PO_4)_3(F,OH,Cl)$] minerals (e.g., bone apatite, synthetic apatite minerals, and rock phosphate) as subsurface reactive barriers (Chen et al. 2006; Yoon et al. 2007). In situ soluble metal and radionuclide can be significantly decreased with the utilization of apatite minerals through various processes such as surface interactions, dissolution-precipitation reactions, and ion exchange. Apatite minerals have been effective in decreasing the soluble concentrations (i.e., a 1000-fold reduction) of contaminants that mostly include Cd, Co, Pb, Sb, Th, and U (Oliva et al. 2011). The hydroxyapatite and vivianite [$Fe_3(PO_4)_2$] nanoparticles (3–10 nm size) can be used as both liquid and solid phosphate amendments.

5.2.8 Soil Amendments

Soil amendments are used for remediation of the degraded soils. Across the world, many types of amendments are used for improving the acidic, saline, and alkaline soils. Peoples are also practicing the chelating agents for enhancing the heavy metal concentration in the soil solution. In which increasing concentration of metal is removed by the hyperaccumulator plants and after this disposal with the proper method is done. In another way, use of organic residues or organic acids produces substances which immobilize the metal concentration or detoxify the metal in the soil without affecting the soil ecosystem (Dotaniya and Meena 2013; Ram and Meena 2014). Such types of amendments are used mostly in the higher contaminated areas of the world.

5.2.9 Landfill Mining

Generation of the huge volume of waste across the living world is increasing with quantum rate, and its scientific disposal is a big challenge for policymakers and researchers. Most of the low economic countries are having the landfill disposal of generated waste. It is cheaper and consumes less manpower but causes a higher risk of contamination to soil and water. Wastes are containing organic and inorganic pollutant leaching down in the soil and water bodies, and their consumption causes many human diseases. With the help of different technologies, precious metals are recovered from the landfill and also used for the physically degraded soils. Compost is also being prepared from the landfill waste after regular turning and periodically monitoring of contaminants. Such type of compost having a significant amount of trace metals causes phytotoxicity and is mostly used in forest plants. However, due to a manifold obstacle, the high cost of processing waste, and technological challenges, it is difficult to landfill mining for metals economically viable.

6 Management Strategies of C and N Metal-Contaminated Soils

Carbon and N dynamics affect crop growth under the contaminated soil. Amount of SOM in the soil reduces the toxicity of metal by conversion of toxic into a nontoxic form. Increasing concentration of Cu in the soil drastically reduces the mineralization rate of OM, and *plant showed* Cu deficiency. Microbial populations and diversity are the key components for governing the C and N availability in the soil. Application of N fertilizers higher than the recommended dose of crop enhances the crop yield (Manna 2013). The C stock in the soil and its annual rate of change can be computed with using single-pool first-order kinetics as mentioned in Eq. 4.

$$oCs/ot = h.A - k.Cs \tag{4}$$

Cs is the SOC, t is the time frame, A represents annual C input to the soil, h is humification rate, and k is the constant.

After repeated application of organic residue, FYM, and other sources, SOC level stabilizes SOC in the soil. The relationship between total SOC and the period of cropping may be computed as per Eq. 5.

$$\mathrm{SOC_t} = \mathrm{SOC_e} + (\mathrm{SOC_0} - \mathrm{SOC_e})\exp(-kt) \quad (5)$$

where, $\mathrm{SOC_t}$ is a value of SOC at a time t; $\mathrm{SOC_e}$ is the value of SOC at equilibrium; $\mathrm{SOC_0}-\mathrm{SOC}$ is the value at $t = 0$; k is the exponential rate of variation (1/year); and t represents time frame as a cropping period. Here k value represents the SOC change toward a new equilibrium. Similar pattern N management also follows the N dynamics in the soil. Increasing the N content in the soil enhances C:N ratio for deciding the mineralization rate of SOC and its uptake toward crop plants.

7 A Future Line of Work

Growing population with higher growth in most of the developing countries is a challenge to feed hungry mouth with healthy food. Poor and limited natural resources have enhanced the amplitude of problem across the globe. Nitrogen and C dynamics play a crucial role in obtaining crop growth and yield in most of the agricultural crops. Higher concentration of HMs in the soil reduces the C and N nutrient mineralization rate by mediating the microbial population and diversity. Addition of organic and inorganic substrates for the immobilization of HMs in the soil for sustainable crop production is a need of today's agriculture. Application of organic crop residue enhances the soil biological activities as well as improves the C and N status in the soil system. It is observed that in the process of decomposition of organic substrates released, a range of low-molecular-weight organic acids mediated soil pH and reduced the toxicity of HMs. Use of different traditional and advanced techniques for management of contaminated soil helps to enhance the crop yield in degraded or physically, chemically degraded soil.

Acknowledgments The authors are highly thankful to Dr. H.M. Meena, Scientist, ICAR-Central Arid Zone Research Institute, Jodhpur, India, for providing motivation and incorporating suggestions during the writing of this manuscript.

References

Angelova V, Ivanova R, Pevicharova G, Ivanov K (2010) Effect of organic amendments on heavy metals uptake by potato plants. https://iuss.org/19th%20WCSS/Symposium/pdf/0660.pdf

Ashoka P, Meena RS, Kumar S, Yadav GS, Layek J (2017) Green nanotechnology is a key for eco-friendly agriculture. J Clean Prod 142:4440–4441

Bolan N, Kunhikrishnan A, Thangarajan R, Kumpiene J, Park J, Makino T, Kirkham MB, Scheckel K (2014) Remediation of heavy metal(loid)s contaminated soils-to mobilize or to immobilize? J Hazard Mater 266:141–166

Buragohain S, Sharma B, Nath JD, Gogaoi N, Meena RS, Lal R (2017) Impact of ten years of bio-fertilizer use on soil quality and rice yield on an inceptisol in Assam, India. Soil Res. https://doi.org/10.1071/SR17001

Chen SB, Zhu YG, Ma YB (2006) The effect of grain size of rock phosphate amendment on metal immobilization in contaminated soils. J Hazard Mater 134(1–3):74–79. https://doi.org/10.1016/j.jhazmat.2005.10.027

Cheng W, Chander K, Inubushi K (2000) Effects of elevated CO2 and temperature on methane production and emission from submerged soil microcosm. In: Methane emissions from major rice ecosystems in Asia. Springer Netherlands, Dordrecht. https://doi.org/10.1007/978-94-010-0898-3_29

Chertov O, Komarov A (1995) On mathematical theory of soil forming processes. I. Theoretical background. II. SOMM-a model of soil organic matter dynamics. III. Basic ideas of a mineral phase modelling. Pushchino Res. Center of Russian Academy of Sciences, Pushchino. Preprint

Chertov OG, Komarov AS (1996) SOMM – a model of soil organic matter and nitrogen dynamics in terrestrial ecosystems. In: Evaluation of soil organic matter models. Springer, Berlin/Heidelberg. https://doi.org/10.1007/978-3-642-61094-3_16

Chu D (2018) Effects of heavy metals on soil microbial community. IOP Conf Ser Earth Environ Sci 113:012009. https://doi.org/10.1088/1755-1315/113/1/012009

Coleman K, Jenkinson DS (1996) RothC-26.3 – a model for the turnover of carbon in soil. In: Evaluation of soil organic matter models. Springer, Berlin/Heidelberg. https://doi.org/10.1007/978-3-642-61094-3_17

Coumar MV, Parihar RS, Dwivedi AK, Saha JK, Lakaria BL, Biswas AK, Rajendiran S, Dotaniya ML, Kundu S (2015a) Pigeon pea biochar as a soil amendment to repress copper mobility in soil and its uptake by spinach. Bioresources 11(1):1585–1595. https://doi.org/10.15376/biores.11.1.1585-1595

Coumar MV, Parihar RS, Dwivedi AK, Saha JK, Rajendiran S, Dotaniya ML, Kundu S (2015b) Impact of pigeon pea biochar on cadmium mobility in soil and transfer rate to leafy vegetable spinach. Environ Monit Assess 188(1):31. https://doi.org/10.1007/s10661-015-5028-y

Dadhich RK, Meena RS (2014) Performance of Indian mustard (*Brassica juncea* L.) in response to foliar spray of thiourea and thioglycollic acid under different irrigation levels. Indian J Ecol 41(2):376–378

Datta R, Anand S, Moulick A, Baraniya D, Pathan SI, Rejsek K, Vranova V, Sharma M, Sharma D, Kelkar A (2017a) How enzymes are adsorbed on soil solid phase and factors limiting its activity: a review. Int Agrophys 31(2):287–302

Datta R, Baraniya D, Wang Y-F, Kelkar A, Meena RS, Yadav GS, Teresa Ceccherini M, Formanek P (2017b) Amino acid: its dual role as nutrient and scavenger of free radicals in soil. Sustainability 9(8):1402

Datta R, Kelkar A, Baraniya D, Molaei A, Moulick A, Meena R, Formanek P (2017c) Enzymatic degradation of lignin in soil: a review. Sustainability 9(7):1163

Dhillion SS, Roy J, Abrams M (1995) Assessing the impact of elevated CO2 on soil microbial activity in a Mediterranean model ecosystem. Plant Soil 187(2):333–342. https://doi.org/10.1007/bf00017098

Dotaniya ML (2018) Remediation of chromium toxicity in wheat by use of FYM and pressmud. In: Rajendiran S, Coumar MV, Meena VD, Kundu S, Saha JK, Patra AK (eds) Oral presentation in the national conference "Organic waste management for food and environmental security"

Dotaniya ML, Datta SC (2014) Impact of bagasse and press mud on availability and fixation capacity of phosphorus in an inceptisol of North India. Sugar Tech 16(1):109–112. https://doi.org/10.1007/s12355-013-0264-3

Dotaniya ML, Datta SC (2015) Phosphorus dynamics mediated by bagasse, press mud and rice straw. In: Biswas DR, Meena HM, Rajendiran S, Meena AL (eds) Agrochimica 59 (4):358

Dotaniya ML, Meena VD (2013) Rhizosphere effect on nutrient availability in soil and its uptake by plants: a review. Proc Natl Acad Sci India Sect B Biol Sci 85(1):1–12. https://doi.org/10.1007/s40011-013-0297-0

Dotaniya ML, Meena BP (2017) Rhizodeposition by plants: a boon to soil health. In: Elanchezhian R, Biswas AK, Ramesh K, Patra AK (eds) Advances in nutrient dynamics in soil plant system for improving nutrient use efficiency. India Publishing Agency, New Delhi, pp 207–224

Dotaniya ML, Pipalde JS (2018) Soil enzymatic activities as influenced by lead and nickel concentrations in a vertisol of Central India. Bull Environ Contam Toxicol 101(3):380–385. https://doi.org/10.1007/s00128-018-2402-9

Dotaniya ML, Datta SC, Biswas DR, Meena BP (2013a) Effect of solution phosphorus concentration on the exudation of oxalate ions by wheat (*Triticum aestivum* L.). Proc Natl Acad Sci India Sect B Biol Sci 83(3):305–309. https://doi.org/10.1007/s40011-012-0153-7

Dotaniya ML, Dasharath P, Meena MH, Jajoria KD, Narolia PG, Pingoliya KK, Meena PO, Kuldeep K, Meena PB, Asha R, Das H, Sreenivasa Chari M, Suresh P (2013b) Influence of phytosiderophore on iron and zinc uptake and rhizospheric microbial activity. Afr J Microbiol Res 7(51):5781–5788. https://doi.org/10.5897/ajmr2013.6461

Dotaniya ML, Das H, Meena VD (2014a) Assessment of chromium efficacy on germination, root elongation, and coleoptile growth of wheat (*Triticum aestivum* L.) at different growth periods. Environ Monit Assess 186(5):2957–2963. https://doi.org/10.1007/s10661-013-3593-5

Dotaniya ML, Datta SC, Biswas DR, Kumar K (2014b) Effect of organic sources on phosphorus fractions and available phosphorus in Typic Haplustept. J Indian Soc Soil Sci 1:80–83

Dotaniya ML, Datta SC, Biswas DR, Meena HM, Kumar K (2014c) Production of oxalic acid as influenced by the application of organic residue and its effect on phosphorus uptake by wheat (*Triticum aestivum* L.) in an inceptisol of North India. Natl Acad Sci Lett 37(5):401–405. https://doi.org/10.1007/s40009-014-0254-3

Dotaniya ML, Kushwah SK, Rajendiran S, Coumar MV, Kundu S, Subba Rao A (2014d) Rhizosphere effect of Kharif crops on phosphatases and dehydrogenase activities in a Typic Haplustert. Natl Acad Sci Lett 37(2):103–106. https://doi.org/10.1007/s40009-013-0205-4

Dotaniya ML, Meena VD, Das H (2014e) Chromium toxicity on seed germination, root elongation and coleoptile growth of pigeon pea (*Cajanus cajan*). Legum Res Int J 37(2):227. https://doi.org/10.5958/j.0976-0571.37.2.034

Dotaniya ML, Thakur JK, Meena VD, Jajoria DK, Rathor G (2014f) Chromium pollution: a threat to environment – a review. Agric Rev 35(2):153. https://doi.org/10.5958/0976-0741.2014.00094.4

Dotaniya ML, Meena VD, Basak BB, Meena RS (2016a) Potassium uptake by crops as well as microorganisms. In: Potassium solubilizing microorganisms for sustainable agriculture. Springer India, New Delhi. https://doi.org/10.1007/978-81-322-2776-2_19

Dotaniya ML, Meena VD, Kumar K (2016b) Impact of biosolids on agriculture and biodiversity. Today and Tomorrow's Printer and Publisher, New Delhi, pp 11–20

Dotaniya ML, Rajendiran S, Meena BP, Meena AL, Dotaniya CK, Meena BL, Jat RL, Saha JK (2016c) Elevated carbon dioxide (CO2) and temperature vis-a-vis carbon sequestration potential of global terrestrial ecosystem. Conservation agriculture. Springer, Singapore. https://doi.org/10.1007/978-981-10-2558-7_9

Dotaniya ML, Meena VD, Lata M, Meena BL (2017a) Climate change impact on agriculture: adaptation strategies. In: Kumar PS, Kanwat M, Meena PD, Kumar V, Alone RA (eds) Climate change & sustainable agriculture. India Publishing Agency, New Delhi, pp 27–38

Dotaniya ML, Meena VD, Rajendiran S, Coumar MV, Saha JK, Kundu S, Patra AK (2017b) Geo-accumulation indices of heavy metals in soil and groundwater of Kanpur, India under long term irrigation of tannery effluent. Bull Environ Contam Toxicol 98(5):706–711. https://doi.org/10.1007/s00128-016-1983-4

Dotaniya ML, Rajendiran S, Coumar MV, Meena VD, Saha JK, Kundu S, Kumar A, Patra AK (2017c) Interactive effect of cadmium and zinc on chromium uptake in spinach grown in vertisol of Central India. Int J Environ Sci Technol 15(2):441–448. https://doi.org/10.1007/s13762-017-1396-x

Dotaniya ML, Aparna K, Dotaniya CK, Singh M, Regar KL (2018a) Role of soil enzymes in sustainable crop production. In: Enzymes in food biotechnology. Elsevier. https://doi.org/10.1016/b978-0-12-813280-7.00033-5

Dotaniya ML, Dotaniya CK, Sanwal RC, Meena HM (2018b) CO2 sequestration and transformation potential of agricultural system. In: Handbook of ecomaterials. Springer. https://doi.org/10.1007/978-3-319-48281-1_87-1

Dotaniya ML, Meena VD, Saha JK, Rajendiran S, Patra AK, Dotaniya CK, Meena HM, Kumar K, Meena BP (2018c) Environmental impact measurements: tool and techniques. In: Handbook of ecomaterials. Springer. https://doi.org/10.1007/978-3-319-48281-1_60-2

Duan X, Zhang G, Rong L, Fang H, He D, Feng D (2015) Spatial distribution and environmental factors of catchment-scale soil heavy metal contamination in the dry-hot valley of Upper Red River in southwestern China. Catena 135:59–69. https://doi.org/10.1016/j.catena.2015.07.006

Facchinelli A, Sacchi E, Mallen L (2001) Multivariate statistical and GIS-based approach to identify heavy metal sources in soils. Environ Pollut 114(3):313–324. https://doi.org/10.1016/s0269-7491(00)00243-8

Fang M, Wong JWC (1999) Effects of lime amendment on availability of heavy metals and maturation in sewage sludge composting. Environ Pollut 106(1):83–89. https://doi.org/10.1016/s0269-7491(99)00056-1

Gadd GM (2000) Bioremedial potential of microbial mechanisms of metal mobilization and immobilization. Curr Opin Biotechnol 11(3):271–279. https://doi.org/10.1016/s0958-1669(00)00095-1

Gadepalle VP, Ouki SK, Herwijnen RV, Hutchings T (2007) Immobilization of heavy metals in soil using natural and waste materials for vegetation establishment on contaminated sites. Soil Sediment Contam Int J 16(2):233–251. https://doi.org/10.1080/15320380601169441

Gisbert C, Ros R, De Haro A, Walker DJ, Pilar Bernal M, Serrano R, Navarro-Aviñó J (2003) A plant genetically modified that accumulates Pb is especially promising for phytoremediation. Biochem Biophys Res Commun 303(2):440–445. https://doi.org/10.1016/s0006-291x(03)00349-8

Gong M, Wu L, Bi X-y, Ren L-m, Wang L, Ma Z-d, Bao Z-y, Li Z-g (2010) Assessing heavy-metal contamination and sources by GIS-based approach and multivariate analysis of urban–rural topsoils in Wuhan, central China. Environ Geochem Health 32(1):59–72. https://doi.org/10.1007/s10653-009-9265-2

Guo G, Wu F, Xie F, Zhang R (2012) Spatial distribution and pollution assessment of heavy metals in urban soils from southwest China. J Environ Sci 24(3):410–418. https://doi.org/10.1016/s1001-0742(11)60762-6

Haase S, Neumann G, Kania A, Kuzyakov Y, Römheld V, Kandeler E (2007) Elevation of atmospheric CO2 and N-nutritional status modify nodulation, nodule-carbon supply, and root exudation of *Phaseolus vulgaris* L. Soil Biol Biochem 39(9):2208–2221. https://doi.org/10.1016/j.soilbio.2007.03.014

Houghton RA (1999) The annual net flux of carbon to the atmosphere from changes in land use 1850-1990*. Tellus B 51(2):298–313. https://doi.org/10.1034/j.1600-0889.1999.00013.x

Hu Y, Cheng H (2013) Application of stochastic models in identification and apportionment of heavy metal pollution sources in the surface soils of a large-scale region. Environ Sci Technol 47(8):3752–3760. https://doi.org/10.1021/es304310k

Impellitteri CA, Allen HE, Yin Y, You SJ, Saxe JK (2000) Soil properties controlling metal partitioning. In: Selim HM, Sparks D (eds) Heavy metals release in soils. Lewis Publishers, Washington DC, pp 149–165

IPCC (2007) IPCC fourth assessment report: climate change 2007. Climate change 2007: working group I: the physical science basis. https://www.ipcc.ch/publicationsanddata/ar4/wg1/en/spmsspm-projections-of.html

Järup L (2003) Hazards of heavy metal contamination. Br Med Bull 68(1):167–182. https://doi.org/10.1093/bmb/ldg032

Jat RL, Jha P, Dotaniya ML, Lakaria BL, Rashmi I, Meena BP, Shirale AO, Meena AL (2018) Carbon and nitrogen mineralization in vertisol as mediated by type and placement method of residue. Environ Monit Assess 190 (7). https://doi.org/10.1007/s10661-018-6785-1

Kabata-Pendia A, Pendias H (1992) Trace elements in soils and plants. CRC Press, Boca Raton/London

Karberg NJ, Pregitzer KS, King JS, Friend AL, Wood JR (2005) Soil carbon dioxide partial pressure and dissolved inorganic carbonate chemistry under elevated carbon dioxide and ozone. Oecologia 142(2):296–306. https://doi.org/10.1007/s00442-004-1665-5

Karmakar R, Das I, Dutta D, Rakshit A (2016) Potential effects of climate change on soil properties: a review. Sci Int 4:51–73

Kelley ME, Brauning SE, Schoof RA, Ruby MV (2002) Assessing oral bioavailability of metals in soil. Ohio: Battelle Press 2:18

Khan M, Zaidi A, Aamil M (2002) Biocontrol of fungal pathogens by the use of plant growth promoting rhizobacteria and nitrogen fixing microorganisms. Indian J Bot Soc 81:255–263

Khan MS, Zaidi A, Wani PA, Oves M (2008) Role of plant growth promoting rhizobacteria in the remediation of metal contaminated soils. Environ Chem Lett. https://doi.org/10.1007/s10311-008-0155-0

Khoudi H, Maatar Y, Brini F, Fourati A, Ammar N, Masmoudi K (2013) Phytoremediation potential of *Arabidopsis thaliana*, expressing ectopically a vacuolar proton pump, for the industrial waste phosphogypsum. Environ Sci Pollut Res 20:270–280

Kumar A (2015) An overview idea on carbon sequestration. Int J Interdisciplinary Res Innov 4:100–105

Kumar S, Dhar H, Nair VV, Bhattacharyya JK, Vaidya AN, Akolkar AB (2016) Characterization of municipal solid waste in high-altitude sub-tropical regions. Environ Technol 37(20):2627–2637. https://doi.org/10.1080/09593330.2016.1158322

Kumar S, Meena RS, Pandey A, Seema (2017) Soil acidity management and an economics response of lime and sulfur on sesame in an alley cropping system. Int J Curr Microbiol App Sci 6(3):2566–2573

Kumar S, Meena RS, Lal R, Singh Yadav G, Mitran T, Meena BL, Dotaniya ML, El-Sabagh A (2018) Role of legumes in soil carbon sequestration. In: Legumes for soil health and sustainable management. Springer Singapore. https://doi.org/10.1007/978-981-13-0253-4_4

Kumararaja P, Shabeer TPA, Manjaiah KM (2016) Effect of bentonite on heavy metal uptake by amaranth (Amaranthus blitumcv. Pusa Kirti) grown on metal contaminated soil. Indian J Hortic 73(2):224. https://doi.org/10.5958/0974-0112.2016.00052.9

Kundu S, Dotaniya ML, Lenka S (2013) Carbon sequestration in Indian agriculture. In: Lenka S, Lenka NK, Kundu S, Rao AS (eds) Climate change and natural resources management. India Publishing Agency, New Delhi, pp 269–289

Kushwah SK, Dotaniya ML, Upadhyay AK, Rajendiran S, Coumar MV, Kundu S, Subba Rao A (2014) Assessing carbon and nitrogen partition in Kharif crops for their carbon sequestration potential. Natl Acad Sci Lett 37(3):213–217. https://doi.org/10.1007/s40009-014-0230-y

Laidlaw MAS, Filippelli GM (2008) Resuspension of urban soils as a persistent source of lead poisoning in children: a review and new directions. Appl Geochem 23(8):2021–2039. https://doi.org/10.1016/j.apgeochem.2008.05.009

Lal R (2001) World cropland soils as a source or sink for atmospheric carbon. Adv Agron. Elsevier. https://doi.org/10.1016/s0065-2113(01)71014-0

Lal R (2003) Global potential of soil carbon sequestration to mitigate the greenhouse effect. Crit Rev Plant Sci 22(2):151–184. https://doi.org/10.1080/713610854

Lehmann J, Rillig MC, Thies J, Masiello CA, Hockaday WC, Crowley D (2011) Biochar effects on soil biota – a review. Soil Biol Biochem 43(9):1812–1836. https://doi.org/10.1016/j.soilbio.2011.04.022

Leyval C, Turnau K, Haselwandter K (1997) Effect of heavy metal pollution on mycorrhizal colonization and function: physiological, ecological and applied aspects. Mycorrhiza 7(3):139–153. https://doi.org/10.1007/s005720050174

Li H, Shi W-y, Shao H-b, Shao M-a (2009) The remediation of the lead-polluted garden soil by natural zeolite. J Hazard Mater 169(1–3):1106–1111. https://doi.org/10.1016/j.jhazmat.2009.04.067

Li H-b, Yu S, Li G-l, Deng H, Luo X-s (2011) Contamination and source differentiation of Pb in park soils along an urban–rural gradient in Shanghai. Environ Pollut 159(12):3536–3544. https://doi.org/10.1016/j.envpol.2011.08.013

Li F, Huang J, Zeng G, Yuan X, Li X, Liang J, Wang X, Tang X, Bai B (2013) Spatial risk assessment and sources identification of heavy metals in surface sediments from the Dongting Lake, Middle China. J Geochem Explor 132:75–83. https://doi.org/10.1016/j.gexplo.2013.05.007

Li F, Huang J, Zeng G, Huang X, Liu W, Wu H, Yuan Y, He X, Lai M (2015a) Spatial distribution and health risk assessment of toxic metals associated with receptor population density in street dust: a case study of Xiandao District, Changsha, Middle China. Environ Sci Pollut Res 22(9):6732–6742. https://doi.org/10.1007/s11356-014-3753-3

Li F, Huang J, Zeng G, Liu W, Huang X, Huang B, Gu Y, Shi L, He X, He Y (2015b) Toxic metals in topsoil under different land uses from Xiandao District, middle China: distribution, relationship with soil characteristics, and health risk assessment. Environ Sci Pollut Res 22(16):12261–12275. https://doi.org/10.1007/s11356-015-4425-7

Li F, Zhang J, Jiang W, Liu C, Zhang Z, Zhang C, Zeng G (2016) Spatial health risk assessment and hierarchical risk management for mercury in soils from a typical contaminated site, China. Environ Geochem Health 39(4):923–934. https://doi.org/10.1007/s10653-016-9864-7

Liang J, Feng C, Zeng G, Zhong M, Gao X, Li X, He X, Li X, Fang Y, Mo D (2017) Atmospheric deposition of mercury and cadmium impacts on topsoil in a typical coal mine city, Lianyuan, China. Chemosphere 189:198–205. https://doi.org/10.1016/j.chemosphere.2017.09.046

Liu J, Chen J, Huang L (2015) Heavy metal removal from MSS fly ash by thermal and chlorination treatments. Sci Rep 5(1). https://doi.org/10.1038/srep17270

Lojkova L, Datta R, Sajna M, Marfo TD, Janous D, Pavelka M, Formanek P (2015) Limitation of proteolysis in soils of forests and other types of ecosystems by diffusion of substrate. In: Amino acids, 2015, vol 8. Springer, Wien, pp 1690–1691

Mandal A, Thakur JK, Sahu A, Bhattacharjya S, Manna MC, Patra AK (2016) Plant–microbe interaction for the removal of heavy metal from contaminated site. In: Plant-microbe interaction: an approach to sustainable agriculture. Springer, Singapore. https://doi.org/10.1007/978-981-10-2854-0_11

Manna MC (2013) Dynamics of soil carbon pools and carbon sequestration. In: Kundu S, Mohanty M (eds) IISS contribution in frontier areas of soil research. Indian Institute of Soil Science, Nabibagh, pp 47–64

Marfo TD, Datta R, Lojkova L, Janous D, Pavelka M, Formanek P (2015) Limitation of activity of acid phosphomonoesterase in soils. In: Amino acids, vol 8. Springer, Wien, pp 1691–1691

Marschner H (1995) Saline soil. In: Mineral nutrition of higher plants. Academic, New York, pp 657–680

McGrath S, Brookes P, Giller K (1988) Effects of potentially toxic metals in soil derived from past applications of sewage sludge on nitrogen fixation by *trifolium repens* L. Soil Biol Biochem 20(4):415–424

Meena VD, Dotaniya ML (2017) Climate change, water scarcity and sustainable agriculture for food security. In: Kumar PS, Kanwat M, Meena PD, Kumar V, Alone RA (eds) Climate change & sustainable agriculture. New India Publishing Agency, New Delhi, pp 123–142

Meena RS, Meena VS, Meena SK, Verma JP (2015a) Towards the plant stress mitigate the agricultural productivity: a book review. J Clean Prod 102:552–553

Meena RS, Yadav RS, Reager ML, De N, Meena VS, Verma JP, Verma SK, Kansotia BC (2015b) Temperature use efficiency and yield of groundnut varieties in response to sowing dates and fertility levels in Western Dry Zone of India. Am J Exp Agric 7(3):170–177

Meena H, Meena RS, Singh B, Kumar S (2016) Response of bio-regulators to morphology and yield of clusterbean [*Cyamopsis tetragonoloba* (L.) Taub.] under different sowing environments. J Appl Nat Sci 8(2):715–718

Meena RS, Kumar S, Pandey A (2017) Response of sulfur and lime levels on productivity, nutrient content and uptake of sesame under guava (*Psidium guajava* L.) based agri-horti system in an acidic soil of eastern Uttar Pradesh, India. J Crop Weed 13(2):222–227

Meena BL, Fagodiya RK, Prajapat K, Dotaniya ML, Kaledhonkar MJ, Sharma PC, Meena RS, Mitran T, Kumar S (2018a) Legume green manuring: an option for soil sustainability. In: Meena R, Das A, Yadav G, Lal R (eds) Legumes for soil health and sustainable management. Springer, Singapore, pp 387–408. https://link.springer.com/chapter/10.1007%2F978-981-13-0253-4_12

Meena RS, Mitran T, Kumar S, Yadav GS, Bohra JS, Datta R (2018b) Application of remote sensing for sustainable agriculture and forest management. Elsevier

Meena RS, Kumar V, Yadav GS, Mitran T (2018c) Response and interaction of *Bradyrhizobium japonicum* and *Arbuscular mycorrhizal* fungi in the soybean rhizosphere. Rev Plant Growth Regul 84:207–223

Molaei A, Lakzian A, Datta R, Haghnia G, Astaraei A, Rasouli-Sadaghiani M, Ceccherini MT (2017a) Impact of chlortetracycline and sulfapyridine antibiotics on soil enzyme activities. Int Agrophys 31(4):499–505

Molaei A, Lakzian A, Haghnia G, Astaraei A, Rasouli-Sadaghiani M, Ceccherini MT, Datta R (2017b) Assessment of some cultural experimental methods to study the effects of antibiotics on microbial activities in a soil: an incubation study. PLoS One 12(7):e0180663

Naser HM, Rahman MZ, Sultana S, Quddus MA, Haoque MA (2017) Remediation of heavy metal polluted soil through organic amendments. Bangladesh J Agric Res 42(4):589–598. https://doi.org/10.3329/bjar.v42i4.35786

Nash KL, Jensen MP, Schmidt MA (1998) Actinide immobilization in the subsurface environment by in-situ treatment with a hydrolytically unstable organophosphorus complexant: uranyl uptake by calcium phytate. J Alloys Compd 271–273:257–261. https://doi.org/10.1016/s0925-8388(98)00066-8

Natarajan A, Hegde R, Naidu L, Raizada A, Adhikari R, Patil S, Rajan K, Sarkar D (2010) Soil and plant nutrient loss during the recent floods in North Karnataka: implications and ameliorative measures. Curr Sci 99(10):1333–1340

Nwoko C, Peter-Onoh C, Onoh GO (2012) Remediation of trace metal contaminated auto-mechanic soils with mineral supplementedorganic amendments. Univers J Environ Res Technol 2(6)

Odukudu FB, Ayenimo JG, Adekunle AS, Yusuff AM, Mamba BB (2014) Safety evaluation of heavy metals exposure from consumer products. Int J Consum Stud 38(1):25–34. https://doi.org/10.1111/ijcs.12061

Okorie A, Entwistle J, Dean JR (2011) The application of in vitro gastrointestinal extraction to assess oral bioaccessibility of potentially toxic elements from an urban recreational site. Appl Geochem 26(5):789–796. https://doi.org/10.1016/j.apgeochem.2011.01.036

Oliva J, De Pablo J, Cortina J-L, Cama J, Ayora C (2011) Removal of cadmium, copper, nickel, cobalt and mercury from water by Apatite II™: column experiments. J Hazard Mater 194:312–323. https://doi.org/10.1016/j.jhazmat.2011.07.104

Oliver DP, Tiller KG, Conyers MK, Slattery WJ, Alston AM, Merry RH (1996) Effectiveness of liming to minimise uptake of cadmium by wheat and barley grain grown in the field. Aust J Agric Res 47(7):1181. https://doi.org/10.1071/ar9961181

Ozcan H, Guvenc S, Guvenc L, Demir G (2016) Municipal solid waste characterization according to different income levels: a case study. Sustainability 8(10):1044. https://doi.org/10.3390/su8101044

Parr JF, Smith S (1969) A multipurpose manifold assembly in evaluating microbiological effects of pesticides. Soil Sci 107:271–276

Park JH, Choppala GK, Bolan NS, Chung JW, Chuasavathi T (2011) Biochar reduces the bioavailability and phytotoxicity of heavy metals. Plant Soil 348(1–2):439–451. https://doi.org/10.1007/s11104-011-0948-y

Pendall E, Leavitt SW, Brooks T, Kimball BA, Pinter PJ, Wall GW, LaMorte RL, Wechsung G, Wechsung F, Adamsen F, Matthias AD, Thompson TL (2001) Elevated CO2 stimulates soil respiration in a FACE wheat field. Basic Appl Ecol 2(3):193–201. https://doi.org/10.1078/1439-1791-00053

Pierart A, Shahid M, Séjalon-Delmas N, Dumat C (2015) Antimony bioavailability: knowledge and research perspectives for sustainable agricultures. J Hazard Mater 289:219–234. https://doi.org/10.1016/j.jhazmat.2015.02.011

Pipalde JS, Dotaniya ML (2018) Interactive effects of lead and nickel contamination on nickel mobility dynamics in spinach. Int J Environ Res 12(5):553–560. https://doi.org/10.1007/s41742-018-0107-x

Post WM, Kwon KC (2008) Soil carbon sequestration and land-use change: processes and potential. Glob Chang Biol 6(3):317–327. https://doi.org/10.1046/j.1365-2486.2000.00308.x

Puga AP, Abreu CA, Melo LCA, Beesley L (2015) Biochar application to a contaminated soil reduces the availability and plant uptake of zinc, lead and cadmium. J Environ Manag 159:86–93. https://doi.org/10.1016/j.jenvman.2015.05.036

Puschenreiter M, Horak O, Friesl W, Hartl W (2005) Low-cost agricultural measures to reduce heavy metal transfer into the food chain – a review. Plant Soil Environ 51(1):1–11. https://doi.org/10.17221/3549-pse

Qiao L (1997) The effects of clay amendment and composting on metal speciation in digested sludge. Water Res 31(5):951–964. https://doi.org/10.1016/s0043-1354(96)00290-4

Rajendiran S, Dotaniya ML, Coumar MV, Panwar NR, Saha JK (2015) Heavy metal polluted soils in India: status and counter measures. JNKVV Res J 49:320–337

Rajendiran S, Basanta Singh T, Saha JK, Vassanda Coumar M, Dotaniya ML, Kundu S, Patra AK (2018) Spatial distribution and baseline concentration of heavy metals in swell–shrink soils of Madhya Pradesh. In: India environmental pollution. Springer, Singapore. https://doi.org/10.1007/978-981-10-5792-2_11

Rajkumar M, Sandhya S, Prasad MNV, Freitas H (2012) Perspectives of plant-associated microbes in heavy metal phytoremediation. Biotechnol Adv 30(6):1562–1574. https://doi.org/10.1016/j.biotechadv.2012.04.011

Ram K, Meena RS (2014) Evaluation of pearl millet and mungbean intercropping systems in arid region of Rajasthan (India). Bangladesh J Bot 43(3):367–370

Rijkenberg MJA, Depree CV (2010) Heavy metal stabilization in contaminated road-derived sediments. Sci Total Environ 408(5):1212–1220. https://doi.org/10.1016/j.scitotenv.2009.11.053

Roane TM, Kellogg ST (1996) Characterization of bacterial communities in heavy metal contaminated soils. Can J Microbiol 42(6):593–603. https://doi.org/10.1139/m96-080

Rodrigues SM, Cruz N, Coelho C, Henriques B, Carvalho L, Duarte AC, Pereira E, Römkens PFAM (2013) Risk assessment for Cd, Cu, Pb and Zn in urban soils: chemical availability as the central concept. Environ Pollut 183:234–242. https://doi.org/10.1016/j.envpol.2012.10.006

Ruddiman WF (2003) The anthropogenic greenhouse era began thousands of years ago. Clim Chang 61(3):261–293. https://doi.org/10.1023/b:clim.0000004577.17928.fa

Rutgers M, Wouterse M, Drost SM, Breure AM, Mulder C, Stone D, Creamer RE, Winding A, Bloem J (2016) Monitoring soil bacteria with community-level physiological profiles using biolog™ ECO-plates in the Netherlands and Europe. Appl Soil Ecol 97:23–35. https://doi.org/10.1016/j.apsoil.2015.06.007

Sabir M, Waraich EA, Hakeem KR, Öztürk M, Ahmad HR, Shahid M (2013) Phytoremediation. In: Soil remediation and plants. Elsevier. https://doi.org/10.1016/b978-0-12-799937-1.00004-8

Saha JK, Selladurai R, Coumar MV, Dotaniya ML, Kundu S, Patra AK (2017) Soil protection policy. In: Environmental chemistry for a sustainable world. Springer Singapore. https://doi.org/10.1007/978-981-10-4274-4_13

Seaman JC, Hutchison JM, Jackson BP, Vulava VM (2003) In situ treatment of metals in contaminated soils with phytate. J Environ Qual 32(1):153. https://doi.org/10.2134/jeq2003.0153

Shahid M, Sabir M, Arif Ali M, Ghafoor A (2014) Effect of organic amendments on phytoavailability of nickel and growth of berseem (*Trifolium alexandrinum*) under nickel

contaminated soil conditions. Chem Speciat Bioavailab 26(1):37–42. https://doi.org/10.3184/095422914x13886890590610
Shamina Imran Pathan TV, Giagnoni L, Datta R, Baldrian P, Nannipieri P, Renella G (2018) Microbial expression profiles in the rhizosphere of two maize lines differing in N use efficiency. Plant Soil:401–413
Sharma A, Talukder G (1987) Effects of metals on chromosomes of higher organisms. Environ Mutagen 9(2):191–226. https://doi.org/10.1002/em.2860090210
Shazia G, Alia N, Iftikhar F, Muhammad I (2015) Reducing heavy metals extraction from contaminated soils using organic and inorganic amendments–a review. Pol J Environ Stud 24:1423–1426
Sihag SK, Singh MK, Meena RS, Naga S, Bahadur SR, Gaurav, Yadav RS (2015) Influences of spacing on growth and yield potential of dry direct seeded rice (*Oryza sativa* L.) cultivars. Ecoscan 9(1–2):517–519
Singh J, Kalamdhad AS (2012) Concentration and speciation of heavy metals during water hyacinth composting. Bioresour Technol 124:169–179
Smičiklas I, Smiljanić S, Perić-Grujić A, Šljivić-Ivanović M, Mitrić M, Antonović D (2014) Effect of acid treatment on red mud properties with implications on Ni(II) sorption and stability. Chem Eng J 242:27–35. https://doi.org/10.1016/j.cej.2013.12.079
Smil V (2000) Phosphorus in the environment: natural flows and human interferences. Annu Rev Energy Environ 25(1):53–88. https://doi.org/10.1146/annurev.energy.25.1.53
Soares MAR, Quina MJ, Quinta-Ferreira RM (2015) Immobilisation of lead and zinc in contaminated soil using compost derived from industrial eggshell. J Environ Manag 164:137–145. https://doi.org/10.1016/j.jenvman.2015.08.042
Sresty TVS, Madhava Rao KV (1999) Ultrastructural alterations in response to zinc and nickel stress in the root cells of pigeonpea. Environ Exp Bot 41(1):3–13. https://doi.org/10.1016/s0098-8472(98)00034-3
Staddon PL, Fitter AH, Graves JD (1999) Effect of elevated atmospheric CO2 on mycorrhizal colonization, external mycorrhizal hyphal production and phosphorus inflow in Plantago lanceolata and *Trifolium repens* in association with the arbuscular mycorrhizal fungus Glomus mosseae. Glob Chang Biol 5(3):347–358. https://doi.org/10.1046/j.1365-2486.1999.00230.x
Stefano DC, Milea D, Porcino N, Sammartano S (2006) Speciation of Phytate ion in aqueous solution sequestering ability toward mercury(II) cation in NaClaqat different ionic strengths. J Agric Food Chem 54(4):1459–1466. https://doi.org/10.1021/jf0522208
Stefano DC, Lando G, Milea D, Pettignano A, Sammartano S (2010) Formation and stability of cadmium(II)/Phytate complexes by different electrochemical techniques. Critical analysis of results. J Solut Chem 39:179–195
Tarnawski S, Hamelin J, Jossi M, Aragno M, Fromin N (2006) Phenotypic structure of Pseudomonas populations is altered under elevated pCO2 in the rhizosphere of perennial grasses. Soil Biol Biochem 38(6):1193–1201. https://doi.org/10.1016/j.soilbio.2005.10.003
The World Bank Annual Report 2015 (2015) World Bank annual report. The World Bank. https://doi.org/10.1596/978-1-4648-0574-5
Thornton I, Farago ME, Thums CR, Parrish RR, McGill RAR, Breward N, Fortey NJ, Simpson P, Young SD, Tye AM, Crout NMJ, Hough RL, Watt J (2008) Urban geochemistry: research strategies to assist risk assessment and remediation of brownfield sites in urban areas. Environ Geochem Health 30(6):565–576. https://doi.org/10.1007/s10653-008-9182-9
USEPA (2008) Landfill methane outreach program: basic information. Available: http://www.epa.gov/lmop/overview.htm. Accessed 6 May 2018
Uchimiya M, Cantrell KB, Hunt PG, Novak JM, Chang SC (2012) Retention of heavy metals in a Typic Kandiudult amended with different manure-based biochars. J Environ Qual 41:1138–1149
Varma D, Meena RS, Kumar S (2017) Response of mungbean to fertility and lime levels under soil acidity in an alley cropping system in Vindhyan Region, India. Int J Chem Stud 5(2):384–389
Varrault G, Bermond A (2011) Kinetics as a tool to assess the immobilization of soil trace metals by binding phase amendments for in situ remediation purposes. J Hazard Mater 192(2):808–812. https://doi.org/10.1016/j.jhazmat.2011.05.097

Verma JP, Jaiswal DK, Meena VS, Meena RS (2015a) Current need of organic farming for enhancing sustainable agriculture. J Clean Prod 102:545–547

Verma SK, Singh SB, Prasad SK, Meena RN, Meena RS (2015b) Influence of irrigation regimes and weed management practices on water use and nutrient uptake in wheat (*Triticum aestivum* L. Emend. Fiori and Paol.). Bangladesh J Bot 44(3):437–442

Wani PA, Khan MS, Zaidi A (2008) Chromium-reducing and plant growth-promoting Mesorhizobium improves chickpea growth in chromium-amended soil. Biotechnol Lett 30(1):159–163. https://doi.org/10.1007/s10529-007-9515-2

Wellman DM, Icenhower JP, Owen AT (2006) Comparative analysis of soluble phosphate amendments for the remediation of heavy metal contaminants: effect on sediment hydraulic conductivity. Environ Chem 3(3):219. https://doi.org/10.1071/en05023

Wellman DM, Pierce EM, Valenta MM (2007) Efficacy of soluble sodium tripolyphosphate amendments for the in-situ immobilisation of uranium. Environ Chem 4(5):293. https://doi.org/10.1071/en07030

Wu S, Shi Y, Zhou S, Wang C, Chen H (2016) Modeling and mapping of critical loads for heavy metals in Kunshan soil. Sci Total Environ 569-570:191–200. https://doi.org/10.1016/j.scitotenv.2016.06.072

Xie Y, Fan J, Zhu W, Amombo E, Lou Y, Chen L, Fu J (2016) Effect of heavy metals pollution on soil microbial diversity and bermudagrass genetic variation. Front Plant Sci 7. https://doi.org/10.3389/fpls.2016.00755

Xiong T, Austruy A, Pierart A, Shahid M, Schreck E, Mombo S, Dumat C (2016) Kinetic study of phytotoxicity induced by foliar lead uptake for vegetables exposed to fine particles and implications for sustainable urban agriculture. J Environ Sci 46:16–27. https://doi.org/10.1016/j.jes.2015.08.029

Yadav GS, Babu S, Meena RS, Debnath C, Saha P, Debbaram C, Datta M (2017a) Effects of godawariphosgold and single supper phosphate on groundnut (*Arachis hypogaea*) productivity, phosphorus uptake, phosphorus use efficiency and economics. Indian J Agric Sci 87(9):1165–1169

Yadav GS, Datta R, Imran Pathan S, Lal R, Meena RS, Babu S, Das A, Bhowmik S, Datta M, Saha P (2017b) Effects of conservation tillage and nutrient management practices on soil fertility and productivity of rice (*Oryza sativa* L.)–rice system in North Eastern Region of India. Sustainability 9(10):1816

Yadav GS, Lal R, Meena RS, Babu S, Das A, Bhomik SN, Datta M, Layak J, Saha P (2017c) Conservation tillage and nutrient management effects on productivity and soil carbon sequestration under double cropping of rice in North Eastern Region of India. Ecol Indic

Yoon JK, Cao X, Ma LQ (2007) Application methods affect phosphorus-induced lead immobilization from a contaminated soil. J Environ Qual 36(2):373. https://doi.org/10.2134/jeq2006.0316

Legumes for Carbon and Nitrogen Cycling: An Organic Approach

Sandeep Kumar, Ram Swaroop Meena, Rahul Datta, Sunil Kumar Verma, Gulab Singh Yadav, Gourisankar Pradhan, Ali Molaei, G. K. M. Mustafizur Rahman, and H. A. Mashuk

Abstract Food security, land degradation, eliminating the threats of climate change, soil sustainability, and crop productivity are the critical challenges in the coming years. Therefore, the sustainability of the agricultural production system is becoming a central component in enhancing food security and environmental sustainability. Legume crops could play a significant role in this perspective by carrying out numerous services in keeping with principles of sustainability. Incorporating legume crops into crop rotation is essential for implementing and integrating the conservation and improvement of soil health, quality, and fertility with diverse aspects of crop and livestock production into the natural farming systems. The source of nutrition to subsequent crops to maintain a continuous nitrogen (N) supply chain greatly varies with regard to the inclusion of legumes in the cropping system. The crop rotation, including pigeon pea (*Cajanus cajan*), improved the total N content in the uppermost soil strata by around 100 μg g^{-1} soils, in comparison

S. Kumar
Department of Agronomy, CCS Haryana Agricultural University, Hisar, Haryana, India

R. S. Meena (✉) · S. K. Verma · G. Pradhan
Department of Agronomy, Institute of Agricultural Sciences (BHU), Varanasi, UP, India
e-mail: Meenars@bhu.ac.in

R. Datta
Department of Geology and Pedology, Mendel University in Brno, Brno, Czech Republic
e-mail: rahul.datta@mendelu.cz

G. S. Yadav
Division of Crop Production, ICAR Research Complex for NEH Region, Lembucherra, India

A. Molaei
Department of Soil Science, Faculty of Agriculture, Ferdowsi University of Mashhad, Mashhad, Iran

G. K. M. M. Rahman · H. A. Mashuk
Department of Soil Science, Bangabandhu Sheikh Mujibur Rahman Agricultural University (BSMRAU), Gazipur, Bangladesh
e-mail: mustafiz@bsmrau.edu.bd

R. Datta et al. (eds.), *Carbon and Nitrogen Cycling in Soil*,
https://doi.org/10.1007/978-981-13-7264-3_10

with 25 μg g^{-1} soils in cropping systems without inclusion of grain legume in the crop rotation. Carefully planned diverse crop rotations diminish the prevalence of insect pests, pathogens, diseases, and weeds. This potential for reducing the attacks by insect pests, diseases, weeds, that positively shape the soil fertility, as a result were reported to boost successive cereal production by 15% to 25%. By means of the complex interactions among the various crop production inputs, the conservation and improvement of soil fertility in crop rotation depends on the long-term integrated approach, in spite of conventional agriculture.

Keywords Crop productivity · Crop rotation · Legumes · Nitrogen fixation · Organic agriculture · Soil fertility

Abbreviations

BNFs	Biological nitrogen fixations
C	Carbon
CO_2	Carbon dioxide
GHGs	Greenhouse gases
IPCC	Intergovernmental panel on climate change
K	Potassium
N	Nitrogen
P	Phosphorus
SOC	Soil organic carbon
SOM	Soil organic matter
UN	United Nations
WUE	Water use efficiency

1 Introduction

In 2012, the United Nations (UN) predicted that the overall population will reach approximately 9.6 billion people worldwide. Therefore, the increased destruction of human activities has considerably altered the chemical make-up of the global biosphere, with enormous implications for the current and future climate. Humans will face diverse inclusive challenges, for example: (i) nutrition and food insecurity; (ii) climate change risks owing to the increasing net discharge of harmful greenhouse gases (GHGs) into the atmosphere; (iii) the growing need for crop production and the good physical condition of the soil; and (iv) the loss of biodiversity and changes in ecosystems (Fliebach et al. 2007). In fact, human activities have been responsible for about 110% of observed warming (ranging from 72% to 146%), with natural factors in isolation leading to a slight cooling over the past 50 years, as pointed out by the Intergovernmental Panel on Climate Change (IPCC) (Hausfather 2017). The effect of the climate change is closely linked with biotic and abiotic stresses, and will have increasingly severe implications for the global food production system (Yadav et al. 2015; Kumar et al. 2016; Ashoka et al. 2017). If in due course no action is taken against climate change, it could cut the probable improvement in food availability by around one third by 2050, along with a per capita average reduction of food availability of 3.2% (kcal day^{-1}), fruit and vegetable intake of 4.0% (14.9 g day^{-1}), and red meat consumption of 0.7% (0.5 g day^{-1}) (Oxford 2016). Projections suggest that the number of people at risk for hunger will increase by 10–20% by 2050 owing to climate change, with 65% of this population in Sub-Saharan Africa (Kumar et al. 2017b). To meet these challenges, a strategy outline is required in which the soil sustainability and crop productivity patterns become essential and vital components. Within this framework, the incorporation of leguminous crops into the production system by way of crop rotation can play a fundamental role by delivering multiple functions with productivity and sustainability principles (Voisin et al. 2014; Jangir et al. 2017; Mitran et al. 2018): (i) at a nutrient production level as an important source of plant proteins (20–25%) (Tharanathan and Mahadevamma 2003); (ii) at a crop production level, because of the ability to fix free atmospheric nitrogen (N) in a form that is available to plants (150 Tg N every year), building soil fertility (Watson et al. 2002; Drinkwater et al. 2017), owing to their function in mitigating GHG emissions (release 5–7 times less GHGs per unit area compared with other crops) (Lemke et al. 2007; Stagnari et al. 2017), improving the soil's organic carbon (SOC) pool (0.88 Mg C ha^{-1} year^{-1}) (Kumar et al. 2018), and encouraging effects on the quantity and quality of the economic produce of successive crops (Kirkegaard et al. 2008; Preissel et al. 2015); (iii) at the crop protection level – disturbing the biological cycles of insect pests and diseases (Van Emden 1991; Nemecek et al. 2008), weeds, and nematodes (McSorley et al. 1994).

Fabaceae is one of the prime families of plants, including 20,000 species cropped per annum, harvested only for dry grain, growing across an extensive range of climates and soil types (Smýkal et al. 2015). Grain legumes are being promoted for utilization as an affordable base of nutritional, protein-rich grains for the human diet (Asif et al. 2013; Temba et al. 2016), animal feed and nourishment (Jezierny et al. 2010), and for their socio-economic impact (Duranti and Gius 1997). Thus, legumes have an important and considerable responsibility in maintaining global food security, ecosystem buoyancy, and support to the various environmental remunerations to the agricultural landscape, improving resource use efficiency, and adding to the balancing of plant protein production (Nemecek et al. 2008; Peoples et al. 2009b; Köpke and Nemecek 2010).Thus, the incorporation of legumes into crop rotation shifts the agricultural production systems away from input-intensive unsustainable systems to more ecologically and economically sustainable ones (Tilman 1998). Indiscriminate and unbalanced application of synthetic nitrogenous fertilizers condenses the grain yields, resulting in steady but unseen economic losses to farmers (Qiao et al. 2012) because of the serious erosion in soil properties and eutrophication from drainage (Vitousek et al. 2009). In China, the loss of mineral N from agroecosystem runoff is projected to reach above 1.74×109 kg year^{-1} (Duan et al. 2000), which is the key contributor to pollution of the water bodies. To moderate the runoff loss of N, a nutrient input source reduction technique was suggested by Yang et al. (2004). In this regard, a sustainably managed agricultural production system not only continues good crop production and plant nutrient supply, but also eliminates the runoff losses of elemental N (Larkin 2008; Kumar et al. 2017a).

The practice of crop rotation can be defined as the sequential production of diverse plant species on the same piece of land (Karlen et al. 1994; Kakraliya et al. 2018; Dadhich et al. 2015). The term "crop rotation" implies that constant monocropping is undesirable and has an adverse effect owing to the increased pressure and stress from insect pests, diseases, and weeds, as this crop rotating system with legumes is the primary way of controlling all the pathogen-causing agents in organic farming (Altieri 1995). The roots of extensive, diverse groups of plants in a rotating system exudate certain bio-chemicals that considerably inhibit the germination and growth rate of latent competitor plants growing in nearby/ surrounding areas. This effect of biochemical/secondary metabolites released by plants on other nearby plants is known as allelopathy, which is exhibited by crop plants such as sunflower (*Helianthus annuus*), rye (*Secale cereal*), maize (*Zea mays*), triticale (*Triticosecale* Wittm.), wheat (*Triticum aestivum*), barley (*Hordeum vulgare*), and vetch (*Vicia sativa*) (Bahadur et al. 2015). Some of the soil fertility-associated factors are also associated with the management of insect pests, soil-borne diseases, leading to increased competition, soil microbial activity, predation, and parasitism in the surrounding regions of the roots – the rhizosphere (Knudsen et al. 1995). Soil-borne pathogens can be affected by a crop rotation span with reduced disease levels among susceptible crops (Clark et al. 1998). In one of their experiments, Phelan et al. (1995) observed that soil organic matter (SOM) management in the past is also interrelated with the vulnerability of crops to ground pests, e.g., European corn borer (*Ostrinia nubilalis*). The inclusion of legumes in the rotation system is very promising and significantly increases soil N supply through atmospheric N fixation and the yields of subsequent crops in the rotation system (Angus et al. 2006; Espinoza et al. 2012). The current agronomic developmental processes, which may absolutely control the production of grain legumes, include the extension of organic farming (Robson et al. 2002), the expansion of reduced/zero tillage (Siddique et al. 2012), or the absence of nonlegume cropping systems (Schäfer 2013). Research work on legume crops can be improved by refining agronomical practices in legume cultivation and by developing efficient breeding programs (Jensen et al. 2011), by utilizing legumes and N provision well (Lithourgidis et al. 2011; Podgorska-Lesiak and Sobkowicz 2011), or by assisting multiple management decisions associated with diverse cropping systems (Reckling et al. 2012).

Crop rotation is of utmost significance in organic farming, where the legumes are the central elements for a successful farming system (Watson et al. 1996). According to Altieri (1995), organic systems are generally described as diverse crop rotation that has the potential for the management of insect pests and diseases. An integrated farming system based on low-input that also provides environmental benefits through a reduction in pesticides, fertilizer and tillage inputs, and crop requirements, because the chemical insecticides, pesticides, and fertilizers are banned in organic farming systems (Jordan et al. 1997). The application of fertilizers and pesticides does not always show the favorable impacts of crop rotation on crops and soil in conventional farming, whereas the crop rotation system is outstanding in organic/natural and low-input farming systems (Bullock 1992). Organic farming improves nutrient cycling and provides environmental benefit, in addition to boosting the quality of the soil (Stockdale et al. 2001). In support of these facts, farmers, researchers, and governmental policy makers are considering crop rotation. The flexible market demands a better perception and understanding of crop rotation that moderate soil quality and harmonize the benefits of rotation with the requirements of farmers.

2 Crop Rotation

Crop rotation with legumes is a central and fundamental process for successful natural and organic farming. This rotational preparation is an annual, on-going procedure that incorporates information, aims, and objectives for many years. It is a type of cyclic rotation wherein the same sequence of crops is repeated continually on the same piece of land, simultaneously progressing the business and management aims of the farmer. Effective crop rotation requires strategic planning for the long term. Sometimes, planning may prove unproductive and unsuccessful because of disturbance by weather, changes in the market demands, labor supply, labor costs, and other factors. However, a lack of planning procedure and management can lead to severe problems, e.g., the soil-borne pathogen due to imbalances in soil nutrient supervision and management. Such problems can require additional and expensive labor, and cause an inability to provide for the demands of market. Rational estimates of legume rotation benefits are needed to guide the design of improved cropping systems and patterns (Carsky et al. 2001), to evaluate the cost-effectiveness of leguminous grain and its involvement in successive crops in the cropping systems (Weitbrecht and Pahl 2000; LMC International 2009), and the farm's economic assessment of the rotational profits of legumes (Beattie et al. 1974). The function of grain legumes as a pre-crop mechanism in cropping systems is increasingly well understood regarding N fixation, several positive environmental bearings (Nemecek et al. 2008; Jensen et al. 2011; Meena et al. 2016a), and their prospective responsibility in organic farming and conservation tillage under crop production (Robson et al. 2002; Luetke-Entrup et al. 2006).

2.1 Need for Crop Rotation

Crop rotation is the practice of growing a sequence of plant species on the same piece of land (Bullock 1992) that is characterized by a cyclic period (Leteinturier et al. 2006) and cheaper resources of advancing soil properties and system productivity (Bagayoko et al. 2000). Likewise, the crop rotating system is also a well thought-out, essential component of an integrated farming system (Stoate et al. 2001), which is inconsistent with monocropping as a sustainable resolution for farming (Leteinturier et al. 2006). It has the potential for various environmental impacts on agriculture through achievements in production and maintaining productivity over many years (Vandermeer et al. 1998) and also for reducing the reliance on external agro-inputs (Bullock 1992) as a result of N delivered by legumes through the biological process of N fixation (Bagayoko et al. 2000). In one of their experiments, Kirkegaard et al. (2008) observed that crop rotation has the potential to reduce the attack of insect pests, diseases, and weeds that positively shape the soil fertility, and as a result successive cereal crop yields increased by 15–25%. The severity of leaf and root disease infection was also reduced in cereal followed by legumes and the yield benefit of legumes was also estimated (Stevenson and van Kessel 1997). Crop rotation remains at the center of all agronomic practices for sustainable production systems and the yield is usually 10–15% higher than monoculture (Cook 1984). The yield profit was also achieved at 49% for Australia and 37% for tropical systems in crop rotation with legumes as a pre-crop followed by cereal crops, without applying any chemical or synthetic fertilizers (Kirkegaard et al. 2008). Experimental results from temperate Europe showed that the economic yield of cereal was more than normal, 17% and 21% respectively in crop rotation monoculture under moderate levels of fertilization (Charles and Vuilloud 2001; Dachler and Köchl 2003). In one of the other experiments, it was observed that rapeseed and potato yields were also higher under crop rotation with legume grains as a pre-crop (Charles and Vuilloud 2001).

A legume-based cropping system also reduces N losses and the leaching of nutrients, and greatly enhances the fraction of crop's remaining carbon (C) that is sequestered in stable SOM (Drinkwater et al. 1998). The inclusion of green manure, forage legumes, and crop rotation along with production systems, does develop the fertility, productivity, and value of soils (Campbell et al. 1992; Biederbeck et al. 1998; Varma et al. 2017; Meena et al. 2018) and reduces dependence on chemical fertilizers (Greenland 1985; Kirkegaard et al. 2008). As a result, environmental conditions and cost-effectiveness concerns move the agricultural system away from input-intensive to more ecologically sustainable production systems (Tilman 1998). This environmental growing condition increases dependence on soil biological processes and on high microbial diversity, as soil microorganisms regulate most of the processes in soil essential for crop production (Paul and Clark 1996) and soil sustainability (Voisin et al. 2014). Thus, crop rotation is designed to (1) earn income, (2) increase soil quality, (3) conserve the ecological system.

2.2 Principles of Crop Rotation

1. Deep-rooted crops should rotate with shallow-rooted crops, e.g., vegetable crops have the ability to extract N from deeper soil layers (Thorup-Kristensen 2001).
2. N-fixing crops should be followed by N demanding crops, e.g., leguminous crops followed by cereal crops (Magdof and Van Es 2000).
3. Crops that are vulnerable to weeds should follow weed-tolerant and -suppressing crops (Lampkin 1990).
4. Autumn- and spring-sown crops should be alternated (Lampkin 1990).
5. Different crops should be suited to various climatic and soil conditions.
6. A balance between cash and forage crops must be maintained (Lampkin 1990).
7. The one annual crop should be grown only once a year (Magdof and Van Es 2000).
8. One crop should not be followed by another closely related species (Lampkin 1990; Magdof and Van Es 2000).
9. Crop sequences that promote healthier crops (Magdof and Van Es 2000) and aid in controlling weeds should be followed (Lampkin 1990).
10. Crops that leave behind a good amount of crop residue in the field after harvest should be adjusted in the crop rotating system (Lampkin 1990; Magdof and Van Es 2000; Thorup-Kristensen 2001), e.g., in general, vegetable crops leave behind additional N in their residue (Thorup-Kristensen 2001).

2.3 Crop Rotation with Cropping Systems

Crop rotational management requires perceptive and multidimensional judgment in the entire field, each individual field, and in the complementary cropping systems in the field. For successful farming, crop rotation planning is a systematic and approachable process, which requires balancing annual and multiyear decisions. For successful farming, planning and implementing are vital components for rotation on an annual and seasonal opportunistic source. Their understanding of the biological ethics of organic farming allows farmers to decide the margins of their fertilizers and pest control management practices. The cash crops and legume crops are a significant source of income. Expert farmers mostly give priority to cultivating their main crops in the most appropriate soils and always focus on crops that are lack the ability to cooperate in improved soil health and a long-standing output (Charles and Johnson 2009). Different types of crops are managed under crop rotation, e.g., leafy vegetables need N, fruiting crops need phosphorus (P), rooting crops need potassium, and legumes are soil builders and provide the cropping system with N, as shown in Fig. 1.

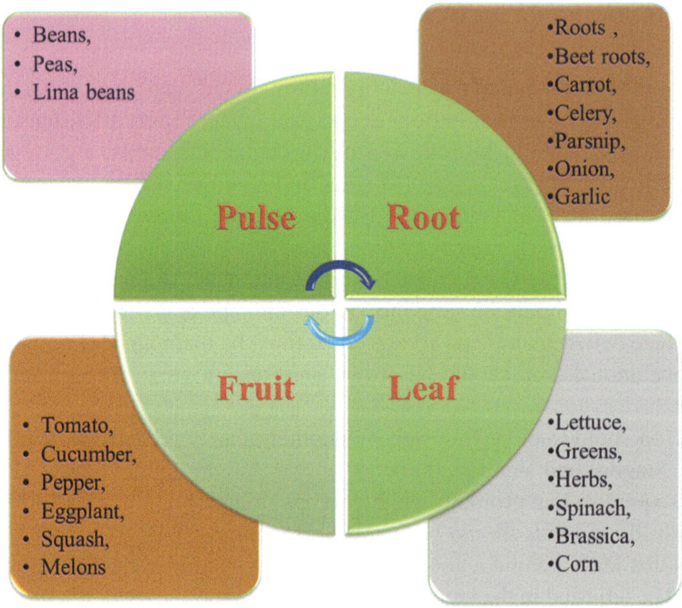

Fig. 1 Crop rotation management in a farm. (Adapted from Reddy 2016)

3 Processes in Crop Rotation

Crop rotation by means of legume crops promotes a physiological process in the soil by enhancing soil formation (Rochester et al. 2001; McCallum et al. 2004), improving soil quality, health, structure, and porosity (McCallum et al. 2004) by increasing the capacity for water retention, and the concentration of soil organic carbon and humus (Jensen et al. 2011) in addition to the maintenance of soil aggregates (Raimbault and Vyn 1991) and is a potential way of increasing the available N supply for cereals at a low cost (Angus et al. 2006; Espinoza et al. 2012; Yadav et al. 2017b). Small grains such as barley, rye, and legumes are liable to encourage soil aggregation with a higher degree of confidence compared with many other cash crops (Angers and Mehuys 1988). In Australia, in a study conducted by Chan and Heenan (1991), the conservation in soil structure was observed in the following order: lupin (*Lupinus angustifolius* L.) > lentil (*Lens culinaris* L.) > canola (*Brassica napus* L.) > pea (*Pisum sativum* L.) > linseed (*Linum usitatissimum* L.) > barley. The leguminous crops not only make a contribution to improving the physical structure of the soil by means of flouting the soil's hardness to expand the air and water circulation in the soil, but also by extracting essential plant nutrients from the deeper soil profile, e.g., pigeon peas possess a strong, deep root system and extract nutrients from deep in the soil. The nutrients accumulate in the deeper soil strata by the decomposition of residual crop, which contributes to nutrient cycling. The experiment conducted by Likoswe (1994) in the Malawi region recorded that the crop rotation along with pigeon peas improved the cumulative N content in the upper soil to around 100 μg g^{-1} soil, compared with 25 μg g^{-1} soils in the crop rotation lacking grain legumes. The build-up of SOM throughout the ley phase also plays an important expressive role in the development of soil's physical structure (Chan and Heenan 1991) and maintenance (Grace et al. 1995). This is a consequence of the invention of agents that bind organic matter, for instance, polysaccharides, decomposition of SOM by soil microorganisms (Hoyt and Hargrove 1986), nutritional effects of organic fertilizers (Fragstein and Schmidt 1999), effects on the right balances of essential plant nutrients (Eriksen et al. 1999), N released through mineralization, and reducing N losses (Wander et al. 1994) through a decrease in the leaching of nutrients (Crews and Peoples 2004) enhance the withholding of N from the system (Reents and Möller 1999) by soil nutrient accessibility (Mikkelsen 1999) and N supply to subsequent crops (Loges et al. 1999) and decreasing C inputs (Drinkwater 1999) by protecting SOC from microbial breakdown and increased net C sequestration (Wander et al. 1994). Consequently, forage legumes had a greater soothing impact on the soil system in comparison with that of monoculture and fallow soil. Finally, legume crops actively enhance the velocity of biological weathering and the mineralization of soil mineral components through increasing the ease of a range of micronutrients (Bormann et al. 1998), increasing both N and P in addition to developing N-scavenging effectiveness (Tilman 1999), controlling decomposition, N mineralization (Wedin and Tilman 1990), management of mycorrhiza (Kahiluoto and Vestberg 1999), humus development in organic ecosystems (Wedin and Tilman 1990; Couteaux et al. 1995), better utilization of cover crops and left-over plant residues for the management of weed and plant diseases (Brandsæter and Riley 1999), and sheep grazing (Jones and Harris 1999) also.

The cropping system along with legumes also develops the water use efficiency (WUE), as legumes have a high leaf area and leaf area index, which leads to the conservation of water (Ogindo and Walker 2005). This in turn leads to an increase in water infiltration rate, air circulation, a reduction in soil erosion (Unger and McCalla 1980), and decreased weed proliferation (Brandsæter and Riley 1999). Cereal–legume crop rotation leads to high water use efficiency as it reduces water evaporation, improves soil moisture conservation in comparison with monocropping of maize (Ghanbari et al. 2010), and as a result obtains a profitable yield (Rana and Rana 2011; Thayamini et al. 2010). The comparative array of legumes and grass cover crops leads to the cumulative water infiltration observed by Obi (1999) in Nigeria: legumes (*Pueraria phaseoloides* L. and *Stylosanthes gracilis* L.) > grasses (*Axonopus compressus* L., *Cynodon plectostachion* L., *Pennisetum polystachion* L., and *Panicum maximum* L.) > bare soil

3.1 Soil Tilth

Tilth in general refers to the physical condition of the soil resulting from tillage. It is related to seed germination, seedling growth, root proliferation, root development, enlargement, fruiting, allowing crops to flourish and consequently increasing the overall yield. In addition, good soil tilth encourages other processes, for example, water infiltration rate and soil airing, which shows a positive impact on both the crop and the environment. These processes are favored by the presence of soil crumbs and stable aggregates, i.e., soil tilth. Crop rotation has a constructive effect on soil tilth subject to crop species being rotated. Other techniques for developing good soil tilth comprise drastically reduced or no/zero-till farming, direct drilling, along with legume crops and the introduction of pulse crops under crop rotation. Thus, zero tillage can be defined as an approach of cultivating crops year after year without affecting the soil system through tillage implementation. The integration of crop rotation along with no-till reduces the impact of rainfall, sediment detachment, sediment transport, surface soil runoff, in addition to losses through soil erosion (Unger and McCalla 1980).

Legumes indirectly improve soil's physical properties and organic C content according to a reduced tillage trial, i.e., legume crop rotations followed by a 2- to 3-year tillage break. Reduced tillage promoted the rapid decomposition of leguminous residues (Nemecek et al. 2008; Alpmann et al. 2013) and enhanced the cycling of nutrients in soil involving lower costs and less energy for cultivation (Hoyt and Hargrove 1986). In Germany, in a study of 75 grain legume producers, 64 respondents plasticized direct seeding followed by legume pre-crop and 30 had reduced or zero-tillage on the whole farm (Alpmann et al. 2013). Faba beans (*Vicia faba*) are directly sown in the rainfed Mediterranean region (López-Bellido et al. 2003) or in areas with low stress from perennial weeds (Köpke and Nemecek 2010; Sihag et al. 2015). By facilitating reduced or zero-tillage systems, the legumes show directly noticeable soil-improving effects on the farm's economic performance. Conservation tillage has been shown to decrease the cost of cultivation more considerably when the no-till was improved by the diversification of the crop rotation with legumes (Lakhran et al. 2017). The cost saving was 21% in a rotation with conservation tillage including a legume, but a cost saving of only 12.5% was made in the tillage system under cereal-dominated rotation in an experiment conducted by Luetke-Entrup et al. (2006).

Likewise, conservation tillage, which includes zero tillage or no-till, practices minimum tillage, and drastically reduced tillage operations are in the habit of growing cereals-based crop rotation. It is accepted not simply for the protection and maintenance of naturally available resources, but also to save time, fuel, wear and tear on agricultural machineries and equipment, and for the ecologically sound and viable qualities of the various techniques. By reducing soil interruption, it promotes a complex decomposition system that improves the stability of the soil environment and the proficiency and continuity of the nutrient cycling chain compared with on-going cropping systems (Francis and Clegg 1990). Tillage reduces the uniformity and multiplicity of the bacterial population in the soil ecosystem by decreasing both the substrate abundance and the uniformity increase in the soil's microbial biomass and diversity, thus confirming them to be more sustainable management systems than conventional tillage and fallowing (Lupwayi et al. 1998). No-tilth soils have a higher bulk density and generally the soil surface is rich in more water-stable aggregates (Edwards et al. 1987), has increased soil water content (Lafond et al. 1992), lower soil temperature, weeds, and insect pest control (Buhler 1995; Cardina et al. 2002), increased disease suppression (Peters et al. 2003) by enhancing the antibiotic abilities of endophytic and root zone bacteria, and as a result consistently increased seed yield (Edwards et al. 1987). Tillage intensity is the most important determinant of increasing SOM in an annual cropping system, and the build-up of SOM could help C sequestration (Sampson and Scholes 2000; Lal et al. 1998) and the C pool equilibrium level of the soil (Cambardella and Elliot 1994), making the soil more resilient. Tillage commonly reduces the organic C equilibrium levels of the soil by increasing the return rate of C that is structurally labile but physically protected (Cambardella and Elliot 1994).

3.2 *Soil Fertility*

Soil fertility is the ability of soil to supply the essential plant nutrients in adequate amounts and in suitable proportions to sustain agricultural plant growth, or we can say that it is the capacity of soil to produce and reproduce (Stocking and Murnaghan 2001). The problem of soil quality and soil fertility status has deteriorated because of population pressure, land constraints, harsh climatic conditions, the continuous use of exhaustive/high yielding crop varieties, with increasing crop intensity, declining traditional soil management practices (Bumb and Baanante 1996), and also because of the over- and imbalanced use of synthetic fertilizers and decreasing use of organic manure (Conway and Pretty 1991). N is the main essential nutrient element that is usually vital for plants in large quantities, and has the quickest and most prominent impact on growth. Most plants take up N in the form of NO_3^- and NH_4^+ from the soil. Air consists of 78% N, but most plants cannot access it. Leguminous crops are therefore special in that they host N-fixing bacteria such as *Rhizobium*, which can convert atmospheric N into an available form that can be used by the host plant. The rotation of cereal legumes not only improves the productivity of cereal crops, but also saves the use of N, i.e., up to 40 kg Nha^{-1}. The easy availability and expanded use of synthetic nitrogenous fertilizer makes it more reactive to N in the atmosphere (Rockström et al. 2009) and also intensifies the risks of NO_3 leaching into water bodies and N_2O emission into the biosphere. N and P are the utmost crop production limiting factors (Gikonyo and Smithson 2003). The highly insoluble Ca and P are bounded by root exudates of legume crops (Arhara and Ohwaki 1989). The leguminous crops have the ability to efficiently dissolve and then utilize the strongly fixed P-Fe/Al on micelle–clay in P-rich acid soils, and possess a moderately low N harvest index (Sanginga et al. 2000). Inorganic or synthetic fertilizers are useful for enhancing the yield production, but these chemical fertilizers are the origin of soil deterioration when applied alone and are also more expensive so they are not affordable to small-scale farmers (McIntire and Fussel 1986). A leguminous crop does not require nitrogenous fertilizer and this effect reduces fertilizer use in subsequent crops in rotation management, reduces fertilizer N requirement

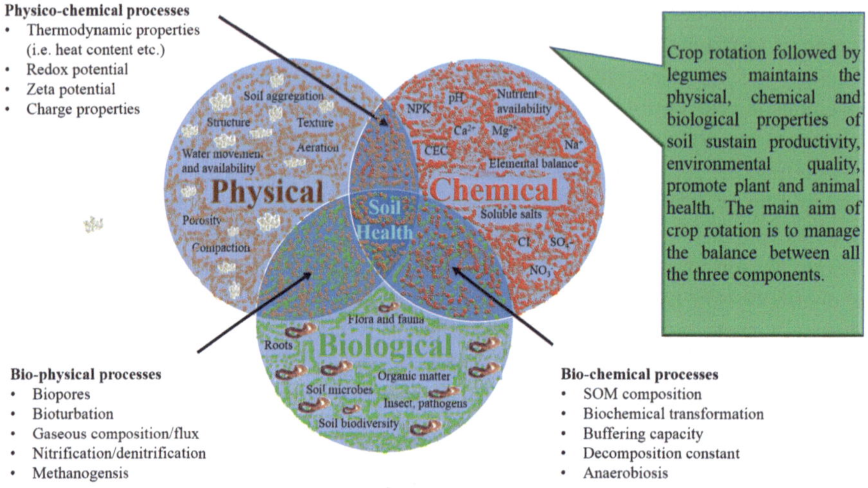

Fig. 2 Crop rotation maintains the balance among these three components

(Greenland 1985), and leads to a significant boost of the obtainable soil N flux (Pierce and Rice 1988). This may be accomplished by adapting grain legumes (Greenland 1985), forage legumes (Giller and Cadisch 1995) or herbaceous cover legumes (Weber 1996). A grain legume crop may save 150–200 kg ha^{-1} of N fertilizer compared with a monocropping of cereal, which left over 20–50 kg ha^{-1}N for the next crop, whereas forage legumes save 150–300 kg ha^{-1} of N fertilizer by fulfilling their own N requirement (Peyraud et al. 2009), together with a residual 20–50 kg ha^{-1}N for the subsequent crop. In case studies across Europe, it has been reported that forage and grain legume-based crop rotation saved on average 88 and 21 kg N fertilizer ha^{-1} compared with rotations without the inclusion of legume plants (Bues et al. 2013). Thus, under this condition of N derived from biological N fixations (BNFs) to the subsequent crops increases crop yield and productivity (Bagayoko et al. 2000; Köpke and Nemecek 2010; Verma et al. 2015).

Crop rotations are very efficient at maintaining and enhancing the value of the soil's physical, biological, and chemical properties, as shown in Fig. 2. The physical characteristics comprise an increase in soil aggregation, granular structure, and friable consistence, and a decrease in the crusting of soil surfaces (Verma and Shekhawat 1991; Bullock 1992), a decrease in soil bulk density, which can significantly obstruct root development and natural nutrient cycling (Tian et al. 2000), and a reduction in the erosion rate (Hoyt and Hargrove 1986; Mielke and Schepers 1986). Crop rotation is a systematic and efficient approach, and includes a combination of leguminous and cash crops, which provides the central means for fertility building and nutrients supply within organic farming and sustainable production systems (Stockdale et al. 2001). The legume in a cropping system can also accelerate soil fertility, health, quality, porosity, and the soil's physical structure (McCallum et al. 2004) and significantly affect the diversity and activity of soil microbial populations in the root atmosphere (Kirkegaard et al. 2008). The special architectural features of the root systems of different legume crops integrated into the rotation also positively moderate and directly act on the development of the soil's structure (Chan and Heenan 1991) through the accumulation of SOM (Grace et al. 1995), resulting from the microbial production of organic binding agents, i.e., polysaccharides decomposing SOM and the entangling impacts of fungal hyphae (Hoyt and Hargrove 1986). Including legumes in crop rotation is a potential way of increasing the available N supply for cereals at a low cost (Angus et al. 2006; Espinoza et al. 2012). The combined impact of all these factors is that post-legume cereal yields are often reported to be 40–80% greater than those achieved in cereal without N fertilizer, representing an additional 450–1000 kg of additional grain per hectare across a range of agro-climatic conditions (Hayat and Ali 2010; Seymour et al. 2012). Soil microbial biomass is a minute fraction of the total amount of carbon in the soil, but is a major flux of labile nutrients and plays a vital role in nutrient cycling and therefore soil fertility in natural and agricultural ecosystems. It covers 2–5% of cumulative soil C and 1–5% of total soil N (Smith and Paul 1990). All organic materials show transformation into the soil by way of the soil's microbial biomass and it also acts as a dynamic flux for a considerable reservoir of C, N, and P (Jenkinson and Ladd 1981). Furthermore, it has the ability to solubilize insoluble Ca-P, fixed P, and occluded P by the root exudates of legumes (Arhara and Ohwaki 1989). Another constructive influence of legume-based crop rotation on soil microbial biomass has been associated with high C inputs and the diversity of plant residues that return back to the soil system, improvement in physical soil, e.g., water-holding/retention capacity, and chemical properties, e.g., soil organic C and humus content (Jensen et al. 2011), and organic N by supplying the crop biomass (Cederberg and Flysjö 2004), soil conservation (Stoop and Staveren 1981), and condensed soil compaction by forming a continuous network of residual root channels and macropores in the subsoil, penetrating hard soil pans (Peoples et al. 2009b). They also supply hydrogen gas in the soil atmosphere as a byproduct of biological N fixation, which enhances the bacterial growth in the surrounding areas of root nodules of legumes (LaFavre and Focht 1983).

3.3 Soil Plant Nutrition

Legume crops have specialized symbiotic N-fixing bacteria, i.e., *Rhizobium* spp. They show mutualistic interaction with the root of the legume crop by converting atmospheric fix N into an available form of N by causing the inflation of root cells to create nodules. The cumulative amount of N fixed by this bacteria–legume symbiosis differs significantly among legume species and their variety, crop duration, climatic conditions, soil type, agronomic interventions, etc. Crop rotations can increase root colonization by mycorrhiza. These mycorrhizae are the beneficial and nonpathogenic fungi that show symbiotic associations with grain legume crops. This valuable and constructive association linking the two organisms could enhance the uptake of water and nutrients by plants and increases N fixation by hastening host nutrition. It is estimated that the extent of N offered by legumes to subsequent crops differs from 50 to in excess of 200 pounds. Mycorrhizal colonization has the affirmative effect on a plant's performance even under harassed environmental conditions, acidic or reclaimed soil, soils deficient in P, and eroded sites (Hamel 2004). In Niger seed (*Guizotia abyssinica*), the monoculture of pearl millet (*Pennisetum glaucum* L.) roots showed 27% colonization by mycorrhizae, whereas 48% of colonization was found in the rotation of millet followed by cowpea (Bagayoko et al. 2000). Legume residues raise mycorrhizal associations in other crop species, as they make available alternative food substrate for macro-/micro-fauna that would otherwise be nourished by mycorrhiza.

The crop rotation shows an effective and complex relationship between PNK and C (Bullock 1992). In an experiment, it was found that rotations together with deep rooted cover crops and legume crops show the distribution of P and potassium (K) from the deep layer to the surface of the soil because plant roots can easily access them (Clark et al. 1998). Crops that have shallow rot systems, such as onions and carrots, may be taken after the harvesting of deep-rooted crops such as maize to help the nutrients to recover. A deep-rooted nutrient exhaustive may be followed by a crop with a shallow root system and a light feeder to scavenge plant nutrients that leach down into the deeper soil horizons after excessive applications of nutrients (Bullock 1992). The light feeder crops include legume sods, grass sods, wheat, barley, and oats, whereas heavy feeder crops are maize, soybean, potato, and vegetables (Clark et al. 1998). Moreover, crop rotations involving legumes improve the concentration of N in soil profile through the biological fixation of free atmospheric N, and thus as a supplementing of synthetic N fertilizer. The quantity of N fixed by legume cover crops greatly differs according to the legume species to be grown, but generally these add 50–200 kg N ha^{-1}(Clark et al. 1998). This N is mineralized by the soil microorganisms with the passage of time. Researchers estimate that 40–75% of all the N enclosed in a leguminous cover crop is available to the soil, which is used by subsequent crops, also subject to environmental conditions (Utomo et al. 1992) may contribute 30–60 kg N ha^{-1} (Utomo et al. 1992) to 110 N ha^{-1}(Tian et al. 2000). In Kenya, Kiiya et al. (2006) reported that the reason for low soil fertility and productivity among small-scale potato farmers is generally the continuous monoculture without a fallow period and a lack of legumes in the cropping sequence owing to small land holdings (Kiiya et al. 2006). Legumes may offer long-lasting advantages in certain soils that again are difficult to transform into monetary output. Typically, legumes in a cropping sequence add significantly more SOM and mineralizable N content than the continuous monocropping of cereal crops. This in turn ensures the improved accessibility of available N fractions to crop plants together with better physical structure of the soil, reduced erosion, and less energetic cultivation (Hoyt and Hargrove 1986). Several studies report that N allied with legume roots and nodules can contribute 30–60% of cumulative N stored by legumes (McNeill and Fillery 2008). In a study of a legume–cereal sequence in a Mediterranean climate in Australia, wheat used 3–10% of below-ground lupin N after being labeled with ^{15}N (McNeill and Fillery 2008). A similar investigation undertaken elsewhere examining a faba bean–wheat sequence suggested that wheat might recover more than twice the amount of N from residual N associated with faba bean-nodulated roots than from shoot residues (Jensen et al. 2010). Such studies clearly demonstrate that the roots and nodules of legumes can be a significant resource of N for subsequent crops.

3.4 Soil Carbon Dioxide

Soil is a major pool of terrestrial C having around 53% of the terrestrial C (Lal 2004). For the previous 100 years, the atmospheric carbon dioxide (CO_2) concentration has reached about 85 ppm (Lal 2004) and ~10% of the atmospheric CO_2every year passes through the soil system (Raich and Potter 1995). This increasing CO_2 concentration in the environment forced scientists of the world to recognize the potential of farming to capture and sequester the atmospheric CO_2 into soil strata for a considerable period of time. Various agronomic interventions are available that moderate the potential of agricultural soils to encourage the sink development of CO_2. In this regard, N fertilizer application and crop rotation are principle factors that have a significant impact on the stock and durability of the soil's C pool (Lal 2004; Meena et al. 2016b). N fertilization influences the soil's C dynamics in two ways. First, enlarged N fertilization will boost plant biomass production in numerous crops, and second, N accessibility is very decisive and significant for microbial action on crop residues with regard to their decomposition (Green et al. 1995). Fogg (1988) observed that increased N fertilization curtailed CO_2 emissions from soil strata. This depression in the soil occurred because of a reduction in the enzymatic activity due to improved N (Fogg 1988; Burton et al. 2004), increased soil acidity (Aerts and de Caluwe 1999), or decreased soil microbial biomass (Soderstrom et al. 1983). Plant biomass, in terms of both below- and above-ground plant parts, is one of the significant resources of C addition into the soil C flux, contributing to soil CO_2 discharges into the surrounding environments. Both N fertilization and crop rotation and can change the sum of plant biomass added to the soil system in a particular year. Increased N fertilizer rates can raise the soil's organic C content in continuing corn–wheat systems (Russell et al. 2005; Agarwal et al. 2017).

3.5 Agroecosystems

Legume plants are thought to be implicated in ecological easing processes in natural and agricultural ecosystems owing to their natural phenomena of BNFs (Table 1). Around the globe, a number of legume species are being cultured to assess protein-rich legume seeds and to gather the uncurtailed plant shoot. In agroecosystems, N-rich legume plants are used to enrich soil in N dynamics (Chalk 1998; Padilla and Pugnaire 2006). Incidentally, numerous species of legumes, including fenugreek

Table 1 Resource and environmental effects of crop rotation with legumes (Reckling et al. 2014)

Process in crop rotation	Leguminous crops	Effect on soil	Effect on agri-food system	Effect on environmental and global eco system
Biological N fixation		Reduced N fertilizer requirement and unwanted losses (Conway and Pretty 1991; Dogliotti et al. 2003)	Reduced CO_2 emissions from industry from industry	Reduced global GHG emissions (Lemke et al. 2007)
Grain protein synthesis		Increased protein supply (Asif et al. 2013; Temba et al. 2016)	Increased protein crop diversity	Reduced pressure on land
N transformations in soil	Reduced N_2O emissions	Effects of NO_3 leaching down into the soil		Curtailed global GHG emanations
Soil development		Enhanced organic matter, water infiltration rate, increased crop yields, and reduced soil erosion (Dogliotti et al. 2003)		
P transformations	Increased mobilization of P in soil	Decreased availability of P to plants		Reduced mining of P rock
Soil C transformations	Soil C balance (Buyanovsky et al. 1994; Balesdent et al. 1988; Cambardella and Elliot 1992)	Increased soil organic matter (Leithold et al. 1997; Jensen et al. 2011), higher and more stable crop yields (Giambalvo et al. 2004; Peoples et al. 2009b; Köpke and Nemecek 2010)		Increased soil C sequestration (Lal et al. 1998; Lal 2004)
Management of insect pests, diseases, nematodes, and weeds		Reduced use of insecticides, pesticides, and weedicides (Dogliotti et al. 2003)		
Species interactions	Enlarged diversity of micro-flora and -fauna increased pollen and nectar provision	Larger population of insects supporting wider wildlife		
Socio-economic	Increased net/annual profit, gross margin, and total income	Annetts and Audsley (2002), Tsakiris and Spiliotis (2006) and Preissel et al. (2015)		

(*Trigonella foenum-graecum* L.), velvet bean (*Mucuna pruriens* Bak.), lupin (*Lupinus angustifolius* L.), vetches (*Vicia sativa* L.), clovers (*Trifolium* sp.), *Crotalaria spectabilis* Roth., or *Sesbania rostrata* Brem. are integrated into the rotation system and also cultivated for green manuring purposes. They also make a contribution to conserving SOM and promoting soil–plant nutrient cycling, moisture retention, and nutrient supply to successive crops, and also promotes the efficient utilization of soil resources spatially and temporally (Corre-Hellou et al. 2006). The efficiency of BNF by legumes is greater in rotation systems than in monocropping (Corre-Hellou et al. 2006) and improved N dynamics of the soil can be beneficial for perennial cover crops such as legume-based grasslands (Rasmunssen et al. 2007), often resulting in the increased economic value of cereal–legume crop rotation compared with the monocropping of nonleguminous crop species, showing a potential effect, i.e., higher in low-input systems than in others, and stable crop yields even under problematic climatic conditions. Jensen (2006) also showed a complementary effect (Loreau and Hector 2001). Legumes grown for forage purposes are extensive and retain the capacity to boost forage production under a wide range of environmental conditions, e.g., subterranean clover (*T. subterraneum* L.), red clover (*T. pratense* L.), white clover (*T. repens* L.), alfalfa (*Medicago sativa* L.), and forage legumes (Frame et al. 1998).

4 Biological Processes with Crop Rotation

4.1 Plant Diseases

Legumes in a rotation system provide N for ensuing nonleguminous crops, but in addition, crop rotation may also provide better control over weeds, disease, insects, and nematodes, as shown in Table 2. The repeated cultivation of similar crops on the same piece of field encourages the infestation and intensification of weeds, insect pests, diseases, and pathogens, and to a susceptible crop because insect pests, pathogens, and weeds are the principal factors in disease spread in field conditions (Benson 1985). The crop rotational design can also majorly affect the prevalence and harshness of several pest and disease problems. The system reduces the spread and multiplication of disease-carrying spores by altering their environmental conditions (Rana and Rana 2011). This rotational system acts as a barrier between the host and the parasite that control insects and diseases by damaging their habitat and prevents the infestation of disease and pathogens, accordingly reducing the requirement for exclusive and hazardous agrochemicals, i.e., fungicides and insecticides (Sekamatte et al. 2003; Datta et al. 2017a). Hence, cereal–legume crop rotation can severely disturb the presence, survival, efficiency, and effectiveness of the natural enemies or predators, such as herbivore insects and pests that depend on plant habitats for feeding and other resources. The changes in the host–plant environment, interaction, and quality directly affect the host plant and indirectly affect their rate of development and their interactions with predators or natural enemies (Rana and Rana 2011).

Table 2 Control of insect pests, diseases, nematodes, and weeds by crop rotation with legume crops

Pest and crops affected	Crop rotation	References
Insect pests		
Hessian fly – wheat	Wheat–soybean/cotton/peanut/velvet bean/rape seed/castor bean/common vetch/French marigold	Reddy (2016)
Colorado potato beetle – potato, tomato	Potato/tomato–rye/wheat/beans	Capinera (2002)
Squash bug – cucurbits	Cucurbits/velvet bean/rape seed/castor bean/common vetch/French marigold	Reddy (2016)
Pepper maggot – pepper	Pepper–velvet bean/rape seed/castor bean/common vetch/French marigold	Reddy (2016)
Wheat stem maggot – wheat	Wheat–leguminous crops	Reddy (2016)
Corn rootworm – corn	Corn–soybean	Leslie and Cuperus (1993) and Orr (2009)
Alfalfa weevil – alfalfa	Alfalfa–corn–cotton–wheat	Canvari and Putnam (2007)
Diseases		
Wheat crown rot fungus (*Fusarium pseudograminearum*)	Wheat–forage legume	Lamprecht et al. (2006)
Stem rot (*Fusarium oxysporum*), black rot (*Ceratocystis fimbriata*), scurf (*Monilochaetes infuscans*)	Sweet potato–grasses/corn	Hamid (2011)
Sclerotinia stem (*Sclerotinia sclerotiorum*) rot – oilseed rape/ canola	Canola/oilseed rape–wheat–barley/ paddy–upland rotation, grasses/corn/ sorghum	Hamid (2011)
Verticillium wilt – eggplant and strawberry	Paddy–upland rotation, eggplant–broccoli/legumes/grains	Hamid (2011)
Kernel smut (*Neovossia horrida*), false smut (*Ustilaginoidea virens*)	Soybean–rice–corn	Brooks (2011)
Take-all – wheat (*Gaeumannomyces herpotrichoides*)	Wheat–faba bean/soybean–wheat	Asefa et al. (2002)
Pea midge and bladder pod Midge	Corn–pea–soybean	Van Emden (1991)
Potato bacterial wilt (*Ralstonia solanacearum*	Potato–common bean	Lemaga et al. (2001)
Bacterial blight – wheat, barley alfalfa	Wheat/barley/alfalfa–legumes	Reddy (2016)
Black shank – tobacco		Reddy (2016)
Northern corn leaf blight – corn	Corn–legumes	Reddy (2016)
Smut – corn		Reddy (2016)
Verticillium wilt – potato, sunflower	Potato–common vetch–field pea, purple vetch–broad bean–bell bean	Reddy (2016)
White mold – peanuts, potato, soybean	Peanuts/potato/soybean–corn/wheat/ barley/oats	Reddy (2016)

(continued)

Table 2 (continued)

Pest and crops affected	Crop rotation	References
Grey leaf spot (*Cercospora zeae-maydis*) corn	Corn–soybean	Kimber and Paull (2011)
Nematodes		
Meloidogyne incognita – velvet bean, rapeseed, common vetch, castor bean, French marigold		Reddy (2016)
M. javanica – velvet bean, sesame, common vetch, French marigold		Reddy (2016)
M. arenaria – velvet bean, sesame, common vetch, castor bean, French marigold		Reddy (2016)
M. hapla – French marigold		Reddy (2016)
Reniform nematode (*Rotylenchulus reniformis*)	Corn–cotton rotation	Bruns et al. (2007)
Weeds		
Decreased weed competition and increased yield and chlorophyll content of rice	Rice–velvet bean/hyacinth bean	Marenco and Santos (1999)
Striga infestation	Cereals–cowpea	Belel et al. (2014)
Integrated weed management	Maize–groundnut and maize–bean	Thayamini et al. (2010)
Weed suppression effect	Wheat–canola–pea	Rana and Rana (2011)

4.2 *Insect Pests*

Crop rotation maintains the cropping land by the cultivation of one crop after another and destroys insect pests' habitat by providing a nonhost environment. The cultivation of the same crop continuously on the same piece of land encourages weeds, insectspests, plant pathogens, and diseases to the vulnerable crop species. Crop rotation with nonhost crops disturbs the life cycle of pests due to habitat or host elimination and many agricultural crops can be saved to some extent from insect pest damage (Van Emden 1991; James et al. 2010). Rotations with leguminous crops are very beneficial because the legumes are likely to supply N to the plant and soil system in addition to disrupting pathogen cycles. Many soil insect pests such as wireworms, termites, scarabaeid grubs, cutworms, and shoot-boring flies and many crop midges such as pea midge and bladder pod midge can be eradicated from fields by crop rotation (Van Emden 1991). In corn to soybean rotation, corn is a grass, whereas soybean is a broadleaf leguminous crop. Both of these crops have different insect pest complexes. Corn rootworm is a vigorous pest and is eradicated by using this rotation (Orr 2009; Leslie and Cuperus 1993). Seran and Brintha (2010) have reported that infestation by corn borer and budworm was more severe under corn monoculture than under a corn–soybean cropping pattern and infestation with corn stalk borers was considerably higher under the monocropping system than under maize–soybean crop rotation (Thayamini et al. 2010).

4.3 Nematodes

Nematodes have a specific or narrow host range with a short life span extent in the soil and are mostly susceptible if nonhost crops are included in the rotation (Peters et al. 2003). Crop rotation has been reported to manage the nematode population in many parts of the world (McSorley et al. 1994). A diversity of crops such as cotton (*Gossypium hirsutum*), soybean (*Glycine max*), sorghum (*Sorghum bicolor*), and sorghum sudangrass (*S. bicolor, S. sudanense*) hybrids have been profitably used in rotation trials to suppress root knot nematodes (McSorley and Gallaher 1993; Rodríguez-Kábana et al. 1991). Bagayoko et al. (2000) in Niger observed that the plant parasitic nematode population showed 70% more profusion in the monoculture of pearl millet than under rotation with cowpea.

5 Role of Crop Rotation in Weed Management

Cereal–legume rotation is one of the conventional cropping sequences that manages or switches weed prevalence at small farms. This is only possible once the competitive nature of the component crops is privileged over the most prevalent weeds (Dimitrios et al. 2010). Despite advances in controlling and managing approaches, weeds have acquired greater intensity and are considered to be the most destructive among all the crop pests, because weed species have the capacity to acclimatize to the novel environment (Sosnoski and Cardina 2006). In an integrated agricultural system, the introduction of an appropriate cropping scheme along with a spatial arrangement such as the integration of legumes into crop rotation and efficient tillage can also make it possible for the crop itself to struggle with weed flora (Avola et al. 2008), but an accepting and perceptive understanding of weed dynamic forces and the effects of interrelated crop and soil factors on the weed life cycle are required (Davis and Liebman 2003). The abundance of weed flora is dependent on integrated managing practices (Marshall et al. 2003). The weed seed reserve also exposes the position of weed distribution and would possibly be considered as a marker for the contact of crop and soil management (Buhler et al. 2001). Still, the association between the size of the weed seed stock inside the soil and the percentage of weeds emerging is very complex (Sjursen 2001). Legume-based crop rotation together with tillage interventions, the usage of chemical pesticides, and various other agronomic approaches have a considerable effect on the weed dynamics and weed seed stock in terms of both size and distribution in different soil horizons (Marshall et al. 2003). The weed flora creates severe difficulties in certain cropping systems, particularly in organic farming systems, where the application of herbicides/weedicides is completely avoided. Verdú and Mas (2007) observed that mulching in crop rotation along with legumes is one of the weed control approaches that also offers additional aids to agricultural sustainability, for instance, soil protection measures avoiding pesticidal contamination and pollutions. Zero tillage accompanied by a rotation system and the use of mulches has the potential to encourage the sustainability of the agricultural system (Carofa et al. 2007; Buragohain et al. 2017; Marfo et al. 2015; Lojkova et al. 2015) because the effects of the design of crop rotation on the density and prevalence of weed flora are heightened by the zero-tillage system (Anderson 2007). Adoption of crop rotation practices reduces the use of weedicides that might modify the size of the weed seed reserve in cultivated soils (Squire et al. 2000) because the crop tolerance to weed species is encouraged by cultural practices (Hald 1999). Crop rotations consisting of two warm-season crops followed by two cool-season crops are more destructive for the growth and development of weed flora (Anderson 2007). Plants with allelopathic effects such as maize, barley, wheat, rye, and triticale promote ecological weed management by decreasing the effect of agricultural practices on environmental quality (Tabaglio et al. 2008; Bahadur et al. 2015; Datta et al. 2017c). A range of mechanical practices may also be useful in diverse cropping systems and rotations that vary with regard to the weed species and their growth stage (Chicouen 2007).

Koocheki et al. (2009) observed that weed composition and density were greatly varied in respect of the species involved in the rotation. In a field of continuous winter wheat, 90% of the total weed population was grasses and sedge, whereas this contribution decreased up to 43% when sugar beet (*Beta vulgaris*) was rotated with wheat. In contrast, the share of broadleaved weeds in the total weed population in continuous winter wheat was 9.4%, which increased up to 55.2% in sugarcane–wheat rotation.

6 Grain–Legume Crop Rotation Benefits for the Cropping System

The importance and effects of grain legumes as a preceding crop are very diverse and unique in nature (Zentner et al. 2002; Seymour et al. 2012). The agronomic pre-crop payback of leguminous grain consists of two mechanisms (Chalk 1998). One is the "N effect," which is caused by BNFs and N processes that make available a longer-term contribution to various crops (Peoples et al. 2009a). Another includes the "break crop effect," which embraces the profit to soil physical structure and SOM (West and Post 2002; Hernanz et al. 2009), P mobilization (Egle et al. 2003), and condensed stress from weeds, insect pests, and plant pathogens–diseases (Robson et al. 2002). The benefits resulting from the N prerequisite are considerably greater under conditions of low N fertilization, and benefits from "break crop effects" are greater in crop rotation dominated by cereal crops, where the reduction in biotic stress arising from plant diseases could be an important factor from the yield increment in grain legumes (Stevenson and van Kessel 1997). Therefore, it was reported by Kirkegaard et al. (2008) in Northern Europe that cereal growing after various "break" crops produces an average 24% higher grain yield in comparison with cereals following continuous monocropping. This effect sustains drastically reduced or zero-tillage practices, where weed and disease management and nonmechanical soil loosening are crucial (Sánchez-Girón et al. 2004; Luetke-Entrup et al. 2006; Molaei et al. 2017a, b), facilitating potential cost investments in the subsequent crop (Von Richthofen et al. 2006). The combination of agronomic effects includes improvement in growing environment conditions, offers manifold ecological benefits to the agricultural system, and increases resource use efficiency (Nemecek et al. 2008; Jensen et al. 2010; Westhoek et al. 2011). Improvement in qualitative characters such as protein and N concentration (López-Bellido et al. 2001; Albrecht and Guddat 2004) increase income by increasing yields in successive crops (Kirkegaard et al. 2008; Köpke and Nemecek 2010).

7 Crop Transition from Conventional to Organic Agriculture

Healthy and vigorous soil is the passion of proponents of organic or natural farming. With respect to the environment and natural processes, this includes the incorporation of various agricultural approaches such as legume-based crop rotation, cover cropping, grass–clover leys, agroforestry, alley farming, animal manuring, integration of crop residues, and livestock farming. Refraining from the use of synthetic and chemical fertilizers and pesticides, organic farming contributes toward a healthy ecosystem, high-quality physical and chemical properties of the soil, and waterbodies. Fields under organic farming normally have superior soils through the preservation and sequestration of organic carbon and humus and the maintenance of biodiversity (Fliebach et al. 2007). Even though organic farming equates to a zero tillage system, a procedure regularly hailed as reducing C losses, it depends on the use of harmful agrochemicals, including factory-made fertilizers, indicating that they are organic to have more N and C (Teasdale 2007; Datta et al. 2014). The organic systems along with reduced tillage also attest to being favorable with regard to the concentration of organic C and biodiversity preservation (Berner et al. 2008). In farming, the major threats stem from intensification and mechanization, i.e., intensive input use, high stocking densities, specialization, greater dependence on agrochemicals, land desertion, adverse climatic stimuli, overgrazing, and the proliferation of invasive alien species. Conventional farming systems have increased the yield per unit area although reducing the labor requirement. The factors responsible for the enhancement in yield consist of the cultivation of new plant varieties with more efficient fertilizer use and many other cultural practices such as higher seed rate and increased crop densities, higher fertilization rate, the efficacious use of pesticides and insecticides, and the extension of the mechanical harvesting of crops (Oelhaf 1978), although many of these approaches diminish naturally available resources, degrade crop quality and superiority, increase the soil–water pollution, and increase the environmental pollution (Hamilton and Helsel 1995). The inclusion of legume crops and composting increases soil productivity and soil quality as they maintain the physical, biological, and chemical components of the soil by storing more organic C (Zinati et al. 2001), the water infiltration rate, the soil's aggregate stability (Dabney et al. 2001), and the cation exchange capacity (Khallel et al. 2000), eliminating the nematode population in the soil and various other pathogenic activities (Gallardo-Lara and Nogales 1987) by improving the plant's tolerance to nematodes (Gallaher and Mcsorley 1993). For the successful organic production system, crop rotation followed by legumes is one of the essential components. By sequential cultivation of diverse crops in the same field, crop assortment in both time and space is obtained and insect pests, nematodes, and weed problems are minimized. The rotational system maintains and/or improves soil structure, organic matter content, and therefore the fertility status of the soil by suppressing root-knot nematodes (*Meloidogyne* spp.). The root-knot nematode can be controlled by incorporating the root-knot-resistant or -tolerant leguminous cover crop (Gallaher and Mcsorley 1993), e.g., velvet bean (*Mucuna deeringiana*), cow pea (*Vigna unguiculata*), or grasses (*Poacaea*), e.g., oats (*Avena sativa*) in a rotation cycle. In agriculture, the concern is more focused on the enduring sustainability of agricultural production and productivity. Mutually, excessive and lower doses of fertilizers than recommended have depressed the management of resources and spoiled the environment. Developed countries particularly tend to over-apply inorganic and organic fertilizers and this has led to water, soil, and environmental contamination (Bumb and Baanante 1996). In developing countries, population stress and pressure, harsh climatic conditions, land constraints, the decline of conventional soil management practices have degraded soil quality and fertility (Bumb and Baanante 1996; Henao and Baanante 1999).

Owing to the fact that agriculture is a soil-based industry that mines numerous plant nutrients from the soil's ecosystem, efficient and effective approaches are required for returning the nutrients back to the soil. When crop rotation is followed, the economic yields of crops are typically 10–15% advanced compared with yields obtained from monoculture (Frick and Johnson 2004). Rotations that include legumes and cereals help to ensure that the land is not exhausted. N is the main nutrient element usually contained in commercial fertilizers used by farmers. It is required by plants in large quantities, and usually has the quickest and most pronounced impact on growth. Plants absorbs N from the soil in the form of NO_3^- and NH_4^+. Air contains 78% N, but most plants cannot access it. Legumes are therefore special in that they host N-fixing bacteria such as rhizobium that can convert atmospheric N into a form that can be used by the host plant. Legumes have been identified for their ameliorative effects on soil since time immemorial. They add substantial amounts of protein-rich plant biomass to the soil system and plant rhizospheric zone, and keep the soil healthy and productive without the application of chemical and synthetic fertilizers (Kakraliya et al. 2017). Thus, crop rotation is a transition form of organic farming, i.e., a novel, innovative, strong and healthy structure of agriculture with respect to the environment and ecosystems. Avast number of studies have convincingly concluded that organic agriculture is more valuable to biodiversity (Oehl et al. 2004; Gabriel et al. 2006; Gabriel and Tscharntke 2007; Gabriel et al. 2010; Yadav et al. 2017a) owing to the adoption of common cultural activities, for instance, crop rotation and the absence of chemical insecticides and pesticides.

8 Intercropping

Intercropping of cereals among legumes is an all-encompassing term, i.e., growing two or more species in same field at the same time (Malezieux et al. 2008) or growing two or more than two crops in secure propinquity, i.e., crops are close up in the same row or bed, or in rows or strips that are adequate for biological relation and interaction. Intercropping includes relay cropping, companion planting, mixed cropping, interseeding, smother cropping, over-seeding, under-seeding, planting polycultures, row crops, cover crops, and mulching with crop residues (Layek et al. 2018; Datta et al. 2017b). Intercropping is basically a performance of eco-function strengthening and intensification that is measured to improve yields of organic or natural farming (Niggli et al. 2008). However, owing to the amplification and intensification of farming systems for the last five decades (Crews and Peoples 2004), annual intercropping is now exceptional in European nations (Malezieux et al. 2008). On the other hand, by introducing cereal–legume intercropping, numerous ecosystem services are provided (Hauggaard-Nielsen and Jensen 2005). They seem to be a renewed concern in cereal–legume intercrops, particularly in natural farming (Malezieux et al. 2008) for the purposes of eco-functional strengthening and intensification. Intercropping has shown a positive role in increasing and balancing the yield (Hauggaard-Nielsen and Jensen 2005) and in increasing the concentration of protein and other nutrients over monocropping (Gooding et al. 2007), mainly in organic farming and low N input systems where N supply can be limited (Corre-Hellou et al. 2006). Intercropping has also been shown to:

(i) Enhance and maintain soil conservation (Anil et al. 1998)
(ii) Promote efficient weed management owing to enlarged surface cover (Dimitrios et al. 2010) and also because of the release of toxic chemicals, i.e., *allelopathy* (Belel et al. 2014)
(iii) Minimize insect pests and diseases (Altieri 1995)
(iv) Offer superior lodging resistance (Anil et al. 1998)
(v) Influence crop production (Liebman and Davis 2000; Hauggaard-Nielsen et al. 2001)
(vi) Efficiently utilize resources (Thayamini et al. 2010).

In comparison, when leguminous grains such as pea (*Pisum sativum*) cultivated in monocropping system are accepted as fragile competitors to weed flora and weed influx has been exposed to harshly bind the N_2 nutrition and grain yield of organically cultivated leguminous grain (Hauggaard-Nielsen et al. 2001). Usually, the weed capability of intercrops to efficiently suppress the weed growth is over-reliant on the type of component crops being selected, their proportion/ratio, the genotype being preferred, their spatial arrangement, plant population or density, and the level of soil moisture and fertility (Dimitrios et al. 2010; Rana and Rana 2011; Belel et al. 2014). Furthermore, grain legumes as sole crops are susceptible to lodging exaggerated by abundant insect pests and diseases, which can be the origin of serious losses of yield in organic farming where pesticide application is prohibited. According to these points of view, intercropping can be an approach to profitably producing organic grain legumes (Hauggaard-Nielsen et al. 2001). Thus, cereal–legume intercrops have the strength to boost total gain yield per unit area and grain superiority, in particular, protein concentration (Gooding et al. 2007).

9 Monoculture

Monoculture is an agricultural practice in which the same crop species is cultivated continuously over the year, i.e., growing or producing the same crop, variety, or the same livestock species, bred on the same piece of land or farming system during one period, whereas polyculture is the alternative to monoculture and the growing of polymorphous crops on the same piece of land at one time. Continuous monoculture or monocropping can bring about more rapid increase in insect pests and plant diseases and then these spread quickly in the entire field once the crop species falls susceptible to it. Under a monocropping pattern, the soil is always deficient in microbial biomass (Moore et al. 2000). The continuous and constant monoculture is undesirable and objectionable because of the enlarged biotic stress liable from weeds, insect pests, diseases, and nematodes, together with complications in maintaining a good fertility level, as shown in Fig. 3.

10 Characteristics and Choice of Crops Commonly Grown in Crop Rotation

The crop choice for crop rotation is a very responsible aspect that includes various factors, such as the previous incidence of the grower, soil quality, soil nature, climatic constraints, market demand, labor costs, and availability. The cultivation system requires a suitable multi-annual rotation together with legume crops, having different rooting depths (UKROFS 2001). Crops with diverse rooting depths should be utilized within the rotation and also within the individual crops, e.g., forage herbs are commonly mixed with several varieties of clover crops and annual grasses to make available different sward structures both above and below the ground. The

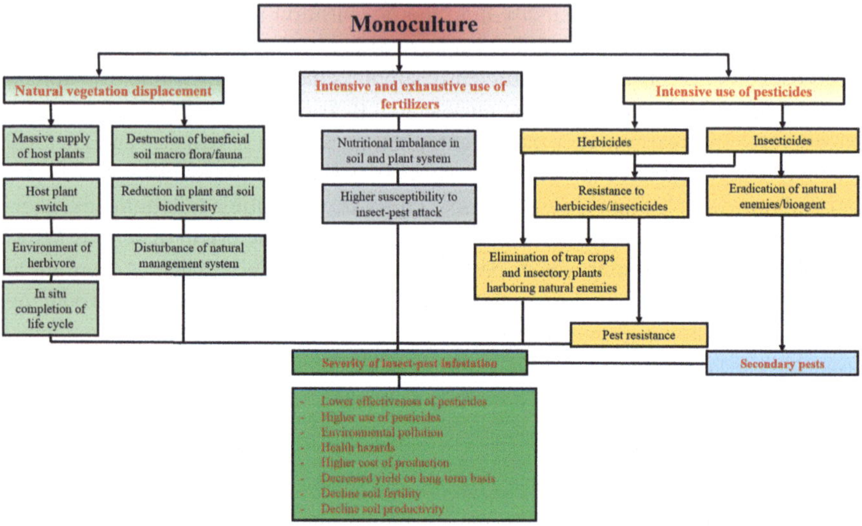

Fig. 3 The problem associated with monoculture. (Modified Source: Reddy 2016)

addition of green manures, cover crops or legume crops in rotation can considerably increase the efficiency of N use. Soil N is immobilized by nonleguminous crop plants that grow up dynamically during the winter season (e.g., rye), otherwise it will leach down into the soil system (Wyland et al. 1995). This N is consequently converted into the available and accessible form by mineralization after merging into the soil. Vigilant responsiveness is mandatory to the timing and method of N assimilation of the cover crop that can match the period of peak crop N demand rate to N mineralization (Rayns et al. 2000). One of the most significant problems in crafting the crop rotations for organic farming is complications in the supervision and management of soil fertility to achieve numerous targets. Although the integration of cover crops or green manures can have favorable influences on N management, there may be the possibility of associated diseases, such as plant pathogens (e.g., *Rhizoctonia solani*) proliferating in plant detritus (Weinhold 1977). In contrast, cover crops and green manures have also been revealed to have the potential for monitoring diseases in vegetables (Abawi and Widmer 2000). The preference of crops and their provision in farming areas is the foundation of agricultural coordination management. These decisions focus on the density concerned in the cropping scheme design and selection at the farm level because of their many involvements at diverse stages of the crop production processes (Navarrete and Bail 2007). Thus, the sketching and planning of cropping patterns are in fact decisive, critical, and vital steps in crop rotation processes and make very significant contributions to the annual and long-term sustainability, efficiency, productivity, prosperity, and profitability of farms. Thus, a suitable cropping arrangement must ensure numerous and conflicting objectives and can take into account a great number of factors and their relations under concern (Nevo and Amir 1991).

11 Future Outlook

By 2030, India needs approximately 32 million tons of pulse grains to meet the feeding requirements of 1.68 billion people, as expected at that time. The global pulse supply is very restricted, as India happens to be a major producer and consumer of pulses. Hereafter, India should produce the required quantity, but also remain competitive to protect indigenous pulse production. The biological N fixation through crop rotation already makes a significant contribution as regards the total amount of N fixed globally, i.e., about 90 million tons year^{-1}, but still need for its development the expansion and widespread application in agriculture, especially to upgrade the world's most exposed livelihoods, cropping systems patterns, and management. Biological N fixation is an effective and essential, complementary, alternative, or cost-effective solution for poor farmers than using industrially manufactured N fertilizers. Grain legumes are the major and richest sources of protein for the survival of the poorest farmer households. When legume production exceeds household necessities, it can be voluntarily traded to earn income, making considerable and express contributions to livelihoods. In addition to the substantial economic benefits and optimistic impact on human health, there is also the potential to cause global ecological settlement by reducing greenhouse gas emissions and water pollution from inorganic N fertilizers. Research on biological N fixation through crop rotation with legumes, particularly an understanding of the molecular genetics concerned with rhizobium–legume symbiosis, has taken significant steps forward by providing new opportunities and possibilities for designing strategies expected to enhance N fixation capacity and legume productivity.

12 Conclusion

Crop rotation with the inclusion of legumes is an essential and central key component of a successful organic farming system. The sequential growth of different crops followed by legumes is a potential way of achieving crop diversity in both space and time that maintains soil structure, soil fertility, soil productivity, and soil organic matter, while suppressing the insect pests, diseases, nematodes, and weed problems. Weed and insect pest management in monoculture is the most crucial factor for crop productivity. Therefore, the crop rotation approach is the primary consideration in the improvement of soil sustainability and crop productivity. The altered soil biology invasion of legumes habitually has a favorable impact on farming systems by accumulation and reutilization of N fixed by BNFs, increasing nutrient absorption, reducing GHG emissions via the reduced use of synthetic N fertilizers, disturbing the life cycles of insect pests and pathogens, refining the physical structure of the soil by adding soil biomass, soil porosity, organic C content, water-holding capacity or moisture retention capacity, and a reduction in soil and water erosion.

References

Abawi GS, Widmer TL (2000) Impact of soil health management practices on soil borne pathogens, nematodes and root diseases of vegetable crops. Appl Soil Ecol 15:37–47

Aerts R, de Caluwe H (1999) N deposition effects on C dioxide and methane emissions from temperate peat land soils. Oikos 84(1):44–54

Agarwal V, Bolch T, Syed TH, Pieczonka T, Strozzi T, Nagaich R (2017) Area and mass changes of Siachen Glacier (East Karakoram). J Glaciol 63(237):148–163

Albrecht R, Guddat C (2004) Welchen Wert haben Körnerleguminosen in der Fruchtfolge. Landesanstalt für Landwirtschaft, Thüringer

Alpmann D, Braun J, Schäfer BC (2013) Fruchtfolgen zielgerichtet auflockern. Land For 13:26–28

Altieri MA (1995) Agroecology: the science of sustainable agriculture. Intermediate Technology Publications, London

Anderson RL (2007) Managing weeds with a dualistic approach of prevention and control. A review. Agron Sustain Dev 27:13–18

Angers DA, Mehuys GR (1988) Effects of cropping on macro-aggregation of a marine-clay soil. Can J Soil Sci 68:723–732

Angus JF, Bolger TP, Kirkegaard JA, Peoples MB (2006) N mineralization in relation to previous crops and pastures. Aust J Soil Res 44:355–365

Anil L, Park J, Philips RH, Miller FA (1998) Temperate intercropping of cereals for forage: a review of the potential for growth and utilization with particular reference UK. Grass Forage Sci 53:301–317

Annetts JE, Audsley E (2002) Multiple objective linear programming for environmental farm planning. J Oper Res Soc 53:933–943

Arhara J, Ohwaki Y (1989) Estimation of available phosphorus in vertisol and alfisol in view of root effects on rhizosphere soil. In: XI Colloquium. Academic Publishers, Wageningen

Asefa T, Tanner DG, Bennie ATP (2002) Effect of stubble management, tillage and cropping sequence on the severity of take-all and eyespot diseases of wheat. Afr Crop Sci J 10:67–79

Ashoka P, Meena RS, Kumar S, Yadav GS, Layek J (2017) Green nanotechnology is a key for eco-friendly agriculture. J Clean Prod 142:4440–4441

Asif M, Rooney LW, Ali R, Riaz MN (2013) Application and opportunities of pulses in food system: a review. Crit Rev Food Sci Nutr 53:1168–1179

Avola G, Tuttobene R, Gresta F, Abbate V (2008) Weed control strategies for grain legumes. Agron Sustain Dev 28:389–395

Bagayoko M, Buerkert A, Lung G, Bationo A, Rornheld V (2000) Cereal/legume rotation effects on cereal growth in Sudano-Sahelian West Africa: soil mineral N, mycorrhizae and nematodes. Plant Soil 218:103–116

Bahadur S, Verma SK, Prasad SK, Madane AJ, Maurya SP, Gaurav VVK, Sihag SK (2015) Eco-friendly weed management for sustainable crop production – a review. J Crop Weed 11(1):181–189

Balesdent J, Wagner GH, Mariotti A (1988) Soil organic matter turnover in long-term field experiments as revealed by carbon-13 natural abundance. Soil Sci Soc Am J 52:118–124

Beattie B, Thompson S, Boehlje M (1974) Product complementarity in production: the by-product case. South J Agric Econ 7:161–165

Belel MD, Halim RA, Rafii MY, Saud HM (2014) Intercropping of corn with some selected legumes for improved forage production. J Agric Sci 6:3

Benson GO (1985) Why the reduced yields when corn follows corn and possible management responses? In: Wilkinson D (ed) Proceeding of the 40th corn sorghum research conference, Chicago. American Seed Trade Association, Washington, DC, pp 161–174

Berner A, Hildermann I, Fließbach A, Pfiffner L, Niggli U, Mäder P (2008) Crop yield and soil fertility response to reduced tillage under organic management. Soil Tillage Res 101:89–96

Biederbeck VO, Campbell CA, Rasiah V, Zentner RP, Wen G (1998) Soil quality attributes as influenced by annual legumes used as green manure. Soil Biol Biochem 30:1177–1185

Bormann BT, Wang D, Bormann FH, Benoit G, April R, Snyder MC (1998) Rapid, plant induced weathering in an aggrading experimental ecosystem. Biogeochemistry 43:129–155

Brandsæter LO, Riley H (1999) Plant residues for weed management in vegetables. In: Olesen JE, Eltun R, Gooding MJ, Jensen ES, Köpke U (eds) Proceeding: designing and testing crop rotations for organic farming. Danish Research Centre for Organic Farming, Foulum

Brooks SA (2011) Influences from long-term crop rotation, soil tillage, and fertility on the severity of rice grain smuts. Plant Dis 95:990–996

Bruns HA, Pettigrew WT, Meredith WR, Stetina SR (2007) Corn yields benefit in rotations with cotton. Online. Crop Management. https://doi.org/10.1094/CM-2007-0424-01-RS

Bues A, Preissel S, Reckling M, Zander P, Kuhlman T, Topp K, Watson CA, Lindström K, Stoddard FL, Murphy-Bokern D (2013) The environmental role of protein crops in the new Common Agricultural Policy. European Parliament, Directorate General for Internal Policies, Policy Department B: Structural and Cohesion Policies, Agricultural and Rural Development IP/B/AGRI/IC/2012-067. 112 pp

Buhler DD (1995) Influence of tillage systems on weed population dynamics and management in corn and soybean in the central USA. Crop Sci 35:1247–1258

Buhler DD, Kohler KA, Thompson RL (2001) Weed seed bank dynamics during a five-year crop rotation. Weed Technol 15:170–176

Bullock DG (1992) Crop rotation. Crit Rev Plant Sci 11(4):309–326

Bumb B, Baanante C (1996) The role of fertilizer in sustaining food security and protecting the environment to 2020. Vision 2020 discussion paper 17. IFPRI, Washington, DC

Buragohain S, Sharma B, Nath JD, Gogaoi N, Meena RS, Lal R (2017) Impact of ten years of bio-fertilizer use on soil quality and rice yield on an inceptisol in Assam, India. Soil Res. https://doi.org/10.1071/SR17001

Burton AJ, Pregitzer KS, Crawford JN, Zogg GP, Zak DR (2004) Simulated chronic NO_3-deposition reduces soil respiration in northern hardwood forests. Glob Chang Biol 10:1080–1091

Buyanovsky GA, Aslam M, Wagner GH (1994) Carbon turnover in soil physical fractions. Soil Sci Soc Am J 58:1167–1173

Cambardella CA, Elliot ET (1992) Particulate soil organic matter changes across a grassland cultivation sequence. Soil Sci Soc Am J 56:777–783

Cambardella CA, Elliot ETC (1994) N dynamics of soil organic matter fractions from cultivated grassland soils. Soil Sci Soc Am J 58:123–130

Campbell CA, Zentner RP, Selles F, Biederbeck VO, Leyshon AJ (1992) Comparative effects of grain lentil-wheat and monoculture wheat on crop production, N economy and N fertility in a Brown Chemozem. Can J Plant Sci 72(109):1–1107

Canveri M, Putnam DH (2007) Managing depleted alfalfa stands: over-seeding and other options. In: Summers CG, Putnam DH (eds) Irrigated alfalfa management for Mediterranean and Desert zones. University of California, Agriculture and Natural Resources, Oakland

Capinera JL (2002) North American vegetable pests: the pattern of invasion. Am Entomol 48:20–39

Cardina J, Herms CP, Doohan DJ (2002) Crop rotation and tillage systems effects on weed seedbanks. Weed Sci 50:448–460

Carofa M, Tourdonneta S, Saulasb P, Flocha DL, Roger-Estradea J (2007) Under sowing wheat with different living mulches in a no till system. II. Competition for light and N. Agron Sustain Dev 27:357–365

Carsky RJ, Singh BB, Oyewole B (2001) Contribution of early season cowpea to late season maize in the savanna zone of West Africa. Biol Agric Hortic 18:303–315

Cederberg C, Flysjö A (2004) Environmental assessment of future pig farming systems – Quantifications of three scenarios from the FOOD 21 Synthesis Work. Swedish Institute for Food and Biotechnology, SIK-rapport Nr 723

Chalk PM (1998) Dynamics of biologically fixed N in legume-cereal rotations: a review. Aust J Agric Res 49:303–316

Chan KY, Heenan DP (1991) Differences in surface soil aggregation under six different crops. Aust J Exp Agric 31:683–686

Mohler C, Johnson SE (2009) Crop rotation on organic farms: a planning manual, NRAES 177. NRAES, Ithaca

Charles R, Vuilloud P (2001) Pois protéagineux et azote dans la rotation. Rev Suisse Agric 33:365–370

Chicouen DE (2007) Mechanical destruction of weeds, a review. Agron Sustain Dev 27:19–27

Clark MS, Horwath WR, Shennan C, Scow KM (1998) Changes in soil chemical properties resulting from organic and low-input farming practices. Agron J 90:662–671

Conway GR, Pretty JN (1991) Unwelcome harvest: agriculture and pollution. Earthscan Publications Ltd., London

Cook RJ (1984) Root health: importance and relationship to farming practices. In: Bezdicek DF (ed) Organic farming, American Society of Agronomy Special Publication No 46. Madison, Wisconsin, pp 111–127

Corre-Hellou G, Fustec J, Crozat Y (2006) Interspecific competition for soil N and its interaction with N_2 fixation, leaf expansion and crop growth in pea-barley intercrops. Plant Soil 282:195–208

Couteaux MM, Bottner P, Berg B (1995) Litter decomposition, climate and litter quality. Trends Ecol Evol 10:63–66

Crews TE, Peoples MB (2004) Legume versus fertilizer sources of N: ecological trade-offs and human needs. Agric Ecosyst Environ 102:279–297

Dabney SM, Delgado JA, Reeves DW (2001) Using winter cover crops to improve soil and water quality. Commun Soil Sci Plant Anal 32:1221–1250

Dachler M, Köchl A (2003) Der Einfluss von Fruchtfolge, Vorfrucht, Stickstoffdüngung und Einarbeitung der Ernterückstände auf Ertrag und Rohproteingehalt von Winterweizen und nachfolgender Sommergerste. Die Bodenkultur 54:23–34

Dadhich RK, Meena RS, Reager ML, Kansotia BC (2015) Response of bio-regulators to yield and quality of Indian mustard (*Brassica juncea* L. Czernj. and Cosson) under different irrigation environments. J App Nat Sci 7(1):52–57

Datta R, Anand S, Moulick A, Baraniya D, Pathan SI, Rejsek K, Vranova V, Sharma M, Sharma D, Kelkar A (2017a) How enzymes are adsorbed on soil solid phase and factors limiting its activity: a review. Int Agrophys 31(2):287–302

Datta R, Kelkar A, Baraniya D, Molaei A, Moulick A, Meena R, Formanek P (2017b) Enzymatic degradation of lignin in soil: a review. Sustainability 9(7):1163

Datta R, Baraniya D, Wang Y-F, Kelkar A, Meena RS, Yadav GS, Teresa Ceccherini M, Formanek P (2017c) Amino acid: its dual role as nutrient and scavenger of free radicals in soil. Sustainability 9(8):1402

Datta R, Vranová V, Pavelka M, Rejšek K, Formánek P (2014) Effect of soil sieving on respiration induced by low-molecular-weight substrates. Int Agrophys 28(1):119–124

Davis AS, Liebman M (2003) Cropping system effects on giant foxtail (*Setaria faberi*) demography: I. Green manure and tillage timing. Weed Sci 51:919–929

Dimitrios B, Panyiota P, Aristidis K, Aspasia E (2010) Weed suppression effects of maize-vegetable inorganic farming. Int J Pest Manag 56:173–181

Dogliotti S, Rossing WAH, van Ittersum MK (2003) ROTAT, a tool for systematically generating crop rotations. Eur J Agron 19(2):239–250

Drinkwater LE (1999) Using plant species composition to restore soil quality and ecosystem function. In: Olesen JE, Eltun R, Gooding MJ, Jensen ES, Köpke U (eds) Proceeding: designing and testing crop rotations for organic farming. DARCOF, Tjele

Drinkwater LE, Schipanski M, Snapp S, Jackson LE (2017) Ecologically based nutrient management. In: Agricultural systems. Academic, Cambridge, MA, pp 203–257. https://doi.org/10.1016/b978-0-12-802070-8.00007-4

Drinkwater LE, Wagoner P, Sarrantonio M (1998) Legume-based cropping systems have reduced carbon and N losses. Nature 396:262–265

Duan S, Zhang S, Huang H (2000) Transport of dissolved inorganic N from the major rivers to estuaries in China. Nutr Cycl Agroecosyst 57:13–22
Duranti M, Gius C (1997) Legume seeds: protein content and nutritional value. J Field Crop Res 53:31–45
Edwards JH, Thrulow DL, Eason JT (1987) Influence of tillage and crop rotation on yields of corn, soybean, and wheat. Agron J 80:76–80
Egle K, Romer W, Keller H (2003) Exudation of low molecular weight organic acids by *Lupinus albus* L., *Lupinus angustifolius* L. and *Lupinus luteus* L. as affected by phosphorus supply. Agronomie 23:511–518
Eriksen J, Askegaard M, Olesen JE, Thorup-Kristensen K (1999) The sulphur balance of organic crop rotations in Denmark. In: Olesen JE, Eltun R, Gooding MJ, Jensen ES, Köpke U (eds) Proceeding: designing and testing crop rotations for organic farming. DARCOF, Tjele
Espinoza S, Ovalle C, Zagal E, Matus I, Tay J, Peoples MB et al (2012) Contribution of legumes to wheat productivity in Mediterranean environments of central Chile. Field Crop Res 133:150–159
Fliebach A, Oberholzer HR, Gunst L, Madar P (2007) Soil organic matter and biological soil quality indicators after 21 years of organic and conventional farming. Agric Ecosyst Environ 118:273–284
Fogg K (1988) The effect of added N on the rate of organic matter decomposition. Biol Rev 63:433–472
Fragstein P, von Schmidt H (1999) External N sources in an organic stockless crop rotation – useful or useless additives? In: Olesen JE, Eltun R, Gooding MJ, Jensen ES, Köpke U (eds) Proceeding: designing and testing crop rotations for organic farming. DARCOF, Tjele
Frame J, Charlton JFL, Laidlaw AS (1998) Temperate forage legumes. CAB International, Wallingford, 327 p
Francis CA, Clegg MD (1990) Crop rotations in sustainable production systems. In: Edwards CA, Lal R, Madden P, Miller RH, House G (eds) Sustainable agricultural systems, Soil and water conservation society. CRC Press, New York, p 130
Frick B, Johnson E (2004) Crop rotations for organic systems. Scott Research Farm, Organic Agriculture Centre of Canada, Truro
Gabriel D, Roschewitz I, Tscharntke T, Thies C (2006) Beta diversity at different spatial scales: plant communities in organic and conventional agriculture. Ecol Appl 16:2011–2021
Gabriel D, Sait SM et al (2010) Scale matters: the impact of organic farming on biodiversity at different spatial scales. Ecol Lett 13(7):858–869
Gabriel D, Tscharntke T (2007) Insect pollinated plants benefit from organic farming. Agric Ecosyst Environ 118:43–48
Gallaher RN, McSorley R (1993) Population densities of *Meloidogyne incognita* and other nematodes following seven cultivars of cowpea. Nematropica 23:21–26
Gallardo-Lara F, Nogales R (1987) Effect of the application of town refuse compost on the soil-plant system: a review. Biol Wastes 19:35–62
Ghanbari A, Dahmardeh M, Siahsar BA, Ramroudi M (2010) Effect of maize (*Zea mays* L.)-cowpea (*Vigna unguiculata* L.) intercropping on light distribution, soil temperature and soil moisture in and environment. J Food Agric Environ 8:102–108
Giambalvo D, Stringi L, Durante G, Amato G, Frenda AS (2004) N efficiency component analysis in wheat under rainfed Mediterranean conditions: effects of crop rotation and N fertilization. Options Méditerr Serie A 60:169–173
Gikonyo EW, Smithson PC (2003) Effects of farmyard manure, potassium and their combinations on maize yields in the high and medium rainfall areas of Kenya. In: Bationo A (ed) Managing nutrient cycles to sustain soil fertility in Sub-Sahara Africa. Academy Science Publishers (ASP), Nairobi, pp 137–149
Giller KE, Cadisch G (1995) Future benefits from biological N fixation in agriculture: an ecological approach. Plant Soil 174:255–277

Gooding MJ, Kasyanova E, Ruske R, Hauggaard-Nielsen H, Jensen ES, Dahlmann C, von Fragstein P, Dibet A, Corre-Hellou G, Crozat Y, Pristeri A, Romeo M, Monti M, Launay M (2007) Intercropping with pulses to concentrate N and sulphur in wheat. J Agric Sci 14:469–479
Grace PR, Oades JM, Keith H, Hancock TW (1995) Trends in wheat yields and soil organic C in the permanent rotation trial at the Waite Agricultural Institute, South Australia. Aust J Exp Agric 35:857–864
Green CJ, Blackmer AM, Horton R (1995) N effects on conservation of C during corn residue decomposition in soil. Soil Sci Soc Am J 59:453–459
Greenland DJ (1985) N and food production in the tropics: contributions from fertilizer N and biological N fixation. In: Kang BT, van der Heide J (eds) N management in farming systems in humid and subhumid tropics. Institute for Soil Fertility International Institute of Tropical Agriculture, Haren, pp 9–38
Hald AB (1999) Weed vegetation (wild flora) of long established organic versus conventional cereal fields in Denmark. Ann Appl Biol 134:307–314
Hamel C (2004) Impact of arbuscular mycorrhizal fungi on N and P cycling in the root zone. Can J Soil Sci 84:383–395
Hamid K (2011) Crop rotations for managing soil-borne plant diseases. Afr J Food Sci Technol 2 (1):001–009
Hamilton PA, Helsel DR (1995) Effects of agriculture on ground-water quality in five regions of the United States. Ground Water 33(2):217–226
Hauggaard-Nielsen H, Ambus P, Jensen ES (2001) Interspecific competition, N use and interference with weeds in pea-barley intercropping. Field Crop Res 70:101–109
Hauggaard-Nielsen H, Jensen ES (2005) Facilitative root interactions in intercrops. Plant Soil 274:237–250
Hausfather Z (2017) Analysis: Why scientists think 100% of global warming is due to humans. Global temperature: CarbonBrief Clear on Climate, December, 13, 2017. https://www.carbonbrief.org/analysis-why-scientists-think-100-of-global-warming-is-due-to-humans
Hayat R, Ali S (2010) N fixation of legumes and yield of wheat under legumes-wheat rotation in Pothwar. Pak J Bot 42:2317–2326
Henao J, Baanante C (1999) Nutrient depletion in the agricultural soils of Africa. 20r20 Vision Brief 62. IFPRI, Washington, DC
Hernanz JL, Sanchez-Giron V, Navarrete L (2009) Soil carbon sequestration and stratification in a cereal/leguminous crop rotation with three tillage systems in semiarid conditions. Agric Ecosyst Environ 133:114–122
Hoyt GD, Hargrove WH (1986) Legume cover crops for improving crop and soil management in the southern United States. Hortic Sci 21:397–402
James B, Atcha-Ahowe C, Godonou I, Baimey H, Goergen H, Sikirou R et al (2010) Integrated pest management in vegetable production: a guide for extension workers in West Africa. International Institute of Tropical Agriculture (IITA). Ibadan, Nigeria 2010:1–120
Jangir CK, Kumar S, Lakhran H, Meena RS (2017) Towards mitigating malnutrition in pulses through biofortification. Trends Biosci 10(17):2999–3002
Jenkinson DS, Ladd JN (1981) Microbial biomass in soil: measurement and turnover. In: Paul EA, Ladd JN (eds) Soil biochemistry, vol 5. Marcell Dekker, New York, pp 415–471
Jensen ES (2006) Grain legume functions in crop rotations. In: Proceedings of the AEP workshop (ed) Grain legumes and the environment: how to assess benefits and impacts. AEP, Zürich, pp 49–54
Jensen ES, Peoples MB, Hauggaard-Nielsen H (2010) Faba bean in cropping systems. Field Crop Res 115:203–216
Jensen ES, Peoples MB, Boddey RM, Gresshoff PM, Hauggaard-Nielsen H, Alves BJR, Morrison MJ (2011) Legumes for mitigation of climate change and the provision of feedstock for biofuels and biorefineries – a review. Agron Sustain Dev 32:329–364
Jezierny D, Mosenthin R, Bauer E (2010) The use of grain legumes as a protein source in pig nutrition: a review. Anim Feed Sci Technol 157:111–128

Jones S, Harris PM (1999) Measurement of N partitioning within different organic systems incorporating strip intercropping, sheep and crop rotation. In: Olesen JE, Eltun R, Gooding MJ, Jensen ES, Köpke U (eds) Proceeding: designing and testing crop rotations for organic farming. DARCOF, Tjele

Jordan VWL, Hutcheon JA, Donaldson GV, Farmer DP (1997) Research into and development of integrated farming systems for less-intensive arable crop production: experimental progress (1989–1994) and commercial implementation. Agric Ecosyst Environ 64:141–148

Kahiluoto H, Vestberg M (1999) Impact of cropping system on mycorrhiza. In: Olesen JE, Eltun R, Gooding MJ, Jensen ES, Köpke U (eds) Proceeding: designing and testing crop rotations for organic farming. DARCOF, Tjele

Kakraliya SK, Jat RD, Kumar S, Choudhary KK, Prakash J, Singh LK (2017) Integrated nutrient management for improving, fertilizer use efficiency, soil biodiversity and productivity of wheat in irrigated rice wheat cropping system in Indo-Gangatic plains of India. Int J Curr Microbiol App Sci 6(3):152–163

Kakraliya SK, Singh U, Bohra A, Choudhary KK, Kumar S, Meena RS, Jat ML (2018) N and legumes: a meta-analysis. In: Meena R, Das A, Yadav G, Lal R (eds) Legumes for soil health and sustainable management. Springer, Singapore

Karlen DL, Wollenhaupt NC, Erbach DC, Berry EC, Swan JB, Eash NS, Jordahl JL (1994) Long-term tillage effects on soil quality. Soil Tillage Res 32:313–327

Khallel R, Jones TE, Knepp AJ, Mann FM, Myers DA, Rogers PM, Serne RJ, Wood MI (2000) Modeling data package for S-SX Field Investigation Report (FIR), RPP-6296, Rev. 0. CH2M Hill Hanford Group, Inc., Richland

Kiiya WW, Mureithi JG, Kiama JM (2006) Improving production of Irish potato (*Solanum tuberosum*, L.) in Kenya: the use of green manure legumes for soil fertility improvement. In: Mureithi JG et al (eds) Development and up scaling of green manure legumes technologies in Kenya. KARI, Nairobi

Kimber RBE, Paull JG (2011) Identification and inheritance of resistance to cercospera leaf spot (*Cercospora zonata*) in germplasm of faba bean (*Vicia faba*). Euphytica 117:419–429

Kirkegaard J, Christen O, Krupinsky J, Layzell D (2008) Break crop benefits in temperate wheat production. Field Crop Res 107:185–195

Knudsen IMB, Debosz K, Hockenhull J, Jensen DF, Elmholt S (1995) Suppressiveness of organically and conventionally managed soils towards brown foot rot of barley. Appl Soil Ecol 12:61–72

Koocheki A, Nassiri M, Alimoradi L, Ghorbani R (2009) Effect of cropping systems and crop rotations on weeds. Agron Sustain Dev 29:401–408

Köpke U, Nemecek T (2010) Ecological services of faba bean. Field Crop Res 115:217–233

Kumar S, Karaliya SK, Chaudhary S (2017a) Precision farming technologies towards enhancing productivity and sustainability of rice-wheat cropping system. Int J Curr Microbiol App Sci 6 (3):142–151

Kumar S, Lakhran H, Meena RS, Jangir CK (2017b) Current needs of sustainable food and forage production to eliminate food and forage insecurity under climate change era. Forage Res 43 (3):165–173

Kumar S, Meena RS, Lal R, Yadav GS, Mitran T, Meena BL, Dotaniya ML, Sabagh A-EL (2018) Role of legumes in soil carbon sequestration. In: Meena R, Das A, Yadav G, Lal R (eds) Legumes for soil health and sustainable management. Springer, Singapore

Kumar S, Sheoran S, Kakraliya SK, Kumar P, Meena RS (2016) Drought: a challenge for Indian farmers in context to climate change and variability. Prog Res – An Int J 11(Special-IX):6243–6246

LaFavre JS, Focht DD (1983) Conservation in soil of H_2 liberated from N_2 fixation by HUP nodules. Appl Environ Microbiol 46:304–311

Lafond GP, Loeppky H, Derksen DA (1992) The effects of tillage systems and crop rotations on soil water conservation, seedling establishment and crop yield. Can J Plant Sci 72:103–115

Lakhran H, Kumar S, Bajiya R (2017) Crop diversification: an option for climate change resilience. Trends Biosci 10(2):516–518

Lal R (2004) Soil C sequestration impacts on global climate change and food security. Science 304:1623–1626

Lal R, Kimble JM, Follett RF, Cole CV (1998) The potential of U.S. cropland to sequester carbon and mitigate the greenhouse effect. Sleeping Bear Press, Chelsea

Lampkin N (1990) Organic farming. Farming Press Books, Ipswich

Lamprecht SC, Marasas WFO, Hardy MB, Calitz FJ (2006) Effect of crop rotation on crown rot and the incidence of *Fusarium pseudograminearum* on wheat in the Western Cape, South Africa. Australas Plant Pathol 35:419–426

Larkin RP (2008) Relative effects of biological amendments and crop rotations on soil microbial communities and soilborne diseases of potato. Soil Biol Biochem 40:1341–1351

Layek J et al (2018) Cereal+Legume intercropping: an option for improving productivity and sustaining soil health. In: Meena R, Das A, Yadav G, Lal R (eds) Legumes for soil health and sustainable management. Springer, Singapore

Leithold G, Huelsbergen KJ, Michel D, Schoenmeier H (1997) In: Diepenbrock W et al (eds) Umweltvertraegliche Pflanzenproduktion. Zeller Verlag, Osnabrueck, pp 43–54

Lemaga BR, Kanzikwera R, Kakuhenzire J, Hakiza J, Manzi G (2001) Effect of crop rotation on bacterial wilt incidence and potato tuber yield. Afr Crop Sci J 9(1):257–266

Lemke RL, Zhong Z, Campbell CA, Zentner R (2007) Can pulse crops play a role in mitigating greenhouse gases from north American agriculture? Agron J 99:1719–1725

Leslie AR, Cuperus GW (1993) Successful implementation of integrated pest management for agricultural crops, vol I, 1st edn. CRC Press, Boca Raton, pp 1–193

Leteinturier B, Herman J, Longueville FD, Quintin L, Oger R (2006) Adaptation of a crop sequence indicator based on a land parcel management system. Agric Ecosyst Environ 112(4):324–334

Liebman M, Davis AS (2000) Integration of soil, crop and weed management in low- external-input farming systems. Weed Res 40:27–47

Likoswe AA (1994) Pigeonpea agronomy annual report: Makoka Research Station. Zomba, Zomba

Lithourgidis AS, Dordas CA, Damalas CA, Vlachostergios DN (2011) Annual intercrops: an alternative pathway for sustainable agriculture. Aust J Crop Sci 5:396–410

LMC International (2009) Evaluation of measures applied under the common agricultural policy to the protein crop sector, Main Report. LMC International, New York/Oxford/Kuala Lumpur

Loges R, Kaske A, Taube F (1999) DiN fixation and residue N of different managed legumes and N uptake of subsequent winter wheat. In: Olesen JE, Eltun R, Gooding MJ, Jensen ES, Köpke U (eds) Proceeding: designing and testing crop rotations for organic farming. DARCOF, Tjele

López-Bellido L, López-Bellido RJ, Castillo JE, López-Bellido FJ (2001) Effects if long-term tillage, crop rotation and N fertilisation on bread-making quality of hard red spring wheat. Field Crop Res 72:197–210

López-Bellido L, López-Bellido FJ, López-Bellido JE, Castillo (2003) Faba bean (*Vicia faba* L.) response to tillage and soil residual N in a continuous rotation with wheat (*Triticum aestivum* L.) under rainfed Mediterranean conditions, Agron J 95: 1253–1261

Loreau M, Hector A (2001) Partitioning selection and complementarity in biodiversity experiments. Nature 412:72–76

Lojkova L, Datta R, Sajna M, Marfo TD, Janous D, Pavelka M, Formanek P (2015) Limitation of proteolysis in soils of forests and other types of ecosystems by diffusion of substrate. In: Amino acids, vol 8. Springer, Wien, pp 1690–1691

Luetke-Entrup N, Schneider M, Stemann G, Gröblinghoff FF et al (2006) Bewertung von neuen Systemen der Bodenbewirtschaftung in erweiterten Fruchtfolgen mit Körnerraps und Körnerleguminosen. Abschlussbericht über die Versuchsjahre 2001–2005, Fachhochschule Südwestfalen, Fachbereich Agrarwirtschaft, Soest (2006), pp. 215

Lupwayi NZ, Rice WA, Clayton GW (1998) Soil microbial diversity and community structure under wheat as influenced by tillage and crop rotation. Soil Biol Biochem 30:1733–1741

Magdof F, Van Es H (2000) Building soils for better crops sustainable soil management. SARE Outreach Publications c/o International Fulfilment Corporation, Brentwood

Malezieux E, Crozat Y, Dupraz C, Laurans M, Makowski D, Ozier-Lafontaine H, Rapidel B, de Tourdonnet S, Valantin-Morison M (2008) Mixing plant species in cropping systems: concepts, tools and models. A review. Agro Sust Dev 29:43–62

Marenco RA, Santos ÁMB (1999) Crop rotation reduces weed competition and increases chlorophyll concentration and yield of rice. Pesqui agropecu Bras 34(10):1881–1887

Marfo TD, Datta R, Lojkova L, Janous D, Pavelka M, Formanek P (2015) Limitation of activity of acid phosphomonoesterase in soils. In: Amino acids, vol 8. Springer, Wien, p 1691

Marshall EJP, Brown VK, Boatman ND, Lutman PJW, Squire GR, Ward LK (2003) The role of weeds in supporting biological diversity within crop fields. Weed Res 43:77–89

McCallum MH, Kirkegaard JA, Green T, Cresswell HP, Davies SL, Angus JF et al (2004) Improved subsoil macro-porosity following perennial pastures. Aust J Exp Agric 44:299–307

McIntire J, Fussel LK (1986) On-farm experiments with millet in Niger. III. Yields and economic analyses. ISC (ICRISAT Sahelian Center), Niamey

McNeill AM, Fillery IRP (2008) Field measurement of lupin below ground N accumulation and recovery in the subsequent cereal-soil system in a semi-arid Mediterranean-type climate. Plant Soil 302:297–316

McSorley R, Dickson DW, de Brito JA (1994) Host status of selected tropical rotation crops to four populations of root-knot nematodes. Nematropica 24:45–51

McSorley R, Gallaher RN (1993) Effect of crop rotation and tillage on nematode densities in tropical corn. J Nematol 25:814–819

Meena BL, Fagodiya RK, Prajapat K, Dotaniya ML, Kaledhonkar MJ, Sharma PC, Meena RS, Mitran T, Kumar S (2018) Legume green manuring: an option for soil sustainability. In: Meena R, Das A, Yadav G, Lal R (eds) Legumes for soil health and sustainable management. Springer, Singapore

Meena H, Meena RS, Singh B, Kumar S (2016a) Response of bio-regulators to morphology and yield of clusterbean [*Cyamopsis tetragonoloba* (L.) Taub.] under different sowing environments. J App Nat Sci 8(2):715–718

Meena RS, Bohra JS, Singh SP, Meena VS, Verma JP, Verma SK, Shiiag SK (2016b) Towards the prime response of manure to enhance nutrient use efficiency and soil sustainability a current need: A book review. J Clean Prod 112:1258–1260

Mielke LN, Schepers JS (1986) Plant response to topsoil thickness on an eroded loess soil. J Soil Water Conserv 41:59–63

Mikkelsen G (1999) Analysing non-replicated data in cropping systems. In: Olesen JE, Eltun R, Gooding MJ, Jensen ES, Köpke U (eds) Proceeding: designing and testing crop rotations for organic farming. DARCOF, Tjele

Mitran T, Meena RS, Lal R, Layek J, Kumar S, Datta R (2018) Role of soil phosphorus on legume production. In: Meena R, Das A, Yadav G, Lal R (eds) Legumes for soil health and sustainable management. Springer, Singapore

Molaei A, Lakzian A, Datta R, Haghnia G, Astaraei A, Rasouli-Sadaghiani M, Ceccherini MT (2017a) Impact of chlortetracycline and sulfapyridine antibiotics on soil enzyme activities. Int Agrophys 31(4):499–505

Molaei A, Lakzian A, Haghnia G, Astaraei A, Rasouli-Sadaghiani M, Ceccherini MT, Datta R (2017b) Assessment of some cultural experimental methods to study the effects of antibiotics on microbial activities in a soil: an incubation study. PLoS One 12(7):e0180663

Moore JM, Klose S, Tabatabai MA (2000) Soil microbial biomass carbon and N as affected by cropping systems. Biol Fertil Soils 31:200–210

Navarrete M, Bail ML (2007) SALADPLAN: a model of the decision-making process in lettuce and endive cropping. Agron Sustain Dev 27(3):209–221

Nemecek T et al (2008) Environmental impacts of introducing grain legumes into European crop rotations. Eur J Agron 28:380–393

Nevo A, Amir I (1991) CROPLOT—an expert system for determining the suitability of crops to plots. Agric Syst 37(3):225–241
Niggli U, Fließbach A, Stolze M, Sanders J, Schader C, Wyss G, Balmer O, Pfiffner L, Wyss E (2008) Gesellschaftliche Leistungen der Biologischen Landwirtschaft, Frick. Forschungsinstitut für Biologischen Landbau (FiBL)
Obi ME (1999) The physical and chemical responses of a degraded sandy clay loam soil to cover crops in southern Nigeria. Plant Soil 211:165–172
Oehl F, Sieverding E, Mäder P, Dubois D, Ineichen K, Boller T, Wiemken A (2004) Impact of long-term conventional and organic farming on the diversity of arbuscular mycorrhizal fungi. Oecologia 138(4):574–583
Oelhaf RC (1978) Organic agriculture: economic and ecological comparisons with conventional methods. Allanheld, Montclair. http://ideas.repec.org/a/eee/jfpoli/v6y1981i3p207-208.html
Ogindo HO, Walker S (2005) Comparison of measured changes in seasonal soil water content by rained maize-bean intercrop and component cropping in semi-arid region in South Africa. Phys Chem Earth 30:799–808
Orr D (2009) Biological control and integrated pest management. In: Integrated pest management: innovation-development process, vol I, 1st edn. Springer, Dordrecht, pp 207–239
Oxford University (2016) Study lays out effects of climate change on food production and health. March 3, 2016. http://www.ox.ac.uk/news/2016-03-03-study-lays-out-effects-climate-change-food-production-and-health
Padilla FM, Pugnaire FI (2006) The role of nurse plants in restoration of degraded environments. Front Ecol Environ 4:196–202
Paul EA, Clark FE (1996) Soil microbiology and biochemistry, 2nd edn. Academic, San Diego
Peoples MB et al (2009a) The contributions of N-fixing crop legumes to the productivity of agricultural systems. Symbiosis 48:1–17
Peoples MB, Unkovich MJ, Herridge DF (2009b) Measuring symbiotic N fixation by legumes. N fixation in crop production. Agron Monogr 52:125–170
Peters RD, Sturz AV, Carter MR, Sanderson JB (2003) Developing disease-suppressive soils through crop rotation and tillage management practices. Soil Tillage Res 72:181–192
Peyraud JL, Le Gall A, Lüscher A (2009) Potential food production from forage legume based-systems in Europe: an overview. Irish J Agric Food Res 48:115–135
Phelan PL, Mason JF, Stinner BR (1995) Soil fertility management and host preference by European corn borer, *Ostrinai nubilalis* (Hübner), on *Zea mays* L.: a comparison of organic and conventional farming. Agric Ecosyst Environ 56:1–8
Pierce FJ, Rice CW (1988) Crop rotation and its impact on efficiency of water and N use. In: Hargrove WL (ed) Cropping strategies for efficient use of water and N, ASA special publication 15. ASA/CSSA/SSSA, Madison, pp 21–42
Podgorska-Lesiak M, Sobkowicz P (2011) Prevention of pea lodging by intercropping barley with peas at different N fertilisation levels. Field Crop Res 149:95–104
Preissel S, Reckling M, Schläfke N, Zander P (2015) Magnitude and farm-economic value of grain legume pre-crop benefits in Europe: a review. Field Crop Res 175:64–79
Qiao J, Yang L, Yan T, Xue F, Zhao D (2012) N fertilizer reduction in rice production for two consecutive years in the Taihu Lake area. Agric Ecosyst Environ 146:103–112
Raich JW, Potter CS (1995) Global patterns of C dioxide emissions from soils. Glob Biogeochem Cycles 9:23–36
Raimbault BA, Vyn TJ (1991) Crop-rotation and tillage effects on corn growth and soil structural stability. Agron J 83:979–985
Rana SS, Rana MC (2011) Cropping system. Department of Agronomy, College of Agriculture, CSK Himachal Pradesh Krishi Vishvavidyalaya, Palampur, p 80
Rasmunssen J, Eriksen J, Jensen ES, Esbensen KH, Høgh-Jensen H (2007) In situ carbon and N dynamics in rye-grass clover mixtures: transfers, deposition and leaching. Soil Biol Biochem 39:804–815
Rayns F, Jackson L, Lennartsson M, Rahn C (2000) Winter cover crops; their relevance for organic horticultural production. In: Proceedings 13th international IFOAM scientific conference, pp 199

Reckling M, Hecker JM, Schläfke N, Bachinger J, Zander P, Bergkvist G, Walker R, Maire J, Eory V, Topp CFA, Rees RA, Toncea I, Pristeri A, Stoddard FL (2014) Agronomic analysis of cropping strategies for each agroclimatic region. Legume Futur Rep 1(4):75

Reckling M, Hecker JM, Bachinger J, Schläfke N, Topp K, Watson C, Zander P (2012) Assessing the economic and agronomic potential of legume-supported crop rotations across Europe using a crop rotation generator. ESA12, Helsinki, Finland, pp 140–141

Reddy PP (2016) Sustainable intensification of crop production. Springer.. ISBN: 978-981-10-2704-2 (Book), Singapore. https://doi.org/10.1007/978-981-10-2704-2

Reents HJ, Möller K (1999) Strategies to avoid nitrate leaching after potato crops by applying different cultivation methods to the following grown cereals. In: Olesen JE, Eltun R, Gooding MJ, Jensen ES, Köpke U (eds) Proceeding: designing and testing crop rotations for organic farming. DARCOF, Tjele

Robson MC, Fowler SM, Lampkin NH, Leifert C, Leitch M, Robinson D, Watson CA, Litterick AM (2002) The agronomic and economic potential of break crops for ley/arable rotations in temperate organic agriculture. Adv Agron 77:369–427

Rochester IJ, Peoples MB, Hulugalle NR, Gault RR, Constable GA (2001) Using legumes to enhance N fertility and improve soil conditions in cotton cropping systems. Field Crop Res 70:27–41

Rockström J, Steffen W, Noone K, Persson A et al (2009) Planetary boundaries: exploring the safe operating space for humanity. Ecol Soc 14(2):32

Rodríguez-Kábana R, Robertson DG, Wells L, Weaver CF, King PS (1991) Cotton as a rotation crop for the management of *Meloidogyne arenaria* and *Sclerotium rolfsii* in peanut. Suppl J Nematol 23:652–657

Russell AE, Laird DA, Parkin TB, Mallarino AP (2005) Impact of N fertilization and cropping system on C sequestration in Midwestern mollisols. Soil Sci Soc Am J 69:413–422

Sampson RN, Scholes RJ (2000) Article 3.4: additional human-induced activities. In: Watson RT et al (eds) Land use, land-use change, and forestry: a special report of the intergovernmental panel on climate change. Cambridge University Press, New York, pp 181–281

Sánchez-Girón V, Serrano A, Hernanz JL, Navarrete L (2004) Economic assessment of three long-term tillage systems for rainfed cereal and legume production in semiarid central Spain. Soil Tillage Res 78:35–44

Sanginga N, Lyasse O, Singh BB (2000) Phosphorus use efficiency and N balance of cowpea breeding lines in a low P soil of the derived savanna zone of West Africa. Plant Soil 220:119–128

Schäfer BC (2013) Mit Fruchtfolgen Witterungsrisiken trotzen! Praxis 1:5–7

Sekamatte BM, Ogenga-Latigo M, Russell-Smith A (2003) Effects of maize-legume intercrops on termite damage to maize, activity of predatory ants and maize yields in Uganda. Crop Prot 22:87–93

Seran TH, Brintha I (2010) Review on maize based intercropping. J Agron 9(3):135–145

Seymour M, Kirkegaard JA, Peoples MB, White PF, French RJ (2012) Break-crop benefits to wheat in Western Australia—insights from over three decades of research. Crop Pasture Sci 63:1–16

Siddique KHM, Johansen C, Turner NC, Jeuffroy MH, Hashem A, Sakar D, Gan Y, Alghamdi SS (2012) Innovations in agronomy for food legumes – a review. Agron Sustain Dev 32:45–64

Sihag SK, Singh MK, Meena RS, Naga S, Bahadur SR, Gaurav YRS (2015) Influences of spacing on growth and yield potential of dry direct seeded rice (*Oryza sativa* L.) cultivars. Ecoscan 9(1–2):517–519

Sjursen H (2001) Change of the weed seed bank during the first complete six-course crop rotation after conversion from conventional to organic farming. Biol Agric Hortic 19:71–90

Smith JL, Paul EA (1990) The significance of soil microbial biomass estimations. In: Bollag JM, Stotzky G (eds) Soil biochemistry, vol 6. Dekker, New York, pp 357–396

Smýkal P, Coyne CJ, Ambrose MJ, Maxted N, Schaefer H, Blair MW et al (2015) Legume crops phylogeny and genetic diversity for science and breeding. Crit Rev Plant Sci 34:43–104

Soderstrom B, Baath E, Lundgren B (1983) Decrease in soil microbial activity and biomasses owing to N amendments. Can J Microbiol 29:1500–1506

Sosnoski LM, Cardina J (2006) Weed seed bank community composition in a 35-yr-old tillage and rotation experiment. Weed Sci 54:263–273

Squire GR, Rodgers S, Wright G (2000) Community-scale seed bank response to less intense rotation and reduced herbicide input at three sites. Ann Appl Biol 136:47–57

Stagnari F, Maggio A, Galieni A, Pisante M (2017) Multiple benefits of legumes for agriculture sustainability: an overview. Chem Biol Technol Agric 4:2. https://doi.org/10.1186/s40538-016-0085-1

Stevenson FC, van Kessel C (1997) N contribution of pea residue in a hummocky terrain. Soil Sci Soc Am J 61:494–503

Stoate C, Boatman ND, Borralho RJ, Carvalho CR, Snoo GR, Eden P (2001) Ecological impacts of arable intensification in Europe. J Environ Manag 63(4):337–365

Stockdale EA, Lampkin NH, Hovi M, Keatinge R, Lennartsson EKM, MacDonald DW, Padel S, Tattersall FH, Wolfe MS, Watson CA (2001) Agronomic and environmental implications of organic farming systems. Adv Agron 70:261–327

Stocking M, Murnaghan N (2001) Hand book for the field assessment of land degradation. FAO, Rome

Stoop WA, Staveren JPV (1981) Effects of cowpea in cereal rotations on subsequent crop yields under semi-arid conditions in Upper Volta. In: Graham PC, Harris SC (eds) Biological N fi xation technology for tropical agriculture. Centro International de Agricultura Tropical (CIAT), Cali, pp 653–657

Tabaglio V, Gavazzi C, Schulz M, Marocco A (2008) Alternative weed control using the allelopathic effect of natural benzoxazinoids from rye mulch. Agron Sustain Dev 28:397–401

Teasdale JR (2007) Strategies for soil conservation in no-tillage and organic farming systems. J Soil Water Conserv 62:144A–147A

Temba MC, Njobeh PB, Adebo OA, Olugbile AO, Kayitesi E (2016) The role of compositing cereals with legumes to alleviate protein energy malnutrition in Africa. Int J Food Sci Technol 51:543–554

Tharanathan RN, Mahadevamma S (2003) Grain legumes—a boon to human nutrition. Trends Food Sci Technol 14:507–518

Thayamini H, Seran TH, Brintha I (2010) Review on maize based intercropping. J Agron 9:135–145

Thorup-Kristensen K (2001) Root growth and soil N depletion by onion, lettuce, early cabbage and carrot. In: Rahn CR, Lillywhite RD, De Neve S, Fink M, Ramos C (eds) Acta Horticultura, vol 563, pp 201–205

Tian G, Kolawole GO, Kang BT, Kirchhof G (2000) N fertilizer placement indexes of legume cover crops in the derived savanna of West Africa. Plant Soil 224:287–296

Tilman D (1998) The greening of the green revolution. Nature 396:211–212

Tilman D (1999) Global environmental impacts of agricultural expansion: the need for sustainable and efficient practices. Proc Natl Acad Sci U S A 96:5995–6000

Tsakiris G, Spiliotis M (2006) Cropping pattern planning under water supply from multiple sources. Irrig Drain Syst 20(1):57–68

UKROFS (2001) UKROFS standards for organic food production. UKROFS, London

Unger PW, McCalla TM (1980) Conservation tillage systems. Adv Agron 33:1–58

Utomo VH, Sitompul SM, Noordwijk MV (1992) Effects of leguminous cover crops on subsequent maize and soybean crop son an ultisol in Lampang. Agriculture 15:44–53

Van Emden HF (1991) Pest control, vol II, 2nd edn. Cambridge University Press, Cambridge, pp 1–128

Vandermeer J, van Noordwijk M, Anderson J, Ong C, Perfecto I (1998) Global change and multi-species agroecosystems: concepts and issues. Agric Ecosyst Environ 67(1):1–22

Varma D, Meena RS, Kumar S (2017) Response of mungbean to fertility and lime levels under soil acidity in an alley cropping system of Vindhyan Region, India. Int J Chem Stud 5(4):1558–1560

Verdú AM, Mas MT (2007) Mulching as an alternative technique for weed management in mandarin orchard tree rows. Agron Sustain Dev 27:367–375

Verma JP, Meena VS, Kumar A, Meena RS (2015) Issues and challenges about sustainable agriculture production for management of natural resources to sustain soil fertility and health: a book review. J Clean Prod 107:793–794

Verma RK, Shekhawat GS (1991) Effect crop rotation and chemical soil treatment on bacterial on bacterial wilt of potato. Indian Phytopathol 449:5–8

Vitousek PM, Naylor R, Crews T et al (2009) Nutrient imbalances in agricultural development. Science 324:1519–1520

Voisin AS, Guéguen J, Huyghe C, Jeuffroy MH, Magrini MB, Meynard JM, Mougel C, Pellerin S, Pelzer E (2014) Legumes for feed, food, biomaterials and bioenergy in Europe: a review. Agron Sustain Dev 34:361–380

Von Richthofen JS, Pahl H, Nemecek T (2006) Economic interest of grain legumes in European crop rotations. Deliverable 3.2 GL-Pro. 58 pp

Wander MM, Traina SJ, Stinner BR, Peters SE (1994) The effects of organic and conventional management on biologically-active soil organic matter pools. Soil Sci Soc Am J 58:1130–1139

Watson CA, Bengtsson H, Løes A-K, Myrbeck A, Salomon E, Schroder J, Stockdale EA (2002) A review of farm-scale nutrient budgets for organic farms in temperate regions. Soil Use Manag 18 (s1):264–273

Watson CA, Ritz K, Younie D, Franklin M (1996) N and soil biomass dynamics in two crop rotations. Asp Appl Biol 47:43–50

Weber G (1996) Legume-based technologies for African savannas: challenges for research and development. Biol Agric Hortic 13:309–333

Wedin DA, Tilman D (1990) Species effects on N cycling: a test with perennial grasses. Oecologia 84:433–441

Weinhold AR (1977) Population of *Rhizoctonia solani* in agricultural soils determined by a screening process. Phytopathology 67:566–569

Weitbrecht B, Pahl H (2000) Lohnt sich der Anbau von Körnerleguminosen? Ökologie Landbau 116:39–41

West TO, Post WM (2002) Soil organic carbon sequestration rates by tillage and crop rotation: a global data analysis. Soil Sci Soc Am J 66:1930–1946

Westhoek H, Rood T, van den Berg M, Janse J, Nijdam D, Reudink M, Stehfest E (2011) The protein puzzle. In: The consumption and production of meat, dairy and fish in the European Union. Netherlands Environmental Assessment Agency (PBL), The Hague

Wyland LJ, Jackson LE, Schulbach KF (1995) Soil-plant N dynamics following incorporation of a mature ryegrass cover crop in a lettuce production system. J Agric Sci 124:17–25

Yadav GS, Babu S, Meena RS, Debnath C, Saha P, Debbaram C, Datta M (2017a) Effects of godawariphosgold and single supper phosphate on groundnut (*Arachis hypogaea*) productivity, phosphorus uptake, phosphorus use efficiency and economics. Indian J Agric Sci 87(9):1165–1169

Yadav GS, Lal R, Meena RS, Datta M, Babu S, Das Layek J, Saha P (2017b) Energy budgeting for designing sustainable and environmentally clean/safer cropping systems for rainfed rice fallow lands in India. J Clean Prod 158: 29–37

Yadav K, Akhtar MS, Panwar J (2015) Rhizospheric plant microbe interactions: key factor to soil fertility and plant nutrition. In: Arora NK (ed) Plant microbe symbiosis: applied facets. Springer, New Delhi, pp 127–145

Yang L, Wang D, Xia L (2004) Characteristics and control approach on agricultural non-point source pollution in Taihu Lake region (in Chinese). China Water Resour 20:29–30

Zentner RP, Wall DD, Nagy CN, Smith EG, Young DL, Miller PR, Campbell CA, McConkey BG, Brandt SA, Lafond GP, Johnston AM, Derksen DA (2002) Economics of crop diversification and soil tillage opportunities in the Canadian prairies. Agron J 94:216–230

Zinati GM, Li YC, Bryan HH (2001) Utilization of compost increases organic C and its humin, humic, and fulvic acid fractions in calcareous soil. Compost Sci Util 9:156–162

Nitrogen-Cycling Communities in Organically Amended Versus Conventionally Managed Agricultural Soil

Lily Pereg and Mary McMillan

Abstract Productivity in agricultural ecosystems is often largely dependent on the input of nitrogen fertilizers such as urea, nitrate, and ammonia, despite the high financial costs and potential detrimental effects on the environment. Soil enrichment with organic matter, such as plant compost, straw, or manure, can enhance soil organic carbon and improve soil quality and increase aggregate stability, biological activity, and microbial diversity, including nitrogen cycles. The abundance of the *nifH* gene, thus nitrogen cycles, was found to be enhanced when organic fertilization was used and correlated mainly to the availability of organic carbon, potentially due to the presence of fuels required to run the energetically expensive nitrogen fixation process. On the other hand, readily available ammonia and nitrate, which are often associated with inorganic fertilization, tend to suppress the soil potential for nitrogen fixation. The impacts of fertilizers on denitrifying microbial communities are complex, due to the great diversity of the denitrifiers and the variation in their abundance in different environments. However, in general, the use of organic fertilizers increases denitrification potential and activity in soils when compared to inorganic fertilization. This is particularly important for increasing the abundance of those denitrifiers containing the *nosZ* genes and capable of the final step of the denitrification process, removing the potent greenhouse gas nitrous oxide by its conversion to dinitrogen. Bacterial and archaeal nitrifiers react differently to variations in soil conditions and to different fertilization management strategies. It is important to note that nitrogen fertilizers in access might turn into an ecosystem hazard where nitrifying microbes convert them to nitrate. This survey of the literature suggests that the addition of organic matter to agricultural soils, even where inorganic nitrogen fertilizers are used, enhances the soil potential for nitrogen cycling and soil sustainability.

Keywords Nitrogen cycle · Nitrogen fixation · Nitrification · Denitrification · Inorganic and organic nitrogen fertilizers · nifH · nirS/K · nosZ · amoA genes

---

L. Pereg (✉) · M. McMillan
School of Science and Technology, University of New England, Armidale, NSW, Australia
e-mail: lily.pereg@une.edu.au; mrookle2@une.edu.au

R. Datta et al. (eds.), *Carbon and Nitrogen Cycling in Soil*,
https://doi.org/10.1007/978-981-13-7264-3_11

Abbreviations

amoA	Gene encoding the monooxygenase enzyme
amoA-arch	Archaeal *amoA* gene
amoA-B	Bacterial *amoA* gene
AOA	Ammonia-oxidizing archaea
AOB	Ammonia-oxidizing bacteria
ATP	Adenosine triphosphate
Av N	Available nitrogen (Kjeldahl method)
C	Carbon
CO_2	Carbon dioxide
DMPP	3,4-Dimethylpyrazole phosphate
IF	Inorganic fertilization
N	Nitrogen
N_2	Dinitrogen gas
N_2O	Nitrous oxide
NH_3	Ammonia
NH_4^+	Ammonium ion
nifH	Gene encoding the dinitrogenase reductase enzyme
nirK	Gene encoding the copper-containing nitrite reductase enzyme
NirK	Copper-containing nitrite reductase
nirS	Gene encoding the cytochrome cd1 nitrite reductase enzyme
NirS	Cytochrome cd1 nitrite reductase
NO	Nitric oxide
NO_2^-	Nitrite
NO_3^-	Nitrate

1 Nitrogen Fertilization and Microbial Communities in Agricultural Soils

Nitrogen (N) is a vital element for all forms of life including plants, with one to four percent of the living material, including proteins, DNA, and RNA, being composed of nitrogen (Woodmansee et al. 1978). Therefore, productivity in agricultural ecosystems is often largely dependent on the input of N fertilizers. The high demand for N in intensive cropping and other agricultural systems is generally met by the addition of fertilizers, such as urea, nitrate, and ammonia, to the soil. However, the effects that such agrochemicals have on soil function are largely unknown. In general, the availability and productivity of agricultural soil is under threat, due to greater urbanization and also intensified farming practices which utilize extensive irrigation, increasing amounts of agrochemicals, and heavy machinery (Berry 1978; Newman et al. 2015; Li et al. 2015). As these intensive agricultural practices result in land degradation, it is increasingly necessary to identify or develop sustainable cropping systems that result in large biomass yields and maintain or improve ecosystem services (Orr et al. 2015), such as soil N cycling.

In agricultural systems, N is taken out from the ecosystem when plants and/or animals are removed. Consequently, there is a need to constantly replace the N supply in these farming systems. This is achieved through the widespread use of N-containing fertilizers, despite the high financial costs and potential detrimental effects on the environment (Ladha et al. 2005; Spiertz 2010). Manufacturing N fertilizers from fossil fuels and transporting manures or other sewage-based fertilizers to agricultural areas are energetically expensive, especially when the amount of fuel required is considered. These processes also result in the release of CO_2 and contribute to the pollution of water bodies through the leaching of nitrate and generation of nitrous oxide, a potent greenhouse gas. These factors contribute to the growing range of environmental concerns around the intensive farming practices and the use of agrochemicals, such as soil compaction and erosion, overuse of land for cropping or grazing, reductions in soil organic matter, depletion of water supplies, and pollution of groundwater and surface waterways through agrochemical runoff (Hirsch and Mauchline 2015). Agrochemical runs off, and soil degradation can both contribute to eutrophication of aquatic habitats. Production practices that result in degraded soil quality may then drive increased use of irrigation and fertilization in an attempt to sustain soil productivity (Tilman et al. 2002; Zalidis et al. 2002). This can, in turn, cause further damage to the soil while also increasing the cost of crop production. While intensive arable farming and use of agrochemicals might negatively impact soil chemical, physical, and biological properties (Caravaca et al. 2002; Bellamy et al. 2005), the enrichment of soil with organic matter, such as plant compost, straw, or manure, can enhance soil organic carbon (SOC) and improve soil quality; increase aggregate stability, biological activity, and microbial diversity (Johnston et al. 2009; Morugán-Coronado et al. 2015; Pérez-Piqueres et al. 2006; García-Orenes et al. 2010, 2013, 2016; Prosdocimi et al. 2016); as well as reduce chemical input, increase plant productivity, and increase the sustainability of organically managed agroecosystems (Macci et al. 2013).

Soil microbial communities are an important component of soil. The structure and function of soil microbial communities vary depending on soil type, pH, temperature, plant cover and rotation, fertilization, tillage management, and water content (Bossio et al. 1998; Saleh-Lakha et al. 2005; Jangid et al. 2008; de Vries et al. 2013; Garbeva et al. 2008; Geisseler and Scow 2014; Quadros et al. 2012; Kibblewhite et al. 2008; Geisseler et al. 2010). Agricultural land management strategies can reduce the diversity and abundance of soil microorganisms and affect various soil properties (Caravaca et al. 2002). For example, decreased availability of water reduces soil organic C, altering the structure of soil microbial communities (Canarini et al. 2016; Bastida et al. 2017). It is widely acknowledged that water availability is essential for the maintenance of soil microbial communities. However, there is a lack of information on how different water management practices impact soil microbial communities (Bastida et al. 2008). In semiarid conditions, where water supplies are limited, the addition of organic matter may support the development of soil microbial communities and increase soil biodiversity (García-Orenes et al. 2010; Frenk et al. 2014; Wafula et al. 2015; Bastida et al. 2017). At the same time, maintaining crop cover on the soil surface, and minimizing soil tillage, can improve the soil physical structure, allowing improved water retention and maintenance of biological activity (Morugán-Coronado et al. 2015). Agrochemicals, including herbicides, fertilizers, fungicides, and insecticides, can affect the soil biota and the structure and function of soil microbial communities and therefore have a significant impact on soil quality (Imfeld and Vuilleumier 2012; Sofo et al. 2012). Soil microbial communities are very sensitive to changes in the soil, including physical disruption of the soil or changes in water or nutrient content. They can thus be considered as possible indicators of variations in the soil environment and of soil quality (Zornoza et al. 2009; Frenk et al. 2014).

An increased understanding of how agricultural management practices influence the structure of soil microbial communities and their overall impact on soil health is markedly important under semiarid conditions (García-Orenes et al. 2013). In semiarid areas, environmental constraints result in the overexploitation of land for food production, and these intensive farming practices result in a decline in soil structure, soil fertility, and a loss of organic matter (Caravaca et al. 2002). Sustainable farming practices, including the application of organic matter to enrich the soil, can enhance SOC and improve soil quality, as observed in arable agriculture following the addition of animal manures (Johnston et al. 2009). Soil organic amendments can also influence diverse soil microbial communities and enhance soil fertility. Organic nutrient sources have been suggested to be beneficial for both soil biodiversity and crop production, as they increase soil organic matter and potentially lead to improvements in a number of soil biological and fertility indicators, including soil aggregation, porosity, and water retention. In southern Spain, a decline in soil organic matter content and an associated loss of soil fertility have been observed as a result of intensive agriculture and the semiarid conditions (Caravaca et al. 2002). In contrast to conventional agriculture in this region, the use of organic amendments promotes the activity of soil microbial communities, increases microbial biodiversity, and improves soil properties. For example, application of oat straw to experimental plots on abandoned agricultural land led to soil restoration (García-Orenes et al. 2010, 2013). Morugán-Coronado et al. (2015) found that management practices including no-tillage, retention of vegetation cover, and application of manure improved soil conditions and increased soil organic matter and soil aggregate stability, leading to increased biological activity. The use of organic fertilization in grapevine production has been trialed over the last 10 years in semiarid regions in Spain. The use of organic amendments has resulted in improved soil quality indicators, including an increase in soil microbial diversity and enhanced soil biological activity, in comparison to traditional farming practices using chemical fertilization (García-Orenes et al. 2016). Earlier, Pérez-Piqueres et al. (2006) reported that the use of organic fertilizers, in the form of compost amendments, manure, and grapevine prunings, promoted an increase in soil microbial activity, enhancing the fertility and productivity of agricultural soils. Interestingly, Castañeda et al. (2015) found that organic rather than conventional management resulted in similar communities in grapevine soils to that of nearby forest in Chile, suggesting that the use of various types of organic amendments in agricultural soils would lead to sustaining soil biological and functional diversity and contribute to conservation of agricultural land. There has been a movement toward implementing organic fertilization systems to help improve soil properties, thereby increasing the productivity of agricultural land, and also allow for a reduction in agrochemical inputs, increasing the sustainability of such agroecosystems (Macci et al. 2013).

2 Microbial Functions in Agricultural Soils: The Nitrogen Cycle

Soil organisms undertake a large number of ecosystem services, such as nutrient cycling. Microorganisms carry out N cycling (Fig. 1), which is one of the most important soil functions (Fitter et al. 2005; Wallenstein and Vilgalys 2005; He et al. 2007). Several soil properties are related to mineralization and immobilization rates of N in soils, including microbial biomass, enzymatic activity, soil respiration rate, and microbial C and N contents (Alef et al. 1988; Hart et al. 1994; Tietema 1998; Barrett and Burke 2000; Bengtsson et al. 2003).

Although our atmosphere is composed of around 78% N (inorganic form), this N source is not available directly to plants and animals. The only organisms that can access this N_2 pool are bacteria and archaea that produce the enzyme nitrogenase. Nitrogenase-producing organisms can reduce the triple bond in atmospheric N_2 to ammonia, converting N into a source which can then be used by other organisms (Dixon and Kahn 2004; Wuebbles 2002). Thus, in nature, the major input of available N into the biosphere is through diazotrophic N fixation by prokaryotes

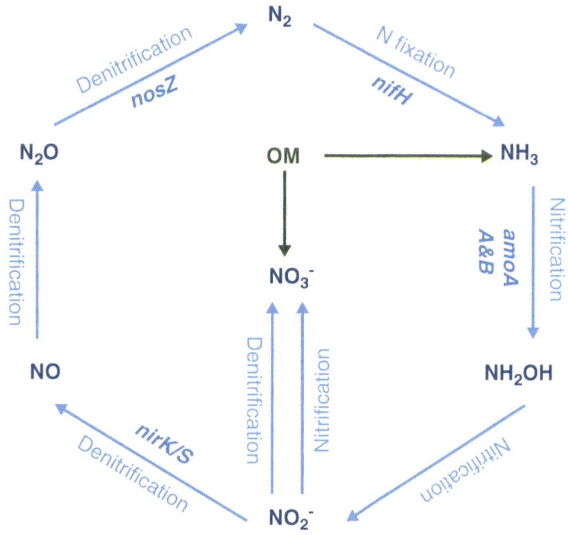

Fig. 1 Nitrogen cycling in the soil. The figure shows the main parts of the N-cycling pathway from N fixation, the main input of biologically available N into the biosphere, to the removal of fixed N from the biosphere. It also shows several genes encoding major enzymes in the N cycle (*nifH*, N fixation; *nirK*, *nirS*, *nosZ*, denitrification; *amoA*, nitrification). A, archaeal *amoA*; B, bacterial *amoA*; OM, organic matter

(Fig. 1). Nitrogenase is the key enzyme in N fixation. The N-reductase subunit of nitrogenase is encoded by the highly conserved gene *nifH* (Coelho et al. 2009). The abundance of the *nifH* gene in an environmental sample gives an indication of the potential for N fixation by bacteria (Coelho et al. 2009). Microbial assimilation of inorganic N is critical for improving soil retention of N (Vinten et al. 2002; Tahovská et al. 2013). Addition of N to the soil is often in forms unavailable for direct uptake by plants, including urea (Witte 2011) or organically bound N in other fertilizers. Therefore, microbial N cycling in soils is key in transforming these N sources into an accessible form of N for plants (Fitter et al. 2005; Wallenstein and Vilgalys 2005; He et al. 2010). In fertilized systems, the hydrolysis of urea, which results in the release of ammonia and carbon dioxide (CO_2), is a significant player in soil N balance. Urea hydrolysis is catalyzed by the bacterial enzyme urease, with the urease alpha subunit being encoded by *ureC* gene (Koper et al. 2004). The abundance of the *ureC* gene can therefore also be used as an indicator for the presence of N cyclers. Since a large portion of soil N is bound in organic matter in the form of proteins and other N-containing macromolecules, mineralization starting with proteolysis is an important step in the release of available N into the biosphere. Proteolysis is a rate-limiting step in soil N cycling. Protease-encoding microbial communities can be studied by quantifying genes encoding the *alkaline* (*apr*) and *neutral* (*npr*) *metallopeptidase*. Using gene abundance assessment, Lori et al. (2018) found that organically managed soils had a more stable N provisioning potential than conventional fertilized soils under drought scenarios, probably facilitated by a distinct and more adaptive proteolytic microbial community.

Nitrification and denitrification, involving ammonia oxidation and nitrate and nitrite reduction to N_2O and N_2, respectively, are major components of the soil N cycle (reviewed by Teixeira and Yergeau 2012). A variety of microbes decompose organic N into NH_4^+ (Zhou et al. 2012), and the preferred N form available for plants, namely, NO_3^-, is produced by nitrifying microorganisms, which oxidize NH_4^+ to NO_2^- and then NO_3^- (Horz et al. 2004; Fierer et al. 2012). Nitrification is an aerobic process which begins with the oxidation of ammonia to nitrite. Nitrite produced in this reaction is then further oxidized to nitrate by nitrite-oxidizing bacteria. The oxidation of ammonia is often the rate-limiting step and is catalyzed by the enzyme ammonia monooxygenase. The abundance of bacterial and archaeal *amoA* genes, which encode the alpha (A) subunit of ammonia monooxygenase, can be measured using molecular techniques and be used to estimate the soil's potential for nitrification. The relative contribution to nitrification by bacterial versus archaeal ammonia oxidizers in soils and other ecosystems is still under debate (reviewed by Teixeira and Yergeau 2012).

The N cycle is considered to be completed when denitrifying microorganism reduces NO_3^- to NO, N_2O, and N_2, returning N to the atmosphere (Braker et al. 1998; Houlton and Bai 2009). Denitrification is a complex anaerobic process involving various enzymes that remove fixed N and convert it to other forms, including gaseous N compounds. N cycling is therefore also important from several other environmental perspectives, including controlling emission of the greenhouse gas nitrous oxide (N_2O). It has been estimated that approximately 5% of soil microbial biomass is denitrifying bacteria (Braker and Conrad 2011), which are functionally diverse and belong to over 60 genera (Philippot 2006). The rate-limiting step in denitrification is the reduction of nitrate (NO_2^-) to nitric oxide (NO), which distinguishes denitrifiers from other nitrate-reducing bacteria. This step is catalyzed by the enzyme nitrate reductase. Denitrifiers may contain one of two functionally and physiologically equivalent types of nitrite reductases: a Cu-containing enzyme (encoded by *nirK*) and a cytochrome cd1 (encoded by *nirS*) (Philippot 2006; Zumft 1997). The size of denitrifying microbial communities has been correlated with denitrification process rates (Throbäck et al. 2007; Hallin et al. 2009; Morales et al. 2010; Szukics et al. 2010; Petersen et al. 2012; Wu et al. 2012; Butterbach-Bahl et al. 2013). Another key reaction in the denitrification process is the reduction of N_2O to N_2. This reaction is catalyzed by the enzyme nitrous oxide reductase, encoded by the gene *nosZ*. The presence of the *nosZ*, a gene in the soil, may suggest that the microbial community present can reduce N_2O to N_2 and influence the balance of the two in the environment (Philippot 2006). Given that denitrifiers account for a significant proportion of soil microbial communities, the response of denitrifying microbes to agricultural management strategies is useful in assessing trends in soil functioning. Community composition has been shown to vary with the use of N and C fertilization (Hallin et al. 2009; Yin et al. 2014; Bastian et al. 2009), crop production practices and cropland use (Reeve et al. 2010; Bissett et al. 2011), as well as cover plant species (Bremer et al. 2007; Hai et al. 2009; Petersen et al. 2012). Our ability to identify agricultural practices that enhance immobilization and transformation of fertilizer N by soils requires an understanding of the impact agricultural management practices, such as the application of N, have on N-cycling microbes. Genes associated with the N cycle have been quantified and assessed in order to elucidate the effects of soil properties and management practices, such as soil geomorphology and land use (Colloff et al. 2008), pasture management (Wakelin et al. 2009), N fertilizers (Okano et al. 2004; Cavagnaro et al. 2008), and tillage (Cavagnaro et al. 2008), on the function of soil microbial communities. The main N-cycling gene targets for molecular analysis of soil N-cycling potential include the denitrification genes nirK, nirS, and nosZ, the nitrification amoA genes (in both bacteria and archaea), and the N fixation gene nifH (reviewed in Teixeira and Yergeau 2012).

3 Nitrogen-Cycling Communities in Soils Under Various Fertilization Managements

3.1 Nitrogen Fixation

It may be anticipated that fertilizers containing N will have some impact on microbial communities involved in N cycling, as it may remove any selection pressure for the ability to carry out N_2 fixation. However, the Rothamsted Broadbalk experiment, which compared the effects of different N inputs over a 170-year period, showed no significant effect on the diversity of the nitrogenase gene *nifH*, despite observed changes in microbial community structure (Ogilvie et al. 2008). Nevertheless, other studies have reported differences. A study conducted in south-east Australia, which compared a number of different soil types and different land management practices, indicated that land use had no influence on *nifH* gene abundance. Instead, the primary factor influencing *nifH* abundance was the amount of microbial biomass carbon (Hayden et al. 2010). In other Australian studies, the use of organic mulch as a soil amendment increased the abundance of the *nifH* gene in cotton soils, and the application of manure helped to maintain the level of *nifH* before and after crop planting. In a Spanish study of grapevine soil, the abundance of the *nifH* gene was significantly increased under treatment with organic fertilizers than when inorganic fertilizer was used (Pereg et al. 2018). Morales et al. (2010) hypothesized that the *nifH* gene, indicating the potential for N fixation, would be found at higher levels in soils where leguminous plants were regularly grown. However, they found that the *nifH* gene was more abundant in soil samples collected from forested sites, or soils from sites used for successional planting, than it was in agricultural soils, including those regularly planted with soybeans. These findings have been attributed to the populations of free-living N fixers present in natural ecosystems. Pereg et al. (2018) also observed that organic treatments which included either leguminous plants or manure resulted in a similar abundance of the *nifH* gene, suggesting that factors other than leguminous plant cover may account for the increased potential for N fixation in organically managed systems when compared to traditional inorganic fertilization practices. Also, in agreement with Morales et al. (2010), Pereg et al. (2018) showed that the increased abundance of *nifH* under organic fertilization regimes correlated strongly with available N (Av N, Kjeldahl method) and total organic carbon (TOC) in grapevine soil. Pereg et al. (2018) suggested that the readily available nitrate and ammonia in inorganic fertilizer may have suppressed the abundance of N fixers, as indicated by the reduced abundance of *nifH* under inorganic fertilization compared with soil treated with slowly released organic N sources. Coelho et al. (2009) have also detected a reduction in free-living diazotroph communities in soil with increasing levels of inorganic N fertilizer. High levels of NH_3 and NO_3^- (available forms of N) found in inorganic fertilizers may repress the synthesis and/or catalytic activity of the enzyme nitrogenase (Bisseling et al. 1978), suppressing N fixation in soil treated with fertilizers and abolishing the competitive advantage that diazotrophs may have in environments poor in available N. The process of N fixation is energetically expensive and requires large amounts of adenosine triphosphate (ATP) and reducing equivalents. Therefore, readily available carbon (C) source is also essential for diazotrophs to fix N (Chan et al. 1994). Results from Pereg et al. (2018) indicated that organic amendments that release organic N and phosphorus gradually into the soil (García-Orenes et al. 2016) support the conservation of N fixer communities in grapevine soils. This is in agreement with earlier studies reporting that phosphorus fertilization stimulates N fixation in soils (Reed et al. 2007) probably due to the high energy requirements of the N fixation process. It should also be noted that other soil parameters will also influence the abundance and activity of N-cycling microbial communities. The seasonal dynamics of these N-cycling communities are also tightly coupled with seasonal changes in labile C and N pools, temperature, oxygen availability, soil compaction, and soil moisture (Meng et al. 2017; Butterbach-Bahl et al. 2013; Rasche et al. 2010; Menneer et al. 2004).

3.2 *Nitrification*

During nitrification, ammonia is converted to nitrite by ammonia oxidizers and then to nitrate by nitrite oxidizers. Ammonia-oxidizing archaea (AOA) are considered to be more abundant than ammonia-oxidizing bacteria (AOB) in the majority of soil types (Leininger et al. 2006). However, it is not yet clear what the relative contributions of these different groups are to soil nitrification and whether or not they can be considered to be functionally interchangeable (Jia and Conrad 2009; Xia et al. 2011). The AOA require less energy and less ammonia than AOB and are therefore likely to be more abundant than AOB in soils that are unfertilized and in natural soils such as forest soils, despite having lower rates of ammonia oxidation than their bacterial counterparts (Martens-Habbena et al. 2009; Tourna et al. 2011). However, AOA has also been found to be more abundant in a variety of arable and agricultural soils (Zhalnina et al. 2013). Acidic soils may also be more likely to support AOA over AOB, as at low pH there is a shift toward ammonium over ammonia, which limits substrate availability and growth of AOB (Hirsch and Mauchline 2015). Although AOA has been found to dominate in arable soils at neutral pH, it appears to be AOB that increase in numbers in response to the application of N. Nitrification increases as N availability increases, and it is assumed that AOB are responsible for these increases in nitrification (Hirsch and Mauchline 2015). Similarly, the size of AOB communities, but not AOA, has been shown to increase with increased N availability in farmed grasslands (Di et al. 2009) and in direct response to application of animal manure (Wakelin et al. 2013), while in contrast, AOA was shown to decline with increasing N in two different arable soils (Bates et al. 2010; Wessén et al. 2011). In acidic soils, however, it is AOA that show a greater response to the application of N fertilizers (Gubry-Rangin et al. 2010). This again reflects the availability of ammonia as a substrate and the different pH optima for the two groups: nitrification by AOB decreases at lower pH (below 7), while conversely, nitrification by AOA decreases as soil pH increases (Hirsch and Mauchline 2015). AOB fix C autotrophically, and ammonia is its only source of usable energy and reductants. AOB isolated from acidic soils are often ureolytic, containing the urease enzyme, and ureolytic AOB can grow at lower pH with urea as an ammonia and CO_2 source (summarized in Koper et al. 2004), two ecologically important traits. Koper et al. (2004) suggested that ureolytic AOB may have an advantage in soils receiving animal wastes or urea fertilizers. The structure of AO microbial communities is also influenced by pH: different groups of AOA and AOB have been found in arable soils maintained at a range of pH from 4.5 to 7.5 (Nicol et al. 2008). The structure and function of these microbial communities seems to be influenced more by the soil properties rather than by the land management strategies used (organic versus conventional), which may explain why significant heterogeneity has been observed in AOA and AOB abundance (Hirsch and Mauchline 2015).

In agreement with Wessén et al. (2010) and Zhang et al. (2010), a study by Pereg et al. (2018) comparing organic versus inorganic grape production indicated that *amoA* genes, involved in ammonium oxidation in both bacteria and archaea, were affected differently by the fertilization practices. Abundance of the bacterial *amoA* (*amoA*-B) gene was lowest in soils treated with organic fertilizer in the form of prunings plus manure (OPM) yet higher in soil treated with prunings and legume cover (OPL) and under inorganic fertilization (IF) (Pereg et al. 2018). IF (20 Kg ha^{-1} N annually) soil was treated with NH_4NO_3 containing NPK, whereas OPL (15 Kg ha^{-1} N annually) was not treated with inorganic fertilizers. Nevertheless, there was a greater abundance of *amoA*-B under both treatments than in OPM-treated (126 Kg ha^{-1} N annually) soil. While NH_3 is produced from degradation of the organic matter in fresh manure, a large proportion of it might be lost by direct conversion into NO_3^- by heterotrophic nitrifiers or by volatilization (Maeda et al. 2011 and references within). In contrast, archaeal *amoA* (*amoA*-arch) genes were evenly distributed under all treatments, OPM, OPL, and IF, in grapevine soil (Pereg et al. 2018). In contrast to Leininger et al. (2006), Di et al. (2009), Hai et al. (2009), and Pereg et al. (2018) found a higher abundance (approx. tenfold) of archaeal than bacterial *amoA* in the soil. Hai et al. (2009) found an even distribution of AOA, but not AOB, in tropical sorghum soils treated with manure or straw with or without urea and, similar to Santoro et al. (2008), concluded that the AOA populations are more stable than AOB populations. AOB and AOA were present in lower numbers in soils treated with ammonium sulfate, compared to other soil treatments (Hallin et al. 2009). Interestingly, treatment with sewage sludge negatively impacted on the size of the AOA communities, but not on the AOB communities. The results of this study confirmed that fertilization regimes could affect not just the abundance of ammonia oxidizers but also the AOB/AOA ratio (He et al. 2007). However, it is not yet clear exactly which conditions favor the dominance of one or the other of the two types of ammonia oxidizers, bacterial and archaeal. Zhang et al. (2017) observed significant effects of pig manure as a fertilizer on nitrification activity and ammonia oxidizer communities. Although in general AOA were obviously more abundant than AOB (hundreds of times greater), it seemed that the AOB community was more sensitive to the shifts in fertilization or soil heterogeneity. This study also found that other factors, such as C- and N-related soil nutrients and enzyme activities, were important factors in shaping AOA and AOB community structures (Hallin et al. 2009). Taken together, these studies also suggest that variations in environmental conditions affect bacterial communities more than their archaeal counterparts. Archaea were found to oxidize ammonia and assimilate C in an agricultural soil, with different groups carrying out either heterotrophic activity or autotrophic CO_2 fixation (Pratscher et al. 2011). Such heterogenicity in ammonia-oxidizing microbes could possibly explain why bacterial and archaeal *amoA* abundance was independent of TOC in grapevine soils as observed by Pereg et al. (2018) and in soils collected from Australian cotton fields.

To summarize, while nitrification takes place in most soils, in soils that are acidic or nutrient-poor, AOA may be responsible for most nitrification, while AOB contribute to a greater extent in neutral and fertilized soils. As a consequence, there may be a delay between the application of fertilizers to nutrient-poor soils and an increase in nitrification, until the AOB populations have the time to adjust and increase. AOA and AOB also show different responses to inhibitors of nitrification. For example, the inhibitor 3,4-dimethylpyrazole phosphate (DMPP) reduces size of AOB communities, but does not impact AOA (Kleineidam et al. 2011); similarly, high levels of sulfadiazine residues in pig manure inhibit growth of AOB to a much greater extent than they inhibit AOA (Schauss et al. 2009). N fertilizers in access might turn into an ecosystem hazard, where nitrifying microbes exist and convert available N to nitrate. Excess of nitrate might leach into, and cause eutrophication of, aquifers, groundwater, lakes, and estuaries (Vitousek et al. 1997; Galloway et al. 2008).

3.3 Denitrification

Denitrification is a useful ability for soil bacteria, allowing anaerobic respiration. Many of the denitrifiers identified are facultative anaerobes, switching to denitrification when soil becomes waterlogged and when organic matter and useable N are readily available. Fungi that contain a mitochondrial gene similar to the bacterial-type nitrite reductase gene can also be involved in denitrification (Kim et al. 2009). Some factors, such as reduced oxygen availability, and changes in nitrate concentration, that influence denitrification have been well-established. The addition of manures and other organic fertilizers to soil tend to increase the activity of denitrifiers, relative to the use of inorganic fertilizers, but the impacts on the structure of microbial denitrifier communities are less clear (Hallin et al. 2009; Philippot et al. 2007). Thompson et al. (2016) showed that the abundance of denitrifiers changed differently than the total bacterial community of soils, suggesting that denitrifier populations are regulated differently from the total bacterial community. However, a long-term Swedish study comparing different fertilizer inputs found that overall denitrification activity was correlated with the total bacterial biomass, with crop yields, and with levels of the *nosZ* gene, which encodes the enzyme nitrous oxide reductase (Hallin et al. 2009). In any case, the abundance, distribution, and diversity of denitrifying microbes appear to be influenced by both soil conditions and management practices, and their activity can be assessed by reference to *nir* gene abundances (Clark et al. 2012; Hallin et al. 2009).

The impacts of inorganic fertilizers on denitrifying microbial communities are complex, due to the great diversity of denitrifiers themselves and the variation in abundance in different environments (Hirsch and Mauchline 2015). However, in general, it has been demonstrated that the use of organic fertilizers increases denitrification activity in soils when compared to inorganic fertilization (Philippot et al. 2007). The Cu-containing enzyme nitrite reductase *nirK* has been reported at higher levels in arable soils and in soils that have been treated with increased N fertilizer inputs (Philippot et al. 2007), while the alternative, *nirS*, has been found to be more abundant in intensively grazed pasture soils with increased soil ammonia levels (Dandie et al. 2011). A study which compared various fertilization managements, including unfertilized bare fallow, unfertilized plots with crop, and plots with crop fertilized with calcium nitrate, ammonium sulfate, solid cattle manure, or sewage sludge, found that the abundance of nitrate reducers (*narG*) and denitrifiers (*nirS*, *nosZ*) were significantly lower in soils treated with sewage sludge than in the soils treated with manure. The positive effects expected by the high organic inputs may have been counteracted by the lower pH in the plots treated with sludge (pH 4.7) compared to the manure-treated plots (pH 6.0) as well as by the higher heavy metal content (Hallin et al. 2009; Bergkvist et al. 2003). Overall, the abundance of denitrifiers was one to two orders of magnitude lower in soils treated with ammonium sulfate, compared with the other treatments. This may also be explained by the lower pH of the soil in these plots compared to the other treatments (Hallin et al. 2009). Other studies have also described the impact of fertilization strategy on denitrifying communities. In Spanish grapevine soils, the abundance of genes involved in denitrification (*nirK*, *nirS*, and *nosZ*) was found to be greater under organic than under inorganic fertilization (Pereg et al. 2018). On the other hand, Zhang et al. (2013) reported different responses of the various denitrification genes (*nirS*, *nirK*, and *nosZ*) to different soil treatments. Clark et al. (2012) reported an overall similar behavior of *nosZ*, *nirK*, and *nirS*, in soil from wheat fields, but found that *nirS* was tenfold less common than the other genes. In contrast, Pereg et al. (2018) found that *nosZ* was approximately fivefold less abundant than *nirS* in the Spanish grapevine soil. Hallin et al. (2009) found nirS/nirK ratio in unfertilized bare fallow soils to be three to ten times higher than in soil from plots with crops. Since the reduction of nitrite by denitrifiers can be performed either by the cytochrome cd1 nitrite reductase (NirS) or by the copper-containing nitrite reductase (NirK), as denitrifying bacteria possess only one form of the enzyme, these findings indicate the habitat created by the presence or absence of plants will select for either NirS- or NirK-type microbes. They suggested that NirS and NirK, although being functionally equivalent, are not ecologically redundant.

The gene *nirS* has been found to dominate over *nirK* in various natural environments, particularly in aquatic environments (Bothe et al. 2000; Nogales et al. 2002; Prieme et al. 2002; Liu et al. 2003; Throbäck et al. 2004; Oakley et al. 2007; Deslippe et al. 2014), as well as in cultured denitrifiers (Zumft 1997), and Thompson et al. (2016) raised a concern about unspecific *nirK* amplification products in PCR assays. Therefore, some researchers shifted their focus from studying both genes to studies on *nirS* alone (Morales et al. 2010; Thompson et al. 2016). Nevertheless, *nirK* is abundant in aerobic, oxygen-rich environments (Desnues et al. 2007; Knapp et al. 2009), and an analysis of the *nirK* PCR primers developed by Henry et al. (2004) and optimized by Pereg et al. (2018) confirmed their suitability for specific *nirK* amplification from soil DNA. Therefore, Pereg et al. (2018) recommended these primers for use in studies estimating the abundance of the denitrifying gene in soil DNA.

Similar to Hallin et al. (2009), Pereg et al. (2018) showed that different N fertilization usage could impact on the size of denitrifying microbial communities.

An organic fertilizer, consisting of manure and plant residues with a relatively high C:N ratio (approx. 80), and six times higher total amount of slow-release N than that of inorganic fertilizer, increased denitrifier abundance in grapevine soils. In another study, such a high C:N ratio was found to correlate with decreased N_2O emissions (Huang 2004). The final step in the N cycle is the reduction of nitrous oxide, releasing N_2 back to the atmosphere. This reaction is catalyzed by nitrous oxide reductase, encoded by the *nosZ* gene, a gene present in the genomes of less than 70% of known denitrifying microbes (Jones et al. 2008). Even when present the gene is not always expressed, resulting in accumulation of N_2O, an important greenhouse gas. It has been reported recently that some microbial species only contain a functional *nosZ* gene (Jones et al. 2013; Sanford et al. 2012), even though they may not possess the other genes involved in the denitrification pathway. Indeed, the abundance of the *nosZ* was higher in the grapevine soils using this organic fertilizer, when compared to inorganic fertilization regimes (Pereg et al. 2018), suggesting a higher abundance of denitrifiers with the ability to reduce N_2O to N_2 and the potential to lower N_2O emissions (Miller et al. 2008). Agricultural management practices that encourage the growth of microorganisms with this functional *nosZ* gene could be important and potentially allow for a reduction in greenhouse gas emissions under conditions which make denitrification inevitable (Hirsch and Mauchline 2015).

4 The Relationship Between Management Strategies and N-Cycling Gene Abundance

One of the fundamental differences between inorganic fertilization and organic matter supplementation is that the first boosts the soil with short-lived nitrate and ammonia in relatively high concentrations, while the latter depends on the slower breakdown of organic matter and thus provides a gradual supply of N and C to the soil. This gradual release of available N and C improves various soil properties (García-Orenes et al. 2016) and supports the maintenance of diazotrophic and denitrifying communities (Pereg et al. 2018). The components of each soil treatment can, therefore, have a significant influence on soil microbial communities. Hartmann et al. (2015) found that systems not receiving manure harbored more dispersed and functionally versatile microbial communities containing oligotrophic organisms that possibly adapted to nutrient-poor environments. On the other hand, systems treated with organic fertilizers harbored specific microbial guilds that degrade complex organic compounds, for example, compost and manure (Hartmann et al. 2015). In a study of Spanish grapevine soils, Pereg et al. (2018) found a significantly high correlation between abundance of *nifH* and TOC, *nosZ*, or *nirS*. The correlation of TOC with *nifH* was particularly high, possibly due to the relatively low available N concentrations at any time in the soil, due to the slow release of available N from organic matter. In contrast to the findings of Pereg et al. (2018) (for grapevine soil,

Spain) and (for cotton in Australia), Morales et al. (2010) found a weak negative correlation between the abundances of the *nifH* and *nirS* genes. The *nirS* gene abundance was not correlated to organic C levels and did not exhibit the same trend as *nosZ* (Morales et al. 2010). Hai et al. (2009) also found that there was an increased abundance of *nifH*, but not of *nirK/S*, in tropical agricultural soils treated with manure, when compared to untreated soil or soil treated with straw. Overall, N fixation and denitrification are opposite reactions. However, some soil microbes are capable of both N fixation and denitrification. For example, various rhizobia species contain *nifH* and *nirK* (Bedmar et al. 2005), while some N-fixing strains of the bacterium *Azospirillum brasilense* contain both *nifH* and cd1-type *nir* gene (Danneberg et al. 1986). Therefore, it is not surprising that the abundance of *nifH* and *nirS/nirK/nosZ* was found to be similar in grapevine soils under various conditions as shown in Pereg et al. (2018). While nitrification is aerobic and denitrification anaerobic, these processes can take place in different micro-niches in soil aggregates and thus run simultaneously where both contribute to soil production of nitrous oxide (Stevens et al. 1997).

Soil with a neutral or slightly basic pH is considered to be optimal for most diazotrophs (Belnap 2001), and soil pH has been considered as a major factor influencing microbial community structure in various studies worldwide (Fierer and Jackson 2006; Noll and Wellinger 2008; Wakelin et al. 2008; Griffiths et al. 2011; Zhalnina et al. 2015). However, often in agricultural managed ecosystems, farmers often manage their soils to achieve particular pH and other soil properties. Therefore, it is not surprising that studies of agricultural soils, such as that carried out in Pago Casa Gran, have shown there was no significant difference in the pH of soil under organic or inorganic fertilization (García-Orenes et al. 2016), so this factor cannot explain the differences observed in N cycler abundance (Pereg et al. 2018). Similar results were observed in a study of Australian cotton soil indicating that factors other than pH are major determinants of N-cycling microbial communities. High NH_4^+ availability and moderate pH favor nitrification in soils, whereas pH close to neutral, high availability of nitrate (electron acceptor), and labile, energy-rich C (Barnard et al. 2005) favor denitrification. There is obviously a great diversity in soil microbial communities across different soils, under different environmental and geographical locations. Moreover, diversity could also reflect temporal factors and technical differences, which may compromise comparisons between studies. When considering N fertilization strategies, higher N fixation than denitrification would be beneficial in order to conserve the newly fixed N in the biosphere. Where *nifH* and *nir/nosZ* genes are both abundant, it is the conditions in the soil that would determine which genes will be fully expressed. Methods such as in situ enzymatic activity, proteomics, and possibly transcriptomics could be employed to assess which conditions in the soil niche would activate particular processes at any given time and allow this to be taken into account when selecting land management strategies.

5 Conclusions

Despite the demonstrated negative environmental impacts, modern agricultural practices, with high levels of agrochemical, mechanization, and modern high-yielding crop varieties, such as cereal production, have increased the productivity of agricultural land and allowed the earth to sustain an increasingly large human population. Since 1700 the global population has increased by at least tenfold. It is currently over seven billion and predicted to grow up to around nine billion people by the year 2050 (Hirsch and Mauchline 2015). This continual population growth places increasing pressure on land for food production. However, we also see aggravated climate change in response to increased industrialization. Food security and the need to feed a growing population must, therefore, be balanced against the negative environmental impacts of intensive crop production (Hirsch and Mauchline 2015). Utilizing N in agricultural systems in ways that consider and help manage microbial communities involved in the N cycle, notably using organic matter supplementation, will play an essential role in ensuring sustainability.

References

Alef K, Beck T, Zelles L, Kleiner D (1988) A comparison of methods to estimate microbial biomass and N-mineralization in agricultural and grassland soils. Soil Biol Biochem 20(4):561–565

Barnard R, Leadley PW, Hungate BA (2005) Global change, nitrification, and denitrification: a review. Glob Biogeochem Cycles 19(1):1

Barrett JE, Burke IC (2000) Potential nitrogen immobilization in grassland soils across a soil organic matter gradient. Soil Biol Biochem 32(11–12):1707–1716

Bastian F, Bouziri L, Nicolardot B, Ranjard L (2009) Impact of wheat straw decomposition on successional patterns of soil microbial community structure. Soil Biol Biochem 41(2):262–275

Bastida F, Zsolnay A, Hernández T, García C (2008) Past, present and future of soil quality indices: a biological perspective. Geoderma 147(3–4):159–171

Bastida F, Torres IF, Romero-Trigueros C, Baldrian P, Větrovský T, Bayona JM, Alarcón JJ, Hernández T, García C, Nicolás E (2017) Combined effects of reduced irrigation and water quality on the soil microbial community of a citrus orchard under semi-arid conditions. Soil Biol Biochem 104:226–237

Bates ST, Berg-Lyons D, Caporaso JG, Walters WA, Knight R, Fierer N (2010) Examining the global distribution of dominant archaeal populations in soil. ISME J 5(5):908–917

Bedmar EJ, Robles EF, Delgado MJ (2005) The complete denitrification pathway of the symbiotic, nitrogen-fixing bacteriumBradyrhizobium japonicum. Biochem Soc Trans 33(1):141–144

Bellamy PH, Loveland PJ, Bradley RI, Lark RM, Kirk GJD (2005) Carbon losses from all soils across England and Wales 1978–2003. Nature 437(7056):245–248

Belnap J (2001) Factors influencing nitrogen fixation and nitrogen release in biological soil crusts. In: Ecological studies. Springer, Berlin Heidelberg

Bengtsson G, Bengtson P, Månsson KF (2003) Gross nitrogen mineralization-, immobilization-, and nitrification rates as a function of soil C/N ratio and microbial activity. Soil Biol Biochem 35 (1):143–154

Bergkvist P, Jarvis N, Berggren D, Carlgren K (2003) Long-term effects of sewage sludge applications on soil properties, cadmium availability and distribution in arable soil. Agric Ecosyst Environ 97(1–3):167–179

Berry D (1978) Effects of urbanization on agricultural activities. Growth Chang 9(3):2–8

Bisseling T, Van Den Bos RC, Van Kammen A (1978) The effect of ammonium nitrate on the synthesis of nitrogenase and the concentration of leghemoglobin in pea root nodules induced by rhizobium leguminosarum. Biochim Biophys Acta Gen Subj 539(1):1–11

Bissett A, Richardson AE, Baker G, Thrall PH (2011) Long-term land use effects on soil microbial community structure and function. Appl Soil Ecol 51:66–78

Bossio DA, Scow KM, Gunapala N, Graham KJ (1998) Determinants of soil microbial communities: effects of agricultural management, season, and soil type on phospholipid fatty acid profiles. Microb Ecol 36(1):1–12

Bothe H, Jost G, Schloter M, Ward BB, Witzel K-P (2000) Molecular analysis of ammonia oxidation and denitrification in natural environments. FEMS Microbiol Rev 24(5):673–690

Braker G, Conrad R (2011) Diversity, structure, and size of N2O-producing microbial communities in soils–what matters for their functioning? In: Advances in applied microbiology. Elsevier, San Diego

Braker G, Fesefeldt A, Witzel K-P (1998) Development of PCR primer systems for amplification of nitrite reductase genes (nirK and nirS) to detect denitrifying bacteria in environmental samples. Appl Environ Microbiol 64(10):3769–3775

Bremer C, Braker G, Matthies D, Reuter A, Engels C, Conrad R (2007) Impact of plant functional group, plant species, and sampling time on the composition of nirK-type denitrifier communities in soil. Appl Environ Microbiol 73(21):6876–6884

Butterbach-Bahl K, Baggs EM, Dannenmann M, Kiese R, Zechmeister-Boltenstern S (2013) Nitrous oxide emissions from soils: how well do we understand the processes and their controls? Phil Trans R Soc B 368(1621):20130122–20130122

Canarini A, Carrillo Y, Mariotte P, Ingram L, Dijkstra FA (2016) Soil microbial community resistance to drought and links to C stabilization in an Australian grassland. Soil Biol Biochem 103:171–180

Caravaca F, Masciandaro G, Ceccanti B (2002) Land use in relation to soil chemical and biochemical properties in a semiarid Mediterranean environment. Soil Tillage Res 68(1):23–30

Castañeda LE, Godoy K, Manzano M, Marquet PA, Barbosa O (2015) Comparison of soil microbial communities inhabiting vineyards and native sclerophyllous forests in Central Chile. Ecol Evol 5(18):3857–3868

Cavagnaro TR, Jackson LE, Hristova K, Scow KM (2008) Short-term population dynamics of ammonia oxidizing bacteria in an agricultural soil. Appl Soil Ecol 40(1):13–18

Chan Y-K, Barraquio WL, Knowles R (1994) N2-fixing pseudomonads and related soil bacteria. FEMS Microbiol Rev 13(1):95–117

Clark IM, Buchkina N, Jhurreea D, Goulding KWT, Hirsch PR (2012) Impacts of nitrogen application rates on the activity and diversity of denitrifying bacteria in the Broadbalk wheat experiment. Phil Trans R Soc B 367(1593):1235–1244

Coelho MRR, Marriel IE, Jenkins SN, Lanyon CV, Seldin L, O'Donnell AG (2009) Molecular detection and quantification of nifH gene sequences in the rhizosphere of sorghum (*Sorghum bicolor*) sown with two levels of nitrogen fertilizer. Appl Soil Ecol 42(1):48–53

Colloff MJ, Wakelin SA, Gomez D, Rogers SL (2008) Detection of nitrogen cycle genes in soils for measuring the effects of changes in land use and management. Soil Biol Biochem 40 (7):1637–1645

Dandie CE, Wertz S, Leclair CL, Goyer C, Burton DL, Patten CL, Zebarth BJ, Trevors JT (2011) Abundance, diversity and functional gene expression of denitrifier communities in adjacent riparian and agricultural zones. FEMS Microbiol Ecol 77(1):69–82

Danneberg G, Kronenberg A, Neuer G, Bothe H (1986) Aspects of nitrogen fixation and denitrification byAzospirillum. Plant Soil 90(1–3):193–202 de Vries FT, Thebault E, Liiri M, Birkhofer K, Tsiafouli MA, Bjornlund L, Bracht Jorgensen H, Brady MV, Christensen S, de Ruiter PC, d'Hertefeldt T, Frouz J, Hedlund K, Hemerik L, Hol WHG, Hotes S, Mortimer SR, Setala H, Sgardelis SP, Uteseny K, van der Putten WH, Wolters V, Bardgett RD (2013) Soil food web properties explain ecosystem services across European land use systems. Proc Natl Acad Sci 110(35):14296–14301

Deslippe JR, Jamali H, Jha N, Saggar S (2014) Denitrifier community size, structure and activity along a gradient of pasture to riparian soils. Soil Biol Biochem 71:48–60

Desnues C, Michotey VD, Wieland A, Zhizang C, Fourçans A, Duran R, Bonin PC (2007) Seasonal and diel distributions of denitrifying and bacterial communities in a hypersaline microbial mat (Camargue, France). Water Res 41(15):3407–3419

Di HJ, Cameron KC, Shen JP, Winefield CS, O'Callaghan M, Bowatte S, He JZ (2009) Nitrification driven by bacteria and not archaea in nitrogen-rich grassland soils. Nat Geosci 2(9):621–624

Dixon R, Kahn D (2004) Genetic regulation of biological nitrogen fixation. Nat Rev Microbiol 2 (8):621–631

Fierer N, Jackson RB (2006) The diversity and biogeography of soil bacterial communities. Proc Natl Acad Sci 103(3):626–631

Fierer N, Lauber CL, Ramirez KS, Zaneveld J, Bradford MA, Knight R (2012) Comparative metagenomic, phylogenetic and physiological analyses of soil microbial communities across nitrogen gradients. ISME J 6(5):1007–1017

Fitter AH, Gilligan CA, Hollingworth K, Kleczkowski A, Twyman RM, Pitchford JW (2005) Biodiversity and ecosystem function in soil. Funct Ecol 19(3):369–377

Frenk S, Hadar Y, Minz D (2014) Resilience of soil bacterial community to irrigation with water of different qualities under Mediterranean climate. Environ Microbiol 16(2):559–569

Galloway JN, Townsend AR, Erisman JW, Bekunda M, Cai Z, Freney JR, Martinelli LA, Seitzinger SP, Sutton MA (2008) Transformation of the nitrogen cycle: recent trends, questions, and potential solutions. Science 320(5878):889–892

Garbeva P, van Elsas JD, van Veen JA (2008) Rhizosphere microbial community and its response to plant species and soil history. Plant Soil 302(1–2):19–32

García-Orenes F, Guerrero C, Roldán A, Mataix-Solera J, Cerdà A, Campoy M, Zornoza R, Bárcenas G, Caravaca F (2010) Soil microbial biomass and activity under different agricultural management systems in a semiarid Mediterranean agroecosystem. Soil Tillage Res 109 (2):110–115

García-Orenes F, Morugán-Coronado A, Zornoza R, Scow K (2013) Changes in soil microbial community structure influenced by agricultural management practices in a mediterranean agro-ecosystem. PLoS One 8(11):e80522

García-Orenes F, Roldán A, Morugán-Coronado A, Linares C, Cerdà A, Caravaca F (2016) Organic fertilization in traditional mediterranean grapevine orchards mediates changes in soil microbial community structure and enhances soil fertility. Land Degrad Dev 27(6):1622–1628

Geisseler D, Scow KM (2014) Long-term effects of mineral fertilizers on soil microorganisms – a review. Soil Biol Biochem 75:54–63

Geisseler D, Horwath WR, Joergensen RG, Ludwig B (2010) Pathways of nitrogen utilization by soil microorganisms – a review. Soil Biol Biochem 42(12):2058–2067

Griffiths RI, Thomson BC, James P, Bell T, Bailey M, Whiteley AS (2011) The bacterial biogeography of British soils. Environ Microbiol 13(6):1642–1654

Gubry-Rangin C, Nicol GW, Prosser JI (2010) Archaea rather than bacteria control nitrification in two agricultural acidic soils. FEMS Microbiol Ecol 74(3):566–574

Hai B, Diallo NH, Sall S, Haesler F, Schauss K, Bonzi M, Assigbetse K, Chotte JL, Munch JC, Schloter M (2009) Quantification of key genes steering the microbial nitrogen cycle in the rhizosphere of sorghum cultivars in tropical agroecosystems. Appl Environ Microbiol 75 (15):4993–5000

Hallin S, Jones CM, Schloter M, Philippot L (2009) Relationship between N-cycling communities and ecosystem functioning in a 50-year-old fertilization experiment. ISME J 3(5):597–605

Hart MJ, Eva A, Zangrilli D, Aaronson SA, Evans T, Cerione RA, Zheng Y (1994) Cellular transformation and guanine nucleotide exchange activity are catalyzed by a common domain on the dbl oncogene product. J Biol Chem 269(1):62–65

Hartmann M, Frey B, Mayer J, Mäder P, Widmer F (2015) Distinct soil microbial diversity under long-term organic and conventional farming. ISME J 9(5):1177–1194

Hayden HL, Drake J, Imhof M, Oxley APA, Norng S, Mele PM (2010) The abundance of nitrogen cycle genes amoA and nifH depends on land-uses and soil types in south-eastern Australia. Soil Biol Biochem 42(10):1774–1783

He J-z, Shen J-p, L-m Z, Zhu Y-g, Zheng Y-m, Xu M-g, Di H (2007) Quantitative analyses of the abundance and composition of ammonia-oxidizing bacteria and ammonia-oxidizing archaea of a Chinese upland red soil under long-term fertilization practices. Environ Microbiol 9 (9):2364–2374

He Z, Deng Y, Van Nostrand JD, Tu Q, Xu M, Hemme CL, Li X, Wu L, Gentry TJ, Yin Y, Liebich J, Hazen TC, Zhou J (2010) GeoChip 3.0 as a high-throughput tool for analyzing microbial community composition, structure and functional activity. ISME J 4(9):1167–1179

Henry S, Baudoin E, López-Gutiérrez JC, Martin-Laurent F, Brauman A, Philippot L (2004) Quantification of denitrifying bacteria in soils by nirK gene targeted real-time PCR. J Microbiol Methods 59(3):327–335

Hirsch PR, Mauchline TH (2015) The importance of the microbial N cycle in soil for crop plant nutrition. In: Advances in applied microbiology. Elsevier

Horz HP, Barbrook A, Field CB, Bohannan BJM (2004) Ammonia-oxidizing bacteria respond to multifactorial global change. Proc Natl Acad Sci 101(42):15136–15141

Houlton BZ, Bai E (2009) Imprint of denitrifying bacteria on the global terrestrial biosphere. Proc Natl Acad Sci 106(51):21713–21716

Huang Y (2004) Nitrous oxide emissions as influenced by amendment of plant residues with different C:N ratios. Soil Biol Biochem 36(6):973–981

Imfeld G, Vuilleumier S (2012) Measuring the effects of pesticides on bacterial communities in soil: a critical review. Eur J Soil Biol 49:22–30

Jangid K, Williams MA, Franzluebbers AJ, Sanderlin JS, Reeves JH, Jenkins MB, Endale DM, Coleman DC, Whitman WB (2008) Relative impacts of land-use, management intensity and fertilization upon soil microbial community structure in agricultural systems. Soil Biol Biochem 40(11):2843–2853

Jia Z, Conrad R (2009) Bacteriarather thanArchaeadominate microbial ammonia oxidation in an agricultural soil. Environ Microbiol 11(7):1658–1671

Johnston AE, Poulton PR, Coleman K (2009) Chapter 1 soil organic matter. In: Advances in agronomy. Elsevier

Jones CM, Stres B, Rosenquist M, Hallin S (2008) Phylogenetic analysis of nitrite, nitric oxide, and nitrous oxide respiratory enzymes reveal a complex evolutionary history for denitrification. Mol Biol Evol 25(9):1955–1966

Jones CM, Graf DRH, Bru D, Philippot L, Hallin S (2013) The unaccounted yet abundant nitrous oxide-reducing microbial community: a potential nitrous oxide sink. ISME J 7(2):417–426

Kibblewhite MG, Ritz K, Swift MJ (2008) Soil health in agricultural systems. Phil Trans R Soc B 363(1492):685–701

Kim SW, Fushinobu S, Zhou S, Wakagi T, Shoun H (2009) Eukaryotic nirK genes encoding copper-containing nitrite reductase: originating from the protomitochondrion? Appl Environ Microbiol 75(9):2652–2658

Kleineidam K, Košmrlj K, Kublik S, Palmer I, Pfab H, Ruser R, Fiedler S, Schloter M (2011) Influence of the nitrification inhibitor 3,4-dimethylpyrazole phosphate (DMPP) on ammonia-oxidizing bacteria and archaea in rhizosphere and bulk soil. Chemosphere 84(1):182–186

Knapp CW, Dodds WK, Wilson KC, O'Brien JM, Graham DW (2009) Spatial heterogeneity of denitrification genes in a highly homogenous urban stream. Environ Sci Technol 43 (12):4273–4279

Koper TE, El-Sheikh AF, Norton JM, Klotz MG (2004) Urease-encoding genes in ammonia-oxidizing bacteria. Appl Environ Microbiol 70(4):2342–2348

Ladha JK, Pathak H, Krupnik TJ, Six J, van Kessel C (2005) Efficiency of fertilizer nitrogen in cereal production: retrospects and prospects. In: Advances in agronomy. Elsevier
Leininger S, Urich T, Schloter M, Schwark L, Qi J, Nicol GW, Prosser JI, Schuster SC, Schleper C (2006) Archaea predominate among ammonia-oxidizing prokaryotes in soils. Nature 442 (7104):806–809
Li J, Deng J, Gu Q, Wang K, Ye F, Xu Z, Jin S (2015) The accelerated urbanization process: a threat to soil resources in eastern China. Sustainability 7(6):7137–7155
Liu X, Tiquia SM, Holguin G, Wu L, Nold SC, Devol AH, Luo K, Palumbo AV, Tiedje JM, Zhou J (2003) Molecular diversity of denitrifying genes in continental margin sediments within the oxygen-deficient zone off the pacific coast of Mexico. Appl Environ Microbiol 69 (6):3549–3560
Lori M, Symanczik S, Mäder P, Efosa N, Jaenicke S, Buegger F, Tresch S, Goesmann A, Gattinger A (2018) Distinct nitrogen provisioning from organic amendments in soil as influenced by farming system and water regime. Front Environ Sci 6
Macci C, Doni S, Peruzzi E, Mennone C, Masciandaro G (2013) Biostimulation of soil microbial activity through organic fertilizer and almond tree association. Land Degrad Dev 27(2):335–345
Maeda K, Hanajima D, Toyoda S, Yoshida N, Morioka R, Osada T (2011) Microbiology of nitrogen cycle in animal manure compost. Microb Biotechnol 4(6):700–709
Martens-Habbena W, Berube PM, Urakawa H, de la Torre JR, Stahl DA (2009) Ammonia oxidation kinetics determine niche separation of nitrifying Archaea and Bacteria. Nature 461 (7266):976–979
Meng H, Wu R, Wang Y-F, Gu J-D (2017) A comparison of denitrifying bacterial community structures and abundance in acidic soils between natural forest and re-vegetated forest of Nanling nature reserve in southern China. J Environ Manag 198:41–49
Menneer JC, Ledgard S, McLay C, Silvester W (2004) The impact of grazing animals on N2 fixation in legume-based pastures and management options for improvement. In: Advances in agronomy. Elsevier
Miller MN, Zebarth BJ, Dandie CE, Burton DL, Goyer C, Trevors JT (2008) Crop residue influence on denitrification, N2O emissions and denitrifier community abundance in soil. Soil Biol Biochem 40(10):2553–2562
Morales SE, Cosart T, Holben WE (2010) Bacterial gene abundances as indicators of greenhouse gas emission in soils. ISME J 4(6):799–808
Morugán-Coronado A, García-Orenes F, Cerdà A (2015) Changes in soil microbial activity and physicochemical properties in agricultural soils in eastern Spain. Span J Soil Sci 5(3):201
Newman L, Powell LJ, Wittman H (2015) Landscapes of food production in agriburbia: farmland protection and local food movements in British Columbia. J Rural Stud 39:99–110
Nicol GW, Leininger S, Schleper C, Prosser JI (2008) The influence of soil pH on the diversity, abundance and transcriptional activity of ammonia oxidizing archaea and bacteria. Environ Microbiol 10(11):2966–2978
Nogales B, Timmis KN, Nedwell DB, Osborn AM (2002) Detection and diversity of expressed denitrification genes in estuarine sediments after reverse transcription-PCR amplification from mRNA. Appl Environ Microbiol 68(10):5017–5025
Noll M, Wellinger M (2008) Changes of the soil ecosystem along a receding glacier: testing the correlation between environmental factors and bacterial community structure. Soil Biol Biochem 40(10):2611–2619
Oakley BB, Francis CA, Roberts KJ, Fuchsman CA, Srinivasan S, Staley JT (2007) Analysis of nitrite reductase (nirK and nirS) genes and cultivation reveal depauperate community of denitrifying bacteria in the Black Sea suboxic zone. Environ Microbiol 9(1):118–130
Ogilvie LA, Hirsch PR, Johnston AWB (2008) Bacterial diversity of the broadbalk 'classical' winter wheat experiment in relation to long-term fertilizer inputs. Microb Ecol 56(3):525–537
Okano Y, Hristova KR, Leutenegger CM, Jackson LE, Denison RF, Gebreyesus B, Lebauer D, Scow KM (2004) Application of real-time PCR to study effects of ammonium on population size of ammonia-oxidizing bacteria in soil. Appl Environ Microbiol 70(2):1008–1016

Orr CH, Stewart CJ, Leifert C, Cooper JM, Cummings SP (2015) Effect of crop management and sample year on abundance of soil bacterial communities in organic and conventional cropping systems. J Appl Microbiol 119(1):208–214

Pereg L, Morugán-Coronado A, McMillan M, García-Orenes F (2018) Restoration of nitrogen cycling community in grapevine soil by a decade of organic fertilization. Soil Tillage Res 179:11–19

Pérez-Piqueres A, Edel-Hermann V, Alabouvette C, Steinberg C (2006) Response of soil microbial communities to compost amendments. Soil Biol Biochem 38(3):460–470

Petersen DG, Blazewicz SJ, Firestone M, Herman DJ, Turetsky M, Waldrop M (2012) Abundance of microbial genes associated with nitrogen cycling as indices of biogeochemical process rates across a vegetation gradient in Alaska. Environ Microbiol 14(4):993–1008

Philippot L (2006) Use of functional genes to quantify denitrifiers in the environment: figure 1. Biochem Soc Trans 34(1):101–103

Philippot L, Hallin S, Schloter M (2007) Ecology of denitrifying prokaryotes in agricultural soil. In: Advances in agronomy. Elsevier

Pratscher J, Dumont MG, Conrad R (2011) Ammonia oxidation coupled to CO2 fixation by archaea and bacteria in an agricultural soil. Proc Natl Acad Sci 108(10):4170–4175

Prieme A, Braker G, Tiedje JM (2002) Diversity of nitrite reductase (nirK and nirS) gene fragments in forested upland and wetland soils. Appl Environ Microbiol 68(4):1893–1900

Prosdocimi M, Jordán A, Tarolli P, Keesstra S, Novara A, Cerdà A (2016) The immediate effectiveness of barley straw mulch in reducing soil erodibility and surface runoff generation in Mediterranean vineyards. Sci Total Environ 547:323–330

Quadros PD, Zhalnina K, Davis-Richardson A, Fagen JR, Drew J, Bayer C, Camargo FAO, Triplett EW (2012) The effect of tillage system and crop rotation on soil microbial diversity and composition in a subtropical acrisol. Diversity 4(4):375–395

Rasche F, Knapp D, Kaiser C, Koranda M, Kitzler B, Zechmeister-Boltenstern S, Richter A, Sessitsch A (2010) Seasonality and resource availability control bacterial and archaeal communities in soils of a temperate beech forest. ISME J 5(3):389–402

Reed SC, Seastedt TR, Mann CM, Suding KN, Townsend AR, Cherwin KL (2007) Phosphorus fertilization stimulates nitrogen fixation and increases inorganic nitrogen concentrations in a restored prairie. Appl Soil Ecol 36(2–3):238–242

Reeve JR, Schadt CW, Carpenter-Boggs L, Kang S, Zhou J, Reganold JP (2010) Effects of soil type and farm management on soil ecological functional genes and microbial activities. ISME J 4 (9):1099–1107

Saleh-Lakha S, Miller M, Campbell RG, Schneider K, Elahimanesh P, Hart MM, Trevors JT (2005) Microbial gene expression in soil: methods, applications and challenges. J Microbiol Methods 63(1):1–19

Sanford RA, Wagner DD, Wu Q, Chee-Sanford JC, Thomas SH, Cruz-Garcia C, Rodriguez G, Massol-Deya A, Krishnani KK, Ritalahti KM, Nissen S, Konstantinidis KT, Loffler FE (2012) Unexpected nondenitrifier nitrous oxide reductase gene diversity and abundance in soils. Proc Natl Acad Sci 109(48):19709–19714

Santoro AE, Francis CA, de Sieyes NR, Boehm AB (2008) Shifts in the relative abundance of ammonia-oxidizing bacteria and archaea across physicochemical gradients in a subterranean estuary. Environ Microbiol 10(4):1068–1079

Schauss K, Focks A, Leininger S, Kotzerke A, Heuer H, Thiele-Bruhn S, Sharma S, Wilke B-M, Matthies M, Smalla K, Munch JC, Amelung W, Kaupenjohann M, Schloter M, Schleper C (2009) Dynamics and functional relevance of ammonia-oxidizing archaea in two agricultural soils. Environ Microbiol 11(2):446–456

Sofo A, Scopa A, Dumontet S, Mazzatura A, Pasquale V (2012) Toxic effects of four sulphonylureas herbicides on soil microbial biomass. J Environ Sci Health B 47(7):653–659

Spiertz JHJ (2010) Nitrogen, sustainable agriculture and food security. Rev Agron Sustain Dev 30 (1):43–55

Stevens RJ, Laughlin RJ, Burns LC, Arah JRM, Hood RC (1997) Measuring the contributions of nitrification and denitrification to the flux of nitrous oxide from soil. Soil Biol Biochem 29 (2):139–151

Szukics U, Abell GCJ, Hödl V, Mitter B, Sessitsch A, Hackl E, Zechmeister-Boltenstern S (2010) Nitrifiers and denitrifiers respond rapidly to changed moisture and increasing temperature in a pristine forest soil. FEMS Microbiol Ecol 72(3):395–406

Tahovská K, Kaňa J, Bárta J, Oulehle F, Richter A, Šantrůčková H (2013) Microbial N immobilization is of great importance in acidified mountain spruce forest soils. Soil Biol Biochem 59:58–71

Teixeira L, Yergeau E (2012) Quantification of microorganisms using a functional gene approach. In: Quantitative real-time PCR in applied microbiology. Caister Academic Press, Norfolk, pp 107–120

Thompson KA, Deen B, Dunfield KE (2016) Soil denitrifier community size changes with land use change to perennial bioenergy cropping systems. Soil 2(4):523–535

Throbäck IN, Enwall K, Jarvis ÃS, Hallin S (2004) Reassessing PCR primers targeting nirS, nirK and nosZ genes for community surveys of denitrifying bacteria with DGGE. FEMS Microbiol Ecol 49(3):401–417

Throbäck IN, Johansson M, Rosenquist M, Pell M, Hansson M, Hallin S (2007) Silver (Ag+) reduces denitrification and induces enrichment of novelnirK genotypes in soil. FEMS Microbiol Lett 270(2):189–194

Tietema A (1998) Microbial carbon and nitrogen dynamics in coniferous forest floor material collected along a European nitrogen deposition gradient. For Ecol Manag 101(1–3):29–36

Tilman D, Cassman KG, Matson PA, Naylor R, Polasky S (2002) Agricultural sustainability and intensive production practices. Nature 418(6898):671–677

Tourna M, Stieglmeier M, Spang A, Konneke M, Schintlmeister A, Urich T, Engel M, Schloter M, Wagner M, Richter A, Schleper C (2011) Nitrososphaera viennensis, an ammonia oxidizing archaeon from soil. Proc Natl Acad Sci 108(20):8420–8425

Vinten AJA, Ball BC, O'Sullivan MF, Henshall JK (2002) The effects of cultivation method, fertilizer input and previous sward type on organic C and N storage and gaseous losses under spring and winter barley following long-term leys. J Agric Sci 139(03):231

Vitousek PM, Aber JD, Howarth RW, Likens GE, Matson PA, Schindler DW, Schlesinger WH, Tilman DG (1997) Human alteration of the global nitrogen cycle: sources and consequences. Ecol Appl 7(3):737–750

Wafula D, White JR, Canion A, Jagoe C, Pathak A, Chauhan A (2015) Impacts of long-term irrigation of domestic treated wastewater on soil biogeochemistry and bacterial community structure. Appl Environ Microbiol 81(20):7143–7158

Wakelin SA, Macdonald LM, Rogers SL, Gregg AL, Bolger TP, Baldock JA (2008) Habitat selective factors influencing the structural composition and functional capacity of microbial communities in agricultural soils. Soil Biol Biochem 40(3):803–813

Wakelin SA, Gregg AL, Simpson RJ, Li GD, Riley IT, McKay AC (2009) Pasture management clearly affects soil microbial community structure and N-cycling bacteria. Pedobiologia 52 (4):237–251

Wakelin SA, Clough TJ, Gerard EM, O'Callaghan M (2013) Impact of short-interval, repeat application of dicyandiamide on soil N transformation in urine patches. Agric Ecosyst Environ 167:60–70

Wallenstein MD, Vilgalys RJ (2005) Quantitative analyses of nitrogen cycling genes in soils. Pedobiologia 49(6):665–672

Wessén E, Hallin S, Philippot L (2010) Differential responses of bacterial and archaeal groups at high taxonomical ranks to soil management. Soil Biol Biochem 42(10):1759–1765

Wessén E, Söderström M, Stenberg M, Bru D, Hellman M, Welsh A, Thomsen F, Klemedtson L, Philippot L, Hallin S (2011) Spatial distribution of ammonia-oxidizing bacteria and archaea across a 44-hectare farm related to ecosystem functioning. ISME J 5(7):1213–1225

Witte C-P (2011) Urea metabolism in plants. Plant Sci 180(3):431–438

Woodmansee RG, Dodd JL, Bowman RA, Clark FE, Dickinson CE (1978) Nitrogen budget of a shortgrass prairie ecosystem. Oecologia 34(3):363–376

Wu L, Osmond DL, Graves AK, Burchell MR, Duckworth OW (2012) Relationships between nitrogen transformation rates and gene abundance in a riparian buffer soil. Environ Manag 50 (5):861–874

Wuebbles D (2002) Atmospheric methane and global change. Earth Sci Rev 57(3–4):177–210
Xia W, Zhang C, Zeng X, Feng Y, Weng J, Lin X, Zhu J, Xiong Z, Xu J, Cai Z, Jia Z (2011) Autotrophic growth of nitrifying community in an agricultural soil. ISME J 5(7):1226–1236
Yin C, Fan F, Song A, Li Z, Yu W, Liang Y (2014) Different denitrification potential of aquic brown soil in Northeast China under inorganic and organic fertilization accompanied by distinct changes of nirS- and nirK-denitrifying bacterial community. Eur J Soil Biol 65:47–56
Zalidis G, Stamatiadis S, Takavakoglou V, Eskridge K, Misopolinos N (2002) Impacts of agricultural practices on soil and water quality in the Mediterranean region and proposed assessment methodology. Agric Ecosyst Environ 88(2):137–146
Zhalnina K, de Quadros PD, Gano KA, Davis-Richardson A, Fagen JR, Brown CT, Giongo A, Drew JC, Sayavedra-Soto LA, Arp DJ, Camargo FAO, Daroub SH, Clark IM, McGrath SP, Hirsch PR, Triplett EW (2013) Ca. Nitrososphaera and Bradyrhizobium are inversely correlated and related to agricultural practices in long-term field experiments. Front Microbiol 4:104
Zhalnina K, Dias R, de Quadros PD, Davis-Richardson A, Camargo FAO, Clark IM, McGrath SP, Hirsch PR, Triplett EW (2015) Soil pH determines microbial diversity and composition in the park grass experiment. Microb Ecol 69(2):395–406
Zhang LM, Offre PR, He JZ, Verhamme DT, Nicol GW, Prosser JI (2010) Autotrophic ammonia oxidation by soil thaumarchaea. Proc Natl Acad Sci 107(40):17240–17245
Zhang X, Liu W, Schloter M, Zhang G, Chen Q, Huang J, Li L, Elser JJ, Han X (2013) Response of the abundance of key soil microbial nitrogen-cycling genes to multi-factorial global changes. PLoS One 8(10):e76500
Zhang Q, Liang G, Myrold DD, Zhou W (2017) Variable responses of ammonia oxidizers across soil particle-size fractions affect nitrification in a long-term fertilizer experiment. Soil Biol Biochem 105:25–36
Zhou J, Xue K, Xie J, Deng Y, Wu L, Cheng X, Fei S, Deng S, He Z, Van Nostrand JD (2012) Microbial mediation of carbon-cycle feedbacks to climate warming. Nat Clim Chang 2(2):106
Zornoza R, Guerrero C, Mataix-Solera J, Scow KM, Arcenegui V, Mataix-Beneyto J (2009) Changes in soil microbial community structure following the abandonment of agricultural terraces in mountainous areas of eastern Spain. Appl Soil Ecol 42(3):315–323
Zumft WG (1997) Cell biology and molecular basis of denitrification. Microbiol Mol Biol Rev 61 (4):533–616

Significance of the Enzymes Associated with Soil C and N Transformation

Anna Piotrowska-Długosz

Abstract Soil enzymes play a crucial role in soil organic matter transformation and nutrient cycling. Enzyme productions are the result of soil microbial community expression and their metabolomic requirement. Understanding the presence and activity of the enzymes of C and N cycles in soil may have important implications on ecosystem disturbances and can help to understand the role of C and N cycling in sustainable soil management and sustaining agricultural productivity. Among the biological features, soil enzymes are often used as a reliable index of changes in the soil status as affected by differentiated natural and anthropogenic factors since they are more sensitive to any changes than other soil variables. As was shown in the reviewed literature, interest in the enzyme systems responsible for C and N transformation in soil is currently still high. This chapter presents a brief overview of earlier and recent findings dealing with the most important soil enzymes involved in the soil C and N cycle, such as cellulase, β-glucosidase, urease, invertase, laccase, peroxidase, proteases, and nitrate reductase. The role of these enzymes in soil C and N transformation, as well as possible changes in enzymatic activity as influenced by differentiated factors, was also analyzed. Moreover, still existing limits related to the methodology adopted to assay soil enzyme activities have been discussed. Additionally, one subchapter is devoted to the relationship between gene abundance and enzymatic activity in soil. The contribution of transcriptomics and proteomics in soil enzymology is still poorly developed probably because there are still some methodological problems in soil proteomics. Moreover, the relationship between enzyme activity and the gene expression in soil is an important aim of research. Finally, further research needs and directions concerning the activity of soil C- and N-cycling enzymes are outlined.

Keywords Nutrient cycling · Cellulase · β-glucosidase · Urease · Invertase · Laccase · Peroxidase · Proteomics

A. Piotrowska-Długosz (✉)
Department of Biogeochemistry and Soil Science, Laboratory of Soil Science and Biochemistry, Faculty of Agriculture and Biotechnology, UTP University of Science and Technology, Bydgoszcz, Poland

R. Datta et al. (eds.), *Carbon and Nitrogen Cycling in Soil*,
https://doi.org/10.1007/978-981-13-7264-3_12

Abbreviations

AA	Ammonification of arginine
Ag_2SO_4	Silver sulfate
amoA and *gdh*	Gene coding the enzyme glutamate dehydrogenase
ANR	Assimilatory nitrate reductase
apr	Alkaline metallopeptidases gene,
bpr or *aprE*	Genes of proteolytic enzymes
C	Carbon
DNA	Deoxyribonucleic acid
EEs	Extracellular enzymes
GlcNAc	*N*-acetyl-β-D-glucosaminide
H_2SO_4	Sulfuric acid
KCl	Potassium chloride
lip A-lip J	Lignin peroxidase genes
LiP H8	Extracellular lignin peroxidase isozyme
MgO	Magnesium oxide
mRNA	Messenger ribonucleic acid
MUB	Modified universal buffer
MUF	4-Methylumbelliferone
N	Nitrogen
NAG	*N*-acetyl D-glucosamine
NAGase	*N*-acetyl-β-D-glucosaminidase
NaOH	Sodium hydroxide
NH_3/NH_4^+	Ammonia/ammonium
NO_2/NO_2^-	Nitrite
NO_3/NO_3^-	Nitrate
npr	Neutral metallopeptidase gene
NR	Nitrate reductase
P	Phosphorus
PAHs	Polycyclic aromatic hydrocarbons
pep Aa, *pepAb*, *pep Ac*, and *pep Ad*	Aspartic protease genes
RT-PCR	Reverse transcription polymerase chain reaction
SOC	Soil organic carbon
sub	Peptidases genes
ureC	Urease-encoding genes

1 Introduction

The carbon (C) and nitrogen (N) cycles represent the most important biogeochemical cycles found in terrestrial ecosystems. Carbon and nitrogen account for 95% of the biosphere and are two of six elements (carbon, hydrogen, oxygen, nitrogen, phosphorus, and sulfur) that are the major components found in plants (Nieder and Benbi 2008; Scharlemann et al. 2014). Most of the organic carbon and nitrogen found in terrestrial ecosystems are related to plant residues and soil organic matter (Schlesinger 1997; Lal 2008), which are mineralized to the simple inorganic forms by a set of extracellular enzymes that are produced mainly by soil microorganisms and less by plant roots (Berg and McClaugherty 2003; Paul 2007).

Soil organic carbon (SOC) represents a significant reservoir of carbon in the global C cycle (Lal 2004). SOC consists of a heterogeneous complex of a wide range of organic materials, including simple molecules (e.g., amino acids, monomeric sugars), polymeric molecules (e.g., lignin, cellulose, nucleic acids, proteins), and plant and microbial residues that consist of simple and composed molecules that are bound together into recognizable cellular structures. SOC is formed mainly from the plant, animal, and microbial residues in various stages of decay (Baldock 2007). Soil organic carbon is one of key drivers of the rhizosphere and bulk soil processes and their functions, which results in better soil structure by influencing aggregate stability, nutrient cycling, and availability as well as infiltration and water storage (Hartemink et al. 2014). The soil SOC provides carbon and energy source for soil microbes and fauna. The content and diversity of the soil microorganisms increase with soil organic carbon increase. Since a lot of transformations in soil are conducted by soil microbial communities, an increase in soil microbial biomass usually enhances plant nutrient availability. The range of soil organic compounds can also promote plant growth and thus enhance plant productivity (Lal 2016; Meena et al. 2016).

All living organisms require nitrogen (N) as a necessary nutrient. In terrestrial ecosystems, N is usually available to plants in a limited range, which results in a strong competition for this element between microbes and plant roots (Vitousek and Howarth 1991). In terrestrial ecosystems, the soil organic nitrogen is mainly derived from the remains of plants and/or microorganisms and less from animals (Kögel-Knabner 2002; Norton and Schimel 2011). Therefore, most of the N entering the soil is in the organic forms, such as proteins, chitin and peptidoglycan, nucleic acids, and other N-containing compounds, all of which first have to be broken down into smaller organic molecules by extracellular depolymerases, which take part in the first step of the organic N degradation (Schimel and Bennet 2004, Geisseler et al. 2010). The small organic compounds that are excreted by enzymes can then be taken up directly or be further decomposed and be taken up by microorganisms as ammonium (Feng et al. 2018). The significant processes in the nitrogen cycle include:

1. Nitrogen has to be fixed and then converted into a usable form (NH_3, NO_3^-) before it can be used by organisms.
2. Mineralization or ammonification: the conversion of amino acids into ammonia.

3. Nitrification: oxidation of ammonium into nitrites and nitrates.
4. Denitrification: the conversion of nitrates into atmospheric nitrogen, N_2.

Mineralization and nitrification are the most significant processes in the soil nitrogen cycle since they mediate plant uptake, nitrate leaching, and trace gas emissions (Norton 2008; Norton and Stark 2011). Soil N mineralization is key process by which organic N is converted into inorganic forms such as ammonium (NH_4^+), nitrite (NO_2), or nitrate (NO_3), and it determines the amount of N that is available to plants. In turn, in the process of nitrification, the reduced N (NH_3/NH_4^+ or organic N) is oxidized into NO_2^- or NO_3^-.

Transformation of carbon- and nitrogen-rich bio-macromolecules that occur in soil is a process which is mainly mediated by a set of extracellular enzymes (EEs) produced by soil microorganisms and plants. The EEs catalyze most of the reactions that are involved in the synthesis and degradation of soil organic matter and assure the supply of the essential energy and nutrient for soil organisms which produce the enzymes (Sinsabaugh et al. 2009; Brzostek and Finzi 2011; Wallenstein et al. 2011). The soil enzyme activities take part in main biological processes related to the SOM quality (e.g., the relative availability of C and N) with the capability of microbes to assimilate nutrients and to use carbon for their own need (Allison et al. 2011). This indicates that changes in the C- or N-cycling enzyme activities may be linked to changes in the availability and/or storage of C and N within the SOM pools (Cenini et al. 2016).

In relation to the above, the aim of this chapter is to point out the present state of our knowledge about the features of the enzymes that take part in soil C and N transformations in relation to other soil properties controlling their function and activity.

2 Potential Role of Soil Enzymes in Maintaining Soil Quality and Fertility

Enzymes are protein-related catalysts whose activities can be measured and quantified in the soil system. Soil enzymes play a key role in the transformation of organic matter and nutrient turnover in a soil ecosystem (Burns 1978; Gianfreda and Ruggiero 2006; Piotrowska-Długosz 2014). The overall soil enzymatic activity consists of various intracellular and extracellular enzymes that are actively secreted by soil microorganisms, such as bacteria and fungi, and they less originate from plants and animals (Gianfreda and Bollag 1996; Rao et al. 2014). With using the current methods in soil enzymology, it is difficult to identify the exact source (origin) of the enzyme as well as the temporal and spatial variability of the enzymatic activity (Gianfreda and Ruggiero 2006). Endoenzymes in the soil are retained in living and proliferating cells, while exoenzymes are produced and secreted by living and proliferating cells, but they act outside of these cells as free enzymes that occurred in a soil solution or as enzymes that remain associated with the external root surface or microbial cell wall. After being secreted outside the cell, enzymes can stay free in a soil solution, or they become rapidly absorbed onto soil mineral and/or organic colloids, mainly clay minerals and humic substances. The activity of free extracellular enzymes occurring in soil solution is rather low compared with those immobilized on soil colloids since they are subjected to many unbeneficial factors shortening their life span (Nannipieri and Gianfreda 1998). Although bound enzymes are resistant to proteolysis and different forms of denaturation (Nannipieri et al. 1996) and absorption protects enzymatic protein against degradation, their activity is significantly lower than that of free enzymes. The activity of absorbed enzymes is, however, the most important part of the overall soil enzymatic activity and is responsible for soil organic matter and nutrient transformation.

Although there are many research and related publications concerning the presence, distribution, and functioning of enzymes in different soil types, measuring their catalytic activity, stability, and changes in their activity caused by various biotic and abiotic agents, etc. (e.g., Burns 1978; Kiss et al. 1998; Burns and Dick 2002; Nannipieri 1994; Tabatabai 1994; Gianfreda and Bollag 1996; Tabatabai and Dick 2002; Gianfreda and Ruggiero 2006; Rao et al. 2014), there are many unresolved questions concerning some topics related to enzymes such as their position in the soil environment; contribution in soil organic matter and nutrient transformation; changes under the influence of natural and anthropogenic factors; possible interaction with organisms that are producers of enzymes in soil; and the elaboration of a universal soil enzyme indicator to access the soil status and health, estimation of the role of soil enzymes in changes of soil environment under the influence of various factors, as well as the determination of the enzymes role in the dynamics of plant nutrient transformation (Burns et al. 2013; Rao et al. 2014). These problems and questions are primarily related to the imperfect methodology that is currently being used in soil enzymology, which is discussed in Sect. 5 below.

It is commonly known that enzymatic activity is a sensitive indicator of soil quality and fertility and can be used for determination of ecosystem responses to management and overall environmental changes as well as sustainability of agricultural ecosystems (Nannipieri et al. 2002; Tabatabai and Dick 2002; Gianfreda and Ruggiero 2006; Varma et al. 2017). Soil enzymes play a key role in the overall process of the transformation of soil organic matter, soil nutrient cycling, and pollutants degradation (Burns 1982; Sinsabaugh et al. 1991). The agricultural importance of soil enzymes has been successively increased since the first statement on soil enzymes was presented many years ago. A positive relationship between the soil enzymes activity and nutrient transformation has been reported in some arable soils. Thus, in a long-term cropping systems using N fertilization, the *N*-acetylglucosaminidase and arylamidase activity were well correlated with the level of nitrogen mineralization (Ekenler and Tabatabai 2002, 2004; Dodor and Tabatabai 2007). Soil enzymatic activity can also be used as an indicator of soil nutrient availability. Many studies have shown that the enzymatic activity of the P and N mineralization is negatively correlated with the available forms of P and N (Muruganandam et al. 2009; Balota and Dias Chaves 2010; Orczewska et al. 2012, Sherene 2017). Similarly, Allison (2005) reported that N fertilization significantly decreased the activity of proteases and chitinase.

3 Enzymes of Soil Carbon and Nitrogen Transformation: A General Overview

The enzymes related to C and N transformations are the most important in the overall process of soil organic matter transformation and energy flow. The main aim of this chapter is to review the current knowledge of the enzymes involved in carbon and nitrogen cycling in soil. Many reviews (e.g., Burns 1982; Tabatabai and Dick 2002; Gianfreda and Bollag 1996; Gianfreda and Ruggiero 2006) dedicated earlier to soil enzymology are important when considering its historical background. More recent research concerning different aspects of enzymatic reactions in soils have been reviewed with special attention paid to the influence of the physical soil properties on enzymatic activity. Finally, future research needs are also specified. Although many of the enzymes that take part in C and N transformation can be determined in soils, only a limited number of enzymes are usually studied. Many of them, like cellulases, are found to act as extracellular enzymes. Others, like urease, are able to catalyze reactions both endo- and extracellularly (Tate III 2002). Earlier, most of the soil-related studies were subjected to meet the agricultural necessity (Dick and Tabatabai 1992; Dick 1994), and determination of soil enzyme activities have mainly been directed toward assessing the quality and quantity of crops and determination of management influence on the enzymes involving in the biogeochemical cycles (e.g., transformation of nutrients in plant biomass/residues, N cycling, and fixation). Special attention has been paid to urease activity due to its great significance in the urea-hydrolyzing. The most often studied enzyme activities involved in N transformations have been related to ammonium formation (amidases, urease), the hydrolysis of proteins (proteolytic enzymes), the loss of N from soil (nitrogen oxide reductases), denitrification (nitrate reductase), and finally N fixation (nitrogenases). In turn, the most important enzymes associated with transformation of carbonaceous compounds are related to the hydrolysis of polysaccharides (e.g., amylases, cellulase complex, xylanases) and hydroxylation of aromatic rings (e.g., laccases and other polyphenol oxidases), which finally lead to either the mineralization or humification of the initial compounds (Dilly et al. 2007) and different lipases and esterases, which catalyze the hydrolysis of the variety of ester linkages in various substrates. Recently, however, these enzyme activities have been considered for evaluation of more broad and universal ecological anxiety, as the impact of human alterations, not only related to agricultural practice, the organic matter, and nutrients transformations in native soil systems such as old-grown forests or barren lands (Rao et al. 2017). For example, some of the C- or N-cycling enzymes (e.g., cellulase, urease) are advantageous in determining the impact of recycling organic wastes (e.g., compost, sewage sludge) on the soil, while determination of laccase and polyphenol oxidase activity is generally related to the breakdown and humification of the xenobiotics with aromatic rings (Gianfreda and Rao 2004; Gómez Jiménez et al. 2011; Meena et al. 2015b; Piotrowska-Długosz 2017).

3.1 Specific Enzymatic Activities of Soil C Transformation

A great number of enzymes are involved in hydrolyzing of C-containing compounds. This group includes mainly enzymes that break down large organic compounds, such as cellulose (cellulases), starch (amylases), chitin (chitinase), and xylan (xylanase). Other enzymes, such as invertase, α- and β-glucosidases, α- and β-galactosidases, and *N*-acetyl-β-glucosaminidase, decompose disaccharides and oligosaccharides into simple sugars. Some enzymes involved in C cycling and the reactions they carried out are characterized below and specified in Table 1.

3.1.1 Cellulase Complex and Glucosidases

The microbial degradation of cellulose, which is the most abundant polysaccharide found in the biosphere, requires the action of at least three groups of enzymes to act synergistically in hydrolyzing the β-1,4 bonds of cellulose to glucose. The catalytic system consists of endo-1, 4-β-glucanase (endocellulase EC 3.2.1.4), exo-1, 4-β-glucanase (cellobiosidase EC 3.2.1.91), and β-glucosidase (Deng and Popova 2011; Phitsuwan et al. 2013). Two glucosidases (EC 3.2.1.20/21) can be identified in soil, α-glucosidase (EC 3.2.1.20) and β-glucosidase (EC 3.2.1.21), which catalyzes the hydrolysis of α-D-glucopyranoside and β-D-glucopyanosides, respectively, and catalyzes the hydrolysis of maltose and cellobiose. Glucosidases are produced by a wide range of microorganisms, animals, and plants (Dodor and Tabatabai 2005). The most often found and determined in soil is β-glucosidase (Deng and Popova 2011; Buragohain et al. 2017), which takes part in the last stage of cellulose decomposition. The most often known reaction carried out by the enzyme is hydrolysis of cellobiose into two molecules of glucose by cleaving the β-glucosidic bonds from the nonreducing terminal ends (Lynd et al. 2002; Jørgensen et al. 2007; Deng and Popova 2011). This process is important since cellobiose is an inhibitor of the cellulolytic enzymes activity (Morais et al. 2004). β-Glucosidase is also responsible for the hydrolysis of β-D-glucopyranoside and many of glycosides, such as phenolic glycosides or flavanone glycosides (Berrin et al. 2003; Acosta-Martinez et al. 2007). Because of wide substrate specificity, the activity of this enzyme is considered to be a good indicator of biomass decomposition in soil (Berrin et al. 2003; Zanoelo et al. 2004). Other known glycosidases are α-galactosidase (EC 3.2.1.23) and β-galactosidase (EC 3.2.1.24), which catalyze the hydrolysis of melibiose and lactose, respectively. These enzymes however not occur in the soil in a significant amount. The importance of glycosidases is related to their participation in soil organic matter mineralization. By hydrolyzing the soil organic carbon and nitrogen compounds, they deliver essential carbon components and nutrients for the growth of heterotrophic microorganisms, thereby increasing soil microbial activity (Dodor and Tabatabai 2005).

Table 1 Soil enzyme activities involved in C cycling

Enzyme common name and EC number	Systematic name	Function/reaction	References
Hydrolases			
Triacylglycerol lipase (lipase), 3.1.1.3	Triacylglycerol acylhydrolase	Triacylglycerol + H_2O = diacylglycerol + a carboxylate	Margesin et al. (1999) and Ebulue et al. (2017)
α-Amylase, 3.2.1.1	4-α-D-glucan glucanohydrolase	Endohydrolysis of (1 → 4)-α-D-glucosidic linkages in polysaccharides containing three or more (1 → 4)-α-linked D-glucose units	Pancholy and Rice (1973), Balota et al. (2004), and Deng and Popova (2011)
Cellulase, 3.2.1.4	4-(1,3;1,4)-β-D-glucan 4-glucanohydrolase	Endohydrolysis of (1 → 4)-β-D-glucosidic linkages in cellulose, lichenin, and cereal β-D-glucans	Phitsuwan et al. (2013)
Endo-1,3(4)-β-glucanase (laminarinase), 3.2.1.6	3-(1 → 3;1 → 4)-β-D-glucan 3 (4)-glucanohydrolase	Endohydrolysis of (1 → 3)- or (1 → 4)-linkages in β-D-glucans when the glucose residue whose reducing group is involved in the linkage to be hydrolyzed is itself substituted at C-3	Machuca et al. (2018)
Endo-1,4-β-xylanase, 3.2.1.8	4-β-D-xylan xylanohydrolase	Endohydrolysis of (1 → 4)-β-D-xylosidic linkages in xylans	Kandeler et al. (1999), Hu et al. (2008), and Štursová and Baldrian (2011)
Chitinase, 3.2.1.14	(1 → 4)-2-Acetamido-2-deoxy-β-D-glucan glycanohydrolase	Random endohydrolysis of *N*-acetyl-β-D-glucosaminide (1 → 4)-β-linkages in chitin and chitodextrins	Adams (2004), Bhattacharya et al. (2007), Brzezińska et al. (2009), and Deng and Popova (2011)
α-Glucosidase, 3.2.1.20	α-D-glucoside glucohydrolase	Hydrolysis of terminal, nonreducing (1 → 4)-linked α-D-glucose residues with release of α-D-glucose	Štursová and Baldrian (2011)
β-Glucosidase, 3.2.1.21	β-D-glucoside glucohydrolase	Hydrolysis of terminal, nonreducing β-D-glucosyl residues with release of β-D-glucose	Lynd et al. (2002), Jørgensen et al. (2007), and Deng and Popova (2011)
α-Galactosidase, 3.2.1.22	α-D-galactoside galactohydrolase	Hydrolysis of terminal, nonreducing α-D-galactose residues in α-D-galactosides, including galactose oligosaccharides, galactomannans, and galactolipids	Zhang et al. (2010) and Machuca et al. (2018)

β-Galactosidase, 3.2.1.23	β-D-galactoside galactohydrolase	Hydrolysis of terminal nonreducing β-D-galactose residues in β-D-galactosides	Zhang et al. (2010) and Bergmann et al. (2014)
β-Fructofuranosidase (invertase), 3.2.1.26	β-D-fructofuranoside fructohydrolase	Hydrolysis of terminal nonreducing β-D-fructofuranoside residues in β-D-fructofuranosides	Ross (1983), Jin et al. (2009), and Deng and Popova (2011)
β-*N*-Acetylhexosaminidase (hexosaminidase), 3.2.1.52	β-*N*-Acetyl-D-hexosaminide *N*-acetylhexosaminohydrolase	Hydrolysis of terminal nonreducing *N*-acetyl-D-hexosamine residues in *N*-acetyl-β-d-hexosaminidase	Suresh et al. (2011)
Oxidoreductases			
L-Ascorbate oxidase (acid oxidase), 1.10.3.3	L-Ascorbate: oxygen oxidoreductase	4 L-ascorbate + O_2 = 4 monodehydroascorbate + 2 H_2O	Mishra et al. (2009)
Laccase, 1.10.3.2	Benzenediol: oxygen oxidoreductase	4 benzenediol + O_2 = 4 benzosemiquinone + 2 H_2O	Shinkafi et al. (2014)
Glucose oxidase, 1.1.3.4	β-D-glucose: oxygen 1-oxidoreductase	β-D-glucose + O_2 = D-glucono-1,5-lactone + H_2O_2	Russel et al. (2014)
Phenol oxidase, 1.14.18	–	Phenol + ½ O_2 = Quinone + H_2O	Sinsabaugh (2010)
Peroxidase, 1.11.1.7	Phenolic donor: hydrogen-peroxide oxidoreductase	2 phenolic donor + H_2O_2 = 2 phenoxyl radical of the donor + 2 H_2O	Sinsabaugh (2010)

3.1.2 Invertase

Invertase (β-D-fructofuranoside fructohydrolase [EC 3.2.1.26]) splits off β-D-fructofuranoside rest from nonreducing end of β-D-fructofuranosides, such as sucrose, raffinose, oligofructose, or inulin (Deng and Popova 2011). The preferable substrate for invertase is sucrose, most commonly occurring in plants soluble sugar, which consists of a molecule of glucose and fructose (Jin et al. 2009). Together with cellulase complex, invertase activity is responsible for the disintegration of plant litter in the soil system (Frankenberger and Johanson 1983; Datta et al. 2017b). Although the invertase activity is generally associated with the heavy fraction of soil (clay minerals and silt), in soil under grasslands, the invertase activity was partially related to light soil fraction (Ross 1983). The soil invertase activity is used as an index for nutrient transformation, energy metabolism, and pollutant degradation (Nannipieri et al. 1990).

3.1.3 *N*-Acetyl-β-D-Glucosidase (Chitinase)

Chitin, the second most often occurring in soil's amino sugar, is an unbranched polymer consisting of *N*-acetyl-D-glucosamine. The sources of this compound in the soil are exoskeletons of insects and arthropods as well as fungal hyphae (Duo-Chuan 2006; Wongkaew and Homkratoke 2009). The chitinases system taking part in the hydrolyzing of chitin consists of endochitinolytic (chitinase, or β-1, 4-poly-*N*-acetylglucosaminidase [EC 3.2.1.14]) and exochitinolytic (*N*-acetyl-β-D-glucosaminidase, NAGase, [EC 3.2.1.52]) enzymes. They hydrolyze the chemical bonds between *N*-acetyl-D-glucosamine particles (Brzezińska et al. 2009; Deng and Popova 2011), but they differ in the way of action. Chitinase randomly hydrolyzes the 1,4-β bonds in chitin, while NAGase hydrolyzes the terminal, nonreducing ends (Webb 1992; Moss 2010) with the free N-acetyl glucosamine (NAG) as the final product of the reaction (Brzezińska et al. 2009). The availability of the NAGase is differentiated in soils and depends on many factors such as soil physical and chemical properties, microorganism wealth, and different substrate quality and quantity (Sinsabaugh et al. 1992). Many soil organisms, such as bacteria, fungi, and plants, are able to produce chitinases (Duo-Chuan 2006; Sihag et al. 2015). Bacterial chitinases mainly degrade chitin in order to use the reaction products as the source of carbon and nitrogen, while in fungi, this group of enzymes also plays a significant role in cell wall development and structure during the active growth (Adams 2004; Bhattacharya et al. 2007). The synthesis of chitin is induced when other labile carbon and nitrogen sources are lacking. That is why chitin is more abundant in areas with low nutrient content (Hanzlikova and Jandera 1993; Brzezińska et al. 2009).

3.1.4 Xylanase

According to Deng and Popova (2011), the xylanase enzyme system (1,4-β-D-xylan xylanohydrolase, EC 3.2.1.8) consists of the following enzymes: β-xylanase, esterase, α-L-arabinofuranosidase, β-xylosidase, α-glucuronidase, acetylxylan, and hydroxycinnamic acid esterases. The enzyme group is classified into tenth and eleventh families of the glycosyl hydrolases and catalyzes the endohydrolysis of β-1, 4-xylosidic linkages in hemicellulose, which, in addition to cellulose, is the second most frequently occurring polysaccharide on earth (Anand et al. 1990; Kandeler et al. 1999; Hu et al. 2008). The final products of hydrolysis are xylose, xylobiose, as well as short chains of various oligomers. The xylanase enzymatic complex is mostly produced by fungi under insufficient quantities of available compounds (Kandeler et al. 1999). According to Hu et al. (2008), xylanse complex plays a significant role in the circulation of organic materials and energy in the soils as well as in seed germination and fruit ripening.

3.1.5 Amylases

Amylases (EC 3.2.11/2), together with cellulases and invertase, are the group of enzymes that are responsible for the rate and course of the decomposition of plant material in soil (Pancholy and Rice 1972). The amylase system includes endo- and exo-amylases that synergistically hydrolyze starch (Deng and Popova 2011; Yadav et al. 2018). Endo-amylases, commonly known as α-amylases, hydrolyze the α-1,4-glycosidic linkages in random. The products of this reaction are dextrins, oligosaccharides, and finally monosaccharides, like glucose. In turn, exo-amylases consist of β- and γ-amylase and hydrolyze the same bonds but solely from the nonreducing ends of the starch molecule, thereby releasing β-maltose and β-D-glucose (Webb 1992; Deng and Popova 2011). Among amylases, the most active in soil is β-amylase which catalyzes the degradation of the so-called heavy fraction of organic material (the large size, recalcitrant carbon) than the light fraction (the small size, rapid transformation) (Ebregt and Boldewijn 1977; Ross 1983). Amylases also occur in plants as intracellular enzymes and can be liberated into soil together with plants residues. Similarly, to other enzymes, amylases are mainly produced by microorganisms, especially by bacteria and fungi (Ebregt and Boldewijn 1977). There are amylases found in some environments with unbeneficial conditions such as acidophilic, alkalophilic, and thermoacidophilic areas (Ebregt and Boldewijn 1977). It was reported that amylase activities in the soil, similar to other extracellular enzymes, were repressed by the presence of clay minerals. Thus, a considerable decrease in the β-amylase activity measured in three different clay fractions collected from soil and from surface layers of two soils from tussock grasslands was found. In these studies the effect of clay minerals on the decreasing of α-amylase activity was in the following sequence: muscovite < allophane < illite < montmorillonite (Ross 1983). In soils that have a higher carbon amount, amylase was more active as compared to the activities of some other enzyme (Pancholy and Rice 1973; Balota et al. 2004).

3.2 *Specific Enzymatic Activities of Soil N Transformation*

The cycling and fate of N in terrestrial ecosystems are critical for many aspects of environmental quality. That is why the enzymes taking part in nitrogen transformation are important in controlling N in this type of ecosystem and for possible use in assessing soil quality and/or degradation (Kandeler et al. 2011). Various enzymes involved in N cycling differ in their response to environmental change like N deposition and N addition (Hungate et al. 2007, Enowashu et al. 2009). That is why it is suggested to test some enzymatic indicators (e.g., proteases, urease, enzymes taking part in ammonium oxidation and denitrification) in order to observe the most important processes in nitrogen transformation. N-transforming enzymes can be divided into extracellular depolymerases (proteases, chitinases, and peptidoglycan hydrolases) that are involved in the decomposition of the main polymers of organic materials entering the soil and enzymes that take part in N mineralization (urease, amino acid oxidase) (Geisseler et al. 2010; Ashoka et al. 2017). Some important enzymes involved in N cycling and their reactions are presented in Table 2.

3.2.1 Extracellular Depolymerases

The group of extracellular depolymerizing enzymes breaks down the complex of organic plant materials and microbial residues into smaller, soluble subunits that can be taken up by microorganisms. Based on the chemical composition of the main sources of organic residues in soil, the most important extracellular depolymerases involved in the hydrolysis of N-containing molecules are proteases, chitinases, and peptidoglycan hydrolases (Geisseler et al. 2010; Kandeler et al. 2011).

3.2.1.1 Proteolytic Enzymes

The protein degradation (proteolysis) is a significant process in N transformation in different ecosystems since it is believed to be a limiting step of N mineralization in soil (Weintraub and Schimel 2005) due to the much slower primary phase of protein mineralization compared to amino acid mineralization (Jan et al. 2009). Earlier, Ladd and Jackson (1982) have discussed the role of protease activities in the process of N mineralization, while the concept of the N mineralization-immobilization in soil system has been reviewed later by Nannipieri and Eldor (2009). Extracellular

Table 2 Soil enzyme activities involved in N cycling

Enzyme common name and EC number	Systematic name	Function/reaction	References
Hydrolases			
Proteases, 3.4 groups		Protein hydrolysis	Rotanova et al. (2004), Nannipieri and Eldor (2009), and Landi et al. (2011)
Arylamidase, 3.4.11.3		Hydrolysis of N-terminal amino acids from peptides, amides, or arylamides	Acosta-Martinez and Tabatabai (2000) and Muruganandam et al. (2009)
L-Asparaginase,3.5.1.1	L-Asparagine amidohydrolase	L-Asparagine + H_2O = L-aspartate + NH_3	Dodor and Tabatabai (2003) and Muruganandam et al. (2009)
L-Glutaminase, 3.5.1.2	L-Glutamine amidohydrolase	L-Glutamine + H_2O = L-glutamate + NH_3	Khorsandi and Nourbakhsh (2007) and Xue et al. (2006)
Urease, 3.5.1.5	Urea amidohydrolase	Urea + H_2O = CO_2 + 2 NH_3	Mobley et al. (1995) and Glibert et al. (2006)
Amidase, 3.5.1.4	Acylamide amidohydrolase	Hydrolysis of C–N bonds other than peptide bonds in linear amides (monocarboxylic acid amide + H_2O = monocarboxylate + NH_3)	Allison et al. (2011), Fraser et al. (2013), and Rasool et al. (2014)
Adenosine deaminase, 3.5.4.4	Adenosine aminohydrolase	Adenosine + H_2O = inosine + NH_3	Sato et al. (1986)
Peptidoglycan hydrolase		Acts on the link between N9 acetylmuramoyl residues and amino acid residues in cell wall glycopeptides	Tabatabai et al. (2010)
Oxidoreductases			
Nitrate reductase (NADH), (assimilatory nitrate reductase) 1.7.1.1	Nitrite: NAD^+ oxidoreductase	Nitrite + NAD^+ + H_2O = nitrate + NADH + H^+	Abdelmagid and Tabatabai (1987), Martens (2005), Singh and Kumar (2008), and Szajdak and Gaca (2010)
Nitrogenase, 1.18.6.1	Reduced ferredoxin: Dinitrogen oxidoreductase (ATP-hydrolyzing)	8 reduced ferredoxin + 8 H^+ + N_2 + 16 ATP + 16 H_2O = 8 oxidized ferredoxin + H_2 + 2 NH_3 + 16 ADP + 16 phosphate	Halsall and Gibson (1991)
Denitrification activity (DEA)		Denitrification of nitric compounds to N_2O and N_2	Šimek et al. (2002)

(continued)

Table 2 (continued)

Enzyme common name and EC number	Systematic name	Function/reaction	References
Transferases and lyases			
L-Aspartase, 4.3.1.1	L-Aspartate ammonia-lyase (fumarate-forming)	L-Aspartate = fumarate + NH_3	Dodor and Tabatabai (2003)
Carboxy-liases 4.1.1 groups (amino acid decarboxylase)		Aspartic acid = alanine	Frankenberger (1983)
		Glutamic acid = γ-aminobutyric acid	
Transaminases, 2.6.1 group		$R_1R_2–CH–NH_2 + R_3R_4CO = R_1R_2–CO + R_3R_4CH–NH_2$	Dar (2010)
Histidine ammonia-lyase (histidase), 4.3.1.3	L-Histidine ammonia-lyase (urocanate-forming)	L-histidine = urocanate + NH_3	Frankenberger (1983) and Burton and McGill (1991)

proteolytic enzymes hydrolyze protein polymers and polypeptides into smaller peptides and finally into amino acids. Some researchers proposed the assay of protease activity as a good way to measure N depolymerization, as proteases are the most responsible for supplying bioavailable N (Schimel and Bennet 2004).

Approximately 30–40% of overall organic nitrogen in soil are derived from proteins and polypeptides (Jones et al. 2009). The main source of various soil proteases is microbes and plants. Among soil microorganisms, the most effective in excretion of proteases are bacteria, such as *Pseudomonas*, *Streptomyces*, and *Bacillus*. Also, fungi, like *Penicillium*, *Pythium*, *and Aspergillus*, secrete numerous proteolytic enzymes, which activity is especially significant in releasing available nitrogen in conditions of its deficiency (Kudryavtseva et al. 2008). Proteolytic enzymes are usually categorized according to the type of reaction they catalyze, the molecular structure, and the type of functional group in the active site (Rotanova et al. 2004; Landi et al. 2011; Dadhich and Meena 2014). Thus, exopeptidases catalyze the hydrolysis of the terminal amino acids of the protein structure, whereas endopeptidases (proteinases) catalyze the hydrolysis peptide linkages between amino acids occurring inside the polypeptide chains. The exopeptidases hydrolyze peptide bonds on both ends of the peptide chains. The exopeptidases involved in removing one, two, and three amino acids from the N-terminal end of the chain is named aminopeptidases, dipeptidyl-peptidases, and tripeptidyl-peptidases, respectively (Landi et al. 2011). As regards the catalytic function, the following carboxypeptidases can be specified: cysteine-type carboxypeptidases, serine-type carboxypeptidases, and metal-carboxypeptidases (Table 3).

The neutral metalloproteases and serine proteases (SUB) are mostly involved in protein decomposition in agricultural soils that was showed by the selective inhibition of various proteases of bacterial origin (Watanabe et al. 2003; Vranova et al.

Table 3 Classification of proteases (peptidases) EC 3.4. according to the EC nomenclature (Webb 1992)

Sub-subclass	Enzyme
3.4.11	Aminopeptidases
3.4.13	Dipeptide hydrolases (dipeptidases)
3.4.14	Dipeptidyl-peptidases and tripeptidyl-peptidases
3.4.15	Peptidyl-dipeptidases
3.4.16	Serine-type carboxypeptidases
3.4.17	Metallocarboxypeptidases
3.4.18	Cysteine-type carboxypeptidases
3.4.19	Omega peptidases
3.4.21	Serine endopeptidases
3.4.22	Cysteine endopeptidases
3.4.23	Aspartic endopeptidases
3.4.24	Metalloendopeptidases
3.4.25	Threonine endopeptidases
3.4.99	Endopeptidases of unknown catalytic mechanism

2013). Endopeptidases are recognized due to the chemical character of the groups that are responsible for their hydrolytic functions (Landi et al. 2011). According to Kalisz (1988) and Page and Di Cera (2008), four different groups of endopeptidases can be distinguished: aspartic-, cysteine-, serine-, and metalloendopeptidases. Trypsin and subtilisin are two significant serine endopeptidases, and thus trypsin is a particular enzyme that breaks down peptides at arginine and lysine amino acids, while subtilisin has a broad spectrum of activity and hydrolyzes the peptide bonds in various peptide amides. Most metalloproteases require some metals (e.g., zinc and cobalt) for their catalysis. A lot of research has been devoted to study endopeptidases, and their pH optimum (e.g., neutral, acidic, and alkaline) has been determined. Moreover, the extracellular proteolytic enzymes generally reveal a wide specificity toward the substrate and can break down a lot of various proteins (Kalisz 1988). Proteolytic enzymes also act intracellularly and are responsible for metabolism regulation and protein transformation within the cells. The protein turnover is crucial for cells to adapt to new environmental circumstances, particularly in a case of nutrients deficiency (Kalisz 1988). Godlewski and Adamczyk (2007) discussed the problem of proteases secretion by plant roots. They stated that different species and cultivars of the same plant growing in a culture medium could vary in the levels of proteolytic activity, which may indicate that they differ in the capacity to excrete proteases. Measuring the proteases activity at different values of culture medium reaction (pH) has pointed out that the produced proteolytic enzyme activity was the highest at pH = 7. The production of proteolytic enzymes varies within root systems; thus, in the apical parts of the root, proteases are more intensively secreted as compared with the mature section of the root. Increased activity of proteases was found in roots of transgenic plants (Eick and Stöhr 2009).

3.2.2 Enzymes Involved in N Mineralization

In the process of depolymerization, the high molecular weight N-containing polymers are breaking down into simplest compounds, like amino acids, amino sugars, or nitrogenous bases. This process is often considered to be a limiting step in soil nitrogen mineralization (Jones et al. 2009; Kemmitt et al. 2008; Wallenstein and Weintraub 2008). Afterward, ammonium is released from those monomers. These two steps are carried out by primarily microbial-derived extracellular enzymes (Burns et al. 2013; Yadav et al. 2017b). Extracellular enzymes degrade complex, N-containing compounds, such as protein, nucleic acids, and cell wall components (Myrold and Bottomley 2008). All these enzymes carry out the hydrolysis of native N compounds or those added to soil and have been used to assess changes in arable soils under different management practices, such as organic and mineral fertilization, tillage practices, and crop rotation (Hallin et al. 2009, Sinsabaugh et al. 2015). Amidohydrolases (e.g., urease, L-asparaginase, L-glutaminase, and amidases) activity is significant in the process of depolymerization of aliphatic and aromatic nitrogen compounds occurring in soil organic matter (Monreal and Bergstrom 2000). Peptidoglycan breaks the linkages between N-9 acetylmuramoyl and amino acids in the cell wall glycopeptides and thus playing the main function in the transformation of microbial biomass nitrogen (Tabatabai et al. 2010). The range of glycosidases are involved in the hydrolysis of amino sugar polymers that are a significant constituent of microorganism's cell walls. In turn, the enzyme *N*-acetyl-β-D-glucosaminidase (NAGase) carries out the hydrolysis of the *N*-acetyl-β-D-glucosamine from the terminal, nonreducing ends of the chitooligosaccharides. This enzyme also takes part in the catabolism of soil chitin polymer (Tabatabai et al. 2010). Chitin and chitodextrins, which are the major constituents in fungi organisms, are substrates for chitinase activity (Alef and Nannipieri 1995).

The relationships between the indicators of nitrogen mineralization and some amidohydrolases activity were shown by Tabatabai et al. (2010). The authors revealed that the activities of some hydrolytic enzymes (e.g., asparaginase, amidase, urease, and glutaminase) were markedly related to the content of nitrogen mineralized at 30 °C. The correlation coefficient values were between 0.35 and 0.61 for L-glutaminase and L-asparaginase, respectively. Similarly, the activities of L-glutaminase and L-asparaginase were notably related to overall nitrogen mineralization in the study of Muruganandam et al. (2009). Moreover, a significant correlation between the range of hydrolytic activities (e.g., urease L-asparaginase, L-glutaminase, amidase) and N mineralization level was presented in other studies (Khorsandi and Nourbakhsh 2007; Xue et al. 2006). Some other authors in turn (Senwo and Tabatabai 1996; Tabatabai et al. 2010; Muruganandam et al. 2009) found that arylamidase, an enzyme which catalyzes the hydrolysis of amino acids from N-terminal peptide chains, as well as from amides and/or arylamides, was significantly related to nitrogen mineralization process. The obtained correlation coefficients ranged between 0.61 ($p < 0.001$) and 0.77 ($p < 0.005$). These data indicated that the activities of some enzymes involved in N mineralization can be used as significant indicators of this process.

3.2.2.1 Urease

There are a lot of studies concerning the soil urease activity because of the importance of urea as a nitrogen fertilizer (Glibert et al. 2006). Urease (EC 3.5.1.5) catalyzes the hydrolysis of urea into two moles of ammonia and one of carbon dioxide. This process is crucial in regulating the N supply to plants after urea fertilization. Urea enters the soil also as a result of the transformation of urine excreted from mammals. Moreover, urea comes from the degradation of the amino acid arginine and of uric acid, which is excreted by birds, reptiles, and insects (Mobley and Hausinger 1989). The wide range of bacteria species, yeasts, fungi, algae, and plants are the main sources of urease activity in soil environment (Mobley and Hausinger 1989; Follmer 2008). Although urease can be released constitutively from some organisms, the production of this enzyme is most commonly regulated by the presence of nitrogen, and its production is inhibited when the producing organism grows in the environment with the sufficient concentration of a suitable N source, like ammonium ions. On the contrary, the presence of urea and some other nitrogen sources activated production of the enzyme (Mobley et al. 1995). It was found that the urease activity in soil was mostly extracellular and was immobilized by soil mineral and organic colloids. According to Pettit et al. (1976) more or less 60% of the urease activity determined in different soils was bound extracellularly, while Klose and Tabatabai (1999) estimated the extracellular proportion of urease activity to be 46%.

A better understanding of the dynamics of urease activity might identify a more effective way of managing N fertilizers (Balota and Chaves 2010; Kumar et al. 2017). Therefore, it is important to detect the set of natural and anthropogenic factors that can modify/reduce/increase the effectiveness of this enzyme activity in an ecosystem. Some of these factors are soil organic matter and nutrient content, agricultural practices such as tillage, mineral and organic fertilization, crop rotation, soil depth, soil pollution with heavy metals, PAHs, soil waste amendments, and weather conditions such as temperatures and rainfall (Yang et al. 2006; Yadav et al. 2017a; Datta et al. 2014). Thus, it has been found that urease activity was very sensitive to higher amounts of heavy metals (Yang et al. 2006). Since urease activity becomes greater with progressive temperature, the fertilizer urea should be applied at a time the temperatures are the lowest. Then the energy of activation is lower, and the loss of nitrogen by the volatilization is minimal. A better understanding of urease properties and its activity would be helpful in urea fertilizer application, particularly in the areas with high rainfall, flooded, and irrigated fields (Bakshi and Varma 2011).

3.2.2.2 Arginine Deaminase Activity

The ammonification of arginine (AA), one of the basic protein amino acids, appears to be a common process in microorganisms (Alef and Kleiner 1986; Singh and Kumar 2008). Arginine ammonification level is significantly correlated with the soil microbial biomass content and other biochemical properties (Alef and Kleiner 1987; Singh and Singh 2005). Ammonification is an important initial stage of organic matter mineralization when proteins and other organic compounds containing amino groups are decomposed by proteolytic enzymes to amino acids which are further deaminated to ammonium ion NH_4^+(Bonde et al. 2001, Lin and Brookes 1999). Bonde et al. (2001) also proposed that arginine ammonification activity provided an index of gross N mineralization in agricultural soils, as this enzyme activity was well correlated with the average rates of gross N mineralization, and there is a similarity between the seasonal variations of gross N mineralization and arginine ammonification activities. The amount of ammonium produced depends on the C/N ratio of the amino acid, with high ammonium production at a low ratio (Ginésy et al. 2017). The arginine deaminase activity is strongly related to respiration and correlated significantly with the carbon content of the soil but is poorly related with soil pH, ammonia content, percentage clay, or the number of microorganisms (Pandey and Singh 2006). Arginine deaminase activity was significantly high in natural, especially forest soil, compared to agricultural soils, thereby indicating continuously higher N inputs in forest stands (Singh and Kumar 2008). The arginine deaminase activity was differentially affected by heavy metals and other pollutants (Kandeler 1996; Guo et al. 2009). The AA activity was markedly stimulated after chlorpyrifos seed and soil treatment. This enhancement might have been due to an increase in the fungal and *Actinomycetes* population after chlorpyrifos application, which might be using this insecticide as an energy source for microorganisms (Hussain et al. 2009). The same results were obtained by Singh and Singh (2005) for diazinon seed and soil management. The AA activity was found to be greater in the control soil samples in comparison to the samples that had been treated with acetamiprid, where at the enzyme activity decreased over 20% shortly after the pesticide was used (Singh and Kumar 2008; Meena et al. 2015a). Recently, the influence of seasonal changes and forest management on the arginine ammonification was determined in the surface, the organic horizon of some spruce forests (Holik et al. 2017). The authors concluded that the AA was the highest in the soil with the most favorable conditions, such as high water content, a generally lower concentration of ammonium N, and a higher population density following a thinning operation.

3.2.2.3 Arylamidase

Arylamidase (a-aminoacyl-peptide hydrolase, EC 3.4.11.2) is the enzyme that catalyzes the separation of amino acids from the N-terminal end of peptides, amides. The enzyme was found in plants, animals, and microorganisms (Hiwada et al. 1980). Although arylamidase activity is important in the beginning stages of the soil amino acids mineralization, not much recent information is available about this enzyme in soil (Muruganandam et al. 2009). Understanding of the role of arylamidase in soil N cycling and the factors (i.e., soil properties, trace elements liming, tillage, and crop residues management practices) that affect the activity of this enzyme will aid in making decision important for the fertility, productivity, and sustainability of soils. Arylamidase activity was highly influenced by tillage and residue placements, and the greatest arylamidae activity was found in treatments of chisel/mulch, moldboard plow/mulch, and no-till/double mulch, whereas the lowest activity was observed in treatments of moldboard plow/normal and no-till/bare (Acosta-Martinez and Tabatabai 2000).

3.2.2.4 Amidase

Amidase (acylamide amidohydrolase, EC 3.5.1.4) is responsible for the production of ammonia (NH_3) and carboxylic acid through the hydrolysis of amides (Fraser et al. 2013). The production of ammonia is an important process in the N cycle. Amidase activity can be increased in the presence of C, but excess C can lead to both an N and P limitation (Allison et al. 2011). Amidase activity has been positively correlated with high molecular weight amide substrates (propionamide) but not with the low weight substrates (formamide) (Fraser et al. 2013). Amidase is secreted by many of microorganisms, plants, and animals, including bacteria *Rhodococcus* (Nawaz et al. 1994) and *Bacillus* (Thalenfeld et al. 1977) genera. Fungal production of amidase was noted in the *Aspergillus* (Benke 1979) and *Fusarium* (Reichel et al. 1991) genera. The wide variety of substrates that are available to amidase indicates the diverse nature and prevalence of these enzymes. The irreversible inhibition of amidase activity by organophosphate insecticides, such as fenitrothion and trichlorfon, has been described by Rasool et al. (2014).

3.2.2.5 Nitrate Reductase (NR) Activity

The dissimilatory nitrate reductase catalyzes the first step of the nitrification process by reducing nitrate (NO_3^-) to nitrite (NO_2^-) (Singh and Kumar 2008; Verma et al. 2015a). Most of the NO_3^- applied to the soil as fertilizer is taken by plants, leachedordenitrified. Not many of this NO_3^- is transformed to ammonium (NH_4^+) by the assimilatory nitrate reductase (ANR) coming from soil microbes (Abdelmagid and Tabatabai 1987; Singh and Kumar 2008). This assumption is based largely on the results showing that, contrary to dissimilatory reduction of NO_3^- to ammonium ions, the assimilatory reduction of NO_3^- is greatly repressed by NH_4^+concentration (McCarty and Bremner 1992) and that the amount of NH_4^+ in soil is usually higher than that needed to inhibit the activity of ANR (Fu and Tabatabai 1989; Šimek et al. 2002; Martens 2005). In the literature, there are different opinions as regards the ways of the inhibition of soil ANR activity by NH_4^+ions. Some researchers have stated that the negative influence of ammonium ions on the assimilatory nitrate reductase activity is caused by the presence of NH_4^+itself and does not depend on the production of ammonium ions by soil microorganisms (Martens 2005), while other studies have concluded that the inhibitory result of NH_4^+ on the soil ANR activity is caused by the occurrence of the glutamine as a result of NH_4^+ assimilated by soil microorganisms (McCarty and Bremner 1992).

4 Genes Encoding the Mineralization Enzymes in Soil

The contribution of transcriptomics and proteomics in soil enzymology is still poorly developed probably because there are still methodological problems in soil proteomics. Linking enzyme activity to gene expression in soil is a challenging task. Detection of the specific enzyme activities does not identify the microbial species directly involved in the measured process, leaving the link between the composition of the microbial community and the production of key enzymes poorly understood (Nannipieri et al. 2002; Krasek et al. 2006; Colloff et al. 2008; Wallenstein and Weintraub 2008). Enzymes in soil may be intracellular or extracellular, wherein these extracellular enzymes are usually absorbed into soil organic (humic substances) and mineral (clay minerals) colloids. Enzymes in the soil can originate mainly from microorganisms but also from plants and animals. Assuming that it is possible to assess that an enzyme comes from microorganisms, there are commonly a lot of microbes that are able to produce the same enzyme. Additionally, there are frequently varied gene-coding enzymes that are able to catalyze the same or similar reactions, thus leading to some functional excess, which increases the ability of the enzyme-producing organism to adapt and cope with a diversity of environmental conditions (Naessens and Vandamme 2003).

Although detecting the presence of selected genes in soil is now possible, only some studies have focused on the connection between gene abundance and enzymatic activity in soil. The chitinase (E.C. 3.2–1.14) activity in soil was the first to be compared using the respective enzyme-encoding genes. An attempt was made by Metcalfe et al. (2003) to cover the whole process of gene expression, chitinase secretion, and its determination in brown forest soil. The community structure was assessed by extracting DNA and cloning and sequencing the PCR products with the application of the primers for A chitinases (bacteria, fungi, virus, animals, and some plant enzymes) for 18 family groups. The activity was measured by either the weight loss of chitin or via an assay using 4-methylumbelliferyl-(GlucNAc)2, which was higher in the soil that had been treated with sludge and was related to many species of actinobacteria. The analysis of the sequence revealed greater changes in the community structure in the case of the sludge and lime treatments.

The degradation of the lingo-cellulose complex is important in maintaining nutrient cycling. A lot of studies have been dealing with enzymes taking part in these processes or their responsible genes, but no many researchers tried to join the enzyme activity to specified gene products. Bogan et al. (1996a) determined the transcripts of the lignin peroxidase genes (*lip A-lip J*) of *Phanerochaete chrysosporium* in soil treated with anthracene at the dose of 400 ppm. Following the mRNA extraction, the occurrence of the *lip* gene transcript was found and determined using competitive RT-PCR. The lip proteins were extracted from the soil, cleaned, and used to a nitrocellulose membrane. The western blotting technique was accomplished using the monoclonal antibodies to *Phanerochaete chrysosporium* LiP H8. Later, Bogan et al. (1996b) were capable to find the transcripts of nine *lip* genes, even in non-sterile soil taken from a polluted site. In soil microbiology, laccase-encoding genes have primarily been used to study the structural and functional diversity of fungi (Theuerl and Buscot 2010). The diversity of fungal laccase-encoding genes was greater in the surface layer than in the deeper soil layers, and there was a great deal of variability in the surface soil (Luis et al. 2004, 2005). The presence of the laccase-encoding genes of basidiomycetes (DNA was extracted, amplified, and cloned with a final sequencing) changed during different seasons, whereas the laccase activity of the phosphate extracts of the soil remained constant through the years (Kellner et al. 2009). Cañizares et al. (2012) related the first link in the disclosure and expression of encoding genes coming from bacterial β-glucosidase (βgluF2/βgluR4 primers) to the relative enzyme activity in soil treated with long-term management practices and found that these genes were overexpressed in the tilled soils, which was probably the response of the bacteria to stress. However, only 50% of the amino acid sequences were matched by the database sequences that were retrieved, which indicates the presence of soil bacteria that have unknown β-glucosidases.

Recent progress in detecting extracellular peptidase-encoding genes and characterizing the diversity of the urease-encoding genes in soil bacteria has supported that molecular methods targeting some of the key enzymes may help us further understand the microbial community active in soil N mineralization (Fuka et al. 2008a). Various proteolytic genes have been determined in the soil, including serine peptidases (*sub*), alkaline metallopeptidases (*apr*), neutral metallopeptidase (*npr*), as well as aspartic protease (*pep Aa, pepAb, pep Ac* and *pep Ad*) (Fuka et al. 2008b; Veening et al. 2008). Sakurai et al. (2007) showed that the structure of the genes expressing bacterial populations (*apr*- and *npr)* was the most significant in characterizing the total protease activity in soil. The latest advances on the *sub* and *npr* genes in cultivation soils have demonstrated that those genes were various and spatially differentiated in soil (Fuka et al. 2008b; Mrkonjic Fuka et al. 2009). Some authors have suggested that the *sub*, *apr*, and *npr* genes plentifulness is frequently related to the obtainable proteolytic activity (Fuka et al. 2008a). A wide range of soil organisms including various populations of bacteria, fungi, as well as plants are able to produce enzyme urease, which hydrolysis is urea into ammonium and carbon dioxide (Tabatabai et al. 2010). Urease-encoding genes (*ureC*) have been found in a range of soil ammonia-oxidizing archaea (*AOA*) and bacteria (*AOB*) (Lu et al. 2012; Lu and Jia 2013). Metagenomic-based studies displayed that the *ureC* gene was widely arranged in various soils and was remarkably positively related to other genes of N-transforming enzymes, such as *amoA* and *gdh* (gene coding the enzyme glutamate dehydrogenase) (Yang et al. 2013, 2014). However, there are confined results related to the variety of the *ureC* genes in soil microbial populations (Singh et al. 2009).

Genes of proteolytic enzymes (e.g., *bpr* or *aprE*) are diversely expressed in different populations of soil microorganisms (Veening et al. 2008). Many environmental factors including carbon, nitrogen, phosphate and calcium concentrations, soil pH, moisture and temperature, the quality and quantity of available substrates sugars, salicylic acid, plant hormones, flavonoids, amino acids, and selected antibiotics increase gene expression (Maunsell et al. 2006; Shivanand and Jayaraman 2009; Verma et al. 2015b; Molaei et al. 2017a, b). The secretion of proteolytic enzymes by soil microbes is constitutive (they are still present in the cells) or inducible (their synthesis is stimulated by the presence of an appropriate inductor) as dependent on various growth stages (Burns 1982). The highest expression of genes coding proteolytic enzymes in bacteria and fungi may be in the initial constant period of their growth, throughout the exponential growth stage, or for the time of the late lag period (Allison and MacFarlane 1990).

5 Methodology of the C- and N-Cycling Enzymes: General Overview and Data Interpretation

Catalase and peroxidase were the first enzymes which activity was determined in soil more than 100 years ago (Skujins 1978). Since that time, many enzyme activities have been found and determined in the soil. However, the amount of enzymes in soil is significantly higher than those that have been determined because of many possible sources of enzymes occurring in soil (microorganisms, plants, fauna) (Ladd 1985). Additionally, by using many assays we are not able to identify the specific enzyme activity; e.g., during the measurement of casein-hydrolyzing activities, there is no possible to identify the specific bonds that are hydrolyzed or products that are released. Many methods of enzyme activities determination in soil have been developed by Tabatabai and coworkers (Tabatabai and Bremner 1969, 1970; Tabatabai and Singh 1976). Later, at least three books devoted to the methodology of soil enzymes were published (Alef and Nannipieri 1995; Scinner et al. 1995; Dick 2011).

Enzymes in soil are willingly determined since the assays are generally simple, accurate, sensitive, and relatively rapid. A range of enzyme activities and a large number of samples can be analyzed over a period of a few days using small quantities of soil. Changes in enzyme activities depend not only on variations of gene expression but are also affected by environmental and anthropogenic factors (Nannipieri 1994; Tabatabai 1994). Therefore, the determination of enzyme activities in soil requires the effective extraction followed by the exact assessment of the remaining substrate or the products formed during the reaction. Because most of the assays for assessment of the resulting product or decreasing substrate are based on colorimetric measurements, it is recommended to use buffers, which generally do not release organic matter from the soil. Interference by the color development originating from soil organic matter constituents is commonly known during colorimetric measurement. Every soil enzyme assay has its own optimum conditions, such as the suitable substrate concentration, the defined buffer pH, and the temperature of the incubation. At a substrate concentration that exceeds the value that limits the reaction rate, the incubation time should warrant a linear decrease of the substrate or product release and ought to be as short as possible in order to show a quantifiable part of the activity. Long-time incubation should be avoided due to the possible microbial proliferation and growth, the activities of intracellular enzymes, and the synthesis of new enzymes (Burns 1978; Burns and Dick 2002).

Current methods in soil enzymology allow us to determine rather the potential than in situ activity. This is due to the fact that the incubation conditions are selected in order to guarantee the highest rate of substrate conversion. Additionally, in enzyme assays, we use soil slurries to limit the spreading restriction. Thus, the assay conditions used in laboratories are dissimilar from those that occur naturally in soil, where pH, temperature, and moisture are rarely optimal and change very often, and the substrate concentration is usually in low concentration (Burns 1978).

The main problem in interpreting measurements of enzyme activities is to distinguish among many components contributing to the overall activity (Burns 1982; Nannipieri 1994; Gianfreda and Bollag 1996). The activity of any particular enzyme in soil depends on enzymes that can have different locations (living cells, dead cells, cell debris, soil solution, adsorbed by inorganic colloids and associated in various ways with humic molecules) (Datta et al. 2017a). In addition, abiotic transformations, the so-called enzyme-like reactions, can contribute to the overall activity (Gianfreda and Ruggiero 2006). The significant fractions of soil enzymatic activity are intracellular enzymes that are present in cells of plants, animals, and microorganisms. Taking into account that visible plant and animal residues are removed during the soil preparation before enzyme determination and that enzymes which have been liberated from dead cells are quickly degraded by microorganisms or inhibited by various unbeneficial factor occurring in soil environment, it can be assumed that enzymes in living and proliferating microbial cells are the most crucial part of intracellular enzymes found in soil. Thus, assessment of the intracellular enzyme activities can provide significant knowledge about the functional diversity of soil microorganisms. As regards the extracellular localization, the free enzymes are assumed to be short-lived as compared to enzymes associated with soil colloids (Burns 1982; Ladd 1985; Nannipieri 1994). With the currently used methodology, it is not possible to separate between enzyme activities originated from different locations. The following factors restrict advances in soil enzymology: (1) inability to isolate extracellular from intracellular enzymatic activity, (2) inability to extract and purify enzymatic proteins from soil, and (3) lack of suitable methods to extract quantitatively products of enzyme reactions occurring in soil (Tabatabai and Dick 2002). Another problem that is connected with the methodology of soil enzymes is the lack of standardization. The methodologies adopted for soil enzyme measurement are not universal and create difficulty in comparing soil enzyme research. Differences between substrates, assay conditions, incubation times, and detection methods (Marx et al. 2001; Burns et al. 2013) contribute to differences in the enzyme readings. The various experimental conditions that are used to determine soil enzyme activities are presented in Table 4.

Generally, there are two groups of methods used for soil enzymatic activity determination. One group is based on measuring of the substrate concentration decreases, and the other group is related to an increase in the concentration of the product released during the reaction. Techniques most often used in soil enzymology are presented below or in Tables 4 and 5.

A. Colorimetric/*spectrophotometric* methods are based on the absorption of light (visible or ultraviolet) by the substrates and/or products released during the enzymatic reactions. The amount of the substrate remaining after reaction or product extracted from soil sample after incubation with an appropriate substrate and buffer, that are incubated at a specific temperature, pH, and time, is determined using colorimeter (or spectrophotometer). Most of the C and N cycle enzymes can be determined using these methods (Tabatabai and Dick 2002).
B. *Distillation – titration* methods are used to determine the amidohydrolases, such as L-asparaginase, L-glutaminase, L-aspartase, amidase, and urease. These methods are based on the incubation of the soil with the proper substrate, and proper pH buffer after which the NH_4^+ produced is assayed. After incubation, the enzymatic reaction is stopped by adding 2MKCl that contains Ag_2SO_4. The defined volume of the obtained mixture is distilled with MgO, and the NH_3 that is released is collected in boric acid that contains the appropriate indicators and is titrated with standard H_2SO_4. The detailed methodology was presented earlier by Tabatabai (1994).

Table 4 Methods for detecting selected activities of N-cycling enzymes in soil

Enzyme, EC number	Substrate and concentration	Buffer and pH	Time and temperature incubation	End product released	Detection/method	References
Urease, EC 3.5.1.5	Urea, 0.72 M	0.1 M borate buffer, pH 10.0	2 h, 37 °C	NH_3–N, CO_2	NH_3N/spectrophotometer (450 nm)	Kandeler and Gerber (1988)
Urease, EC 3.5.1.5	Urea, 0.2%	–	5 h, 37 °C	NH_3–N, CO_2	Urea/spectrophotometer (500–550 nm)	Zantua and Bremner (1975)
Urease, EC 3.5.1.5	Urea, 0.2 M	0.01 M TRIS buffer, pH 9.0	2 h, 37 °C	NH_3–N, CO_2	NH_3N/steam distillation	Tabatabai (1994)
L-Asparaginase, EC 3.5.1.1	L-Asparagine, 0.5 M	0.1 M TRIS buffer, pH 10.0	2 h, 37 °C	L-Aspartic acid, NH_3–N	NH_3N/steam distillation	Frankenberger and Tabatabai (1991a)
L-Glutaminase, EC 3.5.1.2	L-Glutamine, 0.5 M	0.1 M TRIS buffer, pH 10.0	2 h, 37 °C	L-Glutamic acid, NH_3–N	NH_3N/steam distillation	Frankenberger and Tabatabai (1991b)
Amidase, EC 3.5.1.4	Formamide, 0.5 M	0.1 M TRIS buffer, pH 8.5	2 h, 37 °C	Carboxylic acid, NH_3–N	NH_3N/steam distillation	Frankenberger and Tabatabai (1980)
Denitrification enzyme activity	Solution of 1.07 M KNO_3 and 1 mM glucose	–	–	N_2O	N_2O/gas chromatography	Smith and Tiedje (1979)
Arylamidase, EC 3.4.11.2	L-Leucine β-naphthylamide hydrochloride	0.1 M TRIS buffer, pH 8.0	1 h, 37 °C	N-terminal amino acid from peptides, amides and arylamides	*p*-Dimethylaminocinnamaldehyde (red) spectrophotometer (540 nm)	Acosta-Martinez and Tabatabai (2000)

(continued)

Table 4 (continued)

Enzyme, EC number	Substrate and concentration	Buffer and pH	Time and temperature incubation	End product released	Detection/method	References
Nitrate reductase, EC 1.7.99.4	KNO_3, 25 mM	0.19 M NH_4Cl buffer pH 8.5	24 h, 25 °C	NO_2–N	NO_2–N/spectrophotometer (520 nm)	Abdelmagid and Tabatabai (1987)
Proteases, EC 3.4.	Sodium caseinate, 2%	0.05 M TRIS buffer pH 8.1	2 h, 50 °C	Amino acids	Tyrosine/spectrophotometer (700 nm)	Alef and Nannipieri (1995)
Proteases, EC 3.4.	*N*-Benzyl-L-argininamide (BAA), 0.03 M	0.1 M phosphatase buffer pH 7.1	1 h, 40 °C	Amino acids	NH_3N/spectrophotometer (700 nm)	Nannipieri et al. (1980) and Landi et al. (2011)

TRIS tris(hydroxymethyl)aminomethane

Table 5 Methods for detecting selected activities of C-cycling enzymes in soil

Enzyme EC number	Substrate and concentration	Buffer and pH	Time and temperature incubation	End product released	Detection/method	References
CM-cellulase	Carboxymethylcellulose (CMC), 2% w/v	Acetate buffer, 0.1 M, pH 5.5	24 h/50 °C	Glucose	Glucose or reducing sugar/ spectrophotometer (690 nm)	Schinner and von Mersi (1990) and Deng and Tabatabai (1994)
β-Glucosidase, EC 3.2.1.21	*p*-Nitrophenyl-β-D-glucoside, 50 mM	MUB, pH 6.0	1 h/37 °C	Glucose	*p*NP/spectrophotometer (405–415 nm)	Eivazi and Tabatabai (1988)
Invertase, EC 3.2.1.26	Sucrose, 10% w/v	Acetate buffer, 0.1 M, pH 5.0	3 h/50 °C	Glucose and fructose	Reducing sugar/spectrophotometer (690 nm)	Schinner and von Mersi (1990)
Amylase	Soluble starch solution, 2% w/v	Acetate buffer, 0.1 M, pH 5.5	24 h/30 °C	Glucose	Glucose or reducing sugar/ spectrophotometer (620 nm)	Ross (1965) and Al-Turki and Dick (2003)
Xylanase, EC 3.2.1.8	Xylan, 1.2% w/v	Acetate buffer, 0.1 M, pH 5.5	24 h/50 °C	Xylose	Reducing sugar/spectrophotometer (690 nm)	Schinner and von Mersi (1990) and Deng and Tabatabai (1994)
Chitinase, EC 3.2.1.14	Purified chitin, 1.5% w/v	Phosphate buffer, 0.12 M, pH 6.0	16 h/37 °C	*N*-acetyl-β-glucosamine	*N*-Acetyl-β-glucosamine or reducing sugar	Rodriguez-Kabana et al. (1983)
Lipase, EC 3.2.1.3	4-Methylumbelliferyl heptanoate	TRIS, 0.1 M, pH 7.5	10 min/30 °C	Fatty acids and glycerol	4-Methylumbelliferone/ spectrofluorometer (340/ 450 nm)	Cooper and Morgan (1981)

MUB modified universal buffer, *p*NP *p*-nitrophenol, *TRIS* tris(hydroxymethyl)aminomethane

C. *Fluorescence* methods – some enzyme activities in soils have been detected using fluorimetric techniques. One of the earliest techniques of this kind was that proposed by Pancholy and Lynd (1971) for soil lipase activity. The reaction was based on the hydrolysis of the nonfluorescent butyryl ester of 7-hydroxy-4-methylcoumarin to 7-hydroxy-4-umbelliferone, the highly fluorescent compound. The fluorescence methods have been used for assaying the β-glucosidase activity in peat (Freeman et al. 1995) and for assaying β-glucosidase, β-cellobiase, β-*N*-acetylgalactosaminidase, and β-xylosidase in soil (Darrah and Harris 1986). Very important in these methods is to use the small amount of soil sample (milligrams) or to determine the capacity of the soil to adsorb the fluorogenic compounds that are liberated to correct the assay results.
D. A *microplate fluorimetric method* was proposed by Marx et al. (2001) (as a modification of the earlier used fluorimetric techniques) to determine the activity of the hydrolytic enzymes and was based on the application of the methylumbelliferyl (MUB) substrates. Following incubation for a defined period of time at the desired temperature, MUF is quantified by terminating the enzymatic reaction by adding 0.5 M NaOH. The concentration of MUF is measured by a computerized microplate fluorimeter and is expressed as the micromoles MUF that are released per kg^{-1} soil h^{-1} (Deng et al. 2011). This method offers increased sensitivity and the possibility to estimate the kinetic parameter of the enzyme reaction. If it is successful, the advantages of this methodology are (1) speed of operation (less than 1 h), (2) simultaneous analysis of a large number of samples, (3) simultaneous use of a range of MUB conjugates, (4) measurement under standard conditions, and (5) automatic calculation of reaction rates. This method, which requires only milligram quantities of homogenous soil samples, has been used to measure the activities of β-D-glucosidase, β-D-galactosidase, *N*-acetyl-β-D-glucosaminidase, β-cellobiase, and β-xylosidase in a sandy loam and a silty clay loan soil (Marx et al. 2001). Modified, faster, microplate methods for the high-throughput determination of β-glucosidase were discussed recently (Hoehn 2016). The evaluation includes the use of an automated pipetting system and sonication, as well as a reduction in the number of analytical replicates, which permits a higher sample throughput suitable for service laboratory use.

6 Conclusions and Future Challenges

Based on the literature, it can be concluded that interest in the enzyme systems responsible for C and N transformation in soil is currently still high. The problem with the enzyme activities in soil is however related to the imperfect methodologies, which do not allow to measure the actual soil functioning. The methods of soil enzyme activities determination have the following limitations: (1) the methods do not differentiate between the constituents that contribute to the overall soil enzymatic activity, (2) they measure the potential rather than actual enzyme activities, (3) they do not separate the real enzymatic activity from the so-called "enzyme-like" activities, (4) there is no methods standardization, and (5) the currently used methods prevent to detect the origin of the soil enzymes (Gianfreda and Ruggiero 2006). That is why the development of the better assays of enzyme activity in soil is required.

Future research should also investigate if the enzymes are actively expressed and if, after expression, they catalyze the proper reactions without inhibition. The other area in soil enzymology should be devoted to the development of the new methods suitable to assess the genetic potential and gene expression, as well as the direct enzyme activity, showing the potential activity in the soil system. This study field includes research on the shift in the genes that codes for individual enzymatic proteins and the determination of soil factors that influence the expression of the specific enzymes (Sect. 5, Krasek et al. 2006). There is a great need to explain the connection between the genetic diversity and microbial community structure and functioning (Suenaga 2011; van Elsas and Boersma 2011; Meena et al. 2014). Since all of the enzymes have their relative genes, they are an ideal base for research on the relationship between microbial specification and particular processes occurring in the ecosystem. In this regard, progress in the development of "omics" technologies such as proteomics and transcriptomics give a great, although not yet proven, potential to explain a lot of aspects of the regulation of the functioning and ecology of soil enzymes.

References

Abdelmagid HM, Tabatabai MA (1987) Nitrate reductase activity of soils. Soil Biol Biochem 19:421–427

Acosta-Martinez V, Tabatabai MA (2000) Arylamidase activity of soils. Soil Sci Soc Am J 64:215–221

Acosta-Martinez V, Cruz L, Sotomayor-Ramírez D, Pérez-Algería L (2007) Enzyme activities as affected by soil properties and land use in tropical watershed. Appl Soil Ecol 35:35–45

Adams DJ (2004) Fungal cell wall chitinases and glucanases. Microbiology 150:2029–2035

Alef K, Kleiner D (1986) Arginine ammonification, a simple method to estimate microbial activity potentials in soil. Soil Biol Biochem 18:233–235

Alef K, Kleiner D (1987) Applicability of arginine ammonification as indicator of microbial activity in different soils. Biol Fertil Soils 5:148–151

Alef K, Nannipieri P (1995) Methods in soil microbiology and biochemistry. Academic, San Diego

Allison SD (2005) Cheaters, diffusion and nutrients constrain decomposition by microbial enzymes in spatially structured environments. Ecol Lett 8:626–635

Allison C, Macfarlane GT (1990) Regulation of protease formation in Clostridium sporogenes. Appl Environ Microbiol 56(11):3485–3490

Allison S, Weintraub M, Gartner T, Waldrop M (2011) Evolutionary-economic principles as regulators of soil enzyme production and ecosystem function. In: Shukla G, Varma A (eds) Soil enzymology, Soil biology, vol 22. Springer, Berlin/Heidelberg, pp 229–243

Al-Turki AJ, Dick AW (2003) Myrosinase activity in soil. Soil Sci Soc Am J 76:139–145

Anand L, Krishnamurthy S, Vithayathil PJ (1990) Purification and properties of xylanase from the thermophilic fungus, Humicola lanuginosa (Griffon and Maublanc) Bunce. Arch Biochem Biophys 276:546–553

Ashoka P, Meena RS, Kumar S, Yadav GS, Layek J (2017) Green nanotechnology is a key for eco-friendly agriculture. J Clean Prod 142:4440–4441

Bakshi M, Varma A (2011) Soil enzymes: state of art. In: Shukla S, Varma A (eds) Soil enzymology, Soil biology, vol 22. Springer, Berlin/Heidelberg, pp 1–23

Baldock AJ (2007) Principles of nutrient cycling. In: Marchner P, Rengel Z (eds) Nutrient cycling in terrestrial ecosystems, Soil biology, vol 10. Springer, Berlin/Heidelberg, pp 2–35

Balota EL, Dias Chaves JC (2010) Enzymatic activity and mineralization of carbon and nitrogen in soil cultivated with coffee and green manures. R Bras Ci Solo 34:1573–1583

Balota E, Kanashiro M, Filho A, Andrade D, Dick R (2004) Soil enzyme activities under long – term tillage and crop rotation systems in subtropical agro – ecosystems. Braz J Microbiol 35:300–306

Benke AC (1979) A modification of the Hynes method for estimating secondary production with particular significance for multivoltine populations. Limnol Oceanogr 24:168–171

Berg B, McClaugherty CA (2003) Plant litter. Decomposition, humus formation, carbon sequestration. Springer, Berlin/Heidelberg

Bergmann JC, Costa OYA, Gladden JM, Singer S, Heins R, D'Haeseleer P, Simmons BA, Quirino BF (2014) Discovery of two novel β-glucosidases from an Amazon soil metagenomic library. FEMS Microbiol Lett 351(2):147–155

Berrin JG, Czjzek M, Kroon PA, Mclauchlan WR, Puigserver A, Williamson G, Juge N (2003) Substrate (aglycone) specificity of human cytosolic β- glucosidase. Biochem J 373:41–48

Bhattacharya D, Nagpure A, Gupta RK (2007) Bacterial chitinases: properties and potential. Crit Rev Biotechnol 27:21–28

Bogan BW, Schoenike B, Lamar RT, Cullen D (1996a) Manganese peroxidase mRNA and enzyme activity levels during bioremediation of polycyclic aromatic hydrocarbon contaminated soil with *Phanerochaete chrysosporium*. Appl Environ Microbiol 62:2381–2386

Bogan BW, Schoenike B, Lamar RT, Cullen D (1996b) Expression of lip genes during growth in soil and oxidation of anthracene by *Phanerochaete chrysosporium*. Appl Environ Microbiol 62:3697–3703

Bonde TA, Nielsen TH, Miller M, Sørensen J (2001) Arginine ammonification assay as a rapid index of gross N mineralization in agricultural soils. Biol Fertil Soils 34:179–184

Brzezińska MS, Lalke-Porczyk E, Donderski W, Walczak M (2009) Degradation of chitin in natural environment: role of actinomycetes. Pol J Ecol 57:229–238

Brzostek ER, Finzi AC (2011) Substrate supply, fine roots, and temperature control proteolytic enzyme activity in temperate forest soils. Ecology 92(4):892–902

Buragohain S, Sharma B, Nath JD, Gogaoi N, Meena RS, Lal R (2017) Impact of ten years of bio-fertilizer use on soil quality and rice yield on an inceptisol in Assam, India. Soil Res https://doi.org/10.1071/SR17001

Burns RG (1978) Enzyme activity in soil. In: some theoretical and practical considerations. In: Soil enzymes. Academic, New York, pp 1–49

Burns RG (1982) Enzyme activity in soil: location and a possible role in microbial ecology. Soil Biol Biochem 14:423–427

Burns RG, Dick RP (2002) Enzymes in soil: research and development in measuring activities. In: Burns RG, Dick RP (eds) Enzymes in the environment. Activity, ecology and applications. Marcel Dekker, New York, pp 567–596

Burns RG, DeForest JL, Marxsen J, Sinsabaugh RL, Stromberger ME, Wallenstein MD, Weintraub MN, Zoppini A (2013) Soil enzymes in a changing environment: current knowledge and future directions. Soil Biol Biochem 58:216–234

Burton DL, McGill WB (1991) Inductive and repressive effects of carbon and nitrogen on L-histidine ammonia-lyase activity in a black chernozemic soil. Soil Biol Biochem 23:939–946

Cañizares R, Moreno B, Emilio Benitez E (2012) Biochemical characterization with detection and expression of bacterial β-glucosidase encoding genes of a Mediterranean soil under different long-term management practices. Biol Fertil Soils 48:651–663
Cenini V, Fornara D, McMullan G, Ternan N, Carolan R, Crawley MJ, Clément JC, Lavorel S (2016) Linkages between extracellular enzyme activities and the carbon and nitrogen content of grassland soils. Soil Biol Biochem 96:198–206
Colloff M, Wakelin S, Gomez D, Rogers S (2008) Detection of nitrogen cycle genes in soils for measuring the effects of changes in land use and management. Soil Biol Biochem 40:1637–1645
Cooper AB, Morgan HW (1981) Improved fluorimetric method to assay for lipase activity. Soil Boil Biochem 13:307–311
Dadhich RK, Meena RS (2014) Performance of Indian mustard (*Brassica juncea* L.) in response to foliar spray of thiourea and thioglycollic acid under different irrigation levels. Indian J Ecol 41 (2):376–378
Dar GH (2010) Soil microbiology and biochemistry. New India Publishing Agency, Pitam Pura/New Delhi
Darrah PR, Harris PJ (1986) A fluorimetric method for measuring the activity of soil enzymes. Plant Soil 92:81–88
Datta R, Vranová V, Pavelka M, Rejšek K, Formánek P (2014) Effect of soil sieving on respiration induced by low-molecular-weight substrates. Int Agrophys 28(1):119–124
Datta R, Anand S, Moulick A, Baraniya D, Pathan SI, Rejsek K, Vranova V, Sharma M, Sharma D, Kelkar A (2017a) How enzymes are adsorbed on soil solid phase and factors limiting its activity: a review. Int Agrophys 31(2):287–302
Datta R, Kelkar A, Baraniya D, Molaei A, Moulick A, Meena R, Formanek P (2017b) Enzymatic degradation of lignin in soil: a review. Sustainability 9(7):1163
Deng S, Popova I (2011) Carbohydrate hydrolases. In: Dick RP (ed) Methods of soil enzymology, vol 9. Soil Science Society of America, Madison, pp 185–207
Deng SP, Tabatabai MA (1994) Colorimetric determination of reducing sugars in soils. Soil Biol Biochem 26:473–477
Deng S, Kang H, Freeman C (2011) Microplate fluorimetric assay of soil enzymes. In: Dick RP (ed) Methods of soil enzymology, vol 9. Soil Science Society of America, Madison, pp 311–318
Dick RP (1994) Soil enzyme activity as an indicator of soil quality. In: Doran JW, Leman DC, Bezdicek DF, Steward BA (eds) Defining soil quality for a sustainable environment. SSSA, Special Publication 35, Madison, pp 107–124
Dick RP (ed) (2011) Methods of soil enzymology. Soil Science Society of America, Madison
Dick WA, Tabatabai MA (1992) Significance and potential uses of soil enzymes. In: Metting FB Jr (ed) Soil microbial ecology: applications in agricultural and environmental management. Marcel Dekker, New York, pp 95–225
Dilly O, Munch JC, Pfeiffer EM (2007) Enzyme activities and litter decomposition in agricultural soils in northern, central, and southern Germany. J Plant Nutr Soil Sci 170:197–204
Dodor DE, Tabatabai MA (2003) Aminohydrolases in soil as affected by cropping systems. Appl Soil Ecol 24:73–90
Dodor DE, Tabatabai MA (2005) Glycosidases in soils as affected by cropping systems. J Plant Nutr Soil Sci 168:749–758
Dodor DE, Tabatabai MA (2007) Arylamidase activity as an index of nitrogen mineralization in soils. Commun Soil Sci Plant Anal 36(15–16):2197–2207
Duo-Chuan L (2006) Review of fungal chitinases. Mycopathologia 161:345–360
Ebregt A, Boldewijn J (1977) Influence of heavy metals in spruce forest soil on amylase activity, carbon dioxide evolution from starch and soil respiration. Plant Soil 47:137–148
Ebulue MM, Uwakwe AA, Wegwu MO (2017) Soil lipase and dehydrogenases activities in spent engine oil polluted ecosystem. J Bio Innov 6(5):782–789
Eick M, Stöhr C (2009) Proteolysis at the plasma membrane of tobacco roots: biochemical evidence and possible roles. Plant Physiol Biochem 47:1003–1008

Eivazi F, Tabatabai MA (1988) Glucosidases and galactosoidases in soils. Soil Biol Biochem 20:601–606

Ekenler M, Tabatabai MA (2002) β-Glucosaminidase activity of soils: effect of cropping systems and its relationship to nitrogen mineralization. Biol Fertil Soils 36:367–376

Ekenler M, Tabatabai MA (2004) β-Glucosaminidase activity as an index of nitrogen mineralization in soils. Commun Soil Sci Plant Anal 35(7–8):1081–1094

Enowashu E, Poll C, Lamersdorf N, Kandeler E (2009) Microbial enzyme activities under reduced nitrogen deposition in a spruce forest soil. Appl Soil Ecol 43:11–21

Feng J, Turner BL, Wei K, Tian J, Chen Z, Lü X, Wang C, Chen L (2018) Divergent composition and turnover of soil organic nitrogen along a climate gradient in arid and semiarid grasslands. Geoderma 327:36–44

Follmer C (2008) Insights into the role and structure of plant ureases. Phytochemistry 69:18–28

Frankenberger TW (1983) Kinetic properties of L-histidine ammonia-lyase activity in soils. Soil Sci Soc Am J 47:71–80

Frankenberger TW, Johanson BJ (1983) Factors affecting invertase activity in soils. Plant Soil 74:313–323

Frankenberger TW, Tabatabai MA (1980) Amidase activity in soils. 1. Method of assay. Soil Sci Soc Am J 44:282–287

Frankenberger TW, Tabatabai MA (1991a) L-Asparaginase activity of soils. Biol Fertil Soils 11:6–12

Frankenberger TW, Tabatabai MA (1991b) L-Glutaminase activity in soils. Soil Biol Biochem 23:869–874

Fraser FC, Hallett PD, Wookey PA, Hartley IP, Hopkins DW (2013) How do enzymes catalysing soil nitrogen transformations respond to changing temperatures? Biol Fertil Soils 49:99–103

Freeman C, Liska G, Ostle NJ, Jones SE, Lock MA (1995) The use of fluorogenic substrates for measuring enzyme activity in peatlands. Plant Soil 175:147–152

Fu MH, Tabatabai MA (1989) Nitrate reductase activity in soils: effects of trace elements. Soil Biol Biochem 21:943–946

Fuka MM, Engel M, Gattinger A, Bausenwein U, Sommer M, Munch JC, Schloter M (2008a) Factors influencing variability of proteolytic genes and activities in arable soils. Soil Biol Biochem 40:1646–1653

Fuka MM, Engel M, Haesler F, Welzl G, Munch JC, Schloter M (2008b) Diversity of proteolytic community encoding for subtilisin in an arable field: spatial and temporal variability. Biol Fertil Soils 45:185–191

Geisseler D, Horwath WR, Joergensen RG, Ludwig B (2010) Pathways of nitrogen utilization by soil microorganisms – a review. Soil Biol Biochem 42:2058–2067

Gianfreda L, Bollag JM (1996) Influence of natural and anthropogenic factors on enzyme activity in soil. In: Stotzky G, Bollag LM (eds) Soil biochemistry, vol 9. Marcel Dekker, New York/Basel/Hong Kong, pp 123–193

Gianfreda L, Rao MA (2004) Potential of extra cellular enzymes in remediation of polluted soils: a review. Enzym Microb Technol 35:339–354

Gianfreda L, Ruggiero P (2006) Enzyme activities in soil. In: Nannipieri P, Smalla K (eds) Nucleic acids and proteins in soil. Springer, Berlin/Heidelberg, pp 258–311

Ginésy M, Rusanova-Naydenova D, Rova U (2017) Tuning of the carbon-to-nitrogen ratio for the production of L-arginine by *Escherichia coli*. Fermentation 3(4):60

Glibert P, Harrison J, Heil C, Seitzinger S (2006) Escalating worldwide use of urea – a global change contributing to coastal eutrophication. Biogeochemistry 77:441–463

Godlewski M, Adamczyk B (2007) The ability of plants to secrete proteases by roots. Plant Physiol Biochem 45:657–664

Gómez Jiménez RT, Moliterni E, Rodríguez L, Fernández FJ, Villaseñor J (2011) Feasibility of mixed enzymatic complexes to enhanced soil bioremediation processes. Procedia Environ Sci 9:54–59

Guo H, Chen G, Zhaoping LV, Zhao H, Yang H (2009) Alteration of microbial properties and community structure in soils exposed to napropamide. J Environ Sci 21(4):494–502

Hallin S, Jones CM, Schloter M, Philippot L (2009) Relationship between N-cycling communities and ecosystem functioning in a 50-year-old fertilization experiment. ISME J 3:597–605
Halsall DM, Gibson AH (1991) Nitrogenase activity (C_2H_2 reduction) in straw-amended wheat belt soils in response to diazotroph inoculation. Soil Biol Biochem 23:987–998
Hanzlikova A, Jandera A (1993) Chitinase and changes of microbial community in soil. Folia Microbiol 38:159–160
Hartemik AE, Lal R, Gerzabek MH, Jama B, McBratney AB, Six J, Tornquist CG (2014) Soil carbon research and global environmental challenges. Peer J Pre Prints 2:e366v1 https://doi.org/10.7287/peerj.preprints.366v1
Hiwada K, Ito T, Yokoyama M, Kokubu T (1980) Isolation and characterization of membrane-bound Arylamidases from human placenta and kidney. Eur J Biotechnol 10(1):155–165
Hoehn EC (2016) Improved microplate fluorometric soil enzymes assay for β-glucosidase detection. Dissertations and theses in Natural Resources 130, University of Nebraska-Lincoln, USA
Holik L, Kučera A, Rejšek K, Vranová V (2017) Seasonal dynamics of arginine ammonification in forest soils of Norway spruce pure stands under different silvicultural practices. Cent Eur For J 63:66–72
Hu Y, Zhang G, Li A, Chen J, Ma L (2008) Cloning and enzymatic characterization of a xylanase gene from a soil-derived metagenomic library with an efficient approach. Appl Microbiol Biotechnol 80:823–830
Hungate BA, Hart SC, Selmants PC, Boyle SI, Gehring CA (2007) Soil responses to management, increased precipitation, and added nitrogen in ponderosa pine forest. Ecol Appl 17:1352–1365
Hussain S, Siddique T, Saleem M, Arshad M, Khalid A (2009) Impact of pesticides on soil microbial diversity enzymes, and biochemical reactions. In: Sparks D (ed) Advances in agronomy, vol 102. Academic, San Diego, pp 159–200
Jan M, Roberts P, Tonheim S, Jones DL (2009) Protein breakdown represents a major bottleneck in nitrogen cycling in grassland soils. Soil Biol Biochem 41(11):2272–2282
Jin K, Sleutel S, Buchan D, De Neve S, Cai DX, Gabriels D, Jin JY (2009) Changes of soil enzyme activities under different tillage practices in the Chinese Loess Plateau. Soil Till Res 104:115–120
Jones DL, Kielland K, Sinclai FL, Dahlgren RA, Newsham KK, Farrar JF, Murphy V (2009) Soil organic nitrogen mineralization across a global latitudinal gradient. Glob Biogeochem Cycles 23:GB1016
Jørgensen H, Kristensen JB, Felby C (2007) Enzymatic conversion of lignocellulose into fermentable sugars: challenges and opportunities. Biofuels Bioprod Biorefin 1:119–134
Kalisz HM (1988) Microbial proteinases. Adv Biochem Eng Biotechnol 36:1–65
Kandeler E (1996) Influence of heavy metals on the functional diversity of soil microbial communities. Biol Feril Soils 23:299–306
Kandeler E, Gerber H (1988) Short-term assay of urease activity using determination of ammonium. Biol Fertil Soils 6:68–72
Kandeler E, Luxhùi J, Tscherko D, Magid J (1999) Xylanase, invertase and protease at the soil – litter interface of a loamy sand. Soil Biol Biochem 31:1171–1179
Kandeler E, Poll C, Frankenberger Jr. WT, Tabatabai MA (2011) Nitrogen cycle enzymes. In: Dick RP (ed) Methods in soil enzymology. Soil Science Society of America Book Series, 9, pp 211–244
Kellner H, Luis P, Zimdars B, Kiesel B, Buscot F (2009) Diversity of bacterial laccase-like multicopper oxidase genes in forest and grassland Cambisol soil samples. Soil Biol Biochem 40:638–648
Kemmitt SJ, Wright D, Murphy DV, Jones DL (2008) Regulation of amino acid biodegradation in soil as affected by depth. Biol Fertil Soils 44:933–941
Khorsandi N, Nourbakhsh F (2007) Effect of amendment of manure and corn residues on soil N mineralization and enzyme activity. Agron Sustain Dev 27:139–143
Kiss S, Pasca D, Dragan-Bulards M (1998) Enzymology of disturbed soils. Development in soil science, vol 26. Elsevier, Amsterdam

Klose S, Tabatabai MA (1999) Urease activity of microbial biomass in soils. Soil Biol Biochem 31:205–211
Kögel-Knabner I (2002) The macromolecular organic composition of plant and microbial residues as inputs to soil organic matter. Soil Biol Biochem 34:139–162
Krasek M, Gaze WH, Morris NZ, Wellington EMH (2006) Gene detection, expression and related enzyme activity in soil. In: Nannipieri P, Smalla K (eds) Nucleic acids and proteins in soil. Springer, Berlin/Heidelberg, pp 217–255
Kudryavtseva O, Dunaevsky YE, Kamzolkina O, Belozersky M (2008) Fungal proteolytic enzymes: features of the extracellular proteases of xylotrophic Basidiomycetes. Microbiology 77:643–653
Kumar S, Meena RS, Pandey A, Seema (2017) Soil acidity management and an economics response of lime and sulfur on sesame in an alley cropping system. Int J Curr Microb and App Sci 6 (3):2566–2573
Ladd JN (1985) Soil enzymes. In: Vaughan D, Malcom RE (eds) Soil organic matter and biological activity. Matinus Nijhoff, Dordrecht, pp 175–221
Ladd JN, Jackson RB (1982) Biochemistry of ammonification. In: Stevenson FJ (ed) Nitrogen in agricultural soils. American Society of Agronomy, Madisonl, pp 173–228
Lal R (2004) Soil carbon sequestration impacts on global climate change and food security. Science 304(5677):1623–1627
Lal R (2008) Carbon sequestration. Philos Trans R Soc B 363:815–830
Lal R (2016) Soil health and carbon management. Food Energy Secur 5(4). https://doi.org/10.1002/fes3.96
Landi L, Renella G, Giagnoni L, Nannipieri P (2011) Activities of proteolytic enzymes. In: Dick RP (ed) Methods of soil enzymology. Soil Science Society of America, Madison, pp 247–273
Lin Q, Brookes PC (1999) Arginine ammonification as a method to estimate soil microbial biomass and microbial community structure. Soil Biol Biochem 31:1985–1997
Lu L, Jia ZJ (2013) Urease gene-containing Archaea dominate autotrophic ammonia oxidation in two acid soils. Environ Microbiol 15:1795–1809
Lu L, Han WY, Zhang JB, Wu YC, Wang BZ, Lin XG, Zhu JG, Cai ZC, Jia ZJ (2012) Nitrification of archaeal ammonia oxidizers in acid soils is supported by hydrolysis of urea. ISME J 6:1978–1984
Luis P, Walther G, Kellner H, Martin F, Buscot F (2004) Diversity of laccase genes from basidiomycetes in a forest soil. Soil Biol Biochem 36:1025–1036
Luis P, Kellner H, Zimdars B, Langer U, Martin F, Buscoit F (2005) Patchiness and spatial distribution of laccase genes of ecto-mycorrhizal, saprotrophic and unknown basidiomycetes in the upper horizons of a mixed forest Cambisol. Microb Ecol 50:570–579
Lynd LR, Weimer PJ, van Zyl WH, Pretorius IS (2002) Microbial cellulose utilization: fundamentals and biotechnology. Microbiol Mol Biol Rev 66:506–577
Machuca A, Córdova C, Stolpe NB, Barrera JA, Chávez D, Almendras K, Bonilla AM (2018) In vitro sensitivity of forest soil enzymes to temperature increase in Western Patagonia. J Soil Sci Plant Nutr 18(1):202–219
Margesin R, Zimmerbauer A, Schinner F (1999) Soil lipase activity – a useful indicator of oil biodegradation. Biotechnol Tech 13(12):859–863
Martens DA (2005) Denitrification. In: Hatfield JL, Powlson DS, Crosenzweig DS, Scow KM, Singer MJ, Sparks DL (eds) Encyclopedia of Soils in the Environment. Academic, Amsterdam, pp 378–382
Marx MC, Wood M, Jarvis SC (2001) A microplate fluorimetric assay for the study of enzyme diversity in soils. Soil Biol Biochem 33:1633–1640
Maunsell B, Adams C, O'Gara F (2006) Complex regulation of AprA metalloprotease in *Pseudomonas fluorescens* M114: evidence for the involvement of iron, the ECF sigma factor, PbrA and pseudobactin M114 siderophore. Microbiology 152:29–42
McCarty GW, Bremner JM (1992) Regulation of assimilatory nitrate reductase activity in soil by microbial assimilation of ammonium. Proc Natl Acad Sci U S A 89:453–456

Meena RS, Yadav RS, Meena VS (2014) Response of groundnut (*Arachis hypogaea* L.) varieties to sowing dates and NP fertilizers under Western Dry Zone of India. Bangladesh J Bot 43(2):169–173

Meena RS, Dhakal Y, Bohra JS, Singh SP, Singh MK, Sanodiya P (2015a) Influence of bioinorganic combinations on yield, quality and economics of mungbean. Am J Exp Agric 8 (3):159–166

Meena RS, Meena VS, Meena SK, Verma JP (2015b) The needs of healthy soils for a healthy world. J Clean Prod 102:560–561

Meena H, Meena RS, Singh B, Kumar S (2016) Response of bio-regulators to morphology and yield of clusterbean [*Cyamopsis tetragonoloba* (L.) Taub.] under different sowing environments. J App Nat Sci 8(2):715–718

Metcalfe AC, Williamson N, Krasek M, Wellington EMH (2003) Molecular diversity within chitynolytic actinomycetes determined by in situ analysis. Actinomycetologia 17:18–22

Mishra RR, Dangar TK, Rath B, Thatoi HN (2009) Characterization and evaluation of stress and heavy metal tolerance of some predominant Gram negative halotolerant bacteria from mangrove soils of Bhitarkanika, Orissa, India. Afr J Biotechnol 8(10):2224–2231

Mobley HLT, Hausinger RP (1989) Microbial urease: significance, regulation, and molecular characterization. Microbiol Rev 53:85–108

Mobley HLT, Island MD, Hausinger RP (1995) Molecular biology of microbial ureases. Microbiol Rev 59:451–480

Molaei A, Lakzian A, Datta R, Haghnia G, Astaraei A, Rasouli-Sadaghiani M, Ceccherini MT (2017a) Impact of chlortetracycline and sulfapyridine antibiotics on soil enzyme activities. Int Agrophys 31(4):499–505

Molaei A, Lakzian A, Haghnia G, Astaraei A, Rasouli-Sadaghiani M, Ceccherini MT, Datta R (2017b) Assessment of some cultural experimental methods to study the effects of antibiotics on microbial activities in a soil: an incubation study. PLoS One 12(7):e0180663

Monreal CM, Bergstrom DW (2000) Soil enzymatic factors expressing the influence of land use, tillage system and texture on soil biochemical quality. Can J Soil Sci 80:419–428

Morais H, Ramos C, Forgacs E, Jakab A, Cserhati T (2004) Enzyme production of Pleurotus ostreatus evaluated by the three-way principal component analysis. Eng Life Sci 4:165–170

Moss GP (2010) Enzyme nomenclature: Recommendations of the nomenclature committee of the Interantional Union of Biochemistry and Molecular Biology on the nomenclature and classification of enzymes by the reactions they catalyse. www.chem.qmul.ac.uk/iubmb/enzyme (verified 27 May 2011)

Mrkonjic Fuka M, Engel M, Hagn A, Munch JC, Sommer M, Schloter M (2009) Changes of diversity pattern of proteolytic Bacteria over time and space in an agricultural soil. Microb Ecol 57:391–401

Muruganandam S, Israel DW, Robarge WP (2009) Activities of nitrogen-mineralization enzymes associated with soil aggregate size fractions of three tillage systems. Soil Sci Soc Am J 73:751–759

Myrold D, Bottomley PJ (2008) Nitrogen mineralization and immobilization. In: Schepers JS, Raun WR (eds) Nitrogen in agricultural systems. American Society of Agronomy, Crop Science Society of America, Soil Science Society of America, Madison, pp 157–172

Naessens M, Vandamme EJ (2003) Multiple forms of microbial enzymes. Biotechnol Lett 25:1119–1124

Nannipieri P (1994) The potential use of soil enzymes as indicatyors of productivity, sustainability and pollution. In: Pankhurst CE, Doube BM, Gupta VVSR, Grace PR (eds) Soil biota: management in sustainable farming systems. CSIRO, Adelaide, pp 238–244

Nannipieri P, Eldor P (2009) The chemical and functional characterization of soil N and its biotic component. Soil Biol Biochem 41:2357–2369

Nannipieri P, Gianfreda L (1998) Kinetics of enzyme reactions in soil environment. In: Huang PM, Senesi N, Bue J (eds) Structure and surface reactions of soil particles. Wiley, New York, pp 449–479

Nannipieri P, Ceccanti B, Cervelli S, Matarese E (1980) Extraction of phosphatase, urease, protease, organic carbon and nitrogen from soil. Soil Sci Soc Am J 44:1011–1016

Nannipieri P, Grego S, Ceccanti B (1990) Ecological significance of the biological activity in soil. In: Bollag JM, Stotzky G (eds) Soil biochemistry, vol 6. Marcel Dekker, New York, pp 293–355

Nannipieri P, Sequi P, Fusi P (1996) Humus and enzyme activity. In: Piccolo A (ed) Humic substances in terrestrial ecosystems. Elsevier, Amsterdam, pp 293–328

Nannipieri P, Kandeler E, Rugiero P (2002) Enzyme activities and microbiological and biochemical processes in soil. In: Burs RG, Dick RP (eds) Enzymes in the environment. Activity, ecology and application. Marcel Dekker, New York, pp 1–33

Nawaz MS, Khan AA, Seng JE, Leakey JE, Siitonen PH, Cerniglia CE (1994) Purification and characterization of an amidase from an acrylamide-degrading *Rhodococcus* sp. Appl Environ Microbiol 60:3343–3348

Nieder R, Benbi DK (2008) Carbon and nitrogen in the terrestrial environment. Springer, Dordrecht

Norton JM (2008) Nitrification in agricultural soils. American Society of Agronomy, Crop Science Society of America, Soil Science Society of America 677, Madison, WI, USA. In: Nitrogen in Agricultural Systems, Agronomy Monograph, vol 49, pp 173–199

Norton JM, Schimel JP (2011) Nitrogen mineralization-immobilization turnover. In: Huang PM, Li Y, Summers ME (eds) Handbook of soil science, 2nd edn. CRC Press, Boca Raton, pp 8–18

Norton JM, Stark JM (2011) Regulation and measurement of nitrification in terrestrial systems. Methods Enzymol 486:343–368

Orczewska A, Piotrowska A, Lemanowicz J (2012) Soil acid phosphomonoesterase activity and phosphorus forms in ancient and postagricultural black alder [*Alnus glutinosa* (L.) Gaertn.] woodlands. Acta Soc Bot Pol 81(2):81–86

Page MJ, Di Cera E (2008) Serine peptidases: classification, structure and function. Cell Mol Life Sci 65:1220–1236

Pancholy SK, JQ Lynd (1971) Quantification of soil lipase with 4-MUB fluorescence. Agron Abstr, p 83

Pancholy SK, Rice EL (1972) Effect of storage conditions on actives of urease, invertase, amylase and dehydrogenase in soil. Soil Sci Soc Am Proc 36:536–537

Pancholy SK, Rice EL (1973) Soil enzymes in relation to old field succession: amylase, cellulase, invertase, dehydrogenase, and urease. Soil Sci Soc Am Proc 37:47–50

Pandey S, Singh DK (2006) Soil dehydrogenase, phosphomonoesterase and arginine deaminase activities in an insecticide treated groundnut (*Arachis hypogaea* L.) field. Chemosphere 63 (5):869–880

Paul EA (ed) (2007) Soil microbiology, ecology and biochemistry. Elsevier, New York

Pettit NM, Smith ARJ, Freedman RB, Burns G (1976) Soil urease: activity, stability and kinetic properties. Soil Biol Biochem 8:479–484

Phitsuwan P, Laohakunjit N, Kerdchoechuen O, Lay Kyu K, Ratanakhanokchai K (2013) Present and potential applications of cellulases in agriculture, biotechnology, and bioenergy. Folia Microbiol 58:163–176

Piotrowska-Długosz A (2014) Enzymes and soil fertility. In: Gianfreda L, Rao MA (eds) Enzymes in agricultural sciences. OMICS Group eBooks, Foster City, pp 44–79

Piotrowska-Długosz A (2017) The use of enzymes in bioremediation of soil xenobiotics. In: Hashmi M, Kumar V, Varma A (eds) Xenobiotics in the soil environment, Soil biology, vol 49. Springer, Cham, pp 243–265

Rao MA, Scelza R, Gianfreda L (2014) Soil enzymes. In: Gianfreda L, Rao MA (eds) Enzymes in agricultural sciences. OMICS Group eBooks, Foster City, pp 10–43

Rao CS, Grover M, Kundu S, Desai S (2017) Soil enzymes. Encyclopedia of soil science, 3rd edn. https://doi.org/10.1081/E-ESS3-120052906

Rasool N, Reshi ZA, Shah MA (2014) Effect of butachlor (G) on soil enzyme activity. Eur J Soil Biol 61:94–100

Reichel H, Sisler H, Kaufma D (1991) Inducers, substrates, and inhibitors of a propanildegrading amidase of Fusarium oxysporum. Pestic Biochem Physiol 39:240–250

Rodriguez-Kabana R, Godoy G, Morgan-Jones G, Shelby RA (1983) The determination of soil chitinase activity: conditions for assay and ecological studies. Plant Soil 75:95–106

Ross DJ (1965) A seasonal study of oxygen uptake of some pasture soil and activities of enzymes hydrolysing sucrose and starch. J Soil Sci 16:73–85

Ross DJ (1983) Invertase and amylase activities as influenced by clay minerals, soil-clay fractions and topsoils under grassland. Soil Biol Biochem 15:287–293

Rotanova TV, Melnikov EE, Khalatova AG, Makhovskaya OV, Botos I, Wlodawer A, Gustchina A (2004) Classification of ATP-dependent proteases Lon and comparison of the active sites of their proteolytic domains. Eur J Biochem 271:4865–4871

Russel M, Qian Y, Yao J, Choi MF (2014) Potential of glucose measurement in soil and food sample using low molecular weight *O*-(2-hydroxyl)propyl-3-trimethylammonium chitosan chloride nanoparticle-glucose oxidase immobilised on a natural fibre membrane. Int J Environ Anal Chem 94(13):1317–1329

Sakurai M, Suzuki K, Onodera M, Shinano T, Osaki M (2007) Analysis of bacterial communities in soil by PCR–DGGE targeting protease genes. Soil Biol Biochem 39:2777–2784

Sato F, Omura H, Hayano K (1986) Adenosine deaminase activity in soils. Soil Sci Plant Nutr 32:107–112

Scharlemann JPW, Tanner EVJ, Hiederer R, Kapos V (2014) Global soil carbon: understanding and managing the largest terrestrial carbon pool. Carbon Manag 5:81–91

Schimel JP, Bennet J (2004) Nitrogen mineralization: challenges of a changing paradigm. Ecology 85(30):591–602

Schinner F, von Mersi W (1990) Xylanase-, CM-cellulase- and invertase activity in soil, an improved method. Soil Biol Biochem 22:511–515

Schlesinger WH (1997) Biogeochemistry. An analysis of global change. Academic, San Diego

Scinner F, Őhlinger R, Kandeler E, Margesin R (1995) Methods in soil biology. Springer, Berlin/Heidelberg/New York

Senwo ZN, Tabatabai MA (1996) Aspartase activity of soils. Soil Sci Soc Am J 60:1416–1422

Sherene T (2017) Role of soil enzymes in nutrient transformation: a review. Bio Bull 3(1):109–131

Shinkafi SA, Adamu US, Tanyi ST (2014) Production of Laccases by fungi isolated from soil in Sokoto, Nigeria. J Microbiol Res 4(1):24–27

Shivanand P, Jayaraman G (2009) Production of extracellular protease from halotolerant bacterium, *Bacillus aquimaris* strain VITP4 isolated from Kumta coast. Process Biochem 44(10):1088–1094

Sihag SK, Singh MK, Meena RS, Naga S, Bahadur SR, Gaurav YRS (2015) Influences of spacing on growth and yield potential of dry direct seeded rice (*Oryza sativa* L.) cultivars. The Ecoscan 9 (1–2):517–519

Šimek M, Jíšová L, Hopkins DW (2002) What is the so-called optimum pH for denitrification in soil? Soil Biol Biochem 34:1227–1234

Singh DK, Kumar S (2008) Nitrate reductase, arginine deaminase, urease and dehydrogenase activities in natural soil (ridges with forest) and in cotton soil after acetamiprid treatments. Chemosphere 71:412–418

Singh J, Singh DK (2005) Available nitrogen and arginine deaminase activity in groundnut (*Arachis hypogaea* L.) fields after imidacloprid, diazinon, and lindane treatments. J Agric Food Chem 53:363–368

Singh BK, Nunan N, Millard P (2009) Response of fungal, bacterial and ureolytic communities to synthetic sheep urine deposition in a grassland soil. FEMS Microbiol Ecol 70:109–117

Sinsabaugh RL (2010) Phenol oxidase, peroxidase and organic matter dynamics of soil. Soil Biol Biochem 42:391–404

Sinsabaugh RL, Antibus RK, Linkins AE (1991) An enzymic approach to the analysis of microbial activity during plant litter decomposition. Agric Ecosyst Environ 34:43–54

Sinsabaugh RL, Antibus KR, Linkens EA, McClaugherty AC, Rayburn L, Weiland T (1992) Wood decomposition in a first order watershed: mass loss as a function of exoenzyme activity. Soil Biol Biochem 24:743–749

Sinsabaugh RL, Hill BH, Follstad Shah JJ (2009) Ecoenzymatic stoichiometry of microbial organic nutrient acquisition in soil and sediment. Nature 462(7274):795–798
Sinsabaugh RL, Belnap J, Rudgers J, Kuske CR, Martinez N, Sandquist D (2015) Soil microbial responses to nitrogen addition in arid ecosystems. Front Microbiol 6:12
Skujins J (1978) History of abiontic soil enzymes research. In: Burns RG (ed) Soil enzymes. Academic, New York, pp 1–49
Smith MS, Tiedje JM (1979) Phases of denitrification following oxygen depletion in soil. Soil Biol Biochem 11:261–267
Šturosvá M, Baldrian P (2011) Effects of soil properties and management on the activity of soil organic matter transforming enzymes and the quantification of soil-bound and free activity. Plant Soil 338(1–2):99–10
Suenaga H (2011) Targeted metagenomics: a high-resolution metagenomics approach for specific gene clusters in complex microbial communities. Environ Microbiol 14:13–22
Suresh PV, Kumar A, Sachindra NM (2011) Thermoactive β-N-acetylhexosaminidase production by a soil isolate of Penicillium monoverticillium CFR 2 under solid state fermentation: parameter optimization and application for N-acetyl chitooligosaccharides preparation from chitin. World J Microbiol Biotechnol 27(6):1435–1447
Szajdak L, Gaca W (2010) Nitrate reductase activity in soil under shelterbelt and an adjoining cultivated field. Chem Ecol 26:123–134
Tabatabai MA (1994) Soil enzymes. In: Weaver RW, Angle S, Bottomley P, Bezdicek D, Smith S, Tabatabai A, Wollum A (eds) Methods of soil analysis, part 2 – microbiological and biochemical properties. Soil Science Society of America Inc, Madison, pp 775–833
Tabatabai MA, Bremner JM (1969) Use of p-nitrophenyl phosphate for assay of soil phosphatase activity. Soil Biol Biochem 1:301–307
Tabatabai MA, Bremner JM (1970) Arylsulfatase activity of soils. Soil Sci Soc Am Proc 34:225–229
Tabatabai MA, Dick WA (2002) Enzymes in soil. In: Burns RG, Dick RP (eds) Enzymes in the environment. Activity, ecology and applications. Marcel Dekker, New York, pp 227–248
Tabatabai MA, Singh BB (1976) Rhodanase activity of soils. Soil Sci Soc Am J 40:381–385
Tabatabai MA, Ekenler M, Senwo ZN (2010) Significance of enzyme activities in soil nitrogen mineralization. Commun Soil Sci Plant Anal 4:595–605
Tate RL III (2002) Microbiology and enzymology of carbon and nitrogen cycling. In: Burns RG, Dick RP (eds) Enzymes in the environment. Activity, ecology and applications. Marcel Dekker, New York, pp 227–248
Thalenfeld B, Epstein I, Grossowicz N (1977) Relationship between culture density and catabolite repression of an inducible aliphatic amidase in a thermophilic bacillus. Biochim Biophys Acta Gen Subj 497:112–121
Theuerl S, Buscot F (2010) Laccases: toward disentangling their diversity and functions in relation to soil organic matter cycling. Biol Fertil Soils 46:215–225
van Elsas JD, Boersma FGH (2011) A review of molecular methods to study the microbiota of soil and the mycosphere. Eur J Soil Biol 47:77–87
Varma D, Meena RS, Kumar S (2017) Response of mungbean to fertility and lime levels under soil acidity in an alley cropping system in Vindhyan Region, India. Int J Chem Stud 5(2):384–389
Veening JW, Igoshin OA, Eijlander RT, Nijland R, Hamoen LW, Kuipers OP (2008) Transient heterogeneity in extracellular protease production by *Bacillus subtilis*. Mol Syst Biol 4(184). https://doi.org/10.1038/msb.2008.18
Verma JP, Jaiswal DK, Meena VS, Meena RS (2015a) Current need of organic farming for enhancing sustainable agriculture. J Clean Prod 102:545–547
Verma SK, Singh SB, Prasad SK, Meena RN, Meena RS (2015b) Influence of irrigation regimes and weed management practices on water use and nutrient uptake in wheat (*Triticum aestivum* L. Emend. Fiori and Paol.). Bangladesh J Bot 44(3):437–442
Vitousek PM, Howarth RW (1991) Nitrogen limitation on land and in the sea: how can it occur? Biogeochemistry 13:87–115

Vranová V, Rejšek K, Formánek P (2013) Proteolytic activity in soil: a review. Appl Soil Ecol 70:23–32

Wallenstein MD, Weintraub MN (2008) Emerging tools for measuring and modeling the *in situ* activity of soil extracellular enzymes. Soil Biol Biochem 40:2098–2106

Wallenstein M, Allison SD, Ernakovich J, Steinweg JM, Sinsabaugh RL (2011) Controls on the temperature sensivity of soil enzymes: a key driver of in situ enzyme activity rates. In: Shukla G, Varma A (eds) Soil enzymology. Springer, Berlin, pp 245–258

Watanabe K, Sakai J, Hayano K (2003) Bacterial extracellular protease activities infield soils under different fertilizer managements. Can J Microbiol 49:305–312

Webb EC (1992) Enzyme nomenclature. Recommendations of the Nomenclature Committee of the International Union of Biochemistry and Molecular Biology on the Nomenclature and Classification of Enzymes. Academic, San Diego

Weintraub MN, Schimel JP (2005) Seasonal protein dynamics in Alaskan arctic tundra soils. Soil Biol Biochem 37:1469–1475

Wongkaew P, Homkratoke T (2009) Enhancement of soil microbial metabolic activity in tomato field plots by chitin application. AJAFS (Special Issue):S325–S335

Xue D, Yao H, Huang C (2006) Microbial biomass, N mineralization and nitrification, enzyme activities, and microbial community diversity in tea orchard soils. Plant Soil 288:319–331

Yadav GS, Datta R, Imran Pathan S, Lal R, Meena RS, Babu S, Das A, Bhowmik S, Datta M, Saha P (2017a) Effects of conservation tillage and nutrient management practices on soil fertility and productivity of rice (*Oryza sativa* L.)–rice system in north eastern region of India. Sustainability 9(10):1816

Yadav GS, Lal R, Meena RS, Datta M, Babu S, Das A, Layek J, Saha P (2017b) Energy budgeting for designing sustainable and environmentally clean/safer cropping systems for rainfed rice fallow lands in India. J Clean Prod 158:29–37

Yadav GS, Das A, Lal R, Babu S, Meena RS, Saha P, Singh R, Datta M (2018) Energy budget and carbon footprint in a no-till and mulch based rice–mustard cropping system. J Clean Prod 191:144–157

Yang Z, Liu S, Zheng D, Feng S (2006) Effects of cadmium, zinc and lead on soil enzyme activities. J Environ Sci 18:1135–1141

Yang Y, Wu L, Lin Q, Yuan M, Xu D, Yu H, Hu Y, Duan J, Li X, He Z, Xue K, van Nostrand J, Wang S, Zhou J (2013) Responses of the functional structure of soil microbial community to livestock grazing in the Tibetan alpine grassland. Glob Chang Biol 19:637–648

Yang Y, Gao Y, Wang S, Xu D, Yu H, Wu L, Lin Q, Hu Y, Li X, He Z, Deng Y, Zhou J (2014) The microbial gene diversity along an elevation gradient of the Tibetan grassland. ISME J 8:430–440

Zanoelo FF, Polizeli TLTM, Terenzi HF, Jorge JA (2004) β-Glucosidase activity from the thermophilic fungus Scytalidium thermophilum is stimulated by glucose and xylose. FEMS Microbiol Lett 240(2):137–143

Zantua MI, Bremner JM (1975) Comparisons of methods of assaying urease activity in soils. Soil Biol Biochem 7:291–295

Zhang YL, Chen LJ, Sun CX, Wu ZJ, Chen ZH, Dong GH (2010) Soil hydrolase activities and kinetic properties as affected by wheat cropping systems of Northeastern China. Plant Soil Environ 56:526–532

Factors Affecting Soil Microbial Processes

Lucia Zifcakova

Abstract Soil is one of the most abundant environments on the Earth, where microbial processes take place, thus understanding the soil microbial processes in the context of factors influencing their environment is crucial. Soil microbial processes control soil nutrient cycling, foremost carbon cycling; therefore they affect global climate change. Organic and inorganic forms of carbon of natural or anthropogenic origin are sequestered via microbial activity into so-called soil organic matter that can be preserved in the soil for many decades. Soil microbial processes, such as carbon cycling, can be described by models emphasizing either the importance of physicochemical factors or the involvement of microbes. Balancing the carbon intake (e.g., photosynthesis) and output (e.g., decomposition) is one of the most important microbial tasks in the soil. Soil microbial processes are mediated by enzymes and thus are affected by environmental factors affecting enzymatic activities, such as temperature, water content, pH, and seasonality, but also by factors affecting diversity and abundance of microorganisms, such as nutrient availability, amount of soil organic matter, or presence of the symbiotic tree. Some microbial processes, such as N mineralization, are influenced more by abiotic factors (temperature and moisture) than the diversity of the microbial community since many groups of microbes are involved in this redundant process.

Keywords Mineralization · Physicochemical factors · Soil organic matter · Microbial activity

L. Zifcakova (✉)
Evolutionary Genomics Unit, Okinawa Institute of Science and Technology Graduate University, Onna-son, Okinawa, Japan
e-mail: lucia.zifcakova@oist.jp

R. Datta et al. (eds.), *Carbon and Nitrogen Cycling in Soil*,
https://doi.org/10.1007/978-981-13-7264-3_13

Abbreviations

°C	Degrees of Celsius
Al	Aluminum
C	Carbon
Ca	Calcium
CDI	Climate decomposition index
CH_4	Methane
CO_2	Carbon dioxide
Cu	Cuprum
CUE	Carbon use efficiency
DNDC	DeNitrification-DeComposition
DOC	Dissolved organic matter
H	Hydrogen
LIDET	Litter decay study
Mn-peroxidase	Manganese peroxidase
N	Nitrogen
N_2O	Nitrous oxide
NH_4	Ammonium
NO	Nitric oxide
$NO^-{}_2$	Nitrite
NO_3	Nitrate
NPP	Net primary production
NSP	Net secondary production
O	Oxygen
P	Phosphorus
S	Sulfur
SMP	Soil microbial processes
SOM	Soil organic matter

1 Introduction

Soil is defined by Merriam-Webster dictionary as "the superior layer of earth that can be plowed and in which plants grow," but soil is much more than that. Soil ecosystem is composed of various microhabitats that differ in physicochemical gradients and represents discontinuous environmental conditions. Due to its heterogeneity, soil serves as a medium for the growth of plants, microbes, and diverse organisms. The soil is made up of organic remains, so-called soil organic matter (SOM), clay, and rock particles. One of the soil functions is to maintain global biogeochemical cycles, which affects other biotic and abiotic components of ecosystems. Soil processes are retroactively controlled by biotic components, such as plant and microbial communities, and abiotic factors, such as temperature, water content, pH, etc.

The knowledge of the interplay between external factors and soil microbes in simplifying SMP (soil microbial processes) is critical in our understanding of how global climate changes could affect processes of the terrestrial ecosystem. There is still a significant gap in our understanding on how different biotic and abiotic factors and their interaction influence or regulate SMP, which could be due to the complexity of SMP and technical obstacle to study soil microbial community (Hackl et al. 2005; Brockett et al. 2012).

For the purposes of this chapter, we will consider four processes mediated by microbes involved in C and N cycling: SOM degradation, C sequestration, nitrification, and denitrification. These SMP are mediated by microbial enzymes, of which activity and production can be affected by external factors, such as temperature, pH, water, seasonality, and interactions among organisms.

Enzymes are one of the key drivers of soil biological process, such as organic matter degradation, mineralization, or recycling. Activity of hydrolytic enzymes, ligninolytic oxidases, and peroxidases has a direct effect on the transformation rates of soil biopolymers into substrates which are easily available to microorganisms and plants. Thus, studying soil enzyme activities is useful for evaluating the functional diversity of soil microbes, soil organic mass turnover (Kandeler et al. 1999; Yadav et al. 2017; Datta et al. 2014), or fertility of soil.

Soil enzymes are the main indicator of soil quality and health due to their quick response and sensitivity to external environmental factors (Dick 1994; Dick et al. 1996; Datta et al. 2017b). Simultaneous measurement of multiple enzyme activities can be served as a suitable indicator of soil microbial activities (Bolton et al. 1985). Such as β-glucosidase activity, catalyzing the hydrolysis of cellulose to glucose and dehydrogenase activity may be particularly useful enzymes for soil quality monitoring because of their central role in C cycling (Ceccanti et al. 1993; Doi and Ranamukhaarachchi 2009; Pathan et al. 2017).

2 Importance of Soil Organic Matter (SOM) in C Sequestration

One of the most abundant microbial processes in soil mediated by extracellular enzymes is degradation of either plant litter, microbial necromass, or other inputs, including leachates and exudates from different sources. There is a consortium of microorganisms that are degrading and utilizing the majority of C compounds created by NPP (net primary production). NPP turnover supplies energy and forms blocks for heterotrophs to build their biomass termed NSP (net secondary production). NSP could be used in the process of decomposition which takes days to decades and depends on temperature, moisture, and the quality of the live and senesced biomass.

A small fraction of plant (NPP) and heterotrophic decomposer constituents (NSP) are converted into soil organic matter (SOM) that could be persevered for many decades and is an imperative and stable C pool, making up a significant proportion of terrestrial C stocks. Although SOM is mostly a small fraction of the soil, it regulates air and water availability for plant root growth and provides the resistance against wind and water erosion. Organic matter content in different soils ranges from 0.2% to 80%, respectively, in desert and peat soils. In temperate regions, it ranges between 0.4% and 10.0%, with soils of humid region averaging 3–4% and those in semiarid areas 1–3%. Soil C stocks are created in a process called carbon sequestration, during which CO_2 is removed from the atmosphere via photosynthesis and stored soil carbon pool in the form of SOM. The different elements of NPP added to the soil differ significantly as a source of energy and nutrients reflecting their biochemical composition and physical availability to the microorganisms (Wardle and Giller 1996).

The SOM can be separated into two fractions depending on their biological degradability: (1) rapid to medium turnover fraction and (2) recalcitrant fraction with slow turnover. The first one is composed of soluble compounds with small molecular mass and serves as immediate C sources for the soil biota, thus contributing to nutrient cycling. The latter fraction is a complex combination of humic and fulvic acids with different high molecular weight organic molecules attached to soil inorganic particles, represents sequestered C and thus the energy reservoir, and improves soil structure as well (Simpson et al. 2007; Schmidt et al. 2011). Carbon polymers, including hemicellulose, pectins, and cellulose, make up to 50% of NPP inputs in terrestrial environments. These components are structural part of plant cell wall and contain macronutrients, mainly N and P. Cytoplasmic components of plant cells, for example, sugars, organic acids, amino compounds, and proteins, provide energy and essential nutrients for decomposition and form up to 10% dry weight of the plant. After the degradation process, a small fraction of C NPP and NSP is preserved in soil in the form of humic substances by metabolism process or associate with other soil minerals and protected by soil aggregates. SOM thus comprised of distorted decayed plant residues, soil microbes, soil fauna, and by-products of degradation, such as humic substances. Humic substances are results of long oxidation and reduction, causing the material to be increased in C and H but depleted in O content, compared to the original one. During decomposition, N content of humic substances is increased because N compounds react through radical coupling with other compounds, and thus humic substances consist of 50–55% C, 5% H, 33% O, 4.5% N, 1% S, and 1% P. Metals and micronutrients, such as Al, Ca, Zn, and Cu, are also exist but in much smaller amounts. The dominance of aliphatic compounds derived from microbial cell walls in SOM (Schurig et al. 2012) suggests that microbial biomass contribute significantly to stable C pools.

Fungi and bacterial share on total soil biomass is approximately 90% (Rinnan and Bååth 2009), and thus turnover of their necromass is evaluated to contribute as much as 80% to the preservation and accumulation of SOM (Liang and Balser 2010). Throckmorton et al. (2012) hypothesized that cellular biochemistry of different microbes determines the form and amount of C designated to form stable SOM. Martin and Haider (1979) suggested that C stored in SOM is mainly of fungal origin compared to other microbial groups, due to their composite cell walls and pigments that are resistant to decomposition. Rinnan and Bååth (2009) supported this hypothesis by the fact that fungi have higher C-use efficiency (CUE) compared to bacteria that can lead to higher involvement of fungal C to stable SOM. Yet, the overlapping ranges of CUEs for some soil fungi and bacteria question this theory (Six et al. 2006). The composition of the bacterial cell wall is likely to influence its decomposition by soil microorganisms because Gram-positive bacteria contain more peptidoglycan, which is associated with slower decomposition, than Gram-negative bacteria. Though the structure of peptidoglycan in cell wall differs among bacterial species and with growth conditions, what makes it difficult to predict is its decomposability (Vollmer et al. 2008). Due to the limited field-based assessment on comparing bacterial and fungal turnover, our understanding is scarce on the contribution of different microbial groups to SOM (Strickland and Rousk 2010). Throckmorton et al. (2012) reported that the involvement of various cellular biochemistry of main microbial groups contributed evenly to maintaining of SOM, but the results were more dependent on the abundance of microbial groups rather than their unique cellular biochemistries.

Among others, the decomposition processes are regulated by temperature, moisture, soil disturbance, xenobiotics, the quality of SOM as a microbial substrate (Smith and Paul 1990; Smith 1994; Molaei et al. 2017a, b), and microbial community composition (Aber et al. 1990; Couteaux et al. 1995; Fassnacht and Gowerr 1999; Park and Matzner 2003; Pregitzer et al. 2004).

3 Models of Soil Microbial Processes (SMP) Involved in SOM Degradation Are Either Process- or Organism-Oriented

C flux is directly or indirectly controlled by soil organisms through the degradation process. The relative contributions of microbes to CO_2 release vs. C storage in soil are of great interest. The CUE of the organism is the amount of CO_2 lost per unit of energy gained, and environmental conditions can impact CUE (Six et al. 2006). Nutrients in specific ratios or their lack can modify the amount of energy spent to decompose SOM. Nutrient availability, substrate quality, and temperature (del Giorgio and Cole 1998) impact the CUE of soil organisms. Cotrufo et al. (2013) suggested that microbial efficiency should be modeled as a function of substrate characteristics, community structure, and environment. This would widen our understanding of soil microbiota impact on CO_2 flux, SOM retention, C pool composition, and assembly, as well as an improved our knowledge of energy transformations in the microbial community. Multicompartmental models of SOM decomposition dynamics can be either "process-oriented" or "organism-oriented" (Paustian 1994).

Organism-oriented models, also known as "food web models," emphasize diverse functional or taxonomic groups of soil organisms in the description of the flow of organic matter and nutrients (Moore and de Ruiter 2012). Instead of concentrating on the specific organism's activity or group, process-oriented models emphasis on the processes mediating the transformations of organic matter and nutrients.

Most of SOM decomposition dynamics models start by modeling litter decay on the soil surface. The assumption of these models is that plant litter comprises both readily decomposable fraction and recalcitrant fraction composed of cellulose and lignin. Hence, the SOM and litter decay can be divided into different pools based on stabilization mechanisms, bioavailability, and biochemical and kinetic parameters, consisting of small "active" pool with a rapid turnover rate and larger pool with slow turnover rates from decades to thousands of years. Plant lignocellulose ration is positively correlated with plant biomass recalcitrant fraction to decomposition. Many litter decay models work with microbial community as variable and also presume that the majority of plant lignin (>70%) will be transformed into organic material. Plant residue decay is well described by first-order rate kinetics that suggests that inoculation of soil microbes is not limiting factor for degradation rate.

LIDET (litter decay study) (Parton et al. 2007) data was used to assess global ecosystem models. Bonan et al. (2012) have used LIDET data (litter decay data) from study of Parton et al. (2007) to determine global ecosystem models and reported that models should also consider the initial litter N, lignin, labile C content, and effect of climate decomposition index (CDI) to precisely characterize litter decomposition dynamics. The most correlated variable was CDI since it embraces the seasonal patterns of temperature and moisture. From other factors influencing the results of model, the starting N content had a strong impact on N dynamics. Microbial N immobilization during the initial phase of decay (>50% of initial C remaining) was more favorable when N litter content was low (<0.8% N), while high N litter content (>1.5% N) resulted in the immediate release of simultaneous C and N during litter decay.

Many models (Schimel 2001) have coupled soil C decay to microbial biomass and physiology and contain the influence of microbial activity on SOM decay rates as well (Allison et al. 2010). General hypothesis in these models is that extracellular enzymes regulate the decomposition of SOM to dissolved organic matter (DOC) and that DOC availability regulates depolymerization of SOM. Another assumption in these models is that the production of enzymes is equivalent to the amount of microbial biomass. These models utilize Michaelis-Menten equation with the maximum reaction rate, microbial uptake (V_{max}), and half-saturation constant (K_m) being the primary input variables that can represent enzyme reaction rates and microbial uptake of DOC. Other soil environmental factors (water, temperature, soil pH, N, and P) can affect the enzymes production rate and their influence on the decomposition rate of SOM pools (Sinsabaugh and Shah 2012).

Additional relevant factor in substrate-enzyme-microbe models is the ratio of microbial growth to C processing costs known as microbial C-use efficiency (CUE). Substrate-enzyme-microbe models presume that CUE is influenced by the soil environmental variables and the SOM pool, whereas conventional SOM models usually use fixed values for CUE. Thus, conventional SOM models can be improved by counting the influence of soil environmental factors, microbial activity, and enzyme production on CUE.

Nevertheless, how will soil microbial communities react to external variables is also influenced by the type of soil ecosystem. When soil microbial community from the different forests was compared with different climate zones, it was reported that SMP is significantly influenced by soil water content (Brockett et al. 2012). In contrast, it was shown that soil organic carbon one of the main factors affecting soil microbial community function and structure under different types of vegetation (Grayston and Prescott 2005; Yao et al. 2006; Franklin and Mills 2009; Katsalirou et al. 2010). Other studies suggest that soil chemical properties, such as soil C/N ratio (Fierer et al. 2009), nutrient status (Lauber et al. 2008), and soil pH (Rousk et al. 2009), are highly correlated with soil microbial community composition and functioning. Some other studies suggest that composition soil microbial community in forest significantly influenced by the chemistry of plant litter (Ushio et al. 2008; Strickland and Rousk 2010) and spatial pattern of soil properties (Ushio et al. 2010) and these changes in the composition have a direct impact on the functioning of soil microbes.

Moreover, Tilman (1995) reported that biodiversity is regulating the SMP rates and if aboveground species diversity increase could lead to an increase in ecosystem stability. Klironomos et al. (2000) suggested that the presence/absence of arbuscular mycorrhizal fungi changed the relationship of plant biodiversity to aboveground productivity. Diversity and C cycling are significantly correlated to each other during a decrease in diversity, and when diversity increases, C cycling increased. Nevertheless, with high diversity, species-specific traits became more influential than numbers of species. Microbial community functioning can also alter soil chemistry directly via processes, which increase nutrient availability, such as P solubilization and N fixation, and/or alter SOM decomposition rates.

Morris and Blackwood (2015) suggested that availability of a diverse range of organic compounds to varied organisms with a wide range of enzymes could lead to functional redundancy of the microbial community. Experimental studies over the last years have challenged this assumption. Strickland et al. (2009) found differences in C mineralization rates on a community level, using diverse communities, proposing that each combination of microbial communities provided a unique set of metabolic physiologies resulting in different rates. These studies are also supported by metagenomic approaches evaluating metabolic gene diversity (Röling et al. 2010). Changes in the composition of soil microbial community are prone to result in changes of microbial functioning, thus altering SMP (Waldrop and Firestone 2006). For example, increased abundance of microbes producing hydrolytic enzymes that facilitate C acquirement will support the primary metabolism (Cusack et al. 2011), but rise in the production of oxidative enzymes, mainly by saprophytic fungi, will result in higher decomposition of complex compounds (Sylvia et al. 2004). An understanding of the interplay between the function and structure of the microbial community is necessary for estimating the effect of shifts in the structure of microbial community on changes in SMP (Weand et al. 2010). The ability to identify specific soil microbial features driving SOM transformations will expand our mechanistic understanding on how soil C sink and C sequestration work (Lucas et al. 2007; Acosta-Martınez et al. 2010). Soil bacterial community regulates SOM storage in soil by the increase in the C acquisition activity. In contrast, saprophytic fungi are active in SOM turnover because they produce enzymes involved in the oxidation of C compounds.

A wide variety of soil microorganisms are able to produce extracellular enzymes, and some of these enzymes indicate the presence of certain microbial groups (Baldrian 2009). For example, ligninolytic enzymes, such as lignin peroxidase and

Mn-peroxidase, are produced only by saprotrophic Basidiomycetous fungi (Hofrichter 2002; Baldrian 2008; Datta et al. 2017a). Enzymes involved in cellulose and lignin decomposition are the most widely assayed enzymes (Cusack et al. 2011). Other commonly assayed enzymes produced by a wide variety of microorganisms are those involved in the hydrolysis of proteins, chitin, and peptidoglycan, making organic N, S, and P accessible for microorganisms (Caldwell 2005). It was discovered that the relative abundance of particular arbuscular mycorrhizal fungi and Gram-negative bacteria was correlated with activities of hydrolytic enzymes involved in acquisition of C by microorganisms (cellobiohydrolase and β-glucosidase), whereas the relative higher abundance of the saprophytic fungi was associated with the specific activities of enzymes involved in lignin (phenol oxidase and peroxidase) and chitin (N-acetylglucosaminidase) degradation (Colpaert and Laere 1996; Courty et al. 2008, Miller et al. 1998; Burke et al. 2011). On the other hand, Gram-positive bacteria were positively associated with cellobiohydrolases that are involved in cellulose degradation (Waldrop et al. 2000; Bell et al. 2009). It was found that bacteria respond most quickly to additions of simple C compounds such as sugars, starch, and amino acids, while fungi and filamentous bacteria – actinomycetes – dominated when complex C compounds such as lignin and cellulose were added to the beech litter (Moller et al. 1999; Datta et al. 2017c).

Other factors related to the soil fertility, such as SOM content, NH_4, NO_3 and C to N ratio, were correlated with the structure of the soil microbial community. Bacteria are usually found SOM rich soils, while the richness of saprophytic fungi rises with degrade soil (Grayston et al. 2004; Grayston and Prescott 2005; Franklin and Mills 2009; Katsalirou et al. 2010; Wu et al. 2011). The C to N ratio in litter and soil was positively correlated with the occurrence of saprophytic fungi (Högberg et al. 2006; Fierer et al. 2009; You et al. 2014). Abundance of bacteria is high clay soil, while the abundance of saprophytic fungi decreased in clay-rich (Högberg et al. 2006; Lamarche et al. 2007; Fierer et al. 2009; You et al. 2014). Some recent studies suggested that the soil microbial community structure is significantly affected by soil pH (Högberg et al. 2006; Sinsabaugh et al. 2008). On the other hand, it was showed that composition of plant community is a better predictor of variations in microbial community composition than the soil properties, which is mostly due to dependence of litter quality and amount on the plant species, which in turn affects soil physico-chemical properties (Mitchell et al. 2010; Thoms et al. 2010).

4 Interplay Between Photosynthesis and Decomposition

The relative rates of C uptake via photosynthesis vs. C release in the process of autotrophic and heterotrophic respiration = decomposition represent the fluxes in the global C cycle in terrestrial ecosystems. The rates of SMP are significantly impacted by interactions among soil microorganisms and by their interactions with plants. For example, mycorrhizal symbiosis between fungus and plant increases photosynthetic rates, mainly under stress conditions such as water or nutrient restrictions.

Decomposition rates are influenced by competition for resources among decomposers, predation on decomposers, and changes in living conditions.

Decomposition and photosynthesis are key ecosystem processes; thus, individual plant species differentially impact the composition of the soil food webs (Bezemer et al. 2010; Rout and Callaway 2012; Wolfe et al. 2008; Shamina et al. 2018) as well as N turnover rates, which were influenced by plant diversity (Bezemer et al. 2010). Vice versa, the alterations of the microbial community can decrease the flow of nutrients to plants and thus decrease the flow of energy to the microbial community as well.

4.1 *Effect of Seasonality on SMP*

In terrestrial ecosystems, microbial communities are significantly influenced by dominant primary producers – plants. Plants provide not only novel niches for the microbial communities to thrive but most importantly C and N for microbial growth in the form of plant detritus used by saprotrophs, root exudates available for the symbionts, and root-associated microbial communities. Energy input into soil microbial communities highly rely on NPP. Thus the amount of microbial biomass that can be supported in soil depends on plant contributions through root exudates, leaf, or root litter. In temperate zone, photosynthesis associated with the rhizodeposition of easily decomposable C compounds into the soil, either directly or through the root-associated mycorrhizal fungi, is limited to the vegetation period of spring and summer (Ekblad and Högberg 2001). Approximately 30% of the NPP is allocated to roots and soil via root exudates (Beidler et al. 2014). The allocation of C into the soil via plant roots shows several-fold seasonal changes corresponding to the change in intensity of photosynthesis throughout the year (Högberg et al. 2010). Seasonality can be found in plant carbon balance that is positive in the summer due to higher photosynthesis than respiration but negative in winter due to respiration and low photosynthesis (Ryan 1991). Carbon in the form of root exudates derived from plant photosynthates is rapidly consumed by microorganisms, which highlights short-term dynamics in degradation by microbial soil communities (Bellemain et al. 2012). Ten to fifty percent of all assimilated C of plant origin is translocated into mycelia of ectomycorrhizal fungi. Thus they play a role in soil carbon storage, and carbon sink in the boreal forest is driven by these fungi (Orgiazzi et al. 2016).

Seasonality that is represented by changes in plant growth, temperature, and precipitation affects the structure and abundance of microbial community (Högberg et al. 2010; Kaiser et al. 2010; Voriskova et al. 2013). Such seasonal changes in the composition of microbial community and function were observed in the mixed temperate forest (Zhang et al. 2014) and in the Arctic ecosystem (Mundra et al. 2015). One of these changes was the dominance of saprotrophic fungi in spring that was correlated to spring fine root turnover (Satomura et al. 2006), and ectomycorrhizal ones in late summer, when the maximal growth of spruce fine roots occurred (Stober et al. 2000) in temperate (Jumpponen et al. 2010; Voriskova et al. 2013; Wallander et al. 2001) and boreal forests (Davey et al. 2012; Santalahti et al. 2016). Fungal richness and diversity increased more than three times between spring and summer in *Quercus petraea* forest soil (Voriskova et al. 2013). The increase in fungal richness in the Arctic environment was found to correlate with the increase in soil temperature (Mundra et al. 2015), which was showed to be a growth-limiting factor for ectomycorrhizal fungi in these environments (Robinson 2001; Timling and Taylor 2012). Spring snowmelt was correlated to the decrease of fungal biomass and increase of Gram-positive bacteria and Actinobacteria biomass in the soil of Alpine tundra soil, while the Gram-negative bacteria were abundant in summer in the same environment (Bjork et al. 2008). In contrast to the fungal community, bacterial community structure in the soil was responsive to a summer peak of rhizodeposition in a temperate oak forest (Lopez-Mondejar et al. 2015). Seasonal shifts of the relative abundances of individual bacterial groups were found in the alpine soils (Lipson 2007; Kuffner et al. 2012) and were connected to C fluctuations in plant roots. A metaproteomic study in coniferous and deciduous forest showed that the fungi to bacteria ratio have increased in spring compared to winter (Schneider et al. 2012). It was also found that fungi produced more than half the transcribed enzymes involved in SOM degradation, especially in summer in the temperate coniferous forest (Zifcakova et al. 2015, 2017). Enzymes involved in breakdown of complex polysaccharides (endocellulases and endoxylanases) and those decomposing fungal cell wall (N-acetylglucosaminidases) were more active in summer, while cellobiohydrolases involved in cellulose degradation were active in spring (Baldrian et al. 2013). Results of Zifcakova et al. (2017) showed increase in the transcription of enzymes that involved fungal biomass turnover in summer, whereas expression of other compounds such as starch or trehalose is increased during the winter season. Seasonality has a significant influence on gene in soil compared to litter and transcription of the ligninolytic, and cellulolytic enzyme increased during the summer than the winter. Winter communities of microorganisms produced more cellulases and amylases and thus were able to decompose complex carbon substrates, as indicated by decreased mineralization of SOM, while summer communities were able to utilize glucose more effectively since there was the higher availability of dissolved organic carbon in summer than in winter (Koranda et al. 2013). In temperate forest, seasonal differences in the enzyme pools with maxima in summer were found for N-mineralization and denitrification enzymes, but the pool of β-glucosidases enzymes present in most microorganisms did not show any regular seasonal pattern (Rastin et al. 1988; Bohlen et al. 2001) but their transcription varied between summer and winter season (Pathan et al. 2017). Activities of phenol oxidases and peroxidases were highest in late summer, while activities of cellulases and proteases peaked in winter in beech forest soil (Kaiser et al. 2010). The structure of the fungal community of cellulases producers was different in the summer and winter, and it was also suggested that lignin breakdown starts later in summer with the increase of Basidiomycota in metaproteomic data (Schneider et al. 2012). At least in the tundra, the main factor influencing seasonal differences in enzyme activities was temperature (Wallenstein et al. 2009). The total annual enzyme activity in the boreal coniferous forest was 7–32% in winter while 68–93% in summer (Wittmann et al. 2004). Overall, there are evidences not only for the seasonal change in enzyme activities but also for seasonal shifts in abundance of saprotrophic and mycorrhizal fungi also shifts in the bacterial community composition.

4.2 Temperature Dependence of SMP

Correlation of SMP and temperature is complex because individual microbes vary in their optimal temperature, and thus diverse soil microbes may be active at various temperatures. At the top of it, divergent microorganisms have distinctive abilities to adapt to the temperature by changing their physiology, such as membrane fluidity and permeability and structural flexibility of proteins, including enzymes. Due to temperature dependence of enzymatic reactions and biological processes, temperature is one of the key factors affecting SMP. Rise in the temperature by 10 °C will increase the activity of most enzymes by 50 to 100% (Martinek 1969).

The relative temperature sensitivity of microbial activity can be indicated as a Q10 function that is essentially the change in activity proportional to change of temperature about 10 °C and is used to explain the temperature sensitivity of SMP, such as respiration of soil microorganisms. It is generally accepted that microbial activity at 30 °C is twice as high as at 20 °C and activity of soil microbes is usually greatest within 20–40 °C. The metabolic activity of the most microbes decreases drastically around 5 °C referred to as biological zero.

Though activity of microbes is lower at lower temperatures, SMP rates are much higher and more sensitive to temperature changes than predicted from mesophilic temperature range studies. For example, values of Q10 for decomposition of SOM, soil respiration, and N mineralization were quite high, near 8–10, when soil temperature was around 0 °C (Kirschbaum 2013). Microbial activity in SMP during cold periods with dormant plants and barren soil plays a crucial role in the winter losses of soil nutrients, such as N leaching and denitrification, especially during freeze/thaw cycles.

Due to the influence of temperature and moisture on microbes, it is clear that SMP will be modified with climate change; however, it is not yet certain which processes will decrease or increase. It is certain that alterations of nutrient's mineralization rates that are needed by plants and microbes will change ecosystem productivity. Whether the rates of SMP increase or decrease can depend on the changes in temperature and moisture and their impacts on microbial efficiencies but also on the selection of microbial species under the new conditions.

Net flux of CO_2 to the atmosphere is thought to rise over time under most models because microbial decomposition of SOM shall increase with higher temperature and moisture, predominantly in Arctic ecosystems. In the last 100 years, Earth global temperature is increased by 0.5 °C and will be increased by 1 °C–6 °C by 2100, predicted by different model studies. Even though it is only a few degrees' increases, global warming will intensely increase microbial decay rates of the SOM stored in the boreal forests and tundra regions, which contain 30% of the global soil C (Kirschbaum 1995).

Currently, there are studies available that are dealing with the relative contributions of soil microbes to C flux, C-use efficiency, the effect of elevated level of CO_2, and climate change on these fluxes. The main concern is that SOM decomposition is much more accelerated than in NPP representing the C input to SOM. Theory also implies that the decomposition of recalcitrant SOM compounds, such as cellulose and hemicellulose that are usually a rate-limiting step in CO_2 emissions, would become essentially easier at higher temperatures (Davidson and Janssens 2006). More CO_2 atmosphere can positively affect NPP via C fertilization and increased water use efficiency. In addition to losses in soil C, it is anticipated that rising CO_2 will increase emissions of CH_4 and N_2O formed by increased root growth and lowered soil water losses (van Groenigen et al. 2011).

Though, feedback mechanisms that are representative for all biogeochemical fluxes may inhibit the impacts of temperature changes. Soils are complex ecosystems affected by factors, such as change in soil water storage, nutrient cycling, and rainfall patterns that will have an impact on mostly on NPP.

Many of the environmental factors have an influence on decomposition by changing effective SOM (substrate) concentrations at the site of enzymatic reaction, where decomposition occurs. Thus one of the factors to consider in SOM decomposition rates are enzyme affinity levels. Other external factors that are considered in models of the effects of global warming on C cycling are kinetic and thermodynamic properties of extracellular enzymes.

Temperature indeed affects the kinetics of enzymatic reactions but also changes microbial community composition. 5 °C increase in temperate forest soil influences relative abundance of the bacterial community which leads to high bacterial to fungal ration (DeAngelis et al. 2015). Microbial communities react to global warming, and other ecosystem disturbances through resistance, which is facilitated by the plasticity of microbial traits, or via resilience as the community returns back to its initial compositions of species after the stress is gone (Allison and Martiny 2008). Changes in the composition of soil microbial communities are thought to mediate changes in SMP, assuming that a special group of soil microorganisms is different in their functional traits or control a rate-limiting step of SMP (Schimel and Schaeffer 2012). For example, specific microbial groups govern ecosystem functions such as methanogenesis (Bodelier et al. 2000), denitrification (Bakken et al. 2012; Salles et al. 2012), N fixation, and nitrification (Isobe et al. 2011). Changes in the richness of one group of microorganisms that regulating specific processes can have a straightforward impact or influence on the process rate, conversely, some processes occurring at a cruder scale, for example, N mineralization, are more correlated with abiotic factors (temperature and moisture) than composition of microbial community as wide variety of microorganisms is involved in these processes (Hooper et al. 2005).

4.3 Nitrification and Denitrification Models of SMP

Soil macronutrient cycles are strongly connected via microbes' nutrient demands at the time of decomposition, so few of the multicompartmental models of SOM decomposition focusing on C cycle are also able to predict the fluxes of other macronutrients such as N, P, and S. Flows within the N cycle are mainly driven by N fixation (capture of atmospheric N_2 to forms usable for the microbes), mineralization (represented by nitrification and ammonification) of organic N from plant and animal necromass and biomass, and gaseous loses via denitrification and ammonia volatilization. The microbes drive important processes in N cycle, so mutualistic relationships between plants and soil microorganisms, such as *Frankia* (phylum *Actinobacteria*) and *Rhizobium* (class *Alphaproteobacteria*), are very important. In soil systems, where organic N is not yet available, microbial N fixation delivers the initial N source allowing plants to grow. With the increase in plant production, the most N in the ecosystem will originate from the decomposition of plant litter by microorganisms. Such accessible N can either be assimilated by plants or by soil microorganisms via immobilization process when N becomes part of microbial rather than plant cells.

The DNDC (DeNitrification-DeComposition) model simulates plant growth and soil processes (Li et al. 1992) and has few submodules. The nitrification submodule simulates the nitrification rate, the turnover rates of nitrifiers, as well as N_2O and NO productions and is controlled by temperature, moisture, ammonium, and DOC (dissolved organic carbon) concentrations. Denitrification submodule is influenced by soil temperature, moisture, and substrates (DOC, $NO^-{}_3$, $NO^-{}_2$, NO, and N_2O) concentrations and can predict changes in denitrification process, as well as changes in the size of the population of denitrifiers. The fluxes of N_2O and NO^- induced by denitrification are calculated dynamically from soil aeration status, gas diffusion, and substrate limitation. As a source of NO^-production in soils, chemo-denitrification is often considered, and it is dependent on soil pH and nitrite availability. Nitrification occurs mainly in the aerobic fraction of soil, while denitrification is preferred in the anaerobic environment. Denitrification rates can be expressed by Nernst (redox potential) and Michaelis-Menten (enzyme kinetics) equations. When the anaerobic conditions in soil are common and favorable, few processes can happen: (1) more substrates (DOC, $NO^-{}_3$, $NO^-{}_2$, NO, or N_2O) will be allocated to the N pool, (2) rates of sequential denitrification reactions will increase, and (3) the intermediate product gases (N_2O, NO, etc.) will take longer to diffuse from the anaerobic to the aerobic fraction, increasing the rate of N gases being reduced to N_2. The overall effect will be the loss of N from the soil. If N is limiting nutrient in the soil, the microbes will "win" the competition for N between plants and microbes, which will limit the amount of N available for plants and thus decrease NPP and litter quality (van der Heijden et al. 2008). The most of nitrogen found naturally in soil was a product of either N fixation by free-living or symbiotic microbes or of microbial decomposition of organic materials. This does not apply nowadays because anthropogenically generated N is entering soil ecosystems via fertilization and pollutant dispersal and this has resulted in two times increase in the amount of N available for plants. Such nitrogen additions boost soil respiration, reduce microbial biomass, and change enzyme activity in many studied soils, implying significant effect of these N supplements on the soil microbe functions (Ramirez et al. 2012).

4.4 Importance of Soil Water Content in SMP

Another factor influencing not only N and C cycles but all SMP is the soil water content. Soil water influences not only the moisture available to microorganisms and osmotic pressure but also soil aeration status, the solubility of organic materials, and the pH as a function of the soil solution. Physically, water is a transportation agent by mass flow but also a solvent, where enzymatic and chemical reactions happen. Water retention in soil is facilitated by water adsorbing via hydrogen bonding and dipole interactions to soil particles, and thus it is a function of the size of pores in the soil. In soils, where water content is non-limiting, biological activity depends mainly on the temperature, which can be predicted by standard Arrhenius theory. However, when soils dry out, moisture becomes a greater determining factor of SMP than temperature. It is likely that moisture and temperature do not impact the microbial activity in a linear manner, but in complex, nonlinear fashion that reflects the responses of individual microorganisms and their enzyme activities.

Even though the many microbes are capable of tolerating soil stress by accumulation of amino acids and polyols (osmolytes) or altering their outer membrane, soil microbes are significantly affected by rapid dry-wetting cycles and undergo osmotic shocks and induce cell lysis. Following such catastrophic event, there is often a peak in the activity of surviving microbes, called the Birch effect, which is caused by mineralization of the released content of microbial cells.

Further, the lack of soil moisture amplifies the differences in temperature sensitivity of bacterial and fungal community (Briones et al. 2014). Another difference between fungi and bacteria toward the effect of moisture is that bacterial communities respond rapidly to moisture pulses, while the slower-growing fungal community delays in their feedback (Bell et al. 2008; Cregger et al. 2012; Cregger et al. 2014). On the other hand, fungal communities may shift in dominant representatives even with small changes in soil moisture availability ($<30\%$ reduction in water holding capacity), while the representatives of bacterial communities do not change. These observations indicate a higher plasticity of fungal community during wet-dry cycles

(Kaisermann et al. 2015), but soil communities that are adapted to wet-dry cycles or to low water availability will show less functional and compositional changes (Evans et al. 2011). The soil moisture plays a crucial role in S and N cycles as well. For example, sulfur (*Thiobacillus* sp.) and ammonium oxidizers (*Nitrosomonas* sp.) are less tolerant of water stress than are the ammonifiers (*Clostridium* sp., *Penicillium* sp.). Ammonium can pile up in dry soils at the water potentials when ammonification is still possible, but nitrification is restricted, which results in decreased soil pH affecting SMP but also changing the microbial community composition.

4.5 Soil pH as One of the Factors Influencing SMP

A measuring of pH of soil solutions presents a necessary approach allowing to predict of reactions of microbes involved in SMP and enzyme activity in soil. Although pH is easily measured in soil solution, it could be difficult to interpret due to concentrations of cations that are sorbed to the negatively charged soil surfaces and are 10–100 times higher than ones of the soil solution. It has implication for enzyme activity measurements in soil because enzyme sorbed to colloid surfaces in soil have 1–2 pH units' lower optimum as the same and not sorbed enzyme (Marfo et al. 2015 Lojkova et al. 2015).

Although certain microbes can alter pH by acidifying soil in their vicinity to the disadvantage of competitors, the most diverse composition of soil bacterial populations is found near-neutral pH. Acidity, on the other hand, enhances the activity of soil fungi, and it explains why fungi dominate in forested soils, which are acidic, while bacteria usually prevail in rangeland soils and in mildly acidic subhumid to semiarid prairie. Fungi can tolerate low pH and are able to decompose recalcitrant compounds, unlike bacteria, which are thought to be limited by low pH and less enzymatic equipment and have higher requirements for some nutrients and lower tolerance of environmental changes (Allison and Martiny 2008). A pH was found to be the most important factor in determining bacterial community composition (Högberg et al. 2006), and thus *Acidobacteria* and *Alphaproteobacteria* are highly abundant in acidic soils (Bryant et al. 2008; Jones et al. 2009; Baldrian et al. 2011; Shen et al. 2013), but on the other hand, the amount of *Bacteroidetes* and *Actinobacteria* increase with more basic pH (Lauber et al. 2008; Lauber et al. 2009; Jeanbille et al. 2015). In addition, the abundance of *Acidobacteria* in soil is negatively correlated with the dissolved organic carbon availability, which indicates they are slow-growing oligotrophs and are most probably adapted to nutrient limitation (Naether et al. 2012; Garcia-Fraile et al. 2015). *Acidobacteria* were suggested to be very adaptable to environmental modifications due to the high metabolic versatility that allows them to use even highly complex C substrates originated from SOM (Rasche et al. 2010; Naether et al. 2012).

5 Conclusions

Overall, the research suggests the existence of complex interactions between the abiotic and biotic factors that affect the functioning of soil microbial communities in SOM transformations via changes in the allocation of plant-derived C to microbial communities and through modifications of the fungal and bacterial community structure activities.

Particularly, it was found that both soil water and temperature are important drivers of changes in soil microbial community structure (Hackl et al. 2005; Djukic et al. 2010; Brockett et al. 2012). The presence of soil water was positively correlated with the abundance of Gram-negative bacteria, while soil temperature was positively linked with the abundance of saprophytic fungi but negatively with the bacterial community abundance (You et al. 2014). Structure of the soil microbial community was also profoundly affected by SOM, fine root mass, clay content, and C/N ratio. In addition, the relative abundance of Gram-negative bacteria, saprophytic fungi, and actinomycetes was enough to explain most of the variations in the soil enzymes activities involved in SOM transformations (You et al. 2014). The abundance of fungi was found to be associated with activity of enzymes involved in C oxidations, while the abundance of bacteria was linked to activity of extracellular enzymes participating in C transformation (You et al. 2014; Zifcakova et al. 2017). Research findings demonstrate the existence of complex interplay among soil physiochemical properties, soil microenvironment, and plant traits in the decomposition of SOM via regulations in microbial communities. Moreover, external factors that affect the structure of soil microbial communities have also direct/indirect impact on their functioning in soil microbial processes.

References

Aber JD, Melillo JM, McClaugherty CA (1990) Predicting long-term patterns of mass loss, nitrogen dynamics, and soil organic matter formation from initial fine litter chemistry in temperate forest ecosystems. Can J Bot 68(10):2201–2208

Acosta-Martínez V, Bell CW, Morris BEL, Zak J, Allen VG (2010) Long-term soil microbial community and enzyme activity responses to an integrated cropping-livestock system in a semi-arid region. Agric Ecosyst Environ 137(3–4):231–240

Allison SD, Martiny JBH (2008) Resistance, resilience, and redundancy in microbial communities. Proc Natl Acad Sci 105(Supplement 1):11512–11519

Allison SD, Wallenstein MD, Bradford MA (2010) Soil-carbon response to warming dependent on microbial physiology. Nat Geosci 3(5):336–340

Bakken LR, Bergaust L, Liu B, Frostegard A (2012) Regulation of denitrification at the cellular level: a clue to the understanding of N2O emissions from soils. Philos Trans R Soc B Biol Sci 367(1593):1226–1234

Baldrian P (2008) Chapter 2: Enzymes of saprotrophic basidiomycetes, British Mycological Society symposia series, vol 28. Academic, Oxford, pp 19–41

Baldrian P (2009) Microbial enzyme-catalyzed processes in soils and their analysis. Plant Soil Environ 55(9):370–378

Baldrian P, Kolařík M, Štursová M, Kopecký J, Valášková V, Větrovský T, Žifčáková L, Šnajdr J, Rídl J, Vlček Č, Voříšková J (2011) Active and total microbial communities in forest soil are largely different and highly stratified during decomposition. ISME J 6(2):248–258
Baldrian P, Šnajdr J, Merhautová V, Dobiášová P, Cajthaml T, Valášková V (2013) Responses of the extracellular enzyme activities in hardwood forest to soil temperature and seasonality and the potential effects of climate change. Soil Biol Biochem 56:60–68
Beidler KV, Taylor BN, Strand AE, Cooper ER, Schönholz M, Pritchard SG (2014) Changes in root architecture under elevated concentrations of CO2 and nitrogen reflect alternate soil exploration strategies. New Phytol 205(3):1153–1163
Bell C, McIntyre N, Cox S, Tissue D, Zak J (2008) Soil microbial responses to temporal variations of moisture and temperature in a Chihuahuan desert grassland. Microb Ecol 56(1):153–167
Bell CW, Acosta-Martinez V, McIntyre NE, Cox S, Tissue DT, Zak JC (2009) Linking microbial community structure and function to seasonal differences in soil moisture and temperature in a Chihuahuan desert grassland. Microb Ecol 58(4):827–842
Bellemain E, Davey ML, Kauserud H, Epp LS, Boessenkool S, Coissac E, Geml J, Edwards M, Willerslev E, Gussarova G, Taberlet P, Brochmann C (2012) Fungal palaeodiversity revealed using high-throughput metabarcoding of ancient DNA from arctic permafrost. Environ Microbiol 15(4):1176–1189
Bezemer TM, Fountain MT, Barea JM, Christensen S, Dekker SC, Duyts H, van Hal R, Harvey JA, Hedlund K, Maraun M, Mikola J, Mladenov AG, Robin C, de Ruiter PC, Scheu S, Setälä H, Šmilauer P, van der Putten WH (2010) Divergent composition but similar function of soil food webs of individual plants: plant species and community effects. Ecology 91(10):3027–3036
Björk RG, Björkman MP, Andersson MX, Klemedtsson L (2008) Temporal variation in soil microbial communities in Alpine tundra. Soil Biol Biochem 40(1):266–268
Bodelier PLE, Roslev P, Henckel T, Frenzel P (2000) Stimulation by ammonium-based fertilizers of methane oxidation in soil around rice roots. Nature 403(6768):421–424
Bohlen PJ, Groffman PM, Driscoll CT, Fahey TJ, Siccama TG (2001) Plant-soil-microbial interactions in a northern hardwood forest. Ecology 82(4):965
Bolton H, Elliott LF, Papendick RI, Bezdicek DF (1985) Soil microbial biomass and selected soil enzyme activities: effect of fertilization and cropping practices. Soil Biol Biochem 17 (3):297–302
Bonan GB, Hartman MD, Parton WJ, Wieder WR (2012) Evaluating litter decomposition in earth system models with long-term litterbag experiments: an example using the Community Land Model version 4 (CLM4). Glob Chang Biol 19(3):957–974
Briones MJI, McNamara NP, Poskitt J, Crow SE, Ostle NJ (2014) Interactive biotic and abiotic regulators of soil carbon cycling: evidence from controlled climate experiments on peatland and boreal soils. Glob Chang Biol 20(9):2971–2982
Brockett BFT, Prescott CE, Grayston SJ (2012) Soil moisture is the major factor influencing microbial community structure and enzyme activities across seven biogeoclimatic zones in western Canada. Soil Biol Biochem 44(1):9–20
Bryant JA, Lamanna C, Morlon H, Kerkhoff AJ, Enquist BJ, Green JL (2008) Microbes on mountainsides: contrasting elevational patterns of bacterial and plant diversity. Proc Natl Acad Sci 105(Supplement 1):11505–11511
Burke DJ, Weintraub MN, Hewins CR, Kalisz S (2011) Relationship between soil enzyme activities, nutrient cycling and soil fungal communities in a northern hardwood forest. Soil Biol Biochem 43(4):795–803
Caldwell BA (2005) Enzyme activities as a component of soil biodiversity: a review. Pedobiologia 49(6):637–644
Ceccanti B, Pezzarossa B, Gallardo-Lancho FJ, Masciandaro G (1993) Biotests as markers of soil utilization and fertility. Geomicrobiol J 11(3–4):309–316
Colpaert JV, Laere A (1996) A comparison of the extracellular enzyme activities of two ectomycorrhizal and a leaf-saprotrophic basidiomycete colonizing beech leaf litter. New Phytol 134(1):133–141

Cotrufo MF, Wallenstein MD, Boot CM, Denef K, Paul E (2013) The Microbial Efficiency-Matrix Stabilization (MEMS) framework integrates plant litter decomposition with soil organic matter stabilization: do labile plant inputs form stable soil organic matter? Glob Chang Biol 19 (4):988–995

Cou^teaux M-M, Bottner P, Berg B (1995) Litter decomposition, climate and liter quality. Trends Ecol Evol 10(2):63–66

Courty PE, Franc A, Pierrat JC, Garbaye J (2008) Temporal changes in the ectomycorrhizal community in two soil horizons of a temperate oak forest. Appl Environ Microbiol 74 (18):5792–5801

Cregger MA, Schadt CW, McDowell NG, Pockman WT, Classen AT (2012) Response of the soil microbial community to changes in precipitation in a semiarid ecosystem. Appl Environ Microbiol 78(24):8587–8594

Cregger MA, Sanders NJ, Dunn RR, Classen AT (2014) Microbial communities respond to experimental warming, but site matters. Peer J 2:e358

Cusack DF, Silver WL, Torn MS, Burton SD, Firestone MK (2011) Changes in microbial community characteristics and soil organic matter with nitrogen additions in two tropical forests. Ecology 92(3):621–632

Datta R, Vranová V, Pavelka M, Rejšek K, Formánek P (2014) Effect of soil sieving on respiration induced by low-molecular-weight substrates. Int Agrophys 28(1):119–124

Datta R, Kelkar A, Baraniya D, Molaei A, Moulick A, Meena R, Formanek P (2017a) Enzymatic degradation of lignin in soil: a review. Sustainability 9(7):1163

Datta R, Anand S, Moulick A, Baraniya D, Pathan SI, Rejsek K, Vranova V, Sharma M, Sharma D, Kelkar A (2017b) How enzymes are adsorbed on soil solid phase and factors limiting its activity: a review. Int Agrophys 31(2):287–302

Datta R, Baraniya D, Wang Y-F, Kelkar A, Meena RS, Yadav GS, Teresa Ceccherini M, Formanek P (2017c) Amino acid: its dual role as nutrient and scavenger of free radicals in soil. Sustainability 9(8):1402

Davey ML, Heegaard E, Halvorsen R, Ohlson M, Kauserud H (2012) Seasonal trends in the biomass and structure of bryophyte-associated fungal communities explored by 454 pyrosequencing. New Phytol 195(4):844–856

Davidson EA, Janssens IA (2006) Temperature sensitivity of soil carbon decomposition and feedbacks to climate change. Nature 440(7081):165–173

DeAngelis KM, Pold G, TopÃ§uoÄŸlu BD, van Diepen LTA, Varney RM, Blanchard JL, Melillo J, Frey SD (2015) Long-term forest soil warming alters microbial communities in temperate forest soils. Front Microbiol 6:104

del Giorgio PA, Cole JJ (1998) Bacterial growth efficiency in natural aquatic systems. Annu Rev Ecol Syst 29(1):503–541

Dick RP (1994) Soil enzyme activity as indicators of soil quality. In: Doran JW, Coleman DC, Bezdicek DF, Stewart BA (eds) Defining soil quality for a sustainable environment. Soil Science Society of America, Madison, pp 107–124

Dick RP, Breakwell DP, Turco RF (1996) Soil enzyme activities and biodiversity measurements as integrative microbiological indicators. In: Doran JW, Jones AJ (eds) Methods for assessing soil quality. Soil Science Society of America, Madison, pp 247–271

Djukic I, Zehetner F, Mentler A, Gerzabek MH (2010) Microbial community composition and activity in different Alpine vegetation zones. Soil Biol Biochem 42(2):155–161

Doi R, Ranamukhaarachchi SL (2009) Correlations between soil microbial and physicochemical variations in a rice paddy: implications for assessing soil health. J Biosci 34(6):969–976

Ekblad A, Högberg P (2001) Natural abundance of 13C in CO2 respired from forest soils reveals speed of link between tree photosynthesis and root respiration. Oecologia 127(3):305–308

Evans SE, Byrne KM, Lauenroth WK, Burke IC (2011) Defining the limit to resistance in a drought-tolerant grassland: long-term severe drought significantly reduces the dominant species and increases ruderals. J Ecol 99(6):1500–1507

Fassnacht KS, Gowerr ST (1999) Comparison of the litterfall and forest floor organic matter and nitrogen dynamics of upland forest ecosystems in north central Wisconsin. Biogeochemistry 45 (3):265–284

Fierer N, Strickland MS, Liptzin D, Bradford MA, Cleveland CC (2009) Global patterns in belowground communities. Ecol Lett 12(11):1238–1249

Franklin RB, Mills AL (2009) Importance of spatially structured environmental heterogeneity in controlling microbial community composition at small spatial scales in an agricultural field. Soil Biol Biochem 41(9):1833–1840

García-Fraile P, Benada O, Cajthaml T, Baldrian P, Lladó S (2015) Terracidiphilus gabretensis gen. nov., sp. nov., an abundant and active forest soil acidobacterium important in organic matter transformation. Appl Environ Microbiol 82(2):560–569

Grayston SJ, Prescott CE (2005) Microbial communities in forest floors under four tree species in coastal British Columbia. Soil Biol Biochem 37(6):1157–1167

Grayston SJ, Campbell CD, Bardgett RD, Mawdsley JL, Clegg CD, Ritz K, Griffiths BS, Rodwell JS, Edwards SJ, Davies WJ, Elston DJ, Millard P (2004) Assessing shifts in microbial community structure across a range of grasslands of differing management intensity using CLPP, PLFA and community DNA techniques. Appl Soil Ecol 25(1):63–84

Hackl E, Pfeffer M, Donat C, Bachmann G, Zechmeisterboltenstern S (2005) Composition of the microbial communities in the mineral soil under different types of natural forest. Soil Biol Biochem 37(4):661–671

Hofrichter M (2002) Review: lignin conversion by manganese peroxidase (MnP). Enzym Microb Technol 30(4):454–466

Högberg MN, Högberg P, Myrold DD (2006) Is microbial community composition in boreal forest soils determined by pH, C-to-N ratio, the trees, or all three? Oecologia 150(4):590–601

Högberg MN et al (2010) Quantification of effects of season and nitrogen supply on tree belowground carbon transfer to ectomycorrhizal fungi and other soil organisms in a boreal pine forest. New Phytol 187(2):485–493

Hooper DU, Chapin FS, Ewel JJ, Hector A, Inchausti P, Lavorel S, Lawton JH, Lodge DM, Loreau M, Naeem S, Schmid B, Setälä H, Symstad AJ, Vandermeer J, Wardle DA (2005) Effects of biodiversity on ecosystem functioning: a consensus of current knowledge. Ecol Monogr 75(1):3–35

Isobe K, Koba K, Otsuka S, Senoo K (2011) Nitrification and nitrifying microbial communities in forest soils. J For Res 16(5):351–362

Jeanbille M, Buée M, Bach C, Cébron A, Frey-Klett P, Turpault MP, Uroz S (2015) Soil parameters drive the structure, diversity and metabolic potentials of the bacterial communities across temperate beech forest soil sequences. Microb Ecol 71(2):482–493

Jones RT, Robeson MS, Lauber CL, Hamady M, Knight R, Fierer N (2009) A comprehensive survey of soil acidobacterial diversity using pyrosequencing and clone library analyses. ISME J 3(4):442–453

Jumpponen ARI, Jones KL, David Mattox J, Yaege C (2010) Massively parallel 454-sequencing of fungal communities in Quercus spp. ectomycorrhizas indicates seasonal dynamics in urban and rural sites. Mol Ecol 19:41–53

Kaiser C, Frank A, Wild B, Koranda M, Richter A (2010) Negligible contribution from roots to soil-borne phospholipid fatty acid fungal biomarkers 18:2ω6,9 and 18:1ω9. Soil Biol Biochem 42 (9):1650–1652

Kaisermann A, Maron PA, Beaumelle L, Lata JC (2015) Fungal communities are more sensitive indicators to non-extreme soil moisture variations than bacterial communities. Appl Soil Ecol 86:158–164

Kandeler E, Tscherko D, Spiegel H (1999) Long-term monitoring of microbial biomass, N mineralisation and enzyme activities of a Chernozem under different tillage management. Biol Fertil Soils 28(4):343–351

Katsalirou E, Deng S, Nofziger DL, Gerakis A, Fuhlendorf SD (2010) Spatial structure of microbial biomass and activity in prairie soil ecosystems. Eur J Soil Biol 46(3–4):181–189

Kirschbaum MUF (1995) The temperature dependence of soil organic matter decomposition, and the effect of global warming on soil organic C storage. Soil Biol Biochem 27(6):753–760

Kirschbaum MUF (2013) Seasonal variations in the availability of labile substrate confound the temperature dependence of organic matter decomposition. Soil Biol Biochem 57:568–576

Klironomos JN, McCune J, Hart M, Neville J (2000) The influence of arbuscular mycorrhizae on the relationship between plant diversity and productivity. Ecol Lett 3(2):137–141

Koranda M, Kaiser C, Fuchslueger L, Kitzler B, Sessitsch A, Zechmeister-Boltenstern S, Richter A (2013) Seasonal variation in functional properties of microbial communities in beech forest soil. Soil Biol Biochem 60:95–104

Kuffner M, Hai B, Rattei T, Melodelima C, Schloter M, Zechmeister-Boltenstern S, Jandl R, Schindlbacher A, Sessitsch A (2012) Effects of season and experimental warming on the bacterial community in a temperate mountain forest soil assessed by 16S rRNA gene pyrosequencing. FEMS Microbiol Ecol 82(3):551–562

Lamarche J, Bradley RL, Hooper E, Shipley B, Simao Beaunoir AM, Beaulieu C (2007) Forest floor bacterial community composition and catabolic profiles in relation to landscape features in Quebec's southern boreal forest. Microb Ecol 54:10–20

Lauber CL, Strickland MS, Bradford MA, Fierer N (2008) The influence of soil properties on the structure of bacterial and fungal communities across land-use types. Soil Biol Biochem 40 (9):2407–2415

Lauber CL, Hamady M, Knight R, Fierer N (2009) Pyrosequencing-based assessment of soil pH as a predictor of soil bacterial community structure at the continental scale. Appl Environ Microbiol 75(15):5111–5120

Li C, Frolking S, Frolking TA (1992) A model of nitrous oxide evolution from soil driven by rainfall events: 1. Model structure and sensitivity. J Geophys Res Atmos 97(D9):9759–9776

Liang C, Balser TC (2010) Microbial production of recalcitrant organic matter in global soils: implications for productivity and climate policy. Nat Rev Microbiol 9(1):75–75

Lipson DA (2007) Relationships between temperature responses and bacterial community structure along seasonal and altitudinal gradients. FEMS Microbiol Ecol 59(2):418–427

Lojkova L, Datta R, Sajna M, Marfo TD, Janous D, Pavelka M, Formanek P (2015) Limitation of proteolysis in soils of forests and other types of ecosystems by diffusion of substrate. In: Amino acids, vol 8. Springer, Wien, pp 1690–1691

López-Mondéjar R, Voříšková J, Větrovský T, Baldrian P (2015) The bacterial community inhabiting temperate deciduous forests is vertically stratified and undergoes seasonal dynamics. Soil Biol Biochem 87:43–50

Lucas RW, Casper BB, Jackson JK, Balser TC (2007) Soil microbial communities and extracellular enzyme activity in the New Jersey Pinelands. Soil Biol Biochem 39(10):2508–2519

Marfo TD, Datta R, Lojkova L, Janous D, Pavelka M, Formanek P (2015) Limitation of activity of acid phosphomonoesterase in soils. In: Amino acids, vol 8. Springer, Wien, pp 1690–1691

Martin JP, Haider K (1979) Biodegradation of 14C-labeled model and cornstalk lignins, phenols, model phenolase humic polymers, and fungal melanins as influenced by a readily available carbon source and soil. Appl Environ Microbiol 38(2):283–289

Martinek R (1969) Practical clinical enzymology: J. Am Med Tech 31:162

Miller M, Palojärvi A, Rangger A, Reeslev M, Kjøller A (1998) The use of fluorogenic substrates to measure fungal presence and activity in soil. Appl Environ Microbiol 64(2):613–617

Mitchell RJ, Hester AJ, Campbell CD, Chapman SJ, Cameron CM, Hewison RL, Potts JM (2010) Is vegetation composition or soil chemistry the best predictor of the soil microbial community? Plant Soil 333(1–2):417–430

Molaei A, Lakzian A, Datta R, Haghnia G, Astaraei A, Rasouli-Sadaghiani M, Ceccherini MT (2017a) Impact of chlortetracycline and sulfapyridine antibiotics on soil enzyme activities. Int Agrophys 31(4):499–505

Molaei A, Lakzian A, Haghnia G, Astaraei A, Rasouli-Sadaghiani M, Ceccherini MT, Datta R (2017b) Assessment of some cultural experimental methods to study the effects of antibiotics on microbial activities in a soil: an incubation study. PLoS One 12(7):e0180663

Møller J, Miller M, Kjøller A (1999) Fungal–bacterial interaction on beech leaves: influence on decomposition and dissolved organic carbon quality. Soil Biol Biochem 31(3):367–374

Moore JC, de Ruiter PC (2012) Energetic food webs. Oxford University Press, Oxford
Morris SJ, Blackwood CB (2015) Chapter 10 - the ecology of the soil biota and their function in soil microbiology, ecology and biochemistry. Elsevier, pp 273–309
Mundra S, Bahram M, Tedersoo L, Kauserud H, Halvorsen R, Eidesen PB (2015) Temporal variation of *Bistorta vivipara* -associated ectomycorrhizal fungal communities in the High Arctic. Mol Ecol 24(24):6289–6302
Naether A, Foesel BU, Naegele V, Wüst PK, Weinert J, Bonkowski M, Alt F, Oelmann Y, Polle A, Lohaus G, Gockel S, Hemp A, Kalko EKV, Linsenmair KE, Pfeiffer S, Renner S, Schöning I, Weisser WW, Wells K, Fischer M, Overmann J, Friedrich MW (2012) Environmental factors affect acidobacterial communities below the subgroup level in grassland and forest soils. Appl Environ Microbiol 78(20):7398–7406
Orgiazzi A, Bardgett RD, Barrios E, Behan-Pelletier V, Briones MJI, Chotte JL, de Deyn GB, Eggleton P, Fierer N, Fraser T, Hedlund K, Jeffery S, Johnson NC, Jones A, Kandeler E, Kaneko N, Lavelle P, Lemanceau P, Miko L, Montanarella L, Moreira FMS, Ramirez KS, Scheu S, Singh BK, Six J, van der Putten WH, Wall DH (2016) Global soil biodiversity atlas. European Union, Luxembourg
Park J-H, Matzner E (2003) Controls on the release of dissolved organic carbon and nitrogen from a deciduous forest floor investigated by manipulations of aboveground litter inputs and water flux. Biogeochemistry 66(3):265–286
Parton W, Silver WL, Burke IC, Grassens L, Harmon ME, Currie WS, King JY, Adair EC, Brandt LA, Hart SC, Fasth B (2007) Global-scale similarities in nitrogen release patterns during long-term decomposition. Science 315(5810):361–364
Paustian K (1994) Modelling soil biology and biochemical processes for sustainable agriculture research. In: Pankhurst CE, Doube DM, VVSR G, Grace PR (eds) Soil Biota: Management in sustainable farming systems. CSIRO Publishing, Collingwood, pp 182–193
Pathan SI, Žifčáková L, Ceccherini MT, Pantani OL, Větrovský T, Baldrian P (2017) Seasonal variation and distribution of total and active microbial community of β-glucosidase encoding genes in coniferous forest soil. Soil Biol Biochem 105:71–80
Pregitzer KS, Zak DR, Burton AJ, Ashby JA, MacDonald NW (2004) Chronic nitrate additions dramatically increase the export of carbon and nitrogen from northern hardwood ecosystems. Biogeochemistry 68(2):179–197
Ramirez KS, Craine JM, Fierer N (2012) Consistent effects of nitrogen amendments on soil microbial communities and processes across biomes. Glob Chang Biol 18(6):1918–1927
Rasche F, Knapp D, Kaiser C, Koranda M, Kitzler B, Zechmeister-Boltenstern S, Richter A, Sessitsch A (2010) Seasonality and resource availability control bacterial and archaeal communities in soils of a temperate beech forest. ISME J 5(3):389–402
Rastin N, Rosenplänter K, Hüttermann A (1988) Seasonal variation of enzyme activity and their dependence on certain soil factors in a beech forest soil. Soil Biol Biochem 20(5):637–642
Rinnan R, Baath E (2009) Differential utilization of carbon substrates by bacteria and fungi in tundra soil. Appl Environ Microbiol 75(11):3611–3620
Robinson CH (2001) Cold adaptation in Arctic and Antarctic fungi. New Phytol 151(2):341–353
Röling WFM, Ferrer M, Golyshin PN (2010) Systems approaches to microbial communities and their functioning. Curr Opin Biotechnol 21(4):532–538
Rousk J, Brookes PC, Baath E (2009) Contrasting soil pH effects on fungal and bacterial growth suggest functional redundancy in carbon mineralization. Appl Environ Microbiol 75 (6):1589–1596
Rout ME, Callaway RM (2012) Interactions between exotic invasive plants and soil microbes in the rhizosphere suggest that 'everything is not everywhere'. Ann Bot 110(2):213–222
Ryan MG (1991) Effects of climate change on plant respiration. Ecol Appl 1(2):157–167
Salles JF, Le Roux X, Poly F (2012) Relating phylogenetic and functional diversity among denitrifiers and quantifying their capacity to predict community functioning. Front Microbiol 3:209

Santalahti M, Sun H, Jumpponen A, Pennanen T, Heinonsalo J (2016) Vertical and seasonal dynamics of fungal communities in boreal scots pine forest soil. FEMS Microbiol Ecol 92 (11):fiw170
Satomura T, Hashimoto Y, Koizumi H, Nakane K, Horikoshi T (2006) Seasonal patterns of fine root demography in a cool-temperate deciduous forest in central Japan. Ecol Res 21(5):741–753
Schimel J (2001) Biogeochemical models: implicit versus explicit microbiology. In: Schulze E-D, Harrison SP, Heimann M et al (eds) Global biogeochemical cycles in the climate system. Academic, San Diego, pp 177–183
Schimel JP, Schaeffer SM (2012) Microbial control over carbon cycling in soil. Front Microbiol 3:348
Schmidt MWI, Torn MS, Abiven S, Dittmar T, Guggenberger G, Janssens IA, Kleber M, Kögel-Knabner I, Lehmann J, Manning DAC, Nannipieri P, Rasse DP, Weiner S, Trumbore SE (2011) Persistence of soil organic matter as an ecosystem property. Nature 478(7367):49–56
Schneider T, Keiblinger KM, Schmid E, Sterflinger-Gleixner K, Ellersdorfer G, Roschitzki B, Richter A, Eberl L, Zechmeister-Boltenstern S, Riedel K (2012) Who is who in litter decomposition? Metaproteomics reveals major microbial players and their biogeochemical functions. ISME J 6(9):1749–1762
Schurig C, Smittenberg RH, Berger J, Kraft F, Woche SK, Goebel M-O, Heipieper HJ, Miltner A, Kaestner M (2012) Microbial cell-envelope fragments and the formation of soil organic matter: a case study from a glacier forefield. Biogeochemistry 113(1–3):595–612
Shamina Imran Pathan TV, Giagnoni L, Datta R, Baldrian P, Nannipieri P, Renella G (2018) Microbial expression profiles in the rhizosphere of two maize lines differing in N use efficiency. Plant Soil 433:401–413
Shen C, Xiong J, Zhang H, Feng Y, Lin X, Li X, Liang W, Chu H (2013) Soil pH drives the spatial distribution of bacterial communities along elevation on Changbai Mountain. Soil Biol Biochem 57:204–211
Simpson AJ, Simpson MJ, Smith E, Kelleher BP (2007) Microbially derived inputs to soil organic matter: are current estimates too low? Environ Sci Technol 41(23):8070–8076
Sinsabaugh RL, Follstad Shah JJ (2012) Ecoenzymatic stoichiometry and ecological theory. Annu Rev Ecol Evol Syst 43(1):313–343
Sinsabaugh RL, Lauber CL, Weintraub MN, Ahmed B, Allison SD, Crenshaw C, Contosta AR, Cusack D, Frey S, Gallo ME, Gartner TB, Hobbie SE, Holland K, Keeler BL, Powers JS, Stursova M, Takacs-Vesbach C, Waldrop MP, Wallenstein MD, Zak DR, Zeglin LH (2008) Stoichiometry of soil enzyme activity at global scale. Ecol Lett 11(11):1252–1264
Six J, Frey SD, Thiet RK, Batten KM (2006) Bacterial and fungal contributions to carbon sequestration in agroecosystems. Soil Sci Soc Am J 70(2):555
Smith JL (1994) Cycling of nitrogen through microbial activity. In: Hatfield JL, Stewart BA (eds) Soil biology: effects on soil quality. CRC Press, Boca Raton, pp 91–120
Smith JL, Paul EA (1990) The significance of soil microbial biomass estimations. In: Bollag GM, Stotzkey G (eds) Soil biochemistry, vol 6. Marcel Dekker, New York, pp 357–396
Stober C, George E, Persson H (2000) Root growth and response to nitrogen. Ecol Stud 142:99–121
Strickland MS, Rousk J (2010) Considering fungal:bacterial dominance in soils – methods, controls, and ecosystem implications. Soil Biol Biochem 42(9):1385–1395
Strickland MS, Lauber C, Fierer N, Bradford MA (2009) Testing the functional significance of microbial community composition. Ecology 90(2):441–451
Sylvia D, Fuhrmann J, Hartel P, Zuberer D (2004) Principles and applications of soil microbiology, 2nd edn. Prentice Hall, Upper Saddle River
Thoms C, Gattinger A, Jacob M, Thomas FM, Gleixner G (2010) Direct and indirect effects of tree diversity drive soil microbial diversity in temperate deciduous forest. Soil Biol Biochem 42 (9):1558–1565
Throckmorton HM, Bird JA, Dane L, Firestone MK, Horwath WR (2012) The source of microbial C has little impact on soil organic matter stabilisation in forest ecosystems. Ecol Lett 15 (11):1257–1265

Tilman D (1995) Biodiversity: population versus ecosystem stability. Ecology 77(2):350–363
Timling I, Taylor DL (2012) Peeking through a frosty window: molecular insights into the ecology of Arctic soil fungi. Fungal Ecol 5(4):419–429
Ushio M, Wagai R, Balser TC, Kitayama K (2008) Variations in the soil microbial community composition of a tropical montane forest ecosystem: does tree species matter? Soil Biol Biochem 40(10):2699–2702
Ushio M, Kitayama K, Balser TC (2010) Tree species-mediated spatial patchiness of the composition of microbial community and physicochemical properties in the topsoils of a tropical montane forest. Soil Biol Biochem 42(9):1588–1595
van der Heijden MGA, Bardgett RD, van Straalen NM (2008) The unseen majority: soil microbes as drivers of plant diversity and productivity in terrestrial ecosystems. Ecol Lett 11(3):296–310
van Groenigen KJ, Osenberg CW, Hungate BA (2011) Increased soil emissions of potent greenhouse gases under increased atmospheric CO2. Nature 475(7355):214–216
Vollmer W, Blanot D, De Pedro MA (2008) Peptidoglycan structure and architecture. FEMS Microbiol Rev 32(2):149–167
Voříšková J, Brabcová V, Cajthaml T, Baldrian P (2013) Seasonal dynamics of fungal communities in a temperate oak forest soil. New Phytol 201(1):269–278
Waldrop MP, Firestone MK (2006) Seasonal dynamics of microbial community composition and function in oak canopy and open grassland soils. Microb Ecol 52:470–479
Waldrop MP, Balser TC, Firestone MK (2000) Linking microbial community composition to function in a tropical soil. Soil Biol Biochem 32(13):1837–1846
Wallander H, Nilsson LO, Hagerberg D, Baath E (2001) Estimation of the biomass and seasonal growth of external mycelium of ectomycorrhizal fungi in the field. New Phytol 151(3):753–760
Wallenstein MD, McMahon SK, Schimel JP (2009) Seasonal variation in enzyme activities and temperature sensitivities in Arctic tundra soils. Glob Chang Biol 15(7):1631–1639
Wardle DA, Giller KE (1996) The quest for a contemporary ecological dimension to soil biology. Soil Biol Biochem 28(12):1549–1554
Weand MP, Arthur MA, Lovett GM, McCulley RL, Weathers KC (2010) Effects of tree species and N additions on forest floor microbial communities and extracellular enzyme activities. Soil Biol Biochem 42(12):2161–2173
Wittmann C, Kähkönen MA, Ilvesniemi H, Kurola J, Salkinoja-Salonen MS (2004) Areal activities and stratification of hydrolytic enzymes involved in the biochemical cycles of carbon, nitrogen, sulphur and phosphorus in podsolized boreal forest soils. Soil Biol Biochem 36(3):425–433
Wolfe BE, Rodgers VL, Stinson KA, Pringle A (2008) The invasive plantAlliaria petiolata (garlic mustard) inhibits ectomycorrhizal fungi in its introduced range. J Ecol 96(4):777–783
Wu J, Liu Z, Wang X, Sun Y, Zhou L, Lin Y, Fu S (2011) Effects of understory removal and tree girdling on soil microbial community composition and litter decomposition in two Eucalyptus plantations in South China. Funct Ecol 25(4):921–931
Yadav GS, Datta R, Imran Pathan S, Lal R, Meena RS, Babu S, Das A, Bhowmik S, Datta M, Saha P (2017) Effects of conservation tillage and nutrient management practices on soil fertility and productivity of rice (*Oryza sativa* L.)–rice system in north eastern region of India. Sustainability 9(10):1816
Yao H, Bowman D, Shi W (2006) Soil microbial community structure and diversity in a turfgrass chronosequence: land-use change versus turfgrass management. Appl Soil Ecol 34 (2–3):209–218
You Y, Wang J, Huang X, Tang Z, Liu S, Sun OJ (2014) Relating microbial community structure to functioning in forest soil organic carbon transformation and turnover. Ecol Evol 4(5):633–647
Zhang X, Wang W, Chen W, Zhang N, Zeng H (2014) Comparison of seasonal soil microbial process in snow-covered temperate ecosystems of Northern China. PLoS One 9(3):e92985
Žifčáková L, Větrovský T, Howe A, Baldrian P (2015) Microbial activity in forest soil reflects the changes in ecosystem properties between summer and winter. Environ Microbiol 18 (1):288–301
Žifčáková L, Větrovský T, Lombard V, Henrissat B, Howe A, Baldrian P (2017) Feed in summer, rest in winter: microbial carbon utilization in forest topsoil. Microbiome 5(1):122

Carbon and Nitrogen Mineralization Dynamics: A Perspective in Rice-Wheat Cropping System

Kirti Saurabh, Rakesh Kumar, J. S. Mishra, Hansraj Hans, Narendra Kumawat, Ram Swaroop Meena, K. K. Rao, Manoj Kumar, A. K. Dubey, and M. L. Dotaniya

Abstract Rice-wheat cropping system (RWCS), one of the prominent agricultural production systems, at an area of ~26 M ha is confined to the Indo-Gangetic Plains (IGPs) in South Asia and China. Crop residues obtained from field crops are essential sources of nutrition and organic carbon (40% of total dry biomass constituted by C) for the next crops, and hence they not only increase the agricultural productivity but also are responsible for the better quality of soil, water, and air. Perhaps the most important challenge facing exhaustive RWCS in all regions of the world is effective management of post-harvest crop residues. Disposal of wheat residue is easy as it can be used to feed animals. However, due to the presence of high silica content, rice residue is usually burned. Residue burning is the main method of disposal in areas under combined harvesting in the IGPs of eastern India as it reduces cost. However, burning of crop residue (CR) is not eco-friendly as it results in fast degradation of soil organic matter and nutrients and increased CO_2 emission creating intense air pollution as well as global warming. Therefore, exploitation of CR is a crucial element for a sustainable production system, and it has generated much interest in the recent years by reducing the consequence of residue burning and increasing the soil organic matter (SOM) and the nutrient-supplying capacity. CR retention infield can be considered a key element in promoting soil health with increased physical, chemical, and biological properties. In RWCS, residue management can be done by (1) wheat residue retention in rice and its residual effect in succeeding wheat crop, (2) rice straw retention in wheat and its residual impact in following rice, and (3) wheat straw retention in rice and rice straw

---

K. Saurabh · R. Kumar (✉) · J. S. Mishra · H. Hans · K. K. Rao · M. Kumar · A. K. Dubey
Division of Crop Research, ICAR Research Complex for Eastern Region, Patna, India

N. Kumawat
AICRP on Management of Salt Affected Soils and Use of Saline Water in Agriculture, College of Agriculture, Indore, Madhya Pradesh, India

R. S. Meena
Department of Agronomy, Institute of Agricultural Sciences (BHU), Varanasi, UP, India
e-mail: Meenars@bhu.ac.in

M. L. Dotaniya
ICAR-Indian Institute of Soil Science, Bhopal, India

R. Datta et al. (eds.), *Carbon and Nitrogen Cycling in Soil*,
https://doi.org/10.1007/978-981-13-7264-3_14

retention in wheat (cumulative effect). All these crop residue management systems depend on a systematic understanding of the factors that control residue decomposition and their careful application. Significant factors, such as tillage/CR management, influence soil microbial activity and biomass, bulk density, soil moisture content, porosity, soil structure stability, and nutrient-supplying capacity of the soil. Thus, the variations in soil properties consequently bring change in soil C and N dynamics and have an impact on plants' nutrient uptake capacity.

Furthermore, residue quality and quantity are found to affect C and N mineralization rates. Plant remains with higher quality (high N contents; low ratios of C/N; low lignin, cellulose, and polyphenol contents; and lignin/N) show high C decomposition and N mineralization rates. In this way residue retention leads to enhancing nutrient balances and better crop yield. However, there is a requisite to study decomposition and nutrient dynamics in RWCS soil under different residue management system, so that accurate composition of integrated nutrient management (INM) can be developed for this prominent system.

Keywords Carbon · Nitrogen · Crop residues · Residue decomposition · Nutrient dynamics

Abbreviations

AMF	Arbuscular mycorrhizal fungi
CEC	Cation exchange capacity
CR	Crop residue
CRM	Crop residue management
FDA	Fluorescein diacetate
GHG	Greenhouse gas
GM	Green manuring
GWP	Global warming potentials
IGPs	Indo-Gangetic Plains
INM	Integrated nutrient management
MWD	Mean weight diameter
NT	No-tillage
OM	Organic matter
PRQI	Plant residue quality index
RT	Reduced tillage
RWCS	Rice-wheat cropping system
SCS	Soil carbon storage
SMB	Soil microbial biomass
SMBC	Soil microbial biomass carbon
SOC	Soil organic carbon
SOM	Soil organic matter
TN	Total nitrogen

WSA	Water-stable aggregates
ZT	Zero tillage

1 Introduction

Being the most important and major cropping system of IGPs, rice and wheat are harnessing enormous soil fertility; therefore, to maintain the productivity of this RWCS, replenishment of soil nutrients is necessary (Subehia and Sepehya 2012; Kumar et al. 2013, 2017d). The RWCS is a heavy feeder of the nutrients. It has not only resulted in the removal of major nutrients (NPKS) from the soil but also caused a nutrient imbalance, leading to a decline in the soil health and quality. For instance, deficiencies of NPK are most widespread. In the production of 1 ton of wheat grains, the estimated removal of 24.5, 3.8, and 27.3 kg N, P, and K, respectively, is found, whereas there is a removal of 20.1 kg N, 4.9 kg P, and 25.0 kg K with similar production of rice grains (Tandon and Sekhon 1988; Bohra and Kumar 2015). In India, Punjab alone produces ~20.8 Mt and 23.3 Mt of rice and wheat residue annually, respectively (Dhar et al. 2014; Chand et al. 2017). Clearance of CRs by burning is often criticized as it is associated with the making of poor soil and air quality by accelerating the losses of soil organic matter (SOM), increasing CO_2 emission, causing extreme air pollution, and reducing microbial activity in the soil (Choudhary et al. 2013; Bhanwaria et al. 2013). Most of the residue from the wheat field is removed and utilized as animal silage, while most of rice residue (82%) is cleared by burning in the field before growing the next crop (Sodhi et al. 2009). Reddy et al. (2003) studied that integrated use of plant nutrients with mineral fertilizers has either maintained or enhanced soil quality or improved performance of the crop along with the cropping system. Therefore, proper crop residue management (CRM) can increase SOM and nutrient-supplying capacity of the soil, reducing the effects of the environmentally hazardous process, i.e., residue burning, as this leads to the destruction of SOM as well as plant nutrients, i.e., NPKS (Mandal et al. 2004). Addition of CRs in soil is beneficial in many ways as it improves the soil nutrients and water holding capacity by returning the good quantity of carbon to the soil (Jeet et al. 2010), which improves its structure (Granatstein et al. 2017). Residues are a source of many nutrients like K, which are not available in inorganic fertilizers. Therefore, integration of residues along with green manure can be a good option. The use of renewable N and C sources, i.e., green manures and CRs in RWCS of South Asia, may lead to increased grain yield, SOC, and reduced soil bulk density (Naresh et al. 2017).

Apart from cattle feed, the crop residues can be used for numerous purposes like a bed for animals, thatching substance for houses, and fuel energy. The beginning of the mechanized harvesting compelled farmers to burn large quantities of in situ crop residues as these hinder with seeding and tillage operations for the succeeding crops (Ali and Nabi 2016). This practice has a vital role in recycling of nutrients contained in crop residues and organic matter content of the soil. In this way, CRs are wonderful natural resources, not waste materials that require disposal. Narang and Virmani (2001) analyzed factors liable for a sign of a yield plateau in RWCS of IGP of India; in this regard, one important priority job is to develop agro-technology to reutilize massive quantity of rice and wheat residues. Recycling of nutrient in the soil-plant ecology is an essential phenomenon of sustainable agriculture (Kumari et al. 2012, 2014). Though fertilizer application is taking place as a dominant role in the RWCS from the last three decades, crop residues still show an imperative part in the nutrient cycling (Kumari et al. 2010, 2013). The total of N, P, and K nutrient potential was 0.634 Tg with the availability of 37.87 Tg of rice and wheat residues for recycling (Samra et al. 2003). According to Sarkar et al. (1999), RWCS is comprised of one-fourth of the total CR production in India. One ton of rice residue contains nearly 6.1 kg N, 0.8 kg P, and 11.4 kg K, while 1 ton of wheat residue holds ~4.8 kg N, 0.7 kg P, and 9.8 kg K. Many soils under RWCS in IGP region are low in nutrient supplies and organic matter. Exploitative agriculture and declining inputs as organic materials have directed to severe depletion of the soil quality, resulting in a decrease in soil organic matter, soil fertility, and productivity (Kumar et al. 2017a). CRs are the chief source of C inputs and methods that have positive effects on soil physical, chemical, and biological properties (Kumar and Goh 2000). Incorporation of CRs modifies the soil environment, which in turn stimulates the soil microbial population/activity and affects subsequent nutrient transformations (Kumar et al. 2015a). Through this sequence of procedures, application of crop residues controls the efficiency of use of resources like fertilizer and water in RWCS. No single residue management practice is superior under all condition (Kumar and Kumawat 2014). Therefore, there is a need to determine the benefit and adverse effect of residue management options before recommending them to the farmers. Different studies carried out in the last few years, which are related to residue management, and their effects on soil properties (chemical, physical, and biological) following fertilizer management practices in RWCS provide valuable guidelines for efficient management of CRs in RWCS.

2 Availability of Crop Residues in Rice-Wheat Cropping System

India ranks at the top position in the production of several crops (e.g., jute, rice, wheat, sugarcane, cotton, and groundnut). Thus, because of the agricultural potency of the country, CR production is also huge. Every year in India, over 500 Mt of agricultural residues are produced. Residue production has increased significantly with the increased production of rice and wheat. However, distribution and availability of CRs are highly spatiotemporal in nature in India due to diversity in the cropping system and agroclimatic condition (Chauhan 2010). Highest crop residue production was recorded for Uttar Pradesh (60 Mt). Other high crop residue-producing states were Punjab (51 Mt) and Maharashtra (46 Mt) (Fig. 1).

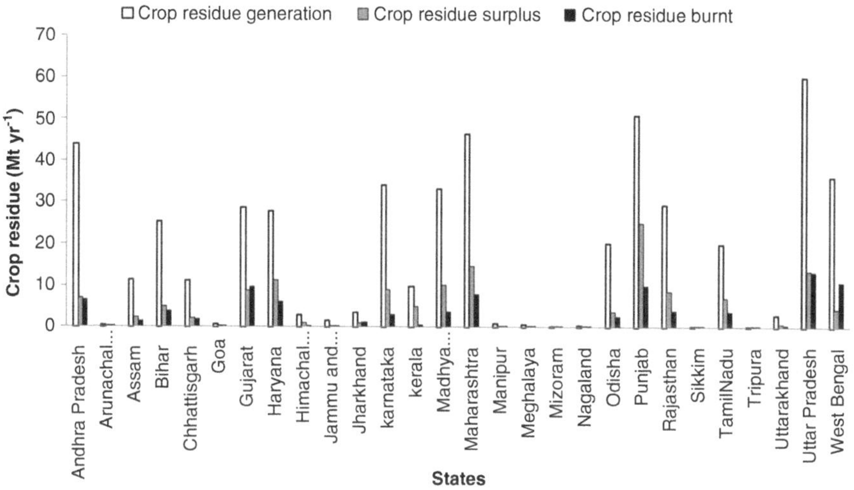

Fig. 1 State-wise data of generation and remaining surplus of crop residues in India; Pathak et al. (2010) and Devi et al. (2017)

The contribution of cereal crops (rice, wheat, maize, millets) is around 70% of total crop residues (352 Mt) in which rice contributes 34% and wheat 22%. Nearly one-fourth of total crop residues produced in the country is contributed by RWCS (Sarkar et al. 1999). Extra CR amount existing in India is projected between the ranges of 84 and 141 Mt year^{-1}, whereas cereal crops contribute ~58% of the 82 Mt of surplus CRs; the amount of residue burned annually is around 70 Mt (44.5 Mt rice straw and 24.5 Mt wheat straw) (Singh and Singh 2003).

3 Problem with Residue Management

Huge quantities of CRs are generated in India by RWCS. Rice and wheat crops in India are mainly harvested through the combine harvester, leaving a huge amount of residues in the field. While ~75% of wheat straw is used as fodder for cattle chopped by cutting machine, still this requires additional task and investment. Due to high silica content, rice straw is considered as poor feed for animals. Because of higher-silica percentage (12–16% vs. 3–5%) and a minor content of lignin (6–7% vs. 10–12%), rice residue differs from other straws (Datta et al. 2017b). Managing rice straw is quite problematic than wheat straw, since there is a very small interval available between rice harvest and wheat sowing and due to the absence of appropriate knowledge for recycling. Other possibilities for CR management include burning, bailing/removal for use as fodder/bedding for animals, in situ integration in the soil, and complete/partial preservation on the surface as mulch using ZT or reduced tillage. After bailing, CRs can be used for paper and ethanol production, bioconversion, and engineering application. Since rice residue has low commercial value due to a shortage of labor, farmers are not in the position to invest in cleaning the field by the use of chopper. In this way, all these practices require other operations which surge the cost. Therefore, farmers in India consider burning as inexpensive and the simplest way of disposing of large tons of residues left after rice crop so that they can grow wheat quickly after rice. At present, 80% or above rice straw of total annual production is being burnt by farmers during the month of October–November (Yadvinder-Singh et al. 2010). Gupta et al. (2004) estimated that the burning of rice straw could cause gaseous emission of about 70% CO_2, 7% CO, 0.66% CH, and 2.09% NO.

Satellite images released by NASA of some northern and northwestern states of India during October and November 2015 exposed the scenario of crop residues burning in Punjab, Haryana, and Uttar Pradesh. These images further revealed the decrease in air quality due to more production of smog, particulate matter, and greenhouse gas. This leads to hazy weather mainly due to the burning of field CRs, which ultimately influences human and animal health in these areas. Therefore, the government of India, Ministry of Agriculture and Farmers' Welfare, formulated the "National Policy for Management of Crop Residues" and circulated it to all states/UTs, to make sure stoppage of burning of CRs, by incentivizing acquisition of recent equipment to curtail leftover CR in the field, in situ management, mixing of residue in soil to increase fertility, multiple uses of crop residue, and formulation of fodder pellets. In situ disposal of CRs through burning resulted in decrease in soil organic matter and valuable nutrients like N and S. Yadvinder-Singh et al. (2010) observed that in Punjab itself annually around 20 Mt of rice and wheat remains are being burned in situ, which results in an 8 Mt of C loss equivalent to a CO_2 load of about 29 Mt per year and around 1×105 tons of N loss; besides this another nutrient like S is also lost and damaged of useful microflora of soil. Disposal of CRs for several other purposes outside of the farm (except for composting and animal feed) possibly has adversarial effects on nutrient supply causing a monetary loss in short term; nevertheless it will have undesirable influence on soil in a long term like quality of water and agriculture sustainability as validated by many studies. There is a requirement of additional nutrient (NPK) fertilization so that the nutrients harnessed by the CR removal can be substituted to maintain the soil fertility in the long term. Some of the farmers in India sell CRs to complement their source of income as cattle feed or as biofuel, whereas others burn/remove it in the absence of marketplace for crop residue; therefore, it befits farmers to use machinery without crop residues on field (Erenstein 2002; Kumar et al. 2016a).

4 Importance of Residue Retention in the Field in Relation to C and N Mineralization

CRs are considered a vital natural resource for conserving and sustaining soil productivity. Addition of CRs to the soil is a useful practice and helps in maintaining and increasing amounts of SOM (Kumar et al. 2015b). Therefore, it is considered that soils have major potential for C storage and to alleviate the atmospheric CO_2 (Nieder and Benbi 2008). CRs can supply essential plant nutrients after mineralization. So, CR addition is suggested as a potential means of sustaining the soil fertility and productivity over long term (Kumar et al. 2018a). Major benefits of residue incorporation are enhancement in physical and biological conditions of soil and prevention of soil degradation. Effective management of CRs is a fundamental constituent of sustainable cropping as it contributes to increasing soil carbon content and nutrient-supplying capacity, reducing harsh effects of residue burning. For recycling CRs, many researchers reported that in situ incorporation and mulching with reduced or NT are the foremost residue management possibilities (Kumar 2015a; Bohra and Kumar 2015; Mishra and Kumar 2017). Ortega et al. (2002) reported that due to the dispersal of CRs incorporated into the soil, there was intense variation in soil with the introduction of NT in agroecosystems. CR management in conservation tillage has a critical effect on soil chemical biological and physical properties. Thus, a better knowledge of about tillage and residue management (surface retained or soil incorporated) is essential. Changes in this condition should affect soil C and N dynamic CR decomposition, and N mineralization dynamics of residue C and N is critical to enumerate probable profits of variations in tillage practices and residue on soil health and agriculture productivity. CRM as practiced in RWCS is of three types: (1) wheat straw management in rice and its residual effect in following wheat, (2) rice straw management in wheat and its residual effect in following rice, and (3) wheat straw management in rice and rice straw management in wheat (cumulative effect) (Kumar et al. 2016b). Residue management affects crop productivity as well as soil fertility and environmental condition so that it plays a key role in the sustainability of cropping systems. Incorporation of CRs offers labile C and N to soils liable to the decomposition rates and synchrony of nutrient mineralization (Murungu et al. 2011).

Mineralization of organic matter indicates a great share in providing essential nutrient to the plant development. Soil husbandry and CRs affect C mineralization and nutrient availability in wetland ecosystems (Kuotsuo et al. 2014). CRM can modify C and N dynamics and subsequently bring significant variations in greenhouse gas emission and crop yields (Liu et al. 2014). For instance, stubble incorporation can stimulate the GHG emissions through increasing C availability for methane-producing microbes in paddy fields and denitrifiers in dryland soils (Yan et al. 2005). Likewise, in comparison to CT, NT and reduced tillage can augment soil C sequestration by reducing soil C breakdown and/or C turnover and, thus, mitigate GHG emission (Ruan and Philip Robertson 2013). Additionally, reduced tillage and residue addition alone or together increase yields by enhancing the fertility of soil (Küstermann et al. 2013). Either rice or wheat crop residues incorporated or surface applied immobilized soil mineral N. Incorporated residues increased SOC and soil aggregate stability significantly by 18% and 55%, respectively (Naresh et al. 2018). CRs incorporated into soil have higher decomposition rate with a quicker mineral N release and more SOM buildup and soil structure improvement than retaining CRs at the soil surface. Compost amendment also considerably lowered specific actions of invertase in macroaggregates and silt + clay fraction, and this effect was more prominent than the addition of fertilizer NPK (Neupane et al. 2011). In contrast, inorganic fertilizer and composting expressively improved specific activities of a cellobiohydrolase in the soil, macroaggregates, and microaggregates (but not in silt + clay fraction) and xylosidase in microaggregates. Increase in SOC in composted soil was hence probably linked with the buildup of lignocellulose and sucrose in macroaggregates, lignocellulose and hemicellulose in microaggregates, and lignin and nonstructural carbohydrates in silt + clay fraction. The C and N mineralization processes have great implication in preserving soil potency and hence agronomic sustainability. Soil management and cropping techniques are known to major elements that affect most of the soil biological properties. Earlier studies have shown that CR retention positively influenced microbial population, activity, and biomass in the soil (Islam and Weil 2000; Pal et al. 2013) and soil moisture content, bulk density, porosity, nutrient distribution, and structure stability (Lobe et al. 2001). These modifications may decrease or increase C and N dynamics and have an impact on nutrient uptake by plants (Kumar and Kumawat 2014). CRs of good quality are, i.e., having high N contents; low lignin, cellulose, and polyphenol contents; and low ratios of C/N and lignin/N that generally result in fast decomposition and N mineralization (Raiesi 1998). Degradation of CRs is directed by N and lignin contents and C/N ratio. However, residue quality is documented as an element affecting C and N mineralization rates. Residue management affects soil nutrient cycling by altering the rate of SOM addition and soil physical and chemical properties, which all affect C and N dynamics in soil (Graham et al. 2002; Shivran et al. 2012; Kumar et al. 2016c). Decomposition of CRs is mostly restricted in wetlands (Qualls and Richardson 2000; Shivran et al. 2013), and decreased decomposition of CRs may bring a reduction in N mineralization and henceforth decrease the rate of N available for plants. Therefore, in wetland soils, management of CRs is decisive. Besides, soil and CRM may interactively stimulate nutrient content that would further have an impact on mineralization of C and N in soil (Salinas-Garcia et al. 2001).

4.1 Soil Organic Nitrogen Mineralization

Rice and wheat residues affect soil organic N mineralization by various processes (Kumar et al. 2017b). Shindo and Nishio (2005) witnessed the rise of organic N mineralization by CRs, potentially caused after enhanced organic matter decomposition present naturally in soil or through hastened turnover of N from microbial biomass (Singh et al. 2017a, b). In contrast, Bradley and Grenon (2006) noticed the negative effect, caused by competition between microorganisms specialized in fresh organic matter decomposition or use of polymerized SOM for energy and nutrient. In addition, surface-applied residue results in cooler and wetter soil environment than open soils (Edwards et al. 2000).

4.2 Organic Nitrogen Mineralization of the Plant Residues

Organic N mineralization from remaining residue can increase soil inorganic N concentration (Kumar et al. 2017c, 2018b). Shindo and Nishio (2005) noted that ~10% of organic N existing in wheat straw was converted into microbial biomass and soil inorganic N content derived from wheat straw ranged between 1.93 and 2.37 mg N/kg. When plant residues are given back to the soil, mineralization of crop residue N contributes to the soil inorganic nitrogen pool. The magnitude of this contribution is governed by the quality of CRs. However, abiotic immobilization of N by CRs can decrease the content of soil mineralizable organic N. Because added inorganic N in CR is transformed into microbial nitrogen, microbial biomass nitrogen, microbial residual nitrogen, and the subsequent nitrogen remineralization rate are enhanced by adding straw residues to the soil (Singh et al. 2017c). However, the effects of CRs on direct inorganic N transformations to soil organic nitrogen remain unknown. There is close interaction between C and N dynamics during the decay of plant stubbles due to the immediate assimilation of C and N by heterotrophic soil microflora involved in the process. CR addition to the soil along with mineral N can act together in a different way. In high C/N crop residues, immobilization related with residue decomposition was increased by adding mineral N. Addition of mineral N also amplified C mineralization during maize residue decomposition and also had no effect on C mineralization (Green et al. 1995). In many cases, high microbial demand for N causes the availability of N (soil + residue N) to be a factor that limits straw decomposition in short term. Rates of C mineralization and N immobilization from high C/N ratio maize straw were ascribed to initial mineral N content of soil (Recous et al. 1995). Similarly, N mineralization from high C/N ratio barley stubble was restricted by the availability of mineral N in the soil to fulfill immobilization requirement (Probert et al. 2005). Influence of N on breakdown of C materials or N availability in soil has been studied in plentiful surveys under laboratory, mainly by changing the initial C/N ratio of the soil residue system by one of the following procedures: (i) incubating one type of residue (maize straw) containing different amounts of N (Scheller and Joergensen 2008), (ii) adding different amounts of mineral N with organic residue (Sall et al. 2007), and (iii) using different types of residues with different initial N content (Bruun et al. 2006). Many studies on N mineralization from high C/N ratio organic materials described the N immobilization under a fixed rate of CR addition by changing the rate of addition of mineral N, though a few have considered both rates of application of residues and mineral N.

5 Decomposition of Crop Residues

Nutrients in the organic form are directly not available to plant and have to undergo decomposition, a process whereby organic nutrient is converted into a mineral form and made available to plant. CR decomposition releases C as CO_2 and nutrients into the mineral nutrient pool. Yadvinder-Singh et al. (2010) conducted a litterbag study to show that rice residue decomposition followed a first-order exponential equation. At the end of the decomposition study, approximately 80% of the rice residue lost its initial weight, and release of total N from buried stubble at maximum tillering stage was ~6 kg N ha^{-1} (15% of initial) in sandy loam and 12 kg N ha^{-1} (27% of initial) in silt loam. CRs are not unwanted but quite a valuable natural resource. About 25% of N and P, 50% of S, and 75% of K are taken up by cereal crops that are allocated in CRs, making them valuable resources of nutrient. Sidhu et al. (2007) assessed the amount of nutrients available in rice and observed that the paddy straw has 39 kg/ha N, 6 kg/ha P, 140 kg/ha K, and 11 kg/ha S. Sidhu and Beri (2005) shared estimated quantity of the nutrients existing in straw, which was burnt, was 106, 65, and 237 thousand tons, respectively, of N, P_2O_5, and K_2O in addition to secondary and micronutrient.

5.1 *Factors Affecting Crop Residue Decomposition and Rate of Release of Nutrients*

CR decomposition is a process controlled by its quality (Goh and Tutua 2004) and environmental factors, and the most important factors are rainfall and temperature. Accordingly, residues with high decomposition and N mineralization rates are usually considered to be of high quality. According to Palm et al. (2001), residues can be divided into four quality classes based on criteria such as N, polyphenol, and lignin content. Similarly, plant residue quality index (PRQI), calculated based on residue C/N ratio, lignin, and polyphenol, was proposed by Tian et al. (1995a). In this index, a higher PRQI value usually specifies greater residue quality, more N mineralization potential, and quicker decomposition. Nutrients in organic form are unavailable to plant, and they have to undergo decomposition, a process whereby organic nutrient is converted into a mineral form made available to the plants. Usually, organic matter accumulates in the soil surface and tends to decrease with depth. Accordingly, decomposition in the field was found to be intensive in topsoil and a decrease with increase in depth (Manjaiah et al. 2000). Purnomo et al. (2000) studied stratification of SOM mineralization in an Australian wheat field. They found that, over the growing season, 32% of total N mineralized in top 20 cm of soil originated from 0 to 2 cm layer, 74% was from 0 to 60 cm, and only 12% was from the soil below 20 cm. To understand the processes which are involved in CR decomposition and nutrient release mainly, N needed to develop efficient soil fertility and management practices; it is important to know factors that govern the processes of decomposition. These processes may be influenced by many factors, such as temperature, moisture, and type of residue, compositions, and placement.

5.1.1 Chemical Composition

The decomposition and mineralization rate of plant materials depend on the chemical composition of the plant tissues. These qualities are the C/N ratio and lignin, cellulose, hemicellulose, and polyphenol content of residues (Palm et al. 2001).

5.1.1.1 N Content and C/N Ratio

Residues having a low content of nitrogen but high in fiber are hard to decay and decompose more slowly than crop residues that are high in N and low in fiber. The leaves of corn, sorghum, soybean, and wheat are all somehow high in N and low in fiber and decompose rapidly (Vigil and Sparks 2004). Earlier studies showed that factor like polyphenol in residue governs the low rates of decomposition of plant materials with narrow C/N ratio. It was stated that the C/N ratio is an important ecological index in the decomposition of OM (Frankenberger and Abdelmagid 1985). In this context, in a lower C/N ratio, easier decomposition of OM would take place and vice versa. In decomposition of OM generally, there are three levels of C/N ratio, the first one with high C/N ratio (>35) and low N contents where N mineralization process is reduced. The second one is medium C/N ratio (20) where mineralization and humification can take place at similar rates and, the last one, low C/N ratio (<10) with high N content, where there is rapid mineralization and releases of N (Duchaufour 1984). There were numerous studies that showed that net N mineralization from different organic materials was correlated with N content. Seneviratne (2000) stated that through decomposition studies of tropical litter from agroforestry that litter with limited N concentration (<1%), N release is controlled by initial N concentration. Nonlegume CRs (wheat, barley, maize, canola, rice, sorghum, sugarcane) with high C/N ratio and/or lignin content may involve the use of fertilizer N to meet microbial N necessities and to ease decomposition. The decomposition dynamics of CRs could be altered by manipulation of quality, principally N, lignin, and polyphenol content, and hence a potentially significant approach to manage N supply in relation to crop demand, thereby improving NUE (Whitbread et al. 2003). When fertilizer N is effective under the thatch layer in the mineral soil with marginal residue disturbance, this reduces the potential for immobilization of the fertilizer N during the decay of wide C/N ratio crop residues (Kumawat et al. 2013b). High N leaf residues from lucerne, medics, pea, and clover can be mineralized relatively fast because of easy decomposition. Crop residues like wheat, rice, sorghum, and maize are composed of high C/N ratio with high lignin content that requires the application of inorganic fertilizer externally to satisfy the microbial N demand which facilitates decomposition. Therefore, the absence of application of fertilizer N can cause immobilization of nutrients by microorganisms when wide C/N ratio-ranged CRs are added (Bhupinderpal-Singh et al. 2006). However, high lignin content may interact with soil or plant residue N, suppressing N availability (Wang et al. 2004) due to the formation of recalcitrant organic N forms.

Intercropping speeds up the decomposition of SOM. Cong et al. (2015) observed 1.5 times increased C decomposition (%) in maize/fava bean intercropping. The maize/fava bean root mix decomposes faster (6%) than estimated from the single litters that were nitrogen rich and hastens decomposition than the litter with less N (Handa et al. 2014). This is likely because allocation of N from high N litter to poor N litter supplies N in the decomposer community, thus accelerating decay of the N-deprived litter (Seastedt 1984).

5.1.1.2 Lignin and Polyphenols

Lignin is a recalcitrant compound, and its relative concentration in residues was reported to increase in initial stages and decline as decomposition proceeds (Berg and Tamm 1991). During decomposition of residues, soluble sugars were reported to be lost rapidly, followed by polysaccharides, cellulose, hemicellulose, and finally lignin (Berg et al. 1982). Mafongoya et al. (1998) reported that N released from leguminous leaves indicated that the (lignin + polyphenol)/N ratio was the utmost vital factor in forecasting N mineralization.

5.1.2 Environmental Factors

Environmental factors are one of the major determinants of the activity of the decomposer community as they directly affect the rate of metabolism (Tanaka 1986). It was observed that in the warm, humid tropics, legumes might be incorporated a few days prior to planting the succeeding crop (Ranells and Wager 1992), while in drier or cooler areas, this may occur as late as 8–9 months prior to planting (Janzen et al. 1990).

5.1.2.1 Temperature

It is evident that N mineralization from slowly decomposable (K = 3.38%; N = 3.4%) CRs is increased by a rise in temperature than from highly decomposable materials (K = 5.36%, N = 1.9%). DeNeve et al. (1996) established a strong correlation between temperature and resistance to degradation. The rise in temperature of 1 °C in areas with an annual mean temperature of 5 °C could eventually lead to a loss of over 10% of SOC, but in the areas with annual mean temperature of 30 °C, the same rise in temperature leads to a loss of only 3% of SOC. This showed that the sensitivity of organic matter decomposition increases with temperature in the temperate region (Kirschbaum 1995). Homann and Grigal (1996) revealed that decomposition of belowground organic materials increased with temperature on cool forest slopes than in a warm field.

5.1.2.2 Moistures

Linn and Doran (1984) stated moisture content is an important factor in controlling decomposition. In their study, they found that the optimum moisture content of the soil for microorganism's activity was 60% water field capacity. Researchers found in most of the soils there is a linear correlation among net N mineralization and moisture content in the range between −0.03 and 4.0 M Pa. It was found that ideal moisture content for net N mineralization corresponded to a soil pore water potential between 0.01 and 0.3 M Pa while that at which no net N mineralization occurred was close to 4.0 M Pa. Although moisture content is an important factor in controlling decomposition, previous studies showed that K was the only nutrient affected by water regime and seemed to increase as water increases due to leaching, because it is not associated with the structural components of the plant cell (Marschner 1995). Parsons et al. (1990) observed higher mass loss (14%) in the first two weeks of decomposition and they attributed large amount of mass loss to high precipitation (>60 mm), which removed most of water-soluble material from leaves during the study period, at the end of the 5-month period, they found that a mass loss of 35% compared with first 2 week. Huang and Schoenau (1997) stated initial litter decomposition rates vary significantly since they were strongly affected by precipitation and moisture.

5.1.3 Soil Condition

In addition to the characteristics of decomposing organic materials in decomposition and release of nutrients, soil characteristics are also very important: soil texture and structure (Lutzow et al. 2002), soil porosity and particle size (Fruit et al. 1999), soil N content (Trinsoutrot et al. 2000), pH (Saggar et al. 1999), cation exchange capacity (Amato and Ladd 1992), sodicity (Nelson et al. 1996), soil fauna (Tian et al. 1995b), specific surface area of clay, nature of clay mineral, and initial soil fertility.

5.1.3.1 Soil Texture

Effects of soil texture in the decomposition of organic matter incorporated into soil seemed to be inconsistent. Skjemstad et al. (1993) stated fine soil particles and organic material interact in the soil to form complexes and microaggregates that render organic substances less susceptible to biodegradation. Pare and Gregorich (1999) comparing between different soil textures showed that after 60 weeks, the proportion of alfalfa N (C/N = 13) mineralized in sandy soil was higher (41%) than clay/loam. In contrast, soybean (C/N = 26) and maize (C/N = 18) residue N were mineralized more in clay or loam soil, 12% and 15%, respectively. They concluded higher mineralization in the fine texture soil was due to their high organic N contents and the presence of clay materials that probably favor mineralization because of their large surface area.

5.1.3.2 Soil pH

Decomposition of residues decreases at low pH due to a decrease in microbial activities (Motavalli et al. 1995). Therefore, the release of N from decomposing plants is influenced by soil pH. The activity of cellulose-degrading enzymes in most terrestrial fungi lies between soil pH 4 and 7 (averaged 5.5), whereas those of their hemicellulose-degrading enzymes ranged from 3.5 to 5.0 (averaged, 4.3) (reported by Wood and Kellogg (1988)).

5.1.3.3 Soil Fauna

Generally, decomposition is thought to proceed through an initial rapid leaching phase followed by a slow stage in which soil fauna activity becomes predominant. For plant materials, decay occurs through initial fragmentation by soil macrofauna (earthworm, millipedes, and termites) with further transformations being accomplished by microbial activity and by enzyme production (Tian et al. 1995b). Each stage basically involves the partial conversion of C to CO and synthesis of microbial tissue. For instance, simple sugars, amino acid, most proteins, and cellulose decompose rapidly (bacterial action), while lignin and microbial melanin decay slowly, mostly through the action of actinomycetes and fungi (Mary et al. 1996; Datta et al. 2017a, c).

5.1.3.4 Soil Nutrients

According to Smethurst and Nambiar (1990), leaf, bark, twigs, stem, and branch material, generally, contain large pools of nutrients. These nutrients were released either by physical leaching or by chemical fragmentation of organic components by soil organisms. Kwabiah et al. (1999) found that release of P was accelerated due to leaching which satisfied the needs of decomposer organisms during decomposition. Forest managers prefer burning of harvest residues as a common practice, as this permits easy access to the site for cultivation, quick nutrient availability, and weed control (O'Connell et al. 2000). Nevertheless, this results in a substantial amount of nutrient loss in the long term, especially N (Raison et al. 1993). If crop residue is reserved, the breakdown of this nutrient-rich material will help to maintain site nutrient capital and will contribute to nutrient supply for the next crop. At the same time, decomposing residues become an important site for immobilization of some nutrients like N (Carlyle et al. 1998). Thus, incorporation of CRs in soil acts as a buffer against nutrient losses by reducing leaching during early stages of plantation development, when root systems are not well developed, have limited spatial coverage, and are unable to utilize available nutrients.

6 Effect of Crop Residues on Soil Properties

Organic matter returned to the soil by decomposing CRs improves soil properties (physical, chemical, and biological), and their interaction affects the nutrient-cycling ability of the soil. Here, we have discussed below the effect of residue management on some important chemical properties like SOC, pH, CEC, structure, texture, porosity, runoff, erosion, compaction, temperature, infiltration, and moisture content and on biomass soil micro- and macrofauna. Higher crop yields will result in the huge production of crop residue after harvest (Fig. 2).

6.1 Chemical Property

6.1.1 Soil Organic Carbon

CRs are the main input in maintaining SOC and nutrients in agricultural soils. Higher amount of SOC helps in achieving better soil health by improving important parameters like the stability of soil aggregate and soil water holding capacity and acts as a soil nutrient pool, therefore, considered as an important indicator of soil quality and agricultural sustainability (Liu et al. 2006). Continuous addition of CRs for long-term determines SOC retention/loss or nutrient cycling as it is governed by the heterotrophic and autotrophic respiration processes of soil in which carbon is released as CO_2. However, anthropogenic activities, such as a change in land use, particularly alteration to agricultural field and paddock, removal of CRs, and cattle grazing, are responsible of the liberation of even larger amounts of C in the air as CO_2 (Dhillon and Wuehlisch 2013). Disturbance in SOC pool by agricultural practices makes the soil a large possible source of emission of N_2O, CH_4, and NOx gasses; decline in soil C can thus lead to degraded soil health and increase

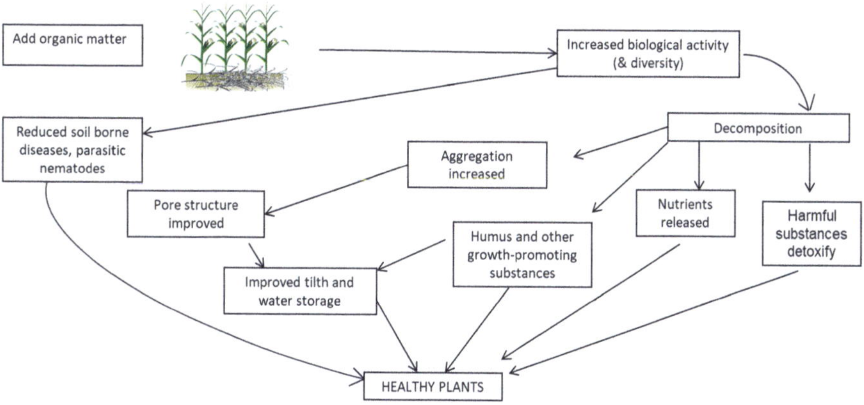

Fig. 2 Effect of SOM on soil properties and plant growth (Oshins and Drinkwater 1999)

stress on food security and sustainable crop production system (Lal 2007). SOC concentration can be increased in soil either by application of the heavy amount of organic matter inputs, by decreasing the rate of decomposition, or both (Shahbaz et al. 2017). The ease of availability of nutrient and carbon stock accumulation depends on labile pool and humus pool of SOM. The labile pool is easily decomposed by microorganisms, while humus pool is recalcitrant in nature and takes more time to decay and thus permits for carbon accumulation in soil. According to Oades (1993), the main reason for the recalcitrant nature of humus pool is because of physical and chemical stabilization which is governed by factors like type of soil, its management, and climatic factor (Lal 2018). The presence of elevated SOC in sandy soils was measured when there was mulch ripping treatment with residue retention compared to clean ripping with residue removed (Chivenge et al. 2007; Meena et al. 2018). Bruun et al. (2015) observed there was around 40% of total SOC stock remaining after 20 years of cultivation of the vertisol and a smaller decrease in the SOC stock of a kaolinite-dominated ultisol. Salinas-Garcia et al. (2001) examined SOC sequestration in two places in Mexico, Apatzingán and Casas Blancas, for 6 long years under rainfed maize production with different conservation tillage and residue treatments. SOC content was found greater (46% in Casas Blancas and 39% in Apatzingán) under all conservation tillage treatments compared to CT and NT with no residue; there might be less contact of CRs applied on surface with soil microorganisms under NT, and therefore decomposition process was steadier as compared to CT where mixing of residue in soil increases the contact with microorganisms (Reicosky et al. 1997; Molaei et al. 2017a, b). Kushwaha et al. (2001) reported a similar result that with elevated rates of decomposition and temperatures, the concentration of SOC and total N noticed maximum under minimum tillage with surface-applied residue compared to incorporation at Varanasi, India. Soil management practices are used to enhance SOC, fertility, and crop productivity around the world (Kumar et al. 2014, 2015a; Kumar 2015b). Memon et al. (2018) designed a field experiment to investigate the impact of different rates of stubble incorporation and tillage methods [reduced tillage (RT) and CT] on crop yield, SOC, total nitrogen (TN), and soil carbon storage (SCS) in RWCS. SOM and SCS were significantly higher by 17.1% and 14.2%, respectively, in RT with 60% straw incorporation.

6.1.2 Soil pH

After the addition of crop residue, there may be changes in soil pH due to the chemical composition of residue and soil properties (Butterly et al. 2011). Xu and Coventry (2003) reported that the pH of the soil is highly affected when residues are high in nitrogen and ash alkalinity, for example, when residues of legume are incorporated in soil compared to lower content residue such as rice and wheat. In another experiment Butterly et al. (2011) concluded that the concentration of organic anions and nitrogen content of the applied residues are responsible for bringing change in soil pH. The change in soil pH with decomposition of organic anions is governed by the following mechanisms: (1) decarboxylation of organic anions that consume H+ anions (if initial soil pH is less than the pKa of weak acid group of the organic carbon, then the H+ will associate with organic anions and vice versa (Bonn and Fish 1991)) and (2) ligand exchange with hydroxyl groups of aluminum and iron oxides. In nitrogen mineralization, initially, there is a change in soil pH due to the ammonification process by consuming H+ and subsequently releasing H+ during nitrification. Therefore, the range of final soil pH depends on the balance of these reactions (Xu and Coventry 2003). Disequilibrium in N cycle can bring the overall acidifying effect of soil. Ogbodo (2011) observed that with the addition of rice/legume residue and burnt rice husk there was a significant improvement in soil pH value when compared with no residue treatment. The OM from residue treatment enhanced soil pH level by increasing the buffering capacity of the soil. Rice husk ash after burning specifically increased soil pH over the other treatments due to its higher concentration of Ca and Mg. Hence, Ca and Mg are found to be most responsible in ameliorating the soil acidity more than other residue treatments. Effect of CR retention in an agricultural field on soil pH is generally confined to topsoil; the effect of incorporation of residue and surface applied on the change in soil pH remained unclear.

6.1.3 Cation Exchange Capacity

Cation exchange capacity (CEC) is one of the soil quality indicators as it is the capacity of the soil to hold cations on their surface for exchange with soil solution. Ogbodo (2011) found significantly higher exchange capacity of soil treated with CRs. Retention of CRs increased SOM content in such plots, which increases the level of elements, their retention, and soil buffer capacity. Hulugalle and Maurya (1991) reported that CR addition increased the CEC and nutrient on treated soil compared to untreated ones. Georgieva (1998) equally reported that burning of forest trash had higher soil fertility and lower acidity. Soil pH-dependent CEC increases after addition of CRs due to increased SOM. Addition of residue in the soil brought change in topsoil layer CEC only (Duiker and Beegle 2006), and the effect of surface-retained and surface-incorporated CRs on CEC remained unclear. Govaerts et al. (2007) observed that after five years, CEC increased in top soil, when CRs where retained compared to soil without residue, but no difference was noticed with increasing soil depth.

6.1.4 Nutrient Availability

CR additions can influence the accessibility of nitrogen to crops. Green manuring (GM) is the most economical means of increasing organic matter content in the soil. Similarly, several scientists proved the increased content of organic matter as a result of GM (Kumawat et al. 2012, 2015). It has been observed to maintain and improve soil structure by addition of SOM and to minimize NPK fixation losses in all types of soil. Dhar et al. (2014) noticed that the maximum amount of available N (206.7 kg ha^{-1}) was recorded by 5 t ha^{-1} straw incorporated 20 days before sowing of wheat along with 5 t ha^{-1} GM as compared to other treatments. GM with the help of nodule fixed atmospheric N in the soil as well as after incorporation of biomass to the soil; it also added a significant amount of N in the soil (Kumawat et al. 2010, 2013a). In this way, rice straw incorporation and GM crop add significantly more nitrogen to the soil. Narayan and Lal (2006) reported an increased amount of 204.5 kg ha^{-1} N in the soil after GM as compared to 170 kg ha^{-1} in the initial soil. Maiksteniene and Arlauskiene (2004) reported that green manure is not rich in P. However, it improves soil physical properties and at the same time stimulates microbiological activity, which makes available P in the soil more readily available to plants. Accumulation of low C/N ratio (e.g., legume) residues resulted in N mineralization, whereas there was temporary immobilization of N due to the presence of cereal residues with a high C/N ratio during the process of decomposition (Govaerts et al. 2006; Kumawat et al. 2017b). Marschner et al. (2015) studied legacy effect, i.e., the impact of former CR C/N ratio on microbial activity, biomass, and nutrient availability after addition of the second residue with same or different C/N ratio. Their results indicated that earlier high C/N residue could decrease net nutrient mineralization from less C/N residue, possibly as a result of nutrient immobilization by microbes decomposing the remaining high C/N residue (Kumawat et al. 2017a). In contrast, initially supplemented low C/N residue could raise nutrient availability after adding high C/N residue, perhaps because nutrients released from the preceding residue still existed in the soil. Changes in microbial community structures could also bring the legacy effect. Retention of CRs on soil surface causes denitrification loss of mineral nitrogen fertilizer due to an increased moisture content of the soil and if incorporation of fertilizers is not done properly. Topsoil concentration of P had been found to be increased after residue addition (Kumawat et al. 2009b, c). This might be due to the transfer of P excavated from lower soil layers (Zibilske et al. 2002). Release of humic molecules and low molecular weight aliphatic acids through the process of CR decomposition was found to block aluminum oxide adsorption sites and subsequently reduce complete adsorption of P. This effect is highly subjective to CR quality, because low C/N ratio legumes are more responsive due to augmented decay rates (Kumawat et al. 2009a).

6.2 Physical Property

6.2.1 Soil Structure

The sustainability of crop production system is highly influenced by soil structure and its ability to resist soil degradation by erosion. Several factors and practices can alter agricultural soil structure such as rainfall, tillage, machinery use, and residue management (Panachuki et al. 2011). Verhulst et al. (2010) identified the various mechanisms involved in improvement of soil structure after retaining CRs in the field: (1) soil aggregation increases through retention of organic matter to the top surface soil, (2) CR retention can protect soil aggregates against runoff from rainwater drops, and (3) CRs present on soil surface can protect soil from compaction/sealing caused by raindrop impact due to less contact. The stability of aggregation and rate of aggregate turnover entirely influenced by the amount and biochemical constituents of CRs reverted to soils. Martens (2000) has found a correlation between water-stable aggregates (WSA), mean aggregate size, and mean weight diameter (MWD) with the biochemical composition of plant residues lignin, phenols, proteins, saccharides, and alkaline extractable humic acids in soil and phenolic acids such as vanillin-vanillic acid in the residue. Corn residues are high in phenols, and its application increases aggregation compared with other crops, although continuous incorporation of corn decreases macroaggregates compared to corn grown in the rotation (Martens 2000). Higher soil aggregate stability was observed under continuous alfalfa (Raimbault and Vyn 1991). When there is soybean cultivation, low aggregation of soil is observed which is attributed to the low concentration of phenols (Martens 2000) along with low residue return to the soil.

6.2.2 Soil Moisture Status

The retention of soil moistures is considered one of the most important benefits of residue covering on the surface in increasing the crop yields under the rainfed condition, where producing the crop is restricted due to nonavailability of the soil moisture. Crop residue retention has been confirmed to regulate the wastage of water through runoff and reduce the evaporation, causative to more soil water holding and elasticity in drought-prone ecology (Verhulst et al. 2011b). CRs also check compaction or sealing of surface; the rate of infiltration for ZT in irrigated condition can be markedly higher compared to traditional cultivation (Verhulst et al. 2011a). Acharya et al. (1998) noticed that distinctly increased soil moisture status in 0–30 cm soil layers in conservation practices with mulching of wild sage and eupatorium as compared to traditional farming. However, mulching contributes to the higher yields of wheat as compared to CT, where the crop residues incorporated. In NER of India, water is insufficient as rainwater runs off slopes. In a study of rice farming in this region, comparing no-till vs. crop residue retention of the surface with minimum tillage showed that both NT and minimum tillage with retention of residue kept more soil moistures. In the upland area of rice-pea rotation with NT, pea performed better with 75% and 50% retention of rice residue. This might be due to improved water retention by covering the residue since no-till without residues exhibits loss of soil moisture to a great extent. Though researchers in many studies noticed the positive effect of maintenance of crop residues on soil moistures and some negative consequences like in semiarid areas with small/frequent rainfall, residue covers intercept precipitation and enhance evaporation (Kozak et al. 2007). On the other hand, crop residues bring about excessive soil moisture and water logging in areas receiving excess rainfall (Rusinamhodzi et al. 2011; Datta et al. 2014; Yadav et al. 2017).

Distinct effect of residue cover on soil surface, on soil moisture, and on crop yield was noted by Thierfelder and Wall (2009). Excessive precipitation and compact runoff contributed to the maximum water accrual which reduced crop yields. However, way out to check the negative impact of waterlogging in areas with excessive precipitation can be the use of permanent raised beds that permit runoff of undue water along with furrow bottoms while holding residues on top bed (Sayre 2004). On the other hand, if precipitation is not extreme, conservation practices with residue retention shrink the threat of failures of crops (Thierfelder and Wall 2009).

6.2.3 Soil Temperatures

The retention of crop residues on soil surface resulted in shrinking in soil temperature at daytime (Verhulst et al. 2010). Under hot humid tropical climate, this may be beneficial because soil temperature is possibly too high for optimum crop development, whereas climates with low-temperature consequence may be harmful to crops. Acharya et al. (1998) noticed that there were lowering in the maximum soil temperatures and increase in the minimum soil temperature in 0–5 cm soil layer in conservation agricultural practices with mulch as compared to CT. This formed a congenial situation for the growth of root attributes. The lowering consequence of retention of the surface crop residue on temperature is in line with results of others as well. Under the cooler climatic condition, reduced soil temperature from crop residue covering can possibly be an inconvenience as it could be dawdling yields of the crops (Verhulst et al. 2010). The consequence of crop residues on the soil temperature is restricted by residues. However, dark-colored crop residues due to more absorption of sunlight resulted in elevated midday temperatures as compared to light-colored residues (Sharratt and Campbell 1994). Under the soil having NT with mulch in northern China, deferred soil heating in early spring delayed sowing and affected germination badly (Wang et al. 2002). Therefore, to tackle this as major issues, Zhang (2011) suggested the use of cold-resistant crops/cultivars for NT with mulching of crop residues. On the other hand, the crop residues possibly are detached from the seed/root zone, in order to uphold the beneficial effect of retention of the crop residues. In spite of differences in seed/root surroundings, no variations in seed emergence after 20 days or in crop yields were reported between CT and NT treatments. Additional study is mandatory to look at whether no-tillage in-row crop residue exclusion/strip tillage leads to higher yields as compared to following the usual no-tillage system with crop residue retention on the soil surface. In most of the climatic condition, removal/burning to a certain extent than recycling of crop residues deteriorates the physical properties of the soil (Prasad and Power 1991). Integration of wheat straw into the soil before rice transplanting over a 5-year cropping period of the rice-wheat system promoted the configuration of the soil aggregates as well as amplified the mean weight diameter of aggregates. However, varied applications of green manuring and crop residues were more effectual than their alone applications. The inclusion of residues alone/in conjunction with green manuring also lowers bulk density of surface soil. Similarly, integration of rice and wheat straw with reference to their burning/removal augmented both infiltration rate and cumulative infiltration. In broad spectrum, the inclusion of crop residues into rice-cultivated soils reduces physical properties like bulk density, penetration resistance, and compaction in the rice-wheat system.

6.3 Biological Properties

Microflora (bacteria, fungi), microfauna (nematodes, protozoa), mesofauna (acarids, enchytrae), and macrofauna (earthworms, termites, large arthropods) together are called as soil biodiversity. Different types of agricultural practices and systems affect the soil diversity in different ways, and the response may be either positive or negative. Addition of fertilizers and manures, which alter the C/N ratio, and tillage practices play a great role in the maintenance of agroecosystems.

6.3.1 Microbial Activity

Important parameters of soil like soil moistures, nutritional availability in agroecosystems, and soil structure are governed by the disintegration of SOM by the soil microorganism. The soil microbial biomass (SMB) can be defined as live part of SOM. It has been projected as another helpful and important sign of soil qualities, as it is a source and pool of organically accessible nutrients and encourages the formation of soil structure and aggregation. The presence of soil microbial population in soil is possibly affected by many ecological factors like soil temperature and moistures (Debosz et al. 1999) and by soil management practices, i.e., crop residue inputs (Govaerts et al. 2007). The maintenance of crop residues is a significant aspect in exciting SMB and microbes' activities in the soil. Lou et al. (2011) compared the treatments with straw retained and straw removed in Northeast China and noted significantly higher SMBC levels, when straw was applied because of enhanced C and N contents, improved soil moistures/porosity, and reduced soil temperature caused by crop residue covers. Integration of CRs in soil surges soil temperature and aeration and proliferates microorganism growth and better contact between them and CR, which resulted in an increase in the rate of decomposition and overall the loss of soil organic carbon (Fontaine et al. 2007). This is dependable with changes in total organic carbon pool due to changes in carbon quantity from residues that are reflected in microbial biomass (Franzluebbers et al. 1999). In the tropical and subtropical condition where temperatures and precipitation is high, residue retention on the surface with no-tillage had increased SOC on topsoil compared to integration (Bayer et al. 2000). This is because of the hindrance of contact between CRs and microorganisms under no-till soil. Thus the rate of decomposition is low. Arbuscular mycorrhizal fungi (AMF) help in recovering the nutritional accessibility to the plants. The AMF is an example of symbiotic relationships between plant roots and fungi, where plant exudates provide glucose to fungi and from fungal hyphae

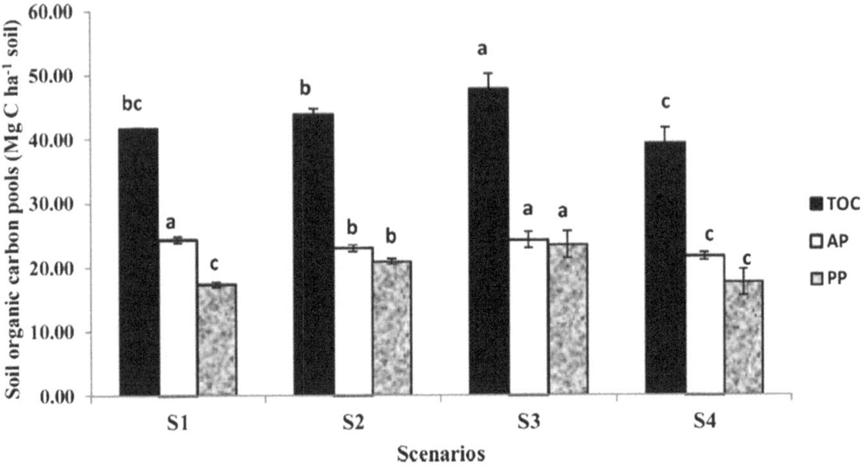

Fig. 3 Total active and passive pool of soil organic carbon in 0–30-cm-depth soil influenced by tillage and crop management practices

phosphate is delivered to plant root. In addition to that, AMF hyphae and the production of glycoprotein, glomalin helps in soil particle bonding, which improves stability of the aggregate (Wright et al. 1999; Marfo et al. 2015; Lojkova et al. 2015). The SOM inputs had an encouraging result on AMF growth and spore population (Emmanuel et al. 2010; Shamina et al. 2018), whereas disturbance of the soil by tillage is recognized to harmfully shock the development of mycorrhizal hyphal (Usuki et al. 2007). Samal et al. (2017) deliberated alteration in microbial properties under long-term scenarios in rice-wheat system and reported that the highest SMBC was recorded in 0–10 cm soil depth S3 (89.32 ± 3.46 μg Cg^{-1} soil). Microbial attributes (MBC, FDA) were improved with an increase in CR carbon addition and, therefore, the buildup in SOC (Fig. 3).

6.3.2 Earthworms

A very important role is played by the larger soil fauna/soil macrofauna, i.e., earthworms in the soil environment. Earthworms carry out numerous significant tasks in the soil. They increase soil structure, water infiltration, nutrient cycling, and subsequently plant growth. They are not the only sign of healthy soil systems, but they also represent a healthy ecosystem. They in a straight line influence carbon and nitrogen cycles by overriding, storing, and cycling nutrients through their biomass, releasing enormous amounts of N through excretion and mortality (Whalen et al. 2000). They add organic matter to the soil through their gut and in their structures like casts, burrows, and middens which ultimately affect soil aggregate stability and have an effect on C and N cycles. Earthworms respond positively to retention of crop residues and minimum tillage practices because of improved soil properties

(Errouissi et al. 2011). Low soil temperature helps in prolonged retention of crop residue on the soil surface which acts as food reservoir to the worms and can lead to increase in earthworm number and their biomass (Chan and Heenan 2006). On the contrary, the process of incorporating the crop residues into the soil through tillage had a harmful impact on earthworm population because it causes physical damage, increases their exposure to predators on surface soil, and destroys their burrows (Chan 2001) Therefore, the CR retention had a varying consequence on the earthworm population, however, depending on their ecological niche (Wuest et al. 2005). On the contrary, in fields with high earthworm numbers, burying the crop residues in soil instead of leaving it on the soil surface can lessen loads, even with superficial tillage, since they feed on the surface (Metzke et al. 2007). The consequence of residues on soil fauna including earthworm can thus vary depending on types of tillage, tillage frequency, plowing depth, residue application method, types of crop residues, amount and quality, etc. The soil microorganisms cooperate significant function in mediating alterations in TOC via mineralization-immobilization of SOM (Breulmann et al. 2014). Progression of residue rotting is chiefly facilitated by the soil microbes and is exaggerated by many another factors, i.e., soil texture, residue quality, and climatic condition (Chen et al. 2014). However, soil microbial diversity reacts in a different way to different stages of residue decay (Marschner et al. 2011). During the 1st stage, there is more abundance of bacterial communities, and fungi dictate the previous stage (Marschner et al. 2011). However, maize straw retention for a longer period (30 years) increased fungal biomass but did not affect bacterial biomass in summer maize-winter wheat system in China (Zhao et al. 2016).

7 Emission of Greenhouse Gasses

Methane (CH_4) and nitrous oxide (N_2O) are considered as important GHGs with global warming potentials (GWP) of 25 and 298 times larger than that of carbon dioxide (CO_2), respectively, on a 100-year horizon. Rice field is identified as the important source of CH_4 emission which contributes about 15–20% of the total methane produced anthropogenically (Aulakh et al. 2001). According to the Kyoto protocol, both CH_4 and N_2O after CO_2 were listed as GHGs which production should be controlled. Thus, it is vital to launch technologies and practices for dropping CH_4 and N_2O emissions from rice and wheat fields while sustaining or increasing rice productivity. Rice-wheat cropping system unavoidably produces huge amounts of straw residues. To increase soil fertility and soil organic carbon storage, crop residue retention in the field is highly recommended. However, the common exercise of incorporating crop stubble uniformly into the surface soil offers a readily available C, which has been proved to significantly augment CH_4 emission and slightly reduce N_2O emission (Zou et al. 2005) from rice fields. Consequently, there may be the opportunity to both sustain soil productivity and mitigate CH_4 and N_2O emissions with appropriate management of fresh CRs. There are other methods of placement of crop residues apart from the common practice of incorporation in topsoil. Ma et al. (2009) studied the effect of five wheat straw application methods (no straw, evenly incorporating, burying straw, ditch mulching, and strip mulching) prior to rice cultivation on CH_4 and N_2O emissions and found that the best management practice for wheat residue addition full or partially is strip mulching onto the field surface, as the method reduced CH_4 emission (by 32% as compared to evenly incorporated) from rice fields without compromise in rice yield. Ditch-buried (DB) method of residue application was reported by Zhu et al. (2012), in which fresh wheat straw was buried in two parallel ditches (20 cm depth) and the remaining area of the field is implemented with shallow rotary tillage (3–5 cm depth). According to earlier studies, DB showed the great potential of C sequestration and decreased the threat of N loss (Yang et al. 2015), showing a new promising way to manage emission of GHG from crop straw addition in a rice-wheat system. Time of incorporation also plays a great role in GHG emission as the incorporation of cereal residues into rice fields before the transplanting can help in minimizing the adverse effect on rice growth and CH_4 emissions. Ma et al. (2007) observed insignificant effect on N_2O emission when wheat straw was integrated before the transplanting of rice due to immobilization of mineral N due to high C/N ratio of the straw. Researchers noticed elevated N_2O emission where CRs were retained on the surface as mulch compared to integration due to more retention of water in soil leading to an anaerobic condition which is favorable for denitrification (Baggs et al. 2003). CR integration in upland cropping system (including wheat) brings no significant production of CH_4 gas. Formation of CH_4 requires anaerobic microsites in the soil for methane-producing bacteria to grow. Therefore, any management practices (irrigation or mulch application) which make soil anaerobic are supposed to increase the threat of CH_4 release. As in submerged condition, any act that results in stubble to decay before becoming anaerobic will minimize the possibility of CH_4 emission. From the perspective of alleviating GHG productions from the wheat crop in RWCS, residues are not main crop management concern. Soil moisture at or near field capacity results in the slight CH_4 formation and N_2O emission, and effect of CR application would be insignificant. Neither surface-applied nor integrated paddy residue into wheat crop would be estimated to have a very substantial influence on CH_4 release in the following rice crop, as the integrated or surface-covered residue would decompose greatly during the upland crop season (Abao et al. 2000).

8 Summary and Conclusions

An important role is played by crop residues in the cycling of essential nutrients for plants despite the principal role of chemical fertilizers in crop production. Crop residue management controls the efficiency with which fertilizer, water, and other resources are used in a cropping system. Due to intensive cropping of rice-wheat system that prevailed in the South Asian region, it is necessary to manage the huge quantity of its residues, which are good source of carbon, nitrogen, and potassium. Several types of research at field and laboratory condition have been done taking different factors since nutrient-supplying capacity and rate of mineralization of organic sources of nutrients vary greatly with soil tillage, residue placement, residue quality, residue quantity, soil type, soil moisture level, temperature, cropping system, etc. Sufficient amount of addition of organic materials to soil is important to improve the soil health as the organic material addition enhances the population of microbes in soil, which are responsible for nutrient transformation resulting in the availability of nutrients, particularly N, P, and S. Therefore, we should incorporate those practices that increase organic matter addition to the soil fertility and crop productivity and also replace the fertilizer nutrients with nutrients supplied from crop residues. After harvesting of the wheat crop and rice transplanting/planting, there is the sufficient availability of time to grow a green manure crop, which has the potential to substitute 50% of the N needs of rice. Therefore, residue quality should be taken into account during soil management. Another important practice is the placement of these residues in the field when the residue is retained on soil surface; this not only lessens runoff and soil loss by erosion but also increases soil physical health and raises SOM concentration, specifically in the surface layer. Furthermore, proliferation in soil microbial biomass and activity after crop residue addition can increase the nutrient-supplying capacity of soil and reduce nutrient losses. Incorporation of CRs by tillage practice may decrease the soil carbon storage by elevation in decomposition process as compared to conservation tillage practice, in this manner making nutrients susceptible to losses (leaching, volatilization, and denitrification), if crop demand and release of nutrients are not matched. Many researchers have reported in their short-term laboratory and field studies that acidic soil can be ameliorated to normal by the application of crop residues. The burning of CRs is beneficial in short term because burnt CRs result in the production of alkalinity of the soil. Burning of CRs could also increase the nutrient-supplying capacity of CRs in the short term by reducing microbial population and their decomposing activity, thus decreasing nutrient immobilization. On the other hand, the effect of long-term residue burning can hasten soil erosion, severely affect soil physical conditions, and increase losses of soil organic matter and related nutrients. The intensity of these ill effects will rest on the frequency and span of CR burning and the soil properties. The quantity of crop residues given back to soil affects soil potency through a sequence of chemical, physical, and biological changes in the soil. Consistent indicators of soil quality changes that show a relationship in line with critical aspects of soil physical, chemical, and biological fertility could offer the basis for emerging sustainable CR management strategies. For instance, the change in the labile fraction of soil organic matter and alteration in soil microbial biomass and functions is a reliable indicator of changes in soil quality parameter due to variation in CR management. Clear-cut evidence as an increase in SOM content and other properties which improves the soil quality and health can be seen only in long-term experiments as these factors depend upon local climatic and edaphic conditions. These kinds of long-term field studies, particularly in tropical regions of the world, should be established and maintained by judiciously selecting the sites considering variations in temperature, moisture, soil mineralogy, and residue management representing different cropping systems across regions.

9 Research Need

This book chapter emphasized the requirement of long-term trials to understand the effects of residue application practices in different regions. It also highlights the practices that involve combinations of interventions at the farm level to design systems that preserve soil health for higher crop production and at the same time provide higher profit to farmers. To distinguish the nutrient-supplying capacity through residues under rice-wheat cropping system in IGPs, there is a need for analysis of yield data of both above- and belowground crop residues. More data needs to be compiled on the effect of management (including soil tillage, residue placement, residue quality, residue quantity, soil type, soil moisture level, temperature, cropping system, etc.) on decomposition and nutrient release rates. Development of technologies for fast in situ decomposition of CRs by irrigation, fertilization, tillage, and application of decomposing microbial consortia for the intensive RWCS is required. Long-term experiments regarding soil property change and nutrient cycling are needed to be performed as several soil properties may clearly be visible after 10 or more years of CR management, and long-term outcomes may vary from those achieved over the short term. Thus, there is a robust necessity to begin long-term experiments at sites judiciously selected for deviations in temperature, moistures, soil mineralogy, and agricultural management covering RWCS in IGPs.

References

Abao EBJ, Bronson KF, Wassmann R, Singh U (2000) Simultaneous records of methane and nitrous oxide emissions in rice-based cropping systems under rain fed conditions. Nutr Cycl Agroecosyst 58:131–139

Acharya CL, Kapur OC, Dixit SP (1998) Moisture conservation for rainfed wheat production with alternative mulches and conservation tillage in the hills of north-west India. Soil Tillage Res 46:153–163

Ali I, Nabi G (2016) Soil carbon and nitrogen mineralization dynamics following incorporation and surface application of rice and wheat residues. Soil Environ 2:207–215

Amato M, Ladd JN (1992) Decomposition of 14C-labelled glucose and legume material in soils. Properties influencing the accumulation of organic residue C and microbial biomass. C Soil Biol Biochem 24:455–464

Aulakh MS, Wassmann R, Rennenberg H (2001) Methane emissions from rice field– quantification, mechanisms, role of management, and mitigation options. Adv Agron 70:193–260

Baggs EM, Stevenson M, Pihlatie M, Regar A, Cook H, Cadisch G (2003) Nitrous oxide emissions following application of residues and fertilizer under zero and conventional tillage. Plant Soil 254:361–370

Bayer C, Mielniczuk J, Amado TJC, Martin-Neto L, Fernandes SV (2000) Organic matter storage in a sandy clay loam Acrisol affected by tillage and cropping systems in southern Brazil. Soil Tillage Res 54:101–109

Berg B, Tamm O (1991) Decomposition and nutrient dynamics of litter in long-term optimum nutrition experiments. Scand J For Res 6:305–321

Berg B, Hannus K, Popoff T, Theander O (1982) Changes in organic chemical components of needle litter during decomposition. Long term decomposition in a Scots pine forest. Can J Bot 60:1310–1319

Bhanwaria B, Ram M, Kumar KN, Kumar R (2013) Influence of fertilizers levels and biofertilizers on growth and yield of linseed (Linum usitatissimum L.) under rainfed condition of South Gujarat. Madras Agric J 100:4–6

Bhupinderpal-Singh, Rengel Z, Jw B (2006) Carbon, nitrogen and sulphur cycling following incorporation of canola residue of different sizes into a nutrient-poor sandy soil. Soil Biol Biochem 38:1591–1597

Bohra JS, Kumar R (2015) Effect of crop establishment methods on productivity, profitability and energetics of rice (*Oryza sativa*)–wheat (*Triticum aestivum*) system. Indian J Agric Sci 85 (2):217–223

Bonn BA, Fish W (1991) Variability in the measurement of humic carboxyl content. Environ Sci Technol 25:232–240

Bradley RL, Grenon F (2006) Evidence that straw does not increase the mobilization of N from decomposing salal (*Gaultheria shallon* Pursh.) leaf litter. Soil Biol Biochem 38:191–194

Breulmann M, Masyutenko NP, Kogut BM, Schroll R, Dorfler U, Buscot F, Schulz E (2014) Short-term bioavailability of carbon in soil organic matter fractions of different particle sizes and densities in grassland ecosystems. Sci Total Environ 498:29–37

Bruun S, Luxhoi J, Magid J, De Neergaard A, Jensen LS (2006) A nitrogen mineralization model based on relationships for gross mineralization and immobilization. Soil Biol Biochem 38:2712–2721

Bruun TB, Elberling B, Neergaard A, Magid J (2015) Organic carbon dynamics in different soil types after conversion of forest to agriculture. Land Degrad Dev 3:272–283

Butterly CR, Kaudal BB, Baldock JA, Tang C (2011) Contribution of soluble and insoluble fractions of agricultural residues to short-term pH changes. Eur J Soil Sci 62:718–727

Carlyle JC, Bligh MW, Nambiar EKS (1998) Woody residue management to reduce nitrogen and phosphorus leaching from sandy soil after clearfelling Pinus radiate plantations. Can J For Res 28:1222–1232

Chan KY (2001) An overview of some tillage impacts on earthworm population abundance and diversity-implications for functioning in soils. Soil Tillage Res 57:179–191

Chan KY, Heenan DP (2006) Earthworm population dynamics under conservation tillage systems in south-eastern Australia. Aust J Soil Res 44:425–431

Chand S, Kumar R, Singh AK (2017) Effects of tillage practices on productivity of wheat under Indo-Gangetic Plains. J AgriSearch 3(209–211):4–03

Chauhan S (2010) Biomass resources assessment for power generation: a case study from Haryana state, India. Biomass Bioenergy 34:1300–1308

Chen L, Zhang J, Zhao B, Yan P, Zhou G, Xin X (2014) Effects of straw amendment and moisture on microbial communities in Chinese fluvo-aquic soil. J Soils Sediments 14:1829–1840

Chivenge P, Murwira H, Giller K, Mapfumo P, Six J (2007) Long-term impact reduced tillage and residue management on soil carbon stabilization: implications for conservation agriculture on contrasting soils. Soil Tillage Res 94:328–337

Choudhary HR, Sharma OP, Singh RK, Singh K, Yadav L, Kumar R (2013) Influence of organic manures and chemical fertilizer on nutrient uptake, yield and profitability of mungbean [Vigna radiate (L.) Wilczek]. Madras Agric J 100(1–3):747–750

Cong WF, Hoffland E, Li L, Janssen BH, van der Werf W (2015) Intercropping affects the rate of decomposition of soil organic matter and root litter. Plant Soil 391:399–411

Datta R, Vranová V, Pavelka M, Rejšek K, Formánek P (2014) Effect of soil sieving on respiration induced by low-molecular-weight substrates. Int Agrophys 28(1):119–124

Datta R, Anand S, Moulick A, Baraniya D, Pathan SI, Rejsek K, Vranova V, Sharma M, Sharma D, Kelkar A (2017a) How enzymes are adsorbed on soil solid phase and factors limiting its activity: a review. Int Agrophys 31(2):287–302

Datta R, Kelkar A, Baraniya D, Molaei A, Moulick A, Meena R, Formanek P (2017b) Enzymatic degradation of lignin in soil: a review. Sustainability 9(7):1163
Datta R, Baraniya D, Wang Y-F, Kelkar A, Meena RS, Yadav GS, Teresa Ceccherini M, Formanek P (2017c) Amino acid: its dual role as nutrient and scavenger of free radicals in soil. Sustainability 9(8):1402
Debosz K, Rasmussen PH, Pedersen AR (1999) Temporal variations in microbial biomass C and cellulolytic enzyme activity in arable soils: effects of organic matter input. Appl Soil Ecol 13:209–218
DeNeve S, Pannier J, Hofman G (1996) Temperature effects on C-and N-mineralization from vegetable crop residues. Plant Soil 1:25–30
Devi S, Gupta C, Jat SL, Parmar MS (2017) Crop residue recycling for economic and environmental sustainability: the case of India. Open Agric 2(1):486–494
Dhar D, Datta A, Basak N, Paul N, Badole S, Thomas T (2014) Residual effect of crop residues on growth, yield attributes and soil properties of wheat under rice-wheat cropping system. Indian J Agric Res 5:373–378
Dhillon RS, von Wuehlisch G (2013) Mitigation of global warming through renewable biomass. Biomass Bioenergy 48:75–89
Duchaufour P (1984) Edafologia: Edafogenesis clarification. Masson, Barcelona, 493 pp
Duiker SW, Beegle DB (2006) Soil fertility distributions in long-term no-till, chisel/disk and moldboard plow/disk systems. Soil Tillage Res 88:30–41
Edwards L, Burney J, Richter G, MacRae A (2000) Evaluation of compost and straw mulching on soil-loss characteristics in erosion plots of potatoes in Prince Edward Island, Canada. Agric Ecosyst Environ 81:217–222
Emmanuel B, Fagbola O, Abaidoo R, Osonubi O, Oyetunji O (2010) Abundance and distribution of arbuscular mycorrhizal fungi species in long-term soil fertility management systems in northern Nigeria. J Plant Nutr 33:1264–1275
Erenstein O (2002) Crop residue mulching in tropical and semi-tropical countries: an evaluation of residue availability and other technological implications. Soil Tillage Res 67:115–133
Errouissi F, Nouira S, Ben Moussa-Machraoui S, Ben-Hammouda M (2011) Soil invertebrates in durum wheat (*Triticum durum* L.) cropping system under Mediterranean semi arid conditions: a comparison between conventional and no-tillage management. Soil Tillage Res 112:122–132
Fontaine S, Barot S, Barré P, Bdioui N, Mary B, Rumpel C (2007) Stability of organic carbon in deep soil layers controlled by fresh carbon supply. Nature 450:277–280
Frankenberger WT, Abdelmagid HM (1985) Kinetic parameters of nitrogen mineralization rates of leguminous crops incorporated into soil. Plant Soil 87:257–271
Franzluebbers A, Haney R, Hons F, Zuberer D (1999) Assessing biological soil quality with chloroform. Can J Soil Sci 79:521–528
Fruit L, Recous S, Richard G (1999) Plant residue decomposition: effect of soil porosity and particle size. In: Effect of mineral-organic-microorganism interactions on soil and freshwater environment. Springer, Boston, pp 189–196
Georgieva LL (1998) Influence of soil tillage, fertilizer application and herbicides on yield of maik grown as second crop for silage. Lasteview dni-Nanki 35:192–195
Goh KM, Tutua SS (2004) Effects of organic and plant residue quality and orchard management on decomposition rates of residues. Commun Soil Sci Plant Anal 35:1532–2416
Govaerts B, Sayre KD, Ceballos-Ramirez JM, Luna-Guido ML, Limon-Ortega A, Deckers L, Dendooven L (2006) Conventionally tilled and permanent raised beds with different crop residue management: effects on soil. C N Dyn Plant Soil 280:143–155
Govaerts B, Mezzalama M, Unno Y, Sayre KD, Luna-Guido M, Vanherck K, Deckers J (2007) Influence of tillage, residue management, and crop rotation on soil microbial biomass and catabolic diversity. Appl Soil Ecol 37(1–2):18–30
Graham MH, Rj H, Jh M (2002) Soil organic matter content and quality: effects of fertilizer applications, burning and trash retention on long-term sugarcane experiment in South Africa. Soil Biol Biochem 34:93–102

Granatstein D, McGuire A, Amara M (2017) Improving soil quality on irrigated soils in the Columbia Basin. Washington State University Extension, Pullman

Green CJ, Blackmer AM, Horton R (1995) Nitrogen effects on conservation of carbon during corn residue decomposition in soil. Soil Sci Soc Am J 59:453–459

Gupta PK, Sahai S, Singh N, Dixit CK, Singh DP, Sharma C, Tiwari MK, Gupta RK, Garg SC (2004) Residue burning in rice–wheat cropping system: causes and implications. Curr Sci 87:1713–1717

Handa IT, Aerts R, Berendse F, Berg MP, Bruder A, Butenschoen O, Chauvet E, Gessner MO, Jabiol J, Makkonen M, McKie BG, Malmqvist B, Peeters ETHM, Scheu S, Schmid B, Van Ruijven J, Vos VCA, Hättenschwiler S (2014) Consequences of biodiversity loss for litter decomposition across biomes. Nature 509:218–221

Homann PS, Grigal DF (1996) Below-ground organic carbon and decomposition potential in a field-forest glacial-outwash landscape. Soil Biol Biochem 23:207–214

Huang WZ, Schoenau JJ (1997) Mass loss measurements and statistical models to predict decomposition of leaf litter in a Boreal Aspen Forest. Commun Soil Sci Plant Anal 28:863–874

Hulugalle NR, Maurya PR (1991) Tillage systems for the West African semi-arid tropics. Soil Tillage Res 20:197–199

Islam KR, Weil RR (2000) Land use effects on soil quality in a tropical forest ecosystem of Bangladesh. Agric Ecosyst Environ 79:9–16

Janzen HH, Bole JB, Biederbeck VO, Slinkard AE (1990) Fate of nitrogen applied as green manure or ammonium fertilizer to soil subsequently cropped with spring wheat at three sites in Western Canada. Can J Soil Sci 70:313–323

Jeet S, Singh D, Kumar R, Kumari A (2010) Yield and physico-chemical properties of soil as affected by seeding rates and nitrogen levels in relation to crop residue management on bed planted soybean (*Glycine max*)-wheat (*Triticum aestivum*) system. Environ Ecol 28 (3B):2063–2067

Kirschbaum MUF (1995) The temperature dependence of soil organic matter decomposition, and the effect of global warming on soil organic storage. Soil Biol Biochem 27:753–760

Kozak JA, Ahuja LR, Green TR, Ma L (2007) Modelling crop canopy and residue rainfall interception effects on soil hydrological components for semi-arid agriculture. Hydrol Process 21:229–241

Kumar R (2015a) Influence of mulching, liming and farm yard manures on production potential, economics and quality of maize (*Zea mays* L.) under rainfed condition of Eastern Himalaya. Bangladesh J Bot 44:3–391

Kumar R (2015b) Productivity, profitability and nutrient uptake of maize (*Zea mays*) as influenced by management practices in North-East India. Indian J Agron 60(2):273–278

Kumar K, Goh KM (2000) Crop residues and management practices: effects on soil quality, soil nitrogen dynamics, crop yields, and nitrogen recovery. Adv Agron 68:197–319

Kumar R, Kumawat N (2014) Effect of sowing dates, seed rates and integrated nutrition on productivity, profitability and nutrient uptake of summer mungbean in Eastern Himalaya. Arch Agron Soil Sci 60(9):1207–1227

Kumar S, Paswan AK, Kumar R, Kumawat N, Singh AK, Singh RS, Singh RK (2013) Effect of organic manures, bio-fertilizers and chemical fertilizers on growth, yield and economics of mustard [*Brassica juncea* (L.) Czern. &Coss.]. Bioinfolct 10(3A):834–838

Kumar R, Chatterjee D, Kumawat N, Pandey A, Roy A, Kumar M (2014) Productivity, quality and soil health as influenced by lime in ricebean cultivars in foothills of northeastern India. Crop J 2:338–344

Kumar R, Deka BC, Kumar M, Ngachan SV (2015a) Productivity, quality and soil health as influenced by organic, inorganic and biofertilizer on field pea in Eastern Himalaya. J Plant Nutr 38:13–2006. https://doi.org/10.1080/01904167.2014.988355

Kumar R, Kumar M, Kumar A, Pandey A (2015b) Productivity, profitability, nutrient uptake and soil health as influenced by establishment methods and nutrient management practices in transplanted rice (*Oryza sativa*) under hill ecosystem of North East India. Indian J Agric Sci 85(5):634–639

Kumar M, Kumar R, Meena KL, Rajkhowa DJ, Kumar A (2016a) Productivity enhancement of rice through crop establishment techniques for livelihood improvement in Eastern Himalayas. Oryza 53(3):300–308

Kumar R, Kumawat N, Kumar S, Kumar R, Kumar M, Sah RP, Kumar U, Kumar A (2016b) Direct seeded rice: research strategies and opportunities for water and weed management. Oryza 53 (4):354–365

Kumar R, Patra MK, Thirugnanavel A, Chatterjee D, Deka BC (2016c) Towards the natural resource management for resilient shifting cultivation system in Eastern Himalayas. In: Bisht J, Meena V, Mishra P, Pattanayak A (eds) Conservation agriculture. Springer, Singapore, pp 409–436. ISBN:978-981-10-2557-0

Kumar M, Rajkhowa J, Meena KL, Kumar R, Zeliang PK, Kikon EL, Rangnamei KL, Namei A (2017a) Effect of nutrient management in lowland rice for improving productivity, profitability and energetics under mid hill altitude of Nagaland. J AgriSearch 4(4):247–250

Kumar S, Devi EL, Sharma SK, Ansari MA, Phurailatpam NTC, Singh TS, Prakash N, Kumar R, Kumawat N, Mandal D, Kumar A (2017b) Rice breeding strategies of North Eastern India for resilience to biotic and abiotic stresses: a review. ORYZA Int J Rice 54(1):1–12

Kumar S, Dwivedi SK, Kumar R, Mishra JS, Singh SK, Prakash V, Rao KK, Bhatt BP (2017c) Productivity and energy use efficiency of wheat (*Triticum aestivum*) genotypes under different tillage options in rainfed ecosystem of middle Indo-Gangetic Plains. Indian J Agron 62:1–31

Kumar S, Kumar R, Mishra JS, Dwivedi SK, Prakash V, Bhakta N, Singh AK, Singh SK, Haris AA (2017d) Evaluation of rice (*Oryza sativa*) cultivars under different crop establishment methods to enhance productivity, profitability and energetics of rice in middle Indo-Gangetic Plains of Eastern India. Indian J Agron 62:3–307

Kumar R, Mishra JS, Kumar R (2018a) Enhancing the productivity of rice fallows area of Eastern India through inclusion of pulses and oilseeds. Indian Farming, August Issue 68(08) (in press)

Kumar S, Kumar R, Mishra JS, Dwivedi SK, Prakash V, Rao KK, Singh AK, Bhatt BP, Singh SS, Haris AA, Kumar V, Srivastava AK, Singh S, Yadav A (2018b) Productivity and profitability of rice (*Oryza sativa*) genotypes as influenced by crop management practices under middle Indo-Gangetic Plains. Indian J Agron 63(1):45–49

Kumari A, Singh ON, Kumar R, Singh AK, Singh R (2010) Effect integrated nutrient management on yield and quality of dwarf pea (*Pisum sativum* L.) vegetable. Science 2:149–152

Kumari A, Singh ON, Kumar R (2012) Effect of integrated nutrient management on growth, seed yield and economics of field pea (*Pisum sativum* L.) and soil fertility changes. J Food Legumes 25(2):121–124

Kumari A, Sinha SP, Kumar R, Kumar A (2013) Effect of vermicompost, biofertilizer and inorganic fertilizers on yield and total nutrient uptake of field pea (*Pisum sativum* L.). J Appl Biol 23:1–4

Kumari A, Singh ON, Kumar R (2014) Root growth, crop productivity, nutrient uptake and economics of dwarf pea (*Pisum sativum* L.) as influenced by integrated nutrient management. Indian J Agric Sci 84(11):1347–1351

Kumawat N, Kumar R, Sharma OP (2009a) Nutrient uptake and yield of mungbean [*Vigna radiata* (L.) Wilczek] as influenced by organic manures, PSB and phosphorus fertilization. Environ Ecol 27(4B):2002–2005

Kumawat N, Sharma OP, Kumar R (2009b) Effect of organic manures, PSB and phosphorus fertilization on yield and economics of mungbean [*Vigna radiata* (L.) Wilczek]. Environ Ecol 27(1):5–7

Kumawat N, Sharma OP, Kumar R, Kumari A (2009c) Response of organic manures, PSB and phosphorus fertilization on growth and yield of mungbean. Environ Ecol 27(4B):2024–2027

Kumawat N, Sharma OP, Kumar R, Kumari A (2010) Yield and yield attributes of mungbean [*Vigna radiate* (L.) Wilczek] as affected by organic manures, PSB and phosphorus fertilization. Environ Ecol 28(1A):332–335

Kumawat N, Singh RP, Kumar R, Kumari A, Kumar P (2012) Response of intercropping and integrated nutrition on production potential and profitability on rainfed pigeonpea. J Agric Sci 4 (7):154–162

Kumawat N, Singh RP, Kumar R (2013a) Productivity, economics and water use efficiency of rainfed pigeonpea + black gram intercropping as influenced by integrated nutrient management. Indian J Soil Conserv 41(2):170–176

Kumawat N, Singh RP, Kumar R, Om H (2013b) Effect of integrated nutrient management on the performance of sole and intercropped pigeonpea (*Cajanus cajan*) under rainfed conditions. Indian J Agron 58(3):309–315

Kumawat N, Singh RP, Kumar R, Yadav TP, Om H (2015) Effect integrated nutrient management on productivity, nutrient uptake and economics of rainfed pigeonpea (*Cajanus cajan*) and black gram (*Vigna mungo*) intercropping system. Indian J Agric Sci 85(2):171–176

Kumawat N, Kumar R, Kumar S, Meena VS (2017a) Nutrient solubilizing microbes (NSMs): its role in sustainable crop production. In: Meena VS, Mishra P, Bisht J, Pattanayak A (eds) Agriculturally important microbes for sustainable agriculture. Springer, Singapore, pp 25–61. https://doi.org/10.1007/978-981-10-5343-6_2. ISBN:978-981-10-5342-9

Kumawat N, Kumar R, Morya J, Tomar IS, Meena RS (2017b) Integrated nutrition management in pigeon pea intercropping systems for enhancing production and productivity in sustainable manner – a review. J Appl Nat Sc 9(4):2143–2151

Kuotsuo R, Chatterjee D, Deka BC, Kumar R, AO M, Vikramjeet K (2014) Shifting cultivation: an 'organic like' farming in Nagaland. Indian J Hill Farming 27(2):23–28

Kushwaha CP, Tripathi SK, Singh KP (2001) Soil organic matter and water-stable aggregates under different tillage and residue conditions in a tropical dryland agroecosystem. Appl Soil Ecol 16:229–241

Küstermann B, Munch JC, Hülsbergen KJ (2013) Effects of soil tillage and fertilization on resource efficiency and greenhouse gas emissions in a long-term field experiment in Southern Germany. Eur J Agron 49:61–73

Kwabiah AB, Vorony RP, Palm CA, Stoskopf NC (1999) Inorganic fertilizer enrichment of soil: effect on decomposition of plant litter under sub humid tropical conditions. Biol Fertil Soils 30:224–231

Lal R (2007) Anthropogenic influences on world soils and implications for global food security. Adv Agron 93:69–93

Lal R (2018) Land use and soil management effects on soil organic carbon dynamics on Alfisols in Western Nigeria. In: Soil processes and the carbon cycle. CRC Press, Boca Raton, pp 123–140

Linn D, Dorna JW (1984) Effect of water-filled pore space on carbon dioxide and nitrogen oxide production in tilled and no tilled soils. Soil Sci Soc Am J 48:1267–1272

Liu X, Herbert SJ, Hashemi AM, Zhang X, Ding G (2006) Effects of agricultural management on soil organic matter and carbon transformation – a review. Plant Soil Environ 52:531–543

Liu C, Lu M, Cui J, Li B, Fang C (2014) Effects of straw carbon input on carbon dynamics in agricultural soils: a meta-analysis. Glob Chang Biol 20:1366–1381

Lobe I, Amelung W, Du Preez CC (2001) Losses of carbon and nitrogen with prolonged arable cropping from sandy soils of the South African Highveld. Eur J Soil Sci 52:93–101

Lojkova L, Datta R, Sajna M, Marfo TD, Janous D, Pavelka M, Formanek P (2015) Limitation of proteolysis in soils of forests and other types of ecosystems by diffusion of substrate. In: Amino acids, 2015, vol 8. Springer, Wien, pp 1690–1691

Lou Y, Liang W, Xu M, He X, Wang Y, Zhao K (2011) Straw coverage seasonal variability of the topsoil microbial biomass and activity. Catena 86(2):117–120

Lutzow M, Leifeld J, Kainz M, Kogel-Knabner L, Munch JC (2002) Indications for soil organic matter quality in soils under different management. Geoderma 105:243–258

Ma J, Li XL, Xu H, Han Y, Cai ZC, Yagi K (2007) Effects of nitrogen fertiliser and wheat straw application on CH4 and N2O emissions from a paddy rice field. Aus J Soil Res 45:359–367

Ma J, Ma E, Xu H, Yagi K, Cai Z (2009) Wheat straw management affects CH4 and N2O emissions from rice fields. Soil Biol Biochem 5:1022–1028

Mafongoya PL, Nair PKR, Dzowela BH (1998) Mineralization of nitrogen from decomposing leaves of multipurpose trees as affected by their chemical composition. Biol Fertil Soils 27:143–148

Maiksteniene S, Arlauskiene A (2004) Effect of preceding crops and green manure on the fertility of clay loam soil. Agron Res 2:87–97

Mandal KG, Misra AK, Hati KM, Bandyopadhyay KK, Ghosh PK, Mohanty M (2004) Rice residue-management options and effects on soil properties and crop productivity. J Food Agric Environ 2:224–231

Manjaiah KM, Voroney RP, Seni U (2000) Soil organic carbon, storage profile and microbial biomass under different crop management systems in a tropical agricultural eco-system. Biol Fertil Soils 32:273–278

Marfo TD, Datta R, Lojkova L, Janous D, Pavelka M, Formanek P (2015) Limitation of activity of acid phosphomonoesterase in soils. In: Amino acids, 2015, vol 8. Springer, Wien, pp 1691–1691

Marschner H (1995) Mineral nutrition of higher plants, 2nd edn. Academic, London

Marschner P, Umar S, Baumann K (2011) The microbial community composition changes rapidly in the early stages of decomposition of wheat residue. Soil Biol Biochem 43:445–451

Marschner P, Hatam Z, Cavagnaro TR (2015) Soil respiration, microbial biomass and nutrient availability after the second amendment are influenced by legacy effects of prior residue addition. Soil Biol Biochem 88:169–177

Martens DA (2000) Plant residue biochemistry regulates soil carbon cycling and carbon sequestration. Soil Biol Biochem 32:361–369

Mary B, Recous S, Darwis D, Robin D (1996) Interactions between decomposition of plant residues and nitrogen cycling in soil. Plant Soil 181:71–82

Meena RS, Mitran T, Kumar S, Yadav GS, Bohra JS, Datta R (2018) Application of remote sensing for sustainable agriculture and forest management. Elsevier

Memon MS, Guo J, Tagar AA, Perveen N, Ji C, Memon SA, Memon N (2018) The effects of tillage and straw incorporation on soil organic carbon status, rice crop productivity, and sustainability in the rice-wheat cropping system of Eastern China. Sustainability 10:961

Metzke M, Potthoff M, Quintern M, Hess J, Joergensen RG (2007) Effect of reduced tillage systems on earthworm communities in a 6-year organic rotation. Eur J Soil Biol 43:209–215

Mishra JS, Kumar R (2017) Conservation tillage and weed management in smallholder agriculture: issues and opportunities. Curr Adv Agric Sci 9(2):186–189

Molaei A, Lakzian A, Datta R, Haghnia G, Astaraei A, Rasouli-Sadaghiani M, Ceccherini MT (2017a) Impact of chlortetracycline and sulfapyridine antibiotics on soil enzyme activities. Int Agrophys 31(4):499–505

Molaei A, Lakzian A, Haghnia G, Astaraei A, Rasouli-Sadaghiani M, Ceccherini MT, Datta R (2017b) Assessment of some cultural experimental methods to study the effects of antibiotics on microbial activities in a soil: an incubation study. PLoS One 12(7):e0180663

Motavalli PP, Palm CA, Parton WJ, Elliott ET, Frey SD (1995) Soil pH and organic C dynamics in tropical forest soils: evidence from laboratory and simulation studies. Soil Biol Biochem 27:1589–1599

Murungu FS, Chiduza C, Muchaonyerwa P, Joergensen RG (2011) Decomposition, nitrogen, and phosphorus mineralization from residues of summer-grown cover crops and suitability for a smallholder farming system. S Afr Commun Soil Sci Plant Anal 42:2461–2472

Narang RS, Virmani SM (2001) Rice-wheat cropping systems of the Indo-Gangetic Plain of India (No. REP-8944CIMMYT)

Narayan D, Lal B (2006) Effect of green manuring on soil properties and yield of wheat under different soil depths in Alfisols under semi-arid conditions in Central India. Bull Nat Inst Ecol 17:31–36

Naresh RK, Panwar AS, Dhaliwal SS, Gupta RK, Kumar A, Rathore RS, Kumar A, Kumar D, Lal M, Kumar S, Tyagi S (2017) Effect of organic inputs on strength and stability of soil aggregates associated organic carbon concentration under rice-wheat rotation in Indo-Gangetic Plain zone of India. Int J Curr Microbiol App Sci 6:1973–2008

Naresh RK, Bhaskar S, Dhaliwal SS, Kumar A, Gupta RK, Vivek R, Vivek R (2018) Soil carbon and nitrogen mineralization dynamics following incorporation and surface application of rice

and wheat residues in a semi-arid area of North West India: a review. J Pharmacogn Phytochem 7:248–259
Nelson PN, Ladd LN, Oades JM (1996) Decomposition of He labelled plant material in a salt-affected soil. Soil Biol Biochem 28:433–444
Neupane MP, Singh RK, Kumar R, Kumari A (2011) Yield performance of baby corn (*Zea mays* L.) as influenced by nitrogen sources and row spacing. Environ Ecol 29(3):1180–1183
Nieder R, Benbi DK (2008) Carbon and nitrogen in the terrestrial environment. Springer, Heidelberg, p 430
O'Connell AM, Grove TS, Mendham D, Rance SJ (2000) O'Connell AM, Grove TS, Mendham D Rance SJ. In: Nambiar EKS, Tiarks A, Cossalter C Ranger J (eds) Site management and productivity in tropical plantation forest. CIFOR, Bogor, pp 61–71
Oades JM (1993) The role of biology in the formation, stabilization and degradation of soil structure. Structure 56:377–400
Ogbodo EN (2011) Effect of crop residue on soil chemical properties and rice yield on an Ultisol at Abakaliki, Southeastern Nigeria. World J Agric Sci 7:13–18
Ortega RA, Peterson GA, Westfall DG (2002) Residue accumulation and changes in soil organic matter as affected by cropping system intensity in no-till dry land agro ecosystems. Agron J 94:944–954
Oshins C, Drinkwater L (1999) An introduction to soil health. [A slide set available at the Northeast Region SARE website: www.uvm.edu/~nesare/slide.html]
Pal V, Singh MM, Kumar R, Verma SS (2013) Response of irrigation scheduling and integrated nutrition on scented rice (*Oryza sativa* L.). Bioinfolet 10(4C):1528–1530
Palm CA, Gachengo CN, Delve RJ, Cadisch G, Giller KE (2001) Organic inputs for soil fertility management in tropical agroecosystems: application of an organic resource database. Agric Ecosyst Environ 83:27–42
Panachuki E, Bertol I, Alves Sobrinho T, Sanches de Oliveira PT, Bicca Rodrigues DBB (2011) Soil and water loss and water infiltration in red latosol under different management systems. Rev Bras Cienc Solo 35:1777–1785
Pare T, Gregorich EG (1999) Soil texture effects on mineralization of nitrogen from crop residues and the added nitrogen interaction. Commun Soil Sci Plant Anal 30:145–157
Parsons WFJ, Taylor BR, Parkinson D (1990) Decomposition of aspen (*Populus tremuloides*) leaf litter modified by leaching. Can J For Res 20:943–951
Pathak H, Bhatia A, Jain N, Aggarwal PK (2010) Greenhouse gas emission and mitigation in Indian agriculture–a review. ING Bull Reg Assess React Nitrogen 19:1–34
Prasad R, Power JF (1991) Crop residue management. Adv Soil Sci 15:205–239
Probert ME, Delve RJ, Kimani SK, Dimes JP (2005) Modelling nitrogen mineralization from manures: representing quality aspects by varying C: N ratio of sub pools. Soil Biol Biochem 37:279–287
Purnomo E, Black AS, Smith CJ, Conyers MK (2000) The distribution of net nitrogen mineralization with in surface soil. I. Field studies under a wheat crop. Aust J Soil Res 38:129–140
Qualls RG, Richardson CJ (2000) Phosphorus enrichment affects litter decomposition, immobilization, and soil microbial phosphorus in wetland mesocosms. Soil Sci Soc Am J 64:799–808
Raiesi F (1998) Impacts of elevated atmospheric CO2 on litter quality, litter decomposability and nitrogen turnover rate of two oak species in a Mediterranean forest ecosystem. Glob Chang Biol 4:667–678
Raimbault BA, Vyn TJ (1991) Crop rotation and tillage effects on corn growth and soil structural stability. Agron J 83:979–985
Raison RJ, O'Connell AM, Khanna PK, Keith H (1993) Effect of repeated fires on nitrogen and phosphorus budgets and cycling processes in forest ecosystems. In: Trabaud L, Prodon R (eds) Fire in Mediterranean ecosystems. CEC, Brussels, pp 347–363
Ranells NN, Wagger MG (1992) Nitrogen release from crimson clover in relation to plant growth stage and composition. Agron J 84:424–430
Recous S, Robin D, Darwis D, Mary B (1995) Soil inorganic N availability: effect on maize residue decomposition. Soil Biol Biochem 12:1529–1538

Reddy BVS, Reddy PS, Bidinger F, Blümmel M (2003) Crop management factors influencing yield and quality of crop residues. Field Crop Res 84(1–2):57–77
Reicosky DC, Dugas W, Torbert H (1997) Tillage-induced soil carbon dioxide loss from different cropping systems. Soil Tillage Res 41:105–118
Ruan L, Philip Robertson G (2013) Initial nitrous oxide carbon dioxide, and methane costs of converting conservation reserve program grassland to row crops under no-till vs. conventional tillage. Glob Chang Biol 19:2478–2489
Rusinamhodzi L, Corbeels M, van Wijk MT, Rufino MC, Nyamangara J, Giller KE (2011) A meta-analysis of long-term effects of conservation agriculture on maize grain yield under rain-fed conditions. Agron Sustain Dev 31:657–673
Saggar S, Parshotam A, Hedley C, Salt G (1999) 14C-labelled glucose turnover in New Zealand soils. Soil Boil Biochem 31:2025–2037
Salinas-Garcia JR, Baez-Gonzalez AD, Tiscareno-Lopez M, Rosales-Robles E (2001) Residue removal and tillage interaction effects on soil properties under rain-fed corn production in central Mexico. Soil Tillage Res 59:67–79
Sall S, Bertrand J, Chotte JL, Recous S (2007) Separate effects of the biochemical quality and N content of crop residues on C and N dynamics in soil. Biol Fertil Soil 43:797–804
Samal SK, Rao KK, Poonia SP, Kumar R, Mishra JS, Prakash V, Mondal S, Dwivedi SK, Bhatt BP, Naik SK, Choubey AK, Kumar V, Malik RK, Mc DA (2017) Evaluation of long-term conservation agriculture and crop intensification in rice-wheat rotation of Indo-Gangetic Plains of South Asia: carbon dynamics and productivity. Eur J Agron 90:198–208
Samra JS, Singh B, Kumar K (2003) Managing crop residues in the rice–wheat system of the Indo-Gangetic Plain. In: Improving the productivity and sustainability of rice–wheat systems: issues and impacts, (improving the pro). American Society of Agronomy, Crop Science Society of America, and Soil Science Society of America, Charlotte, pp 173–195
Sarkar A, Yadav RL, Gangwar B, Bhatia PC (1999) Crop residues in India. Tech Bull Project Directorate for Cropping System. Research, Modipuram, India
Sayre KD (2004) Raised-bed cultivation. In: Lal R (ed) Encyclopedia of soil science. Marcel Dekker Inc, New York
Scheller E, Joergensen GR (2008) Decomposition of wheat straw differing in nitrogen content in soils under conventional and organic farming management. J Plant Nutr Soil Sci 171:886–892
Seastedt T (1984) The role of microarthropods in decomposition and mineralization processes. Annu Rev Entomol 29:25–46
Seneviratne G (2000) Litter quality and nitrogen release in tropical agriculture: synthesis. Biol Fertil Soils 31:60–64
Shahbaz M, Kuzyakov Y, Heitkamp F (2017) Decrease of soil organic matter stabilization with increasing inputs: mechanisms and controls. Geoderma 304:76–82
Shamina IP, Vetrovsky T, Giagnoni L, Datta R, Baldrian P, Nannipieri P, Renella G (2018) Microbial expression profiles in the rhizosphere of two maize lines differing in N use efficiency. Plant Soil 433:401–413
Sharratt BS, Campbell GS (1994) Radiation balance of a soil-straw surface modified by straw color. Agron J 86:200–203
Shindo H, Nishio T (2005) Immobilization and remineralization of N following addition of wheat straw into soil: determination of gross N transformation rates by 15N-ammonium isotope dilution technique. Soil Biol Biochem 37:425–432
Shivran RK, Rokadia P, Kumar R (2012) Phosphorus and sulphur nutrition with P-solublizing bacterial inoculation enhanced the quality and yield of soybean (Cultivar JS-335). Madras Agric J 99(1–3):68–72
Shivran RK, Kumar R, Kumari A (2013) Influence of sulphur, phosphorus and farm yard manure on yield attributes and productivity of maize (*Zea mays* L.) in humid south eastern plains of Rajasthan. Agric Sci Dig 1:9–14

Sidhu BS, Beri V (2005) Experience with managing rice residues in intensive rice-wheat cropping system in Punjab. In: Conservation agriculture: status and prospects. Centre for Advancement of Sustainable Agriculture, New Delhi, pp 55–63
Sidhu HS, Manpreet-Singh, Humphreys E, Yadvinder-Singh, Balwinder-Singh, Dhillon SS, Blackwell J, Bector V, Malkeet-Singh, Sarbjeet-Singh (2007) The happy seeder enables direct drilling of wheat into rice stubble. Aus J Exp Agric 47:844–854
Singh B, Singh Y (2003) Management of crop residues. In: Rice-wheat cropping system in the Indo-Gangetic Plains. Indian Council of Agricultural Research/Punjab Agricultural University, New Delhi/Ludhiana, pp 286–301
Singh AK, Singh AK, Choudhary AK, Kumari A, Kumar R (2017a) Towards oilseeds sufficiency in India: present status and way forward. J AgriSearch 4(2):80–84
Singh AK, Singh AK, Kumar R, Prakash V, Sundaram PK, Yadav SK (2017b) Indian cereals saga: standpoint and way forward. J AgriSearch 4(1):1–9
Singh SK, Abraham T, Kumar R, Kumar R (2017c) Response of crop establishment methods and split application of nitrogen on productivity of rice under irrigated ecosystem. Environ Ecol 35 (2A):859–862
Skjemstad JO, Janik LJ, Head MJ, McClure SG (1993) High energy ultraviolet photo-oxidation: a novel technique for studying physically protected organic matter in clay and silt-sized aggregates. J Soil Sci 44:485–499
Smethurst PJ, Nambiar EKS (1990) Effects of slash and litter management on fluxes of N and tree growth in a young P. radiate plantation. Can J For Res 20:1498–1507
Sodhi GPS, Beri V, Benbi DK (2009) Soil aggregation and distribution of carbon and nitrogen in different fractions under long-term application of compost in rice–wheat system. Soil Tillage Res 2:412–418
Subehia SK, Sepehya S (2012) Influence of long-term nitrogen substitution through organics on yield, uptake and available nutrients in a rice-wheat system in an acidic soil. J Indian Soc Soil Sci 60:213–217
Tanaka DL (1986) Wheat residue loss for chemical and stubble-mulched fallow. Soil Sci Soc Am J 50:434–440
Tandon HLS, Sekhon GS (1988) Potassium research and agricultural production. Fertilizer Development and Consultation Organization, New Delhi
Thierfelder C, Wall PC (2009) Effects of conservation agriculture techniques on infiltration and soil water content in Zambia and Zimbabwe. Soil Tillage Res 105:217–227
Tian G, Brussaard L, Kang BT (1995a) Breakdown of plant residues with contrasting chemical compositions under humid tropical conditions: effects of earthworms and millipedes. Soil Biol Biochem 27:277–280
Tian G, Brussaard L, Kang BT (1995b) An index for assessing the quality of plant residues and evaluating their effects on soil and crop in the (sub-) humid tropics. Appl Soil Ecol 2:25–32
Trinsoutrot J, Rccous S, Bentz B, Lincres M, Chenoby D, Nicolardot B (2000) Biochemical quality of crop residues and carbon and nitrogen mineralization kinetics under nonlimiting nitrogen conditions. Soil Sci Am J 64:918–926
Usuki K, Yamamoto H, Tazawa J (2007) Effects of previous cropping and tillage system on growth of maize and symbiotic association with arbuscular mycorrhizal fungi in central region of Japan. Jpn J Crop Sci 76:394–400
Verhulst N, Govaerts B, Verachtert E, Mezzalama M, Wall PC, Chocobar A, Deckers J, Sayre KD (2010) Conservation agriculture, improving soil quality for sustainable production systems? In: Lal R, Stewart BA (eds) Advances in soil science: food security and soil quality. CRC Press, Boca Raton, pp 137–208
Verhulst N, Carrillo-Garcia A, Moeller C, Trethowan R, Sayre KD, Govaerts B (2011a) Conservation agriculture for wheat-based cropping systems under gravity irrigation: increasing resilience through improved soil quality. Plant Soil 340:467–479

Verhulst N, Nelissen V, Jespers N, Haven H, Sayre KD, Raes D, Deckers J, Govaerts B (2011b) Soil water content, maize yield and its stability as affected by tillage and crop residue management in rainfed semi-arid highlands. Plant Soil 124:347–356

Vigil MF, Sparks D (2004) Factors affecting the rate of crop residue decomposition under field conditions. Conservation tillage Fact Sheet# 3–95. US Department of Agriculture, Washington, DC

Wang J, Liu X, Zhang F (2002) The effect of different soil mulch materials on the growth and yield of rice. Acta Ecol Sin 22:922–929

Wang WJ, Baldock JA, Dalal RC, Moody PW (2004) Decomposition dynamics of plant materials in relation to nitrogen availability and biochemistry determined by NMR and wet-chemical analysis. Soil Biol Biochem 36:2045–2058

Whalen JK, Parmelee R, Subler S (2000) Quantification of nitrogen excretion rates for three lumbricid earthworms using 15N. Biol Fertil Soils 32:347–352

Whitbread A, Blair G, Konboon Y, Lefroy R, Naklang K (2003) Managing crop residues, fertilizers and leaf litters to improve soil C, nutrient balances, and the grain yield of rice and wheat cropping systems in Thailand and Australia. Agric Ecosyst Environ 100:251–263

Wood WA, Kellogg ST (1988) Miomass. Part A: Cellulose and hemicellulose. In: Abelson JN, Simon MI (eds) Methods in enzymology, vol 160160. Academic, San Diego, pp 3–774

Wright S, Starr J, Paltineanu I (1999) Changes in aggregate stability and concentration of glomalin during tillage management transition. Soil Sci Soc Am J 63:1825–1829

Wuest S, Caesar-TonThat T, Wright SF, Williams J (2005) Organic matter addition, N, and residue burning effects on infiltration, biological, and physical properties of an intensively tilled silt-loam soil. Soil Tillage Res 84:154–167

Xu RK, Coventry DR (2003) Soil pH changes associated with lupin and wheat plant materials incorporated in a red-brown earth soil. Plant Soil 250:113–119

Yadav GS, Datta R, Imran Pathan S, Lal R, Meena RS, Babu S, Das A, Bhowmik S, Datta M, Saha P (2017) Effects of conservation tillage and nutrient management practices on soil fertility and productivity of rice (*Oryza sativa* L.)–rice system in north eastern region of India. Sustainability 9(10):1816

Yadvinder-Singh, Gupta RK, Singh J, Singh G, Singh G, Ladha JK (2010) Placement effects on rice residue decomposition and nutrient dynamics on two soil types during wheat cropping in rice–wheat system in northwestern India. Nutr Cycl Agroecosyst 88:471–480

Yan X, Yagi K, Akiyama H, Akimoto H (2005) Statistical analysis of the major variables controlling methane emission from rice fields. Glob Chang Biol 11:1131–1141

Yang HS, Yang B, Dai YJ, Xu MM, Koide RT, Wang XH, Liu J, Bian XM (2015) Soil nitrogen retention is increased by ditch-buried straw return in a rice–wheat rotation system. Eur J Agron 69:52–58

Zhang F (2011) The effects of no-tillage practice on soil physical properties. Afr J Biotechnol 10:17645–17650

Zhao S, Li K, Zhou W (2016) Changes in soil microbial community, enzyme activities and organic matter fractions under long-term Straw return in north-central China. Agric Ecosyst Environ 216:82–88

Zhu L, Sun QF, Liu CX, Wang XH, Bian XM (2012) Effects of burial of rice straw in furrows on soil environment of wheat field. J Ecol Rural Environ 28:399–403

Zibilske LM, Bradford JM, Smart JR (2002) Conservation tillage induced changes in organic carbon, total nitrogen and available phosphorus in a semi-arid alkaline subtropical soil. Soil Tillage Res 66:153–163

Zou JW, Huang Y, Jiang JY, Zheng XH, Sass RL (2005) A 3-year field measurement of methane and nitrous oxide emissions from rice paddies in China: effects of water regime, crop residue, and fertilizer application. Glob Biogeochem Cycles 19:2021

The manufacturer's authorised representative in the EU is Springer Nature Customer Service Centre GmbH, Europaplatz 3, 69115 Heidelberg, Germany. If you have any concerns regarding our products, please contact ProductSafety@springernature.com

Printed and bound by CPI Group (UK) Ltd, Croydon, CR0 4YY
30/09/2026
02225452-0018